World Top Secret: Our Earth IS Hollow!

The Scientific, Scriptural and Historical Evidence that Our Earth Is Hollow!

Rodney M. Cluff

World Top Secret:
Our Earth IS Hollow!

The Scientific, Scriptural and Historical Evidence
that Our Earth Is Hollow!

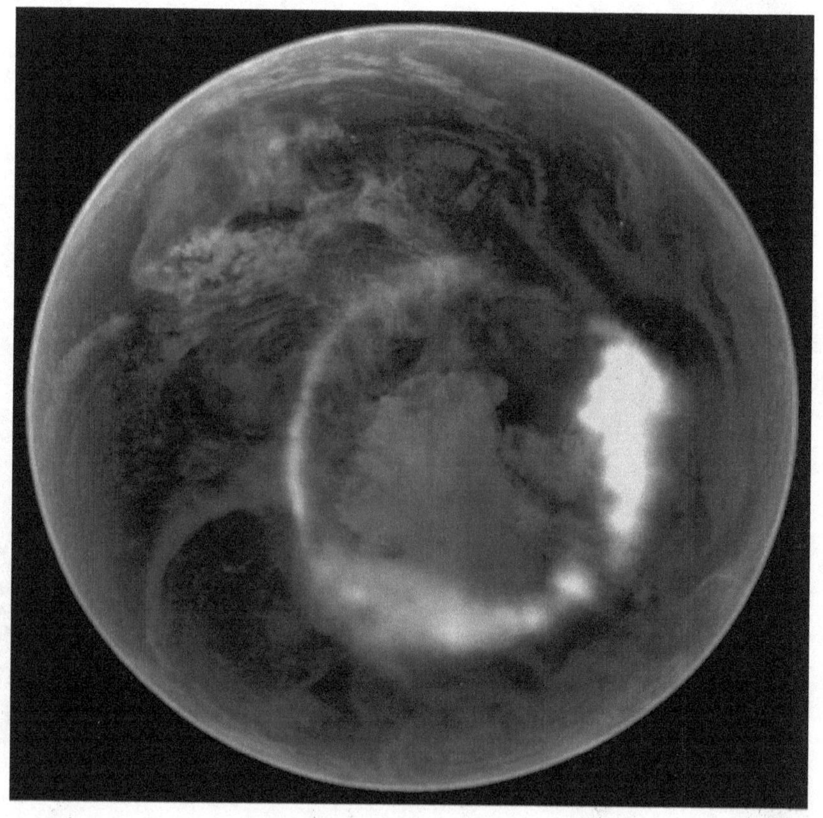

by

Rodney M. Cluff

https://ourhollowearth.com/

DEDICATED

To those who love truth that is stranger than fiction.

World Top Secret:
Our Earth IS Hollow!

CONTENTS

FROM THE AUTHOR

As a young man, I had two favorite subjects, science and religion. In my study, it became my conviction that ultimately science and religion will become one and the same, science being the study of God's creation; and religion consisting of the revelations of God to mankind. Both are ultimately manifestations of the truth of all things given to man by God in His infinite kindness and love to bring about the happiness of His children.

It is from the Book of Mormon, a text of scripture written by ancient American prophets of God that I gained the desire to obtain the object of both true religion and true science: **the search for the truth of all things**. The ancient American prophet of the Book of Mormon concluded this book of scripture with a perfect scientific test anyone can perform on that book to know it is of God. Moroni wrote 421 A.D.:

"Behold, I would exhort you that when ye shall read these things, if it be wisdom in God that ye should read them, that ye would remember how merciful the Lord hath been unto the children of men, from the creation of Adam even down until the time that ye shall receive these things, and ponder it in your hearts.

"And when ye shall receive these things, I would exhort you that ye would ask God, the Eternal Father, in the name of Christ, if these things are not true; and if ye shall ask with a sincere heart, with real intent, having faith in Christ, he will manifest the truth of it unto you, by the power of the Holy Ghost.

"And by the power of the Holy Ghost ye may know the truth of all things." (Moroni 10:3-5)

It is by application of this scientific test to that book of ancient American scripture that I came to a knowledge that it is of God, because God did answer my prayer and let me know by the power of the Holy Ghost of its truthfulness. Millions of Latter-day Saints have performed this same test and received the same answer of the divinity of that book. Therefore, it could be said that Mormonism is a scientific religion.

The acquisition of this one precious truth has given me the impulse to discover the ultimate: the truth of all things. And my search has not been in vain. In fact, my search is a much more efficient one because my hits in the dark are much more infrequent when I have the power of the Holy Ghost to lighten the way to the next truth.

Thus my search has been an exciting one and I hope some of the things I have uncovered concerning this earth of ours will be as exciting to you as it has been to me.

Would you like to receive a free copy of The Book of Mormon? Then click here:

The Book of Mormon | Come unto Christ

ABOUT THE AUTHOR

    RODNEY M. CLUFF, author of World Top Secret: Our Earth IS Hollow! was born and raised in the American Mormon colony of Colonia Juarez in Mexico's northern state of Chihuahua. He became interested in the Hollow Earth Theory at the age of 16 while working on a New Mexico farm where the farm manager's son told the workers of the theory. He thought: What an ideal place for the Lord to hide the Ten Lost Tribes of Israel!

    After graduating from high school, Mr. Cluff served a full-time mission for the Church of Jesus Christ of Latter-day Saints in Mexico where he met his wife, Maria Enriqueta Flores Morales. One year after the release from his mission, they were married in the Mesa, Arizona Temple and now have five lovely children, fifteen lovelier grandchildren, and now two great-grandchildren!

    They moved to Phoenix, Arizona, where one day Mr. Cluff noticed an advertisement of Raymond Bernard's book, **The Hollow Earth** in a tabloid newspaper. He sent for it and thereby began many years of study and writing which has led to the present work. After 16 years with the Arizona Department of Health Services working as an Information Technology Specialist, Mr. Cluff retired and now lives with his wife Enriqueta in the retirement community of Sun City, Arizona, and continues his research into evidences for hollow planets as a hobby.

    He firmly believes: OUR EARTH IS HOLLOW! Backed with scientific evidence, including satellite photos of the polar holes, analysis of the observations of polar explorers, analysis of earthquake data and much more -- coupled with evidence from the scriptures and history that the Lost Tribes of Israel are now

FOUND within the Hollow of Our Earth, he presents his argument in favor of the Hollow Earth Theory.

It is his hope that someday, he may have the privilege of visiting his cousins of the Lost Ten Tribes in the North Countries of our Hollow Earth! The author's own ancestry is of Israelite origin, of the Tribes of Ephraim, Manasseh and Judah, and can be traced back to the Exile of the Ten Tribes from Palestine when they were carried away captive into Assyria in 721 B.C.

The Ten Tribes were held captive for thirty-four years by the Assyrians, but then escaped over the Caucasus Mountains sometime before Babylon conquered Assyria in 605 B.C. They made their home in the region of the Crimea and the Steppe of Russia just north of the Black Sea up until the first century B.C. While there they were ruled by an illustrious leader named Odin. The Roman armies threatened to conquer the region so his ancestors, because of their fierce love of freedom and independence, determined to migrate. From their custom of burying their dead in burial mounds, their migrations have been traced from the Black Sea up the valley of the river Dnieper in Russia to the Baltic Sea and from thence to northern Germany and Scandinavia.

One branch of these people became known as the Sakae or Saxons (a contraction of "Sons of Issac") and settled in Northern Germany. Shortly after the Romans left the British Isles in the fourth century A.D., certain Celtic tribes of the British Isles invited the Engles, Saxons, and Jutes (who had previously raided the east coast of England as pirates) to bring their bands over and help defeat other Celts. From the eighth to the eleventh century they were known as the Scandinavian Vikings. They became the most volatile sea power and military force in Europe. They often attacked coastal areas with fleets that ran into the hundreds of ships and highly organized armies of several thousand. The French became weary of being looted each harvest season and so they invited the Vikings to accept a large section of France and raise their own crops. The Norsemen agreed and the territory became known as Normandy, or loved of the Norsemen.

The Author's Clough forefathers were of the Vikings who settled Normandy in France. They came to England with William the Conqueror in 1066 A.D. In the distribution of lands among his officers, a large estate fell to one CLOUGH in Yorkshire. This estate has been transmitted from father to son until the present time and is known as the Esquire Clough Estate, and is situated about 26 miles from the old city of York.

ABOUT THE AUTHOR

In the year 1635, fifteen years after the first Pilgrims immigrated to America, at about the age of 19 to 21, John Clough with his brother sailed from London, England on the Clipper ship The Elizabeth. Upon arriving in America, John Clough settled in Massachusetts. One of his descendants, David Cluff, changed the spelling of his last name when he joined The Church of Jesus Christ of Latter-day Saints (commonly known as the Mormons) in 1830. The author is a descendant of this David Cluff, whose ancestry can be traced back through the Saxons and Vikings to the House of Israel. In fact, he has been able to trace his ancestry back to Adam and Eve, the first parents of the human race, with the help of Michel L. Call's book, ROYAL ANCESTORS OF SOME L.D.S. FAMILIES, and FamilySearch.org, in which he finds that he is the 119th generation from Adam. He also calculated the generation spans between each generation from Adam to his eldest son and find that they add up to 6004 years.

In 2003, the author was contacted by Steve Currey of the Expedition Company of Provo, Utah who asked him to help plan an expedition to Our Hollow Earth. They put together the Voyage to Our Hollow Earth Expedition with plans to take an airliner into the Arctic to pin-point the location of the North Polar Opening, and then follow that with a sea expedition chartering the Russian Nuclear Icebreaker, the Yamal.

Unfortunately, in 2006, three months before the departure date of the over flight of the Arctic, Steve discovered he had six inoperable brain tumors, and died two months later. His family then canceled their involvement in the expedition and returned all expedition members' monies.

Many expedition members expressed interest to continue the quest to reach Inner Earth, and so an expedition member, Dr. Brooks Agnew offered to be their new expedition leader. Dr. Agnew formed a new expedition company, put up a nice expedition website for the new North Pole Inner Earth Expedition, but after 7 years promoting the expedition, he resigned as expedition leader September 9, 2013. The largest stockholder of his start-up electric car company, Vision Motor Cars, threatened to remove all his money from his company if he continued as our expedition leader.

Currently, Mr. Rodney Cluff encourages all interested parties to do their own expedition to Our Hollow Earth, providing the best available evidence and instructions on how to get there in this book and its future updated editions.

Then in January 2020, Physicist, Dr. Brooks Agnew reposted his website and projected a new date to carry out his North Pole Inner Earth Expedition. He is chartering the new Russian nuclear icebreaker, the Arktika to carry expedition members into the Arctic in search of the North Polar Opening. Since the itinerary shows the expedition is going to the Magnetic North Pole which was located at 86.448 N 175.34585 E in 2019, and is now located one-quarter of the way into the polar opening, scientists using a gyroscope aboard the Arktika should be able to detect the curvature into the earth towards the North Polar Opening, which my latest evidence indicates is located centered near Northland, Russia at 85 N, 130 E.

You can now subscribe to Dr. Agnew's Hollow Earth TV channel to keep informed of the progress and to help finance the expedition.

If you would like to purchase a copy of the book, World Top Secret: Our Earth IS Hollow! -- the result of 35 plus years research in support of the Hollow Earth / Hollow Planets Theory, you may do so on Amazon.com or on Rodney M. Cluff's website at:

https://www.ourhollowearth.com/

FOREWORD

In 1981, I took my family to Alaska to look for evidence of the hollow in our earth. While there, I made two friends whom I found had a deep desire to journey to Our Hollow Earth via the North Polar Opening. We spent the summer of 1981 with Fred M. Sandelin, who 12 years previously had gone to Alaska to see if he could find a way to go to the North Countries. He had a very profitable commercial Salmon fishing business which he hoped to use to eventually buy a float plane with which to make a flight north. He asked me to go fishing with him. We had a successful season and I earned $7,000 in two months fishing plus an offer for a summer job every year if I wanted it.

From Fred I learned that while he was working on the DEW line (the early warning radar stations in the far north to warn of missile attack across the pole from the Soviet Union) he saw some pictures taken in the North Countries of the Hollow Earth showing the Giant people and vegetation that exist there. He worked at Prudhoe Bay for the oil companies as an electric generator repairman and says that the North Slope for over a hundred miles to the south has no trees on it. Yet on the coast the Eskimos gather driftwood that has drifted in from the North -- that land beyond the pole where the Eskimo says the sun never sets.

My other friend, John Gagné, whom I also met in Fairbanks, Alaska, had come to Alaska years ago and tried to drum up some support for expeditions to the Hollow Earth without much success. He had three interesting incidents to relate. While a student at BYU, he met a girl who was a good friend of the Admiral Byrd family. Since John was at that time a member of an expedition preparing to go in search for Noah's Ark on Mount Ararat and had written the script for the movie commentary, In Search Of Noah's Ark, he asked her if she would ask the Byrd family if they would let him examine Byrd's writings, hoping his credentials would influence them favorably. But when she returned to Virginia for Christmas vacation, and asked them, she was told that they do not let ANYONE look at Byrd's writings. They had them under lock and key.

In 1991, my wife and I attended a UFO Conference in Phoenix, Arizona where we met Harley Byrd, who claimed to be the nephew of Admiral Byrd. In the conference newsletter, an article about Harley Byrd said that he worked for the CIA. So I believe he was there to tell us lies and disinformation about the

Hollow Earth. He gave the workshop at the conference on the Hollow Earth. Harley had a copy of the supposed Diary of Admiral Byrd's flight beyond the poles and told us a story of how he obtained it, but I caught him in his lie. He declared that at the Admiral's state funeral, his aunt was standing by the casket and happened to pass her hand over the Admiral's body when she felt some papers in his jacket. So at the opportune moment when Harley was handing her the U.S. flag to cover the body, she retrieved the papers from the Admiral's jacket and passed it to Harley beneath the flag.

I asked him if I could buy the copy of the diary that he had in his hand. He wanted $10 for it, so I wrote him a check, but as I was looking through it, I told him it looked very similar to a copy of Byrd's diary I had received several years before from Bruce Walton of Provo, Utah. Harley seemed surprised and asked, *"Do you know Bruce Walton?"* To which I replied, *"Yes, we have been friends for years."* And with that he strangely took the diary back and gave me back my check without saying another word.

The diary has now been published and is available from Inner Light Publications and Amazon.com. It consists of a few short pages describing the Admiral's flight log on the February 1947 flight to that Land Beyond the Poles. Dennis Crenshaw, hollow earth researcher, after examining it, claimed the diary is fake because he found some verbiage in it from James Hilton's book, Lost Horizon. When Bruce had sent me a copy of the diary in the late '80's, he said he had obtained it from Tawani W. Shoush of the International Society for a Complete Earth, and on the first page, it had Tawani's address at Rt 1 Box 63, Houston, Missouri 85483. Bruce told me he thought the diary was a fabrication. Nevertheless, the diary is amazingly similar to a story told me by John Gagné in Alaska.

John told me that a few years before as he was working as a radio commentator in Alaska's state capital, Juneau, he was out one weekend with some buddies when they saw a UFO. It was night and they were looking up at the stars when a bright white light lit up, above a nearby mountain. Presently it turned bright red and zipped off into space. Back on the radio he encouraged people to call in if they had seen the UFO that weekend and share the experience. One lady, Sylvia Darvell, even came to talk to him privately. She had been involved in Alaskan politics from way back and said that it was through this involvement that she had come in contact with Admiral Byrd and they had become close friends. She said that after Byrd's flight into the Arctic beyond the poles that he came to her and confided information which he said he was afraid to tell the world for fear he would be condemned as insane.

She said that Byrd had told her that in his flight beyond the North Pole, he had come to open ocean and then to land covered with lush vegetation and that he then came to cities of a people *"large in stature."* According to Sylvia, they actually landed up there and met the people of the Hollow Earth who received them well. He said they were friendly and highly advanced in the sciences; that they possessed advanced transportation systems such as monorail trains between their cities and craft which since have become known as FLYING SAUCERS.

John said that when he first moved to Fairbanks to find a way to go to the North Countries that he met an agent of the United States Foreign Secret Service. John figured that since the United States government has kept the discovery of Our Hollow Earth a secret that perhaps this agent knew about it. He therefore proceeded to try and convince the agent to give him information about the polar opening in hopes of figuring out the best way to go up there. But the agent persistently denied he knew anything about it. However, it wasn't too long afterward that John received a call from the agent with instructions to meet him at a certain place after which John followed him into a dark corner of a bar. And in a very secretive manner, the agent offered to take John to the Hollow Earth via float plane. But he wanted 3 million dollars to take him! Needless to say, John couldn't come up with the money.

Recently, a friend has told me about an article his friends at work were talking about in which reports from the pilots of Airlines which fly daily over the Arctic say they have seen a subtropical land in the region of the North Pole covered with lush vegetation. Incidentally, Fred, my friend in Fairbanks who has worked on the Distant Early Warning radar line and watched the Airline planes come in over the polar area on the radar, noticed that none ever passed over the area described in this book as the North Polar Opening. All flew on either side of it. The fact is, that if they tried to fly over the Polar Opening, they wouldn't be able to because it is too big. If they did try to fly over it, the airplane would follow the curvature of the earth at the polar lip into the Hollow interior of our planet.

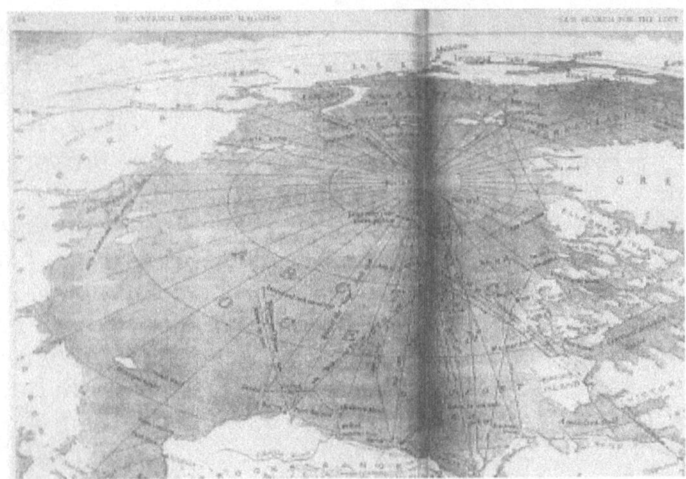

It is the opinion of this author that this actually happened August 12, 1937, to the so called *"missing Soviet fliers"* as described in Vilhjalmur Stefansson's book, UNSOLVED MYSTERIES OF THE ARCTIC, in which a flight tried to pass directly over the Arctic from Moscow, Russia to Fairbanks, Alaska. Radio transmissions from the ill-fated flight were received, but grew fainter and fainter until they were received no more. Subsequent rescue flights were unable to locate them. Moscow called off the search after 7 months. Most likely, the missing Soviet fliers inadvertently flew through the North Polar Opening and into Our Hollow Earth where they crashed or decided to stay because it is a paradise there.

Jack West commented on the theory of a Hollow Earth as a good place for the Lord to hide the Lost Ten Tribes of Israel in his tapes, THE SECOND COMING OF CHRIST. Located towards the end of the first side of the 2nd tape, this Mormon leader had this to say:

> *"Now if you quote me, I'll probably deny it, because I can't give you book and page. But wouldn't it be thrilling if those stories that we keep hearing are true? Wouldn't it be thrilling if the earth IS hollow instead of a molten state down there and that there ARE people inside the earth?"*

> *"Wouldn't it be thrilling if that book is based on fact? And this wonderful fellow lived close to me. He lived in Glendale in southern California just a few miles from where I live. And when he was about to pass away and go meet his maker, Olaf Jansen...told his story in almost a life time to his dearest friend...His friend went there and here he had documented evidence, everywhere you turn, of a trip*

that he and his father made to the inside of the earth. They were Norwegian fishermen. In a book called, THE SMOKY GOD, if you want to make a note of it. You read it and see what you think!"

"You see," he continues, "I heard that broadcast as many of you did, by Admiral Byrd. For he testified, 'We've flown hundreds of miles north of the north pole, every inch of the way, over beautiful forest land and green hills and lovely blue waters, and we've seen giant animals down there in the woods.' And then they had to return because they were almost out of gas. He went to the South Pole. Got more gas this time; went still farther south of the SOUTH pole, every inch of the way he testified, over blue waters and beautiful wooded areas and green hills."

"Now, I don't know the answer. All I know is that the greatest number of sightings of UFO's have been near the north and south poles. Wouldn't that be fascinating if some of them are coming from down inside? Yes, they are way ahead of us they all testify. Get this book called, THE HOLLOW EARTH. It's a scientific book this time; gives evidence all over the place that the government purposely quieted down and hushed up that story of Admiral Byrd that I heard both of his international broadcasts as many of you did on international radio hookup when he told about these stories. Believe me, HE WAS NOT OFF HIS ROCKER! I believe with all my heart that they literally did exactly what he said they did!"

"...what about Peary when they got that bird that saved their lives? It lit on their equipment and they caught it. And when they opened it up, it had GREEN GRASS in its craw. And yet they didn't know of any green grass within THOUSANDS of miles of where they were...What about THE GOLD OF THE GODS and the story of von Daniken. I know him personally. Erich von Daniken, who wrote, CHARIOTS OF THE GODS and now GOLD OF THE GODS. Get it if you don't have it. It's thrilling. What about these thousands of miles of tunnels underneath the earth starting with Ecuador. We don't know where they really start. But they found them up at the north end of Ecuador clear down into Chile. They haven't come to the ends of them at either end and yet there is a wind blowing through there all the time and the legend is that some of these tunnels go down spiraling down inside the earth to a beautiful land inside the earth."

"I don't know the answer," Jack West says, who is also an amateur archaeologist. *"It's interesting. I've done a lot of research on it and I'm beginning to believe that it might be a possibility. Brigham Young taught as though the Ten Tribes were quite close to us; that they certainly didn't have to come from another planet when they came back."* (Cassette tapes, THE SECOND COMING OF CHRIST, published in 1978 by Sounds of Zion, Box 7332, Murray, Utah 84107)

It is the opinion of this author that the Hollow Earth Theory answers a host of questions that heretofore have been complete mysteries with no solutions in sight. At last for the person with an OPEN mind and an INQUIRING heart, may a more complete picture come into view. It is, therefore, with confidence that I present the ideas in this book as TRUTHS which time will prove to be correct and valid theories.

I give my whole-hearted thanks to all who have encouraged me to write this book; my wife, Queta (Keta), my sister, LaVerne, my friends in Alaska; my mother, who together with my father at first rejected the very thought of a Hollow Earth but subsequently accepted it as THE place the Lord has hidden the LOST TRIBES and encouraged me to copyright and publish my book.

PREFACE

What you are about to read is so absolutely FANTASTIC, INCREDIBLE and UNKNOWN to you that you will think it impossible that it could be the truth -- BUT IT IS!

DID YOU KNOW?...

--That our supposed solid earth is really HOLLOW with polar openings into the interior!

--That the home of the LOST but now FOUND! Ten Tribes of Israel is the inside surface of OUR HOLLOW EARTH!

--That the capital city of the Ten Tribes of Israel is the true location of the GARDEN OF EDEN inside our earth!

--That the physical location of PARADISE where the righteous spirits of the dead go is a sun in the center of Our Hollow Earth which gives LIGHT to the plant, animal and human life within our earth!

--That greater civilizations than any you have ever heard of, presently live in giant lighted CAVERNS of the earth's crust!

--That not only is our earth HOLLOW, but our moon is hollow as are ALL the planets and even the sun!

--That the richest, perhaps most populated, most powerful and most highly advanced nation of this earth is INSIDE the earth!

--That the true home of the FLYING SAUCERS is the Israelite nation inside our earth, and that they also come from other planets of our solar system, and from other stars in our galaxy!

--That the majority of the flying saucers sighted around the world are the military of the Hollow Earth Nation of Israelites and are operating a DEFENSIVE against the INTERNATIONAL ILLUMINIST-COMMUNIST-JESUIT CONSPIRACY that controls our world!

--That the International Illuminist-Communist-Jesuit Conspiracy, which is in control of the U.S. government 60-80%, is currently operating an OFFENSIVE against that Israelite Nation inside Our Hollow Earth with intentions of subjugating it under One World Order rule!

--That the Illuminist-Communist-Jesuit stranglehold upon the world will ultimately be broken with the help of the Ten Tribe Nation of the Hollow Earth and their flying saucers!

--Why the governments of the world hold the Hollow Earth, its inhabitants and their flying saucers as the WORLD TOP SECRET?

INTRODUCTION

Perhaps the ideas that will be developed here may be considered by some as speculation. But all truth seekers, that is, true scientists, are invited to take these theories and search for evidence which will prove them to be true. Science is the search for truth. The search begins with the formulation of a theory based upon an accumulation of evidences which are not sufficient to establish a fact, the fact being a manifestation of the ultimate truth.

Many scientists have obtained the truth of their theories, because their theories were valid. Such a scientist was Christopher Columbus, who with evidence accumulated by himself and others reached the conclusion, or in other words, formulated a theory that the earth is round and not flat as many at that time supposed. His theory was a valid one, because when he set out to prove it true, he found the sufficient evidence to prove it a fact, that the earth was truly round, and that a man could go west and get to the same place as the man who went east. Although the earth was not actually circumnavigated until after Columbus' day, nevertheless, it was he who accumulated the evidence sufficient enough to prove to the world of his day that the earth was indeed round and that it could be circumnavigated.

The Ferdinand Magellan expedition was the first to circumnavigate the Earth by way of the Pacific Ocean beginning in Spain in 1519 and ending in Spain by September 1522 proving the Earth to be round. Then starting in 1979, the Transglobe Expedition with Sir Ranulph Fiennes and Charles R. Burton circumnavigated the earth's polar axis by the 29th of August 1982. Since the launch of the first satellites orbiting the earth in 1957, at the beginning of 2019, the United Nations Office for Outer Space Affairs reported in their Index of Objects Launched into Outer Space, that there were 4,987 satellites orbiting the planet earth. Stuff In Space gives a pretty good visual idea of how many satellites and debris are now orbiting earth. If you rotate the earth to its north and south poles, you can see an interesting dearth in polar orbiting satellites at the poles on Stuff In Space.

Columbus had a valid theory which proved to be true, that the earth was indeed round. But there have been many scientists who have formulated theories which upon accumulating more evidence have been found to be false. One such theory, proven false by many scientists, and still propagated as though it were a religion, is the so-called theory of organic evolution -- formulated and made popular by Charles Darwin. This theory, which states that higher life forms evolved from lower life forms, directly contradicts the highly established fact that species can propagate only after their own kind. And if two species mate, they either have no offspring, or their offspring such as the mule, cannot have offspring. And the fact that although many species have become extinct throughout history, there are species living today which have not changed in the least detail from the beginning of the earth's creation, such as the ant, and the cockroach; whose prehistoric fossils found in the earliest strata are identical to those of today.

If life truly did evolve throughout eons of time as Darwin maintained, it is only logical that the fossil record should be replete with transitional life forms in the process of evolving from one species to the next. Yet NO transitional forms have been found in the fossil record. The fact that animal, plant and human remains are fully developed in the earliest strata of the Earth in which they are found, proves that there was no gradual development or evolution of life forms throughout eons of time. Life forms are continually found to be suddenly placed on the earth fully developed in their highest forms.

Would it not be a more profitable search for truth where theories are based upon scientific evidence PLUS inspiration from God and supported by the revealed word of God recorded in the scriptures and not just upon the unreliable imagination of man? Darwin imagined that the origin of the species happened by chance and evolved into higher, more complex species through millions of years of time. He, himself admitted before he died that his theory could be false and that God could have created the world and all life in it after all.

Darwin's theory has been proven false. Many excellent works disproving evolution can be obtained from the scientists at the _Institute for Creation Research_, PO Box 59029, Dallas, Texas 75229, 1806 Royal Ln, Dallas, Texas 75229. A superb example is a book by Dr. Edward F. Blick, A SCIENTIFIC ANALYSIS OF GENESIS. Other recommended literature can be obtained from _The Evolution Protest Movement_, 110 Havant Road, Hayling Island, Hants, Polloll, England. A Creation Science magazine is published by 500 creation scientists, all with graduate degrees in science, of the _Creation Research Society_.

Christopher Columbus, whose theory that the earth is round, said he was inspired of the Holy Ghost to go and prove it true. Columbus spoke of the source of his theory. Said he,

> "...our Lord unlocked my mind, sent me upon the sea, and gave me fire for the deed. Those who heard of my enterprise called it foolish, mocked me, and laughed. But who can doubt but that the Holy Ghost inspired me?" (Jacob Wasserman, COLUMBUS, DON QUIXOTE OF THE SEAS, p. 20)

Columbus' theory has even been proven true in fulfillment of scriptural prophecy. Nephi, the great ancient American prophet wrote 600 years before Christ, of Columbus' discovery of America. In a vision of the future, Nephi wrote,

> "And I looked and beheld a man among the Gentiles, who was separated from the seed of my brethren by the many waters; and I beheld the Spirit of God that it came down and wrought upon the man; and he went forth upon the many waters, even unto the seed of my brethren, who were in the promised land (America)." (1 Nephi 13:12, THE BOOK OF MORMON)

Columbus' theory that the earth is round is also supported by scripture. Alma said to the anti-Christ, 74 years before Christ's birth,

> "The scriptures are laid before thee, yea, and all things denote there is a God, yea, even the earth and all things that are upon the face of it, yea, and its MOTION, yea, and also all the PLANETS which move in their regular form do witness that there is a Supreme Creator." (Alma 30:44)(Emphasis MINE throughout this book)

In another place in the Book of Mormon, this motion of the earth that Alma referred to indicated that the Nephites understood that it was the rotation of a round earth. Helaman wrote,

> "Yea, by the power of his voice doth the whole earth shake; yea, by the power of his voice, do the foundations rock, even to the very center. Yea, and if he say go back, that it lengthen out the day for many hours--it is done; And thus, according to his word the earth goeth back, and it appeareth unto man that the sun standeth still; yea, and behold, this is so; FOR SURELY IT IS THE EARTH THAT MOVETH AND NOT THE SUN." (Helaman 12:11-15)

And in our own day, the Lord revealed to Joseph Smith the modern American Prophet that the earth rotates:

"The earth ROLLS upon her wings, and the sun giveth his light by day, and the moon giveth her light by night, and the stars also give their light, as they ROLL upon their wings in their glory." (DOCTRINE AND COVENANTS 88:45)

In the Bible, Isaiah, speaking of God, wrote, *"It is he that sitteth upon the circle of the earth..."* (Isaiah 40:22) The word *"circle"* originating from the Hebrew word *"khug"* is interpreted by some Hebrew scholars to mean *"sphericity"* or *"roundness."*

Thus we see that Columbus' theory was inspired of God, is supported by scripture, and confirmed by exploration. On the other hand, the scriptures do not support the theory of organic evolution, which in fact, contradicts the written word of God.

1. The theory of evolution maintains that the earth came into existence millions of years ago. This contradicts the scriptures which state that the earth was created (organized) in six days of the Lord, one day of the Lord being 1,000 of our earth years. (Abraham, 3:4, The Pearl of Great Price)

Some creation scientists believe that the seven days of the earth's creation period were the same as our current 24 hour days. If God said that he created the earth in six days, and rested on the seventh, then he did just that! But whose days? Earth days, or the days on the planet where God resides?

Now if you're God and you are creating an earth, how are you going to measure time? You would measure it according to the time on the planet where you live. Right? So how long is a day on the planet where God resides?

Peter taught that a day of the Lord is as a 1,000 earth years,

"But, beloved, be not ignorant of this one thing, that one day is with the Lord as a thousand (Earth) *years, and a thousand* (Earth) *years as one day* (of the Lord)." (2 Peter 3:8)

Now, why would you believe in the *"literal"* six days of Creation and then say that Peter is not speaking literally? Peter is plainly saying that a *"day"* of the Lord is 1,000 earth years. Peter goes on to say in verse 10 that,

"...the day of the Lord will come as a thief in the night; in the which the heavens shall pass away with a great noise, and the elements shall melt with fervent heat, the earth also and the works that are therein shall be burned up..."

So WHEN is the Day of the Lord?

The Day of the Lord is 7 days after the last Day of the Lord -- the day when the Lord rested from his labors of creation. A day of the Lord being 1,000 earth years, Adam fell 4,000 years before Christ, so the year 2,000 would be 6,000 years since the Fall -- and begins the Millennium -- the 7th day, the Day of the Lord. So the Lord still counts his days as equal to 1,000 earth years!

Therefore the creation period was 7,000 years, NOT seven 24-hour days. Neither was it millions of years as evolutionists would have us believe. Adam and Eve were expelled from the presence of God in the Garden of Eden 4,000 years before Christ and in the year 2000, there had been about 2,000 years since Christ's birth. Hence:

7,000 Creation Period

4,000 Adam to Christ

<u>2,000</u> from Christ to the year 2000

13,000 years had passed from the beginning of the Earth's creation to the year 2000. (<u>DOCTRINE OF SALVATION</u>, Joseph F. Smith, Vol. I., pp. 78-80)

After reading the foregoing, an Atheist from Canada wrote,

"I happen to be an atheist. I do not believe in superstition and all religions are founded on myths. On page 15 of your book you said that the creation began 13,000 years ago. Ever hear of the dinosaurs? Some of these lived 130 to 170 million years ago. They are not a myth. I have seen their skeletons. They have several at Elk Island Park in Calgary, Alberta. Where were these animals before your creation? drifting around in space? They have found human bones which are dated at 30,000 to 40,000 years of age. Your creation theory is just so much bull shit."--John Sandbekken, Lot 65, R.R. 2, Riva Ridge Est., Penticton, BC V2A 6J7

Scientists also say that the dinosaur became extinct long before man came on the scene of history. However, along the Puluxy River near Glenrose, Texas were discovered human footprints alongside those of a dinosaur embedded in the sedimentary stone. The human tracks were 16 inches long, 9 inches across and the stride was 6 ft. Its height was probably about 10-15 feet tall. As the tracks were uncovered, it became clear that the man had been walking but broke into a run when the meat-eating three-toed dinosaur commenced its attack and chased the human. Now, this surely is solid stone-proof that man lived in the times of the dinosaurs which contrary to popular scientific opinion were not millions of years ago but only a few thousand.

Consider also that dinosaur bones have been discovered with blood vessels with red blood cells and heme in the marrow of their bones, which would not be possible if the dinosaurs had died millions of years ago. (Discovery by Paleontologist Mary Schweitzer, in an article titled "Dinosaur Shocker," by Helen Fields, reported in Smithsonian Magazine, smithsonian.com)

One method of geological dating is under certain suspicion. That is Carbon-14 dating. This method of dating is based upon the assumption that the rate of the formation of carbon-14 in the upper atmosphere is equal to the rate of decay. Dr. Libbi, who invented the method, reasoned that the formation rate must be equal to the rate of decay because it would only take 30,000 years for this equilibrium to be established, and he says everyone "knows" that the earth is over 30,000 years old. Yet the latest studies indicate that the rate of formation IS NOT EQUAL to the rate of decay -- positive proof that the earth is younger than 30,000 years and that this dating method is invalid. For example, penguins living in the Antarctic today have yielded 3,000 year old carbon-14 ages when tested. Seals killed recently gave ages of 1,000 years. (THE GREAT DINOSAUR MISTAKE, by Kelly L. Seagraves, 1975, Beta Books, California) Other methods of geological dating are equally flawed as documented in Sylvia Baker's booklet, BONE OF CONTENTION, IS EVOLUTION TRUE?

On the other hand, a study described in Dr. Blick's book, A SCIENTIFIC ANALYSIS OF GENESIS, of several thousand fossils using Dr. Libbi's Carbon-14 dating method corrected to account for the actual rate of formation and decay of Carbon-14 in the upper atmosphere, dated the majority of the fossils at around 5,000 years ago -- indicating very strongly that the fossils were created by Noah's flood in 2,345 B.C. and not by millions of years of depositions. The layers of sediments were laid down by the waters of Noah's flood. And in those layers are where the fossil bones are found. If Evolution is correct, then why aren't fossils being created today?

George F. Dodwell, the government astronomer for South Australia, and Fellow of the Royal Astronomical Society of Great Britain, in his unpublished book, THE OBLIQUITY OF THE ECLIPTIC, describes evidence of the cause of Noah's flood from astronomic sites throughout the world. After graphing historical observations made by ancient astronomers from Greece, India, Chinese, Arabia, Egypt and medieval Europe from ancient times up through the middle ages to modern times, and examining ancient astronomic sites such as the solar temples of Egypt, the ancient British monument at Stonehenge, and the solar temple of Tiahuanaco, near Lake Titicaca in the Andes Mountains of Bolivia, astronomer George F. Dodwell determined with a graphic curve of the astronomical observations of the obliquity of the earth's inclination to its ecliptic about the sun from ancient times to the present -- that in the year 2,345 B.C. the earth was suddenly tipped on its axis to about 26.5 degrees by the passage of some planet-size comet that devastated the ancient world with Noah's flood, and which over the years since then has gradually returned back to earth's present 23.45 degree inclination to the earth's ecliptic plane about the Sun. See his graph here:

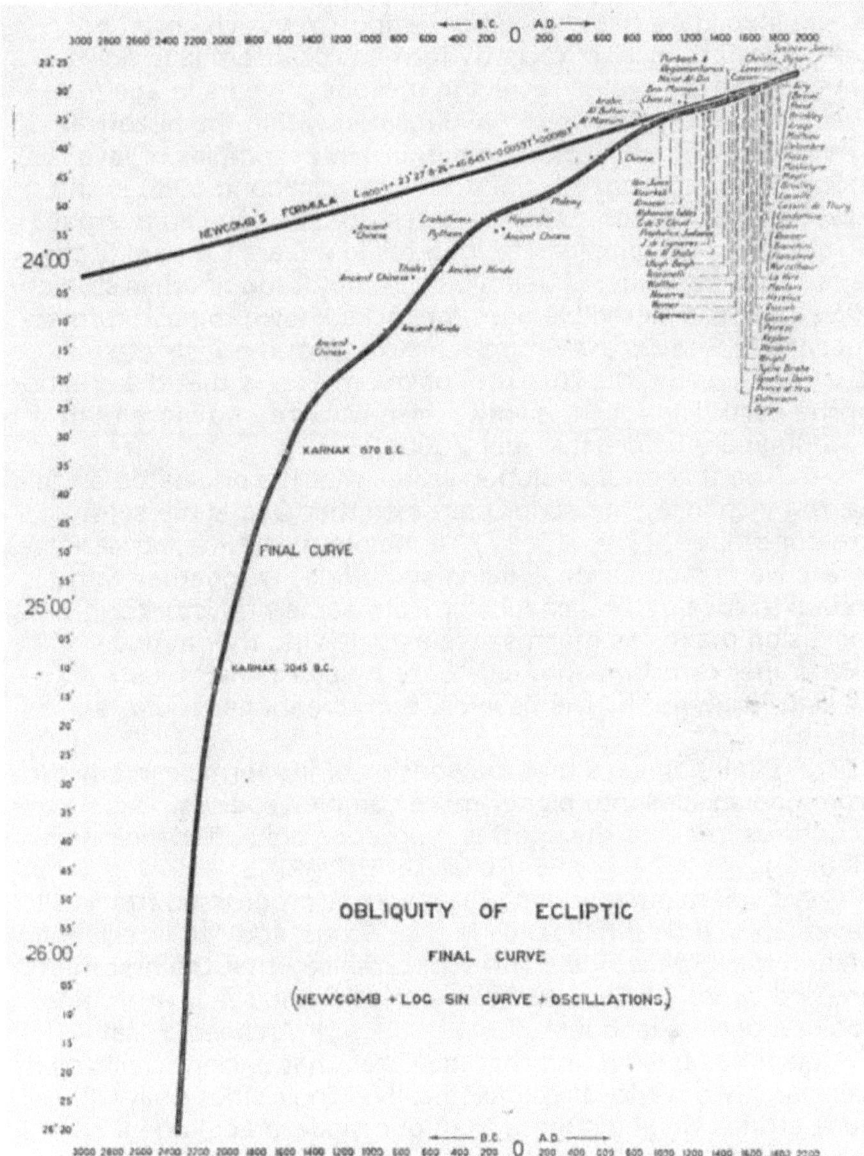

The Obliquity of the Ecliptic by George F. Dodwell

Other scientific methods of dating are under certain suspicion. Consider the Potassium-Argon dating method for lava flows. This method of dating rocks is based on the supposition that there is no Argon in the rocks resulting from radioactive decay of Potassium into Argon when the rock was formed as lava. To find argon in lava rock upon its formation would invalidate this method of dating. Yet such argon has been found to exist in recently formed rock from lava flows, which when lab tested results in millions of years in age for these lava rocks known to have formed within the historical knowledge of man. For example, in 1992, samples of lava stone extruded from Mt. Saint Helens volcano in 1980 in the State of Washington, USA, were tested that gave an average age of the rock samples of 1.056 million years. In reality the lava stone was only 12 years old. A multitude of other studies have reported incredible ages for rocks known to have formed recently. (See *Excess Argon*... article from the Institute for Creation Research). The truth of the matter is that the creation of the Earth is a recent event, which occurred no more than 13,000 years before the year 2000.

2. The theory of evolution states that life originated on this Earth by chance. The scriptures state that God is the sole creator of life. (2 Nephi 2:13) To maintain that the intricate life forms we find on earth all living symbiotically together came into existence by chance luck or from some prehistoric big-bang explosion makes as much sense as believing that a modern skyscraper can come into existence by all the parts just *"falling"* together by themselves. Both creations require a Creator.

3. Evolution says that the species of life forms can advance from one species into higher more complex species. The scriptures say that the species reproduce only after their own kind. (Moses 2:22-24, PEARL OF GREAT PRICE)

4. Evolution maintains that man has progressed from cavemen of little intelligence in the *"Stone Age,"* to intelligent men in the *"Space Age."* The scriptures say that the first men on earth were intelligent, literate people who read, wrote and spoke a perfect language. (Moses 6:5, 6) Archaeological evidence has been found that indicates that ancient civilizations were highly advanced technologically. Their cities built with huge stones weighing more than our modern equipment can move are found all over the world, many of which are submerged off shore on the continental shelves indicating that the ocean levels during their time were much lower than now. See books by Archaeologist Jonathan Gray of New Zealand (now deceased).

5. Evolution denies that Adam was the first man. It insists that cavemen existed thousands of years before intelligent men evolved. The scriptures plainly state that Adam was the first

man on this earth. (Moses 1:34)(I Corinthians 15:45) Adam's wife, Eve, was "the mother of all living." (Genesis 3:20)

6. Evolution maintains that death has existed for millions of years ever since life first appeared on the earth. The scriptures say that before Adam brought death into the world, all things were immortal. (2 Nephi 2:22, 1 Corinthians 15:21, 22)

7. Evolution maintains that the survival of the fiercest (or fittest) is the law of progression. The scriptures plainly state that the peaceful man, obedient to the laws of God, brings prosperity and long life, whereas the fierce wicked man brings about his complete destruction. Concerning the people of America, the Lord told the Brother of Jared about 2,000 years before Christ,

> *"Behold, this is a land which is choice above all other lands; wherefore he that doth possess it shall serve God or shall be swept off; for it is the everlasting decree of God. And it is not until the fullness of iniquity among the children of the land, that they are swept off."* (Ether 2:10)

The Book of Mormon is a historical record of two such civilizations that were, in fact, literally swept off the face of the land of America when they became ripe in iniquity. One, the Jaredite nation whose people came to America in submarines from the Tower of Babel were destroyed in a civil war about 500 B.C. The Nephite nation which originated from two families who came to America in a ship from Jerusalem in 600 B.C. was destroyed in 421 A.D. In fact, both civilizations were destroyed by conspiracies. The Book of Mormon called them *"secret combinations."* Today, once again, civilization in America is being threatened by a conspiracy of the Super Rich -- the very ones who promote the godless theory of Organic Evolution.

The truth concerning the Theory of Organic Evolution is that it is a doctrine of the Devil. It is a fact that the first thing the Communists teach a nation that they have taken over is the Theory of Organic Evolution. It denies the truth revealed by God and persuades men not to believe in Christ.

Some professing Christians believe that they can be consistent in their belief in Jesus Christ and still believe in the Theory of Organic Evolution. They are kidding themselves. Joseph F. Smith, Jr., in his book, MAN, HIS ORIGIN AND DESTINY, shows how this cannot be.

The very fact that Evolution promotes the belief that death existed from the very beginning that life evolved on this planet contradicts the biblical teaching that Adam brought death into this world for disobeying God. The Apostle Paul wrote,

<h1 style="text-align:center">INTRODUCTION</h1>

> *"For as in Adam ALL die, even so in Christ shall ALL be made alive."* (1 Corinthians 15:22)

By partaking of the forbidden fruit which changed their bodies from immortality to mortality, Adam and Eve brought death into the world 4,000 years before Christ's birth. They fell from the presence of God when they were expelled from the Garden of Eden to our surface world.

Since God saves all the works of his hands, in order for fallen man to be able to come back into His presence, a Savior had to be provided. Why? Because fallen man cannot do it alone. This contradicts evolutionist belief, which maintains that we ARE alone, and will constantly evolve on our OWN! Evolution confidently asserts there is no need for a God or a Savior, for after all, this earth and all life on it came into existence on its OWN -- by chance, so they say!

Therefore, if you believe in Evolution you have to admit that Adam did not bring death into this world! And as Joseph F. Smith asserts in his book, if Adam did not bring death into the world, then you will have to admit that there is NO need for a Savior, who the Bible maintains will save us from that death that Adam brought upon us, when Christ resurrects all of us from the grave! There is just NO way around this for any compromise. If you believe in the Theory of Organic Evolution, you have no need to believe in our Savior, Jesus the Christ! That is why Communists put Christians to death! There is just no such thing as a Christian who is a Communist, or a Communist who is a Christian. Communists believe in the Theory of Organic Evolution where there is no need for a God to save us from death and sin. If there is no God, then there is no SIN, for sin is the breaking of God's law. Communists and evolutionists are a law unto themselves, and so they act like animals which they think they are! But we're NOT! We are children of God, and so need to act like it.

Good counsel is given us by the prophets by which to judge all theories. Moroni wrote, 400 years after Christ's visit to the Americas after his resurrection,

> *"For behold, my brethren, it is given unto you to judge, that ye may know good from evil; and the way to judge is as plain, that ye may know with a perfect knowledge, as the daylight is from the dark night."*

> *"For behold, the Spirit of Christ is given to every man, that he may know good from evil; wherefore, I show unto you the way to judge; for every thing which inviteth to do good, and to persuade to believe in Christ, is sent forth by*

the power and gift of Christ; wherefore ye may know with a perfect knowledge it is of God."

"But whatsoever thing persuadeth men to do evil, and believe not in Christ, and deny him, and serve not God, then ye may know with a perfect knowledge it is of the devil; for after this manner doeth the devil work, for he persuadeth no man to do good, no, not one; neither do his angels; neither do they who subject themselves unto him." (Moroni 7:15-17)

Hence we see we have here two men propounding theories, which have had immense influence upon the people of the world. One, his theory based upon the imaginations of his mind, unsupportable by scientific evidence, contradictory to the revealed word of God, and propagated as a doctrine of the devil -- persuading men not to believe in Christ; the other, his theory based upon inspiration of the Holy Ghost, proved to be true by scientific observation, fulfilling scriptural prophecy: which theory proved to be valid?

The objective of science should be to find out the truth. It should also seek for evidence that proves what God has said to be true. In doing so, all theories searching for the truth must be supported by the revealed word of God, and by valid scientific observation, or they shall come to not.

To discover the truth, it should be the first law of a scientist to base his search for the truth on the scriptures and inspiration of the Holy Ghost. If a theory contradicts the scriptures, the revealed word of God, he may know of a surety his theory is false. For God is the greatest of all scientists and cannot lie.

In MORMON DOCTRINE, Apostle Bruce R. McConkie, wrote concerning the importance of basing the search for the truth on revelation from God,

"The Bible, Book of Mormon, Doctrine and Covenants, and Pearl of Great Price (are the standard works of the Church of Jesus Christ of Latter-day Saints). These four volumes of scripture are the standards, the measuring rods, the gauges by which all things are judged. Since they are the will, mind, word, and voice of the Lord (Doctrine & Covenants 68:4) *they are true; consequently, all doctrine, all philosophy, all history, and all matters of whatever nature with which they deal are truly and accurately presented. THE TRUTH OF ALL THINGS IS MEASURED BY THE SCRIPTURES...and one truth never contradicts another."* (MORMON DOCTRINE, Bruce R. McConkie, pp. 690, 691)

The second law of a scientist should be obedience to the commandments of God, so that he will be worthy to receive truth from God. Christ said to the prophet Joseph Smith on May 6, 1833,

> *"The Spirit of truth is of God. I am the Spirit of truth, and John bore record of me, saying; He received a fullness of truth, yea, even of all truth;"*
>
> *"And no man receiveth a fullness unless he keepeth his commandments. He that keepeth HIS commandments receiveth truth and light, until he is glorified in truth and knoweth all things."* (D&C 93:26-28).

The third law of a scientist should be to base his theory on valid scientific observations, not falsehoods, and to test his theory with unrelenting persistence to discover any possible contradictions.

For example, one test that could be conducted to prove the earth is NOT hollow, would be to dig a hole to the center of the earth. If the earth is hollow, the hole will discover a more rapid decrease in the acceleration of gravity with depth than if the earth has matter to its center. A gravimeter lowered down the hole would discover that the center of gravity is located in the earth's shell rather than in the center of the earth.

One such experiment was carried out in the Greenland Icehole Experiment with results being reported in the February 27, 1989 Physical Review Letters journal, p. 986. What the experiment discovered was that at a depth of 2.037 kilometers from the surface of the hole, their gravimeter measured 167.42 LESS miliGals in the acceleration of gravity than if the center of gravity were in the center of the earth, proving that the center of gravity is closer to the surface, which would be the case if the earth is hollow.

In the proposed hole to the center of the earth, the hole will not be able to reach the center of the earth because it will end at the inner surface of the earth's shell. The hole may well pay for itself by possibly finding gold, silver, diamonds or other precious ores in the process of digging the hole. After digging the hole and installing an elevator, many people would pay for the privilege to go to the center of the earth, even if it turns out that the earth is not full of matter, but is instead hollow.

Many object to measuring the truth of theories by the scriptures, saying that they are mysteries of God and shouldn't be dwelt upon. But God says,

> *"Seek not for riches, but for wisdom, and behold, the mysteries of God shall be unfolded unto you, and then shall*

you be rich. Behold, he that hath eternal life is rich...And if thou wilt inquire, thou shalt know mysteries which are great and marvelous; therefore thou shalt exercise thy gift, that thou mayest find out mysteries, THAT THOU MAYEST BRING MANY TO THE KNOWLEDGE OF THE TRUTH, YEA, CONVINCE THEM OF THE ERROR OF THEIR WAYS." (D&C 6:7, 11)

To this end I dedicate this book: To bring many to a knowledge of the truth about the frozen northland, about our earth and of those who live in it, that those who would accept these evidences might know of mysteries which are great and marvelous which have been kept hidden from the beginning of creation; that it might become evident that the Hollow Earth Theory IS a valid theory based upon scientific fact and IS supported by the revealed word of God recorded by His prophets in books of holy scripture; that in obtaining this knowledge we might develop in ourselves a greater appreciation for the fabulous creations of God and His love for His children because He CARES for those who love Him; that the Hollow Earth theory might be considered with enough validity by enough people that an expedition might soon be fitted out to establish without guile or disguise the truth to all the world about the frozen northland and its EDEN hidden somewhere beyond the ice; and last of all, that this work might help, in some small way, to create an awareness of the need for all earth's inhabitants to repent and obey that God which gave them life, or be destroyed at the hands of a secret Godless Conspiracy, whose foundation is the Devil.

CHAPTER ONE
The World's Top Secret!:
The Greatest Geographical Discovery In History!

There is a saying that the truth is stranger than fiction. Certainly this is so concerning the greatest geographical discovery in modern history, which is, that our assumed solid earth is in reality HOLLOW with openings near the poles, and that the inner surface of our planet is populated by a highly advanced civilization of humans.

Confirming that our earth is indeed hollow, in February 1927, Richard E. Byrd of the United States Navy flew north from Alert Base on the northern shore of Ellesmere Island, Canada beyond the north pole on a flight of 1,700 miles towards 87.7 N Latitude, 142.2 E Longitude through a giant hole in the Arctic Ocean and into the hollow interior of our planet. After flying over the Arctic Ocean past the ice, he came to the interior continent of Our Hollow Earth which he saw was covered with vegetation, lakes and rivers and even saw a prehistoric-type mammoth wandering around alive on the terrain below. Some people on the ground even waved to him as he circled them in his airplane before heading back to Alert Base. (From The Hollow Earth file that Col. Billie F. Woodard read while stationed at Area 51, WORLDS BEYOND THE POLES by F. Amadeo Giannini, and GENESIS FOR THE SPACE RACE, by John B. Leith, p. 144.)

John B. Leith states in his book that it was during Byrd's May 9, 1926 flight to the North Pole the previous year in competition with Roald Amundsen's first crossing of the polar sea by dirigible that Byrd's co-pilot Floyd Bennett had first fired Byrd's imagination about the Earth being hollow with polar openings. After passing Amundsen's slower flying dirigible, they had arrived at the North Pole. They then circled around for several miles to make sure they had taken in the pole. Then at one point, they discovered the surface of the ocean devoid of ice and the horizon seemed to enlarge itself beyond the 85th parallel. As they continued in that direction, their compass became erratic, a tail wind increased, the sun's position sank lower on the horizon, and their navigation computations became more uncertain. The Earth's surface seemed to begin dipping down into a deep depression within the Arctic Ocean disappearing into an unending hole down into the Earth. At this point, Byrd decided to turn back and head for their base camp at Spitzbergen. Before reaching their base camp, they had resolved to return the next year to explore this hole into the earth that they had found.

The next year, in February 1927, the United States Navy sponsored Bryd's flight through the North Polar Opening from Alert Base, Ellesmere Island, Northern Canada to the Inner Continent of Our Hollow Earth. John B. Leith, as an intelligence officer of the U.S. government during and after World War II, read this account of Byrd's first flight to Our Hollow Earth through the North Polar Opening in the United States National Archives. He learned that upon Byrd's return from the Arctic, President Calvin Coolidge did not want Byrd's discovery publicized to the world. The photos and flight log of Byrd's flight beyond the Pole into the hollow interior of the Earth were sealed and placed in a vault in the Library of Congress. It was such a fantastic discovery that our Earth is hollow, that President Coolidge did not think anyone would believe it. It was later classified as the "The White Hole Project," and it wasn't until the second year of World War II that the US government fully realized the significance and truth that our earth is hollow when a couple of other pilots also discovered the North Polar Opening. One of those pilots that discovered the North Polar Opening was the famous aviator, Charles A. Lindbergh. He discovered the North Polar Opening while searching out possible northwest passages with his airplane in northern Canada in about 1931. See: Charles A. Lindbergh Discovered the North Polar Opening)

Amadeo Giannini reported that before Byrd's flight beyond the North Pole in February 1927, Byrd had announced over his radio, *"I'd like to see that land BEYOND THE POLE. That area beyond the Pole is the CENTER OF THE GREAT UNKNOWN."*

In 1929, Byrd flew his airplane through both the north and south polar openings. As Byrd was entering the South Polar

Opening, he looked up and could see the other side of the opening, the continent in the sky above his head, and exclaimed, *"THAT ENCHANTED CONTINENT IN THE SKY, LAND OF EVERLASTING MYSTERY!"*

The Operation High Jump expedition of 1946 and 1947 led by Admiral Richard E. Byrd of the United States Navy converged on Antarctica consisting of 4,700 men, with 12 supply ships, warships, icebreakers, and a submarine, with an aircraft carrier with 25 airplanes. Upon departing from the United States, Defense Secretary James Forrestal gave Admiral Byrd his last secret instructions which were to search out where thousands of Germans had escaped to after World War II. The White House called it the *"greatest polar expedition in history."*

In February 1946, Admiral Richard E. Byrd flew his Falcon airplane launched from an aircraft carrier stationed in the ocean north of McMurdo Base in the Antarctic. He flew half way through the South Polar Opening. Overland continually, they went from 300 miles per hour in his propeller airplane before entering the opening which slowed to 50 miles per hour at the half-way point as he met warm air rising up out of the opening that slowed them down. They also went from minus 60 degrees Fahrenheit before entering the opening to plus 60 degrees Fahrenheit at the half-way point before he had to turn back for lack of fuel. At the half-way point through the polar opening, Byrd calculated the opening to be about 125 miles in diameter. (GENESIS FOR THE SPACE RACE, by John B. Leith, Chapter X)

The next year in February 1947, Admiral Byrd flew again through the South Polar Opening, this time with several other planes. After passing through the polar opening, they were approached by flying saucers from the Hollow Earth. Byrd, thinking the saucers were enemy German flying saucers, fired upon them, and so the saucers returned fire with their lasers. Several of Byrd's planes were knocked out of the sky and Byrd was told by radio to leave and go back the way he had come through the South Polar Opening and never return.

Before Byrd's plane could return to the aircraft carrier near Antarctica, flying saucers came up out of the ocean and attacked the Operation High Jump fleet. The Hollow Earth flying saucers knocked several airplanes out of the air that were launched from the carrier, and set their destroyer on fire with laser beams. A nearby Russian ship filmed the battle. (The Russians were still our allies at the end of World War II. You can watch the film of this UFO War in Antarctica: Part One, Part Two, and Part Three, which can also be found at the bottom of my Other Links page on my website).

The fight only lasted a few minutes, and there were some men that died. Some airplanes were downed by the Hollow Earth flying saucers before the saucers dived back beneath the waves. As result, Operation High Jump was terminated six months earlier than they had expected to remain in the Antarctic and returned to the United States. (GENESIS FOR THE SPACE RACE, Chapter XI)

On March 5, 1947, the prestigious Chilean newspaper, EL MERCURIO of Santiago, interviewed Admiral Byrd, saying,

"Admiral Byrd declared today that it was imperative for the United States to initiate defense measures against the possible invasion of the country by hostile aircraft operating from the polar regions.

The Admiral stated, 'I don't want to frighten anyone unduly but it is a bitter reality that in the case of a new war the continental United States will be attacked by aircraft flying in from one or both poles.'"

Upon Byrd's return from that land beyond the South Polar Opening, Byrd announced by radio that, *"The present expedition has opened up A VAST NEW LAND."* (WORLD'S BEYOND THE POLES, F. Amadeo Gianinni)

So great was this fabulous discovery by Admiral Richard E. Byrd, that news of it was quickly suppressed. Shortly after the initial announcement of Admiral Byrd's flights through the South Polar opening, U.S. Navy Intelligence clamped down on any further publication of the greatest geographical discovery in history. Henceforth, our Hollow Earth has been the WORLD'S TOP SECRET!

The subsequent ouster of Forrestal as Secretary of Defense and his untimely death on May 22, 1949 was suspiciously called a *"suicide,"* because he wanted to tell the American people what really happened in the Antarctic with Operation High Jump.

The secret that our Earth is HOLLOW has been hidden purposely from the world. A secret society of the world's richest and most influential men, called the Order of the Illuminati, an international conspiratorial organization, has kept the discovery of our Hollow Earth a secret. This secret society was founded May 1, 1776 at Ingolstadt, Bavaria by Jesuit Priest, Adam Weishaupt, as detailed in Professor John Robison's book, PROOFS OF A CONSPIRACY, first published in 1798.

The World's Top Secret!:
The Greatest Geographical Discovery in History!

The Order of the Illuminati has as its goal to take over the whole world through Socialism and Communism with the objective of eventually forming a one world government. These super-rich of America and Europe come close to controlling the world today. They have giant foundations, repositories of practically tax-free money, with which they push their goals into reality. Through its control of big business and its control of world governments including the United States government, which it controls 60 to 80%, the Illuminati has successfully influenced the scientific community into opposition to any idea that would indicate that the earth is a hollow spheroid, or that anything in space is hollow.

But, no matter how much the scientific community rejects the Hollow Earth Theory, rejection cannot take away the evidence. The evidence that our earth IS hollow and, in fact, every planet in the heavens is hollow, is there for anyone who may desire to know for himself.

A belief in itself does not make a truth. A truth must be demonstrable. If you want to know, for example, if New York City exists, you may do one of two things. One, you may go there personally and see the city for yourself; or Two, you can accept the reliable witness of someone who has been there. In this way the truth of New York City's existence is plainly demonstrated.

Evidence does exist to prove that the Hollow Earth Theory is valid. However, so abundant is the evidence, once a person sets out to earnestly find it, that it cannot properly be called a *"theory,"* but comes so very close to that point of acquiring the truth that it becomes a fact almost. Of course, the truth is always there and never changes, but a fact becomes a fact when the truth is established in the minds of men with sufficiently substantial evidence that it becomes a reality.

To become a reality, the Hollow Earth Theory must be demonstrable. First, we may accept the reliable witness of those who have seen the evidence of the hollow in our earth. And Second, if we doubt that evidence, we may repeat those observations and even make more observations of our own and thus see for ourselves that the earth, contrary to all popular opinion, is indeed HOLLOW! After all, didn't the whole world believe at one time that the earth was flat? Or that the sun revolved around the earth instead of the earth around the sun?

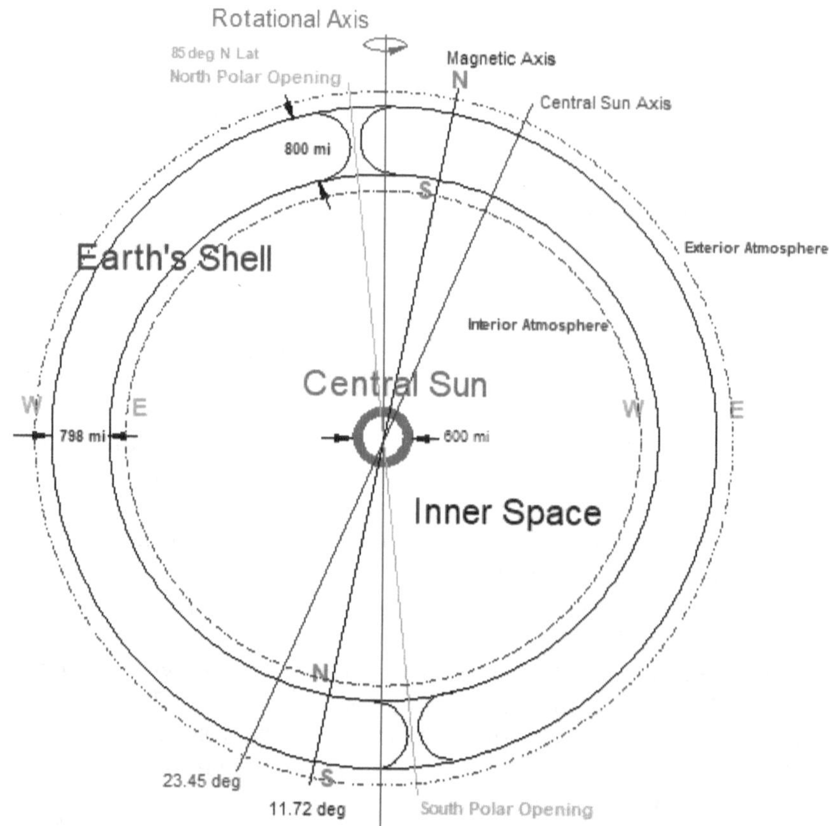

Our Hollow Earth

The Hollow Earth Theory states that our supposed solid earth is in reality HOLLOW, with polar openings that lead into a hollow interior illuminated by a small inner sun. My estimate of the size of the polar openings is that they are about 925 miles in diameter measured at the perimeter of the polar opening, and from that perimeter the earth gradually curves in to a 125 mile diameter polar lip where the sides are closest together and that the distance around the curvature of the polar openings from the outer to inner surfaces of the planet is 1,258 statute miles.

I have concluded that the polar openings are definitely NOT located centered over the polar axis of the earth. However, since the earth most probably was created in rotation, then the polar openings would have formed at the polar axis. There is evidence that earth has been tipped on its axis since creation by the passage of planet-sized comets so that the polar openings are not now centered over the polar rotational axis of the earth.

I am firmly convinced the polar openings DO exist. However, they definitely are NOT the size that Marshall B. Gardner, in his book, A JOURNEY TO THE EARTH'S INTERIOR, thought they were.

Estimates on the thickness of the earth's crust are 300 to 1,800 miles. Explorer Olaf Jansen, in his book, THE SMOKY GOD, estimated the earth's shell to be 300 miles from the outside surface to the inside surface. Marshall B. Gardner figured the earth's shell to be 800 miles thick with 1,400 mile diameter polar openings. The Inner Earth Guide in John Uri Lloyd's book, ETIDORPHA, who lived in the caverns of the earth's crust, reported that the shell is 800 miles thick with the center of gravity 700 miles down. (ETIDORPHA, p. 193) I accept the Inner Earth Guide's estimate in ETIDORPHA as the most likely correct estimate for the thickness of the earth's shell, which has been confirmed by Ret. Col. Billie F. Woodard who has traveled through the 800 mile thick shell of the planet on trips from Area 51 to the capital city of Our Hollow Earth in Hollow Earth tunnel trains.

Jan Lamprecht, in his 1998 book, HOLLOW PLANETS, considered that maybe the orthodox science estimate of the discontinuity inside the earth where P-waves suddenly decrease velocity at a depth of 1,800 miles is perhaps the inner surface of the earth's shell. This, however, would require polar openings more like straight tunnels into the earth. Polar openings with a rounding curvature into the interior through a 1,800 mile thick shell would require polar openings so large they would not fit in the Arctic Ocean and would engulf the entire Antarctic continent. Such a polar opening would be difficult, or impossible, to hide.

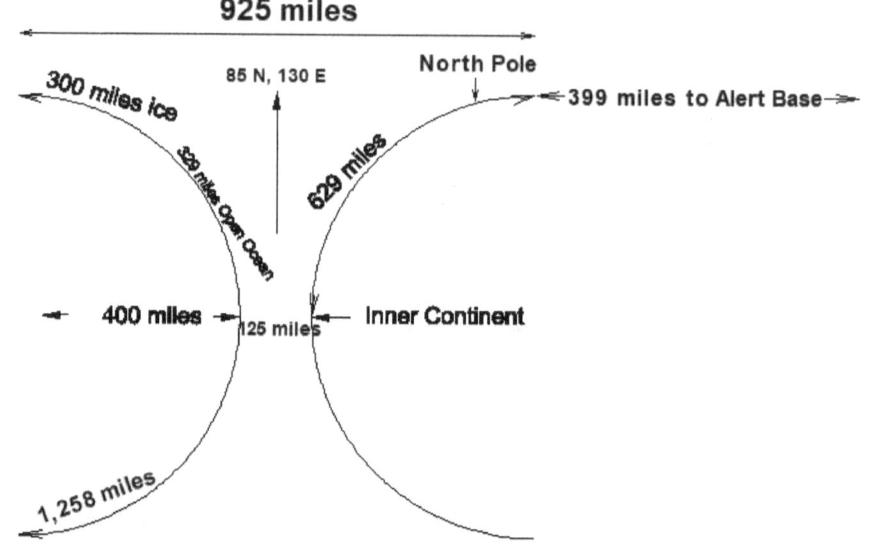

North Polar Opening

It would seem most probable that the polar openings have a rounding curvature into the interior, instead of a tunnel-like shape, and perhaps are about 125 miles wide at the polar *"neck,"* -- the place where the sides of the opening are closest to each other, as reported by Admiral Byrd. My estimate is that the polar openings begin 462.5 statute miles from the center of the opening centered over 85 North and South Latitude and gradually curve into the earth so that at the neck they are 125 miles in diameter.

As reported by the Inner Earth Guide in ETIDORPHA, an earth shell thickness of 800 miles, or about 10% of the earth's diameter is the most realistic estimate of the distance from the outer surface to the inner surface of our Hollow Earth. However, the fact that the acceleration of gravity increases towards the poles indicates that the shell at the equator is THINNER than at the poles.

Orthodox science claims this increase of gravity towards the poles is caused because the surface of the planet at the poles is closer to the center of the earth. More likely, however, it is caused by the greater amount of mass in the thicker polar shell that exerts a greater acceleration of gravity at the poles than the thinner shell at the equator.

We can calculate the thickness of the earth's shell at the equator, assuming that the thickness of the earth's shell at the poles is 800 miles, as given to us by the Inner Earth Guide in ETIDORPHA. The surface acceleration of gravity at the equator is 978 cm/sec^2, as compared to 983 cm/sec^2 at the poles, a difference of 5 cm/sec^2, which cannot be accounted for by centrifugal force at the equator of 3 cm/sec^2. Therefore, the difference of 2 cm/sec^2 has to be accounted for by a thinner shell at the equator than at the poles.

Using this information, we must add the centrifugal force at the equator to the acceleration of gravity at the equator of 978 cm/sec^2 + 3 cm/sec^2 = 981 cm/sec^2 to obtain the acceleration caused by the mass of the earth's shell at the equator. Orthodox Science claims that centrifugal force is a fictitious force, while it is NOT, because they do not believe in the ether of space. Centrifugal force at the equator is an acceleration of an ether flow outwards from the earth, caused by the rotation of the earth, slowing down the acceleration of gravity towards the earth caused by the mass in the equator's shell. If centrifugal force at the equator did not exist, (for example, if the earth stopped rotating) then the acceleration of gravity at the equator would be 981 cm/sec^2. We can then calculate the thickness of the earth's shell at the equator as x = 981 * 800 / 983, which is 798 miles, two miles less than at the poles.

The interior sun of the Earth is calculated by Marshall B. Gardner in his book, to be about 600 miles in diameter. Since space is considered to begin above Earth's atmosphere at about 63 miles above the Earth's outer surface, and since I have calculated a 42.8% less inner surface gravity of the Earth, this would give us 27 miles above the inner surface for the extent of the inner atmosphere. So if we allow 27 miles as the extent of the atmosphere above the inner surface and taking the 800 mile estimate of the Earth's polar shell, and 600 miles for the diameter of the inner sun, that leaves us with: 7,899.8 polar Earth diameter minus 1600 polar thickness of Earth's shell minus 600 miles of the inner sun minus 54 miles of the inner atmosphere equals **5,645.8 miles** inside our Earth that consists of SPACE! Incredible? To scientists who have been fed the Illuminati's line and have swallowed it whole, unknowingly, the Hollow Earth Theory is impossible. But the truth is that the theory is based upon more truth than opinion. All that is lacking is for the theory to be put to the test, and with undoubting scientific investigators, proceed to extract the correct measurements of Our Hollow Earth.

However, the scientific community does consider the Hollow Earth Theory as so much hearsay. Representative of the viewpoint of the scientific community is an article by John M. Prytz in Flying Saucers Magazine. In his article titled, *"The Hollow Earth Hoax,"* Prytz says:

"Believers in the existence of UFOs must have also some ideas as to what they are, why they are here, who made them, and where do they come from."

"In regards to the latter part, where do they come from, and providing one subscribes to the artificial intelligence theory, the only two possible answers are terrestrial sources and extraterrestrial sources. This is the same as saying the source of UFOs is somewhere in the Universe! That is a pretty broad range of places to pick from. However, we can reduce the list slightly and at the same time in doing so tell you at least one place UFOs are not from--that's Earth."

"If UFOs have the Earth as their origin, then either the UFOs must be made and flown by earthlings, or else some other terrestrial theory must be put forth. It is unanimously agreed by practically everyone, believer, skeptic, and non-believer alike that UFOs are not secret weapons or craft of this nation or any other nation. That leaves the other, of course, it's the hollow Earth theory."

"Now belief in the hollow earth theory has declined over the past years, but not fast enough for something that is so obviously wrong. It is rare that one writing about UFOs can make an ABSOLUTE and positive statement, but here is one that can, should, and will be made--the hollow earth theory is not only improbable, but as impossible as the definition of that word allows--and such a strong word too, but rightly chosen under the circumstances, when used with all the factual evidence at hand against it."

"This is not speculation why the hollow earth might not exist, but factual evidence explaining why the hollow earth cannot exist. No speculation or common sense will be used, but rather factual accurate observations of astronomy, geology, physics, oceanography, all of which together disprove totally the notion of a hollow earth. Nor will the lines of evidence be so technical as to put one over on someone not specialized, but basic lines of evidence the nature of which can be found in high school texts, and indeed, even in elementary school general science books. Hence, all the information can be understood and easily

checked should one not be so prejudiced to do so of course. The hollow earth theory will be looked at from two points of view, that of the hollow earth itself, and that of the polar openings." ("The Hollow Earth Hoax," *Flying Saucers* Magazine, June 1970, p. 34)

Let us pause here to comment upon Prytz's source of evidence with which he will try to disprove the Hollow Earth Theory. We must remember that if we are to accept the testimony of others in obtaining evidence, their witness must be reliable. Prytz states his source of evidence and bases his rebuttals to the Hollow Earth Theory on *"evidence"* laid down even in high school and grade school texts. These authoritative reference books are, indeed, as unreliable as are their perpetrators.

Prytz either ignores or does not know that the Illuminati substantially controls the government school system of the United States. Certainly, back at the turn of the 20th century when John Dewey succeeded in imposing government education on the nation, perhaps the citizens did not know that the tenth plank of Karl Marx's Communist Manifesto for the destruction of a capitalist nation was to provide *"free education to all children in public schools."* If the citizens of the United States had known that the Illuminati had gained control over the government, surely they would not have allowed the government to obtain control over the schools.

As the result, text books have been rewritten to conform with the Illuminati's viewpoint. And Dr. Felix Wittmer shows in his book, <u>CONQUEST OF THE AMERICAN MIND</u>, how the

"Columbia Teachers College have in the course of twenty years turned thousands and thousands of teachers into missionaries of the collectivist creed." (Meador Publishing Co. Boston, 1956, p. 39)

Columbia University and many others came to be controlled outright through grants by the super-rich members of the Illuminati. In Rene A. Wormser's book, <u>FOUNDATIONS: THEIR POWER AND INFLUENCE</u>, he shows how the Illuminati's foundations have become a major influence in education in the United States through grants to Universities, through the influence of the Progressive Education Association and the financing and promotion of socialist textbooks. On pages 209-210, Wormser shows how history books were rewritten for the schools in America to keep Americans from learning the truth about the Illuminati.

In addition to this influence on American education, the Illuminati have exerted substantial CONTROL over the educational system of the U.S. ever since the Illuminati came into control of the government of the United States at the beginning of the 20th century. Their control has increased in the government until today it is estimated that they control the government between 60-80%. This control over the government is made possible by the election of Illuminati financed Presidents of the United States. These presidents, in turn, appoint to high government positions members of the Illuminati's most influential organization in America, The Council on Foreign Relations (C.F.R.). In Gary Allen's book, NONE DARE CALL IT CONSPIRACY, he lists 110 members of the C.F.R. appointed to high government positions in the Nixon administration. On page 22 of the April, 1982 issue of American Opinion Magazine, Alan Stang points out that President Reagan *"has appointed seventy-two of these C.F.R. members to top federal posts...it is fair to say that the Council on Foreign Relations controls the federal government under Reagan. Indeed, it is fair to say that the C.F.R. IS the government."* Subsequent administrations have also been filled primarily with C.F.R. members.

Since the Judicial body of the U.S. government is staffed by presidential appointment with Congressional approval, the Supreme Court is also staffed with a majority of supporters of the Illuminati. And in the Congress are some 150 congressmen (in 1977) bought by Big Labor (A.F.L.-C.I.O.) which is controlled by the Illuminati. Also, Susan Huck reveals in her article *"Buying Congress,"* in the July-August, 1977 issue of *American Opinion Magazine*, that

> *"The Rockefeller-financed National Committee for an Effective Congress is a political front for the Far Left which has helped to buy the seats of 200 members of the current (94th) Congress."*

This control of the Illuminati over the U.S. government gives them control over the government school system to a substantial degree. And even though there is no federal school system, the Illuminati have exerted their influence through federal grants, national educators associations, Federal laws and regulations, the *"free"* school lunch program, and especially through the influence of rewritten school texts that conform to the Illuminati's' viewpoint. This then puts these texts under the suspicion of promoting false theories and ignoring the truth. John Prytz, who takes his evidence from school texts, therefore could very well be basing his evidence against the Hollow Earth Theory on the Illuminati's *"line."* And as we shall see later on, the Illuminati has vital reasons for keeping the discovery of Our Hollow Earth a secret.

From this knowledge of the Illuminati's control over education we also find it conceivable that the Conspiracy also exerts substantial control over the scientists who work in the Educational System, thus pressuring them into an anti-Hollow Earth stand. Those few who would venture too close to disclosing the true origin of FLYING SAUCERS from the interior of Our Hollow Earth, mysteriously take their lives in their hands, or at least run the risk of losing their jobs.

Consider these incidents from Brinsley Le Poer Trench's book, SECRET OF THE AGES, UFO's FROM INSIDE THE EARTH. Trench writes,

"Veteran ufologists will recall the sensational closing down of Albert K. Bender's International Flying Saucer Bureau back in 1953."

"At that time, Bender had quite a large international movement. He had been in close touch with Edgar Jarrold, another ufologist, who headed the Australian Flying Saucer Bureau, and they had been working together on a theory linking the UFOs with Antarctica."

"Mr. Bender had written an article for publication in his magazine which divulged the secret of the saucers. That article was never published. Three 'Men in Black' visited Bender at his home. They so frightened Bender that for a long time he gave up all UFO research. This incident has been fully described by Gray Barker in his book, THEY KNEW TOO MUCH ABOUT FLYING SAUCERS."

"Bender's colleagues questioned him afterwards for some time and with his permission tape recorded his answers. One of the questions put to him was: 'How did you find out about it? Can't you tell me just where you got your theory?'"

"Bender's answer was: 'All I can say is this: It was something that I was thinking about for a long time. I went into the fantastic and came up with the answer." (SECRET OF THE AGES, pp. 170, 171)

The *"Men in Black"* who caused Bender to desist from publishing his theory of the origin of UFOs, which most likely was the Hollow Earth Theory, are mysterious persons who consistently warn people not to publicize knowledge they happen to receive about UFOs. However, it is the belief of this author, that *"Men in Black,"* called such because they wear all black clothing, are member-agents of a satanic church of Lucifer (which is also part of the International Communist-Illuminist-Jesuit Conspiracy) and, in fact, work against the purposes of the Ufonauts. Their commandments are exactly opposite those of God and a livelihood is provided for those who covenant to do anything they are ordered to do. Their purpose in covering up knowledge of UFOs and our Hollow Earth is to prevent the expansion of the political Kingdom of God to the surface world of our planet.

The Illuminati has quieted such persons as Albert K. Bender who come close to revealing the origin of the Flying Saucers. The *"Men in Black"* generally try to *"scare"* their contacts into not disclosing their information about the Flying Saucers. If that does not work, then the Illuminati's agents may KILL those who would go too far. These victims of the Illuminati are scientists, who after much study and observation, come up with the true origin of UFOs.

Le Poer Trench cites some such incidents of the Illuminati's dirty work:

"Consider the case of the late Dr. Morris K. Jessup, a professional astronomer. He was, also, a prominent ufologist and wrote several fine books on the subject. Jessup believed that the majority of the UFOs emanated from what he called the binary Earth-Moon system. He considered the UFOs came from installations inside the interior of the moon and from inside the earth, and that they had bases in the oceans, too."

"On 20 April, 1959, Dr. Morris K. Jessup was found dead. He had apparently committed suicide by inhaling carbon monoxide, presumably by connecting a hose to the exhaust of his station wagon and introducing it inside the vehicle."

"Did Jessup know too much?"

"A very prominent scientist and leading ufologist was the late Dr. James E. McDonald, senior physicist, Institute of Atmospheric Physics, and Professor, Department of Meteorology, University of Arizona."

"Dr. McDonald became a very outstanding speaker and writer on UFOs, and was very critical of the U.S. Air Force's handling of the UFO situation. Keel wrote that Dr. McDonald ... 'privately discussed in his last years, the possibility that alien beings were not only present on this

planet but were systematically taking over top posts in the government and military.'"

"On 13 June, 1971, his body was found in the desert north of Tucson, Arizona. He, too, had apparently done away with himself."

"Did McDonald know too much?"

"Then there is the case of Professor Rene Hardy. He was a scientist of world repute; a prodigious inventor with over 250 patents to his name, in the fields of electronics, radio, television, ultrasonics, and optics. He was interested in ufology, parapsychology and interstellar navigation, among many other subjects."

"On 12 June, 1972, the Professor was found dead with a bullet in his head, and a revolver in his hand, just two days before he was to have announced a highly important discovery concerning space phenomena. There was no reason for his apparent suicide." (SECRET OF THE AGES, pp. 170-175)

That the U.S. government is engaged in such gangsterism is hard to take, but isn't that the way with all truth? The method used to operate the cover-up is by murdering those who would disclose the truth.

Even the Hollow Earth Theory's most recent advocates are no longer around. Ray Palmer, editor of SEARCH and FLYING SAUCERS magazines was being harassed physically and over the phone by callers from the US military for his convincing arguments in favor of the Hollow Earth Theory. Then all of a sudden, on a trip, Palmer became mysteriously ill, was rushed to a hospital and died within hours. Palmer's 1978 death was a real victory for the Illuminati. Many of his magazine readers wrote in expressing their suspicion concerning his strange death.

Then there is Raymond Bernard, author of the book, THE HOLLOW EARTH. Some say he was last heard of in Brazil. He supposedly took an expedition into the Matto Grosso of Brazil's dense western jungle looking for legendary subterranean super-advanced cavern cities in the late 60's and NEVER returned. Some believe he actually found the cities. Of course, how do we know for sure? Hollow Earth researcher Brownley says this is all hogwash. He says he talked to Bernard's former secretary who said he died of pneumonia here in the United States. But still, HE is no longer with us -- another victory for the Illuminati's cover-up. In 1986, another Hollow Earth publisher, Gray Barker, died soon after publishing and reissuing a number of out-of-print hollow earth books.

One of them was a compilation of Ray Palmer's SEARCH Magazine articles on the Hollow Earth which Barker titled, LANDS BEYOND THE POLES. Although Barker and Palmer could have died of some complication of old age, yet their demise leaves a void among faithful promoters of the Hollow Earth Theory. Hollow Earth researcher Bruce Walton after publishing his book on Mount Shasta seems to have drifted out of the hollow earth research scene after having a devastating automobile accident. Nevertheless, his GUIDE TO INNER EARTH, published by Amazon.com, continues to interest researchers into the hollow earth theory.

In 2003, I was contacted by Steve Currey of the Expedition Company of Provo, Utah who asked me to help him put together an expedition to our Hollow Earth. I helped him put together a nice website, https://www.voyagehollowearth.com/ for our expedition called the Voyage to Our Hollow Earth Expedition with plans to charter an airliner into the Arctic to pin-point the location of the North Polar Opening, and then follow that with a sea expedition chartering the Russian Nuclear Icebreaker, the Yamal.

Unfortunately, in 2006, three months before the departure date of the over flight of the Arctic, Steve discovered he had six inoperable brain tumors, and died two months later. His family cancelled their involvement in the expedition and returned all expedition members' monies.

With the demise of Steve Currey, an expedition member, Dr. Brooks Agnew, with a PhD degree in Physics, offered to be our expedition leader. He created a nice expedition website and renamed the expedition, the North Pole Inner Earth Expedition. A New York film company offered to fund the expedition with the Yamal, but then mysteriously disappeared. Their phone was disconnected, and when Dr. Agnew sent people to their address in New York City their offices were found empty. After seven years of promoting the North Pole Inner Earth Expedition, Dr. Agnew resigned as our expedition leader in September 2013 citing losses to his start-up business, Vision Motor Cars as the main reason for resigning. Basically, it was financial blackmail, because the largest stockholder of his company threatened to withdraw all his money from Vision Motor Cars because Dr. Agnew was involved as leader of our expedition to the hollow earth.

In 2008, I was contacted by Retired Col. Billie Faye Woodard. He expressed an interest in joining our expedition to go to the Hollow Earth and said he had some important information we needed – the coordinates of the North Polar Opening, which he said he had obtained from the Hollow Earth file he had access to while working 11.5 years at the top secret facility, Area 51, in Nevada, USA. So we visited him in Pahrump, Nevada, after which we put together an expedition plan to the Inner Continent by seaplane. We located a bush pilot in Alaska, who was a retired airline pilot for Alaska Airlines, Terry Smith.

Terry agreed to hire a couple of pilots to take us in his Albatross seaplane as close as we could get to the Inner Continent through the North Polar Opening. Unfortunately, before we could come up with the money to pay for the flight to the Inner Continent, Terry Smith crashed his airplane into a mountain in Alaska while taking an Alaskan ex-Senator and 4 others fishing along with two teenagers. The two teenagers survived the crash. Five people including Terry Smith perished in the crash. See: https://www.adn.com/alaska-news/article/gci-crash-grassroots-search-and-rescue-led-horrible-crash-scene/2010/08/12/

Just when there appeared to have occurred a void in hollow earth research by the demise of some of its top promoters, the 1990's saw a resurgence of interest in the Hollow Earth theory. Dennis Crenshaw published a Hollow Earth newsletter for several years and had a nice website at https://thehollowearthinsider.com/ (now defunct). Danny Weiss consolidated his International Society for a Complete Earth with a nice website at https://www.hollowearthresearch.org intent on reenacting Admiral Byrd's flight into the North Polar Opening. Jan

Lamprecht of South Africa published a 600 page tome on the most recent scientific evidences for hollow planets. He also had a nice website. The Internet has assisted in a renewed interest in hollow planets research with email, websites and forums.

Nevertheless, incidents continue to surface indicating that the Illuminati is opposed to the widespread knowledge that our earth is HOLLOW and is the origin of the majority of the FLYING SAUCERS seen around the world. Such knowledge would destroy the Illuminati's/Jesuit's hold upon the governments and economies of the world. Widespread knowledge of the Illuminati would cause the people to rise up and oust them from the United States government which the Illuminati uses to finance and extend their control of the world.

Additionally, they do not want flying saucer technology to get out to the common people. They want a monopoly on flying saucer technology, which they have obtained by downing a few flying saucer craft with high-powered radar, and advanced particle beam weapons and plan on using this technology to consolidate their power over the nations of Earth. They also do not want the free energy technology that powers the flying saucers to get out to the people because it would destroy their energy monopolies around the world.

The Hollow Earth Nation is daily carrying on reconnaissance of the Illuminati's activities via their flying saucers. Their knowledge of the Illuminati, if communicated to the people would be disastrous for the conspiracy, which can only operate successfully in secret through lies and deceit in their drive to establish world dictatorship over us. For this reason, the Illuminati and their Jesuit controllers want the discovery of our Hollow Earth kept a secret.

The Hollow Earth Nation, on the other hand, it seems would want the discovery that our earth is hollow kept a secret in order to preserve their homeland from wholesale immigration and infiltration of their land by wicked and lawless people. In the opinion of this author, it will be only when people of our outer earth repent of their wicked ways and begin to live the commandments of God that the Hollow Earth Nation will permit people on the earth's surface to visit their homeland or to migrate there. It becomes imperative then, that those of us who love freedom alert our friends and neighbors to the onslaught of the Godless, Satanic forces of the Jesuits and their Order of the Illuminati and its instrument of world dictatorship, the International Communist Conspiracy, and at the same time, prepare ourselves for ultimate contact and alliance with the highly advanced and righteous nation inside our Hollow Earth.

The Hollow Earth Theory is intimately involved in the Conspiratorial Theory of History as uncovered by the late Dr. Cleon Skousen, of the National Center For Constitutional Studies, in his books, THE NAKED CAPITALIST, and THE NAKED COMMUNIST; Gary Allen, a member of the patriotic John Birch Society, in his book, NONE DARE CALL IT CONSPIRACY, and Robert L. Preston, in his book, WAKE-UP AMERICA IT's LATER THAN YOU THINK!, all of whom published their books in 1972. Skousen's book, THE NAKED CAPITALIST, is a commentary on Dr. Carol Quigley's book, TRAGEDY AND HOPE, A HISTORY OF CIVILIZATION IN OUR OWN TIME. Dr. Quigley was an insider of the Illuminati who offered their proposal for a world government under their control as the HOPE of the world and a TRAGEDY to citizens unwilling to accept it. A more recent book, THE HIDDEN HAND, by Ralph Epperson, a University of Arizona professor, is also an excellent treatise on the Conspiracy Theory of History.

For years, I have had certain people tell me that the Jesuit Order of the Catholic Church was at the very top of world control. It seems that whenever I would investigate strange events in the world, there would be a report of Jesuit involvement. It had seemed strange to me that the Vatican and Italy were allies of the Third Reich under Hitler. Then when the communists rose to such heights of power in the Soviet Union, it seemed strange that a Pope would come from the Communist country of Poland. And now, with the ascent of Pope Francis, and the report that he is a Jesuit, it all starts to make sense. Recently, when I read Johnny Cirucci's book, ILLUMINATI UNMASKED: EVERYTHING YOU NEED TO KNOW ABOUT THE "NEW WORLD ORDER" AND HOW WE WILL BEAT IT, in which he lays out all the involvement of the Jesuit Order throughout the world, I became even more convinced that the Jesuit Order is at the top of world control, including the Catholic Church, all the governments of the world, the Communist movement and even the Illuminati. The founder of the Illuminati, Adam Weishaupt was a Jesuit Priest. So now it makes sense that the Jesuit Order is at the top of world control, as illustrated by the recent visit of Pope Francis to America in 2015 to speak to Congress and the United Nations, as well as his visit to Russia where he met with their Christian church leader.

This all confirms what the scriptures say, that the Great and Abominable Church dominates our outer world today, at least until Christ returns in glory to reign as King of Kings for a thousand years. Both the Book of Mormon and the Bible refer to this Great and Abominable Church that dominates our outer world. The ancient American prophet Nephi saw a vision, of which he wrote, as recorded in 1 Nephi 14:

17 *"And when the day cometh that the wrath of God is poured out upon the mother of harlots, which is the great and abominable church of all the earth, whose founder is the devil, then, at that day, the work of the Father shall commence, in preparing the way for the fulfilling of his covenants, which he hath made to his people who are of the house of Israel."*

And in the Bible, John, the Apostle, was shown in vision the Great and Abominable Church over all the earth, and called it, *"Mystery Babylon the Great, the Mother Of Harlots and Abominations of the Earth,"* as recorded in Revelation 17:

1 *"And there came one of the seven angels which had the seven vials, and talked with me, saying unto me, Come hither; I will shew unto thee the judgment of the great whore that sitteth upon many waters:*

2 With whom the kings of the earth have committed fornication, and the inhabitants of the earth have been made drunk with the wine of her fornication.

3 So he carried me away in the spirit into the wilderness: and I saw a woman sit upon a scarlet coloured beast, full of names of blasphemy, having seven heads and ten horns.

4 And the woman was arrayed in purple and scarlet colour, and decked with gold and precious stones and pearls, having a golden cup in her hand full of abominations and filthiness of her fornication:

5 And upon her forehead was a name written, MYSTERY, BABYLON THE GREAT, THE MOTHER OF HARLOTS AND ABOMINATIONS OF THE EARTH."

Without a knowledge of the Jesuits, their Illuminati and their conspiracy to establish a one world government as a military dictatorship of the UNTIED NATIONS, it is impossible to understand why the discovery that our earth is hollow and is the origin of most of the FLYING SAUCERS seen around the world and why they are here, has been kept secret from the whole world. With this knowledge now revealed, you may now KNOW the enemy which is within our borders!

The idea that our earth is a solid sphere is not a fact, but a theory treated as fact in school texts. Arnold de Azevedo, in his PHYSICAL GEOGRAPHY, wrote about the world beneath our feet, concerning which scientists know nothing beyond a few miles deep, propounding FALSE theories, hypothesis and conjectures to hide their ignorance:

"We have below our feet an immense region whose radius is 6,290 kilometers, which is COMPLETELY UNKNOWN, challenging the conceit and competence of scientists."

Thus with a knowledge of the Illuminati's control of the governments of the world and their educational systems, economies and even big industry including the scientific community in that control or influence, and additionally, with the knowledge of the Jesuit control of the Illuminati, we can begin to understand the absolute rejection by scientists of the Hollow Earth Theory. But rejection does not take away the evidence. The evidence is there for all to see who would take the pains to study it out or to go see for themselves.

In Dr. Steven Greer's recent 2017 book, UNACKNOWLEDGED is a document by a W.B. Smith of the Canadian Department of Transportation, titled TOP SECRET, from Ottawa, Ontario, dated November 21, 1950, in which W.B. Smith, wrote that the Canadian government was working on flying saucer technology. In the course of attempting to discover how flying saucers work, he made inquiries with top government officials in the United States government who told him that the subject was the most classified subject in the United States government, even higher than the H-bomb, that flying saucers exist and their modus operandi is under investigation by a small group headed by Dr. Vannevar Bush. (UNACKNOWLEDGED, pages 723, 724)

So if in the 1950's the subject of Flying Saucers and how they operate was the Top Secret of the United States and Canadian governments (and continues to be Top Secret, according to Dr. Greer), certainly where they come from would be even a higher secret, which is why I have called this the World's Top Secret: that Our Earth IS hollow and is where most of the flying saucers seen around the world are coming from.

It is my prediction that discovery of our Hollow Earth WILL be established as fact before the world shortly, and this knowledge will DESTROY the International Illuminist-

Communist Conspiracy and their bosses in the Jesuit Order with their false and Godless theories, including ORGANIC EVOLUTION, and including their thrust toward world domination, because the political Kingdom of God, which is presently established within Our Hollow Earth will expand to our surface world and unite itself with the spiritual Kingdom of God -- The Church of Jesus Christ of Latter-day Saints in preparation for the Second Coming of our Lord and Savior, our Creator, Jesus Christ who will establish a world government with Himself as King!

CHAPTER TWO
Our Earth IS Hollow! -- The Scientific Evidence

Therefore, let us take the evidence of the scientific community as presented by John M. Prytz in his article, "The Hollow Earth Hoax," and see how well his evidence recorded in high school texts, even in elementary texts, stands up against the first-hand observations of trained observers.

Number One item of Prytz's *factual evidence explaining why the hollow earth (and its polar openings) cannot exist,"* states:

> *"The North polar area is covered with water, commonly known as the Arctic Ocean. It has an area of 3,622,200 square miles and an average depth of 4,362 feet. This Arctic Ocean is the name for water north of the continental land masses in the region of the Arctic Circle, and is often covered with pack ice. If any polar opening were present, the water under the force of gravity would drain into the hole, like the water which drains into a sink drain. Therefore, either the hollow earth would be FLOODED TO CAPACITY, or else water would still be draining down the hole, not only causing a gigantic whirlpool, but also lowering the level of all the world's oceans--such a drop has not been noticed."*

Our answer to Mr. Prytz is that the North Polar Opening does indeed exist. However, the Arctic Ocean does not "drain" into the "hole," because the earth's principal center of gravity is not at its center, as would be the case if the earth were a solid sphere.

A look at my drawing of Our Hollow Earth, which is a center section through the earth's polar axis, we see that the earth's shell is approximately 800 miles from the outer to the inner surface. As we previously determined, the earth's atmosphere extends from the inner surface upwards 27 miles towards an Inner Sun of 600 miles in diameter, thus giving us some 5,645 miles inside our earth of SPACE – a near vacuum. Any student of physics knows that gravity is caused by mass. Empty space alone does not produce the force of gravity.

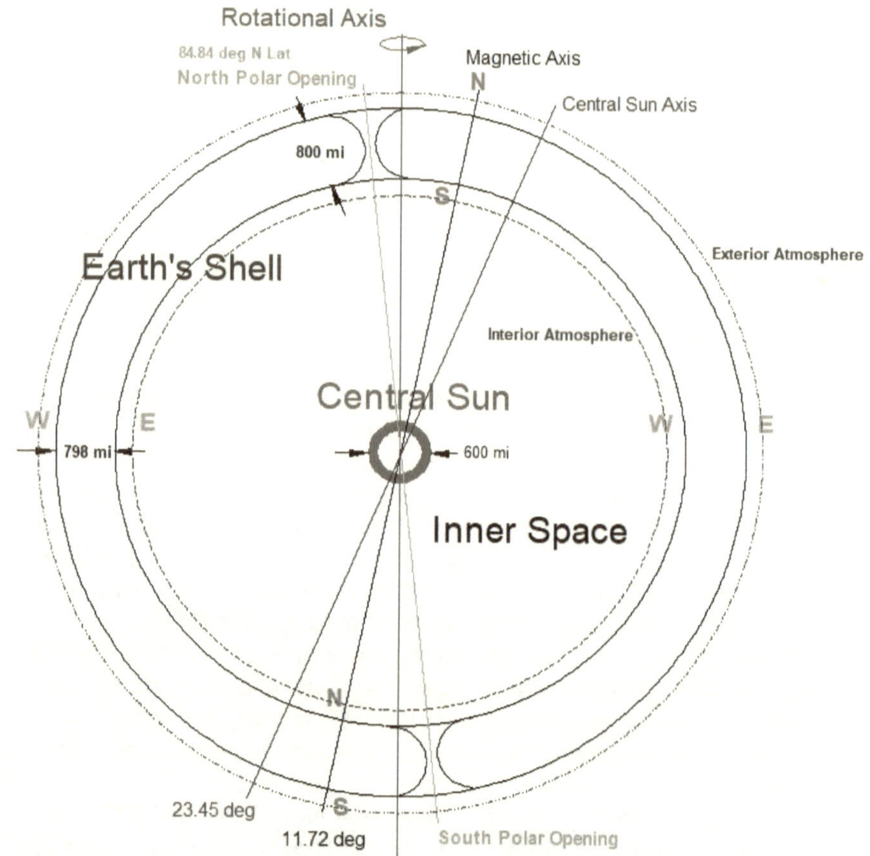

Our Hollow Earth

Granted, an inner sun would contain some of the mass of the earth, but very little, in comparison to the shell. Because our earth is hollow and not full of matter as orthodox science claims, a small center of gravity is located in the inner sun, but because the preponderance of the mass in a hollow earth is located in its shell, the primary center of gravity is located in the shell, somewhere between the outer and inner surfaces, and describes a sphere -- a central sphere of gravity. Because of a higher density in the earth's shell toward the inner surface, the central sphere of gravity is located 700 miles down from the outer surface in the earth's 800 mile thick shell -- according to the Inner Earth guide in ETIDORPHA. (ETIDORPHA, p. 193)

Explorer Olaf Jansen, who supposed the earth's shell to be 300 miles thick, explained the earth's center of gravity thus:

"Sir James Ross claimed to have discovered the magnetic pole at about seventy-four degrees latitude. This is wrong--the magnetic pole is exactly one-half the distance through the earth's crust. Thus, if the earth's crust is three hundred miles in thickness, which is the distance I estimate it to be, then the magnetic pole is undoubtedly one hundred and fifty miles below the surface of the earth, it matters not where the test is made. And at this particular point one hundred and fifty miles below the surface, gravity ceases, becomes neutralized; and when we pass beyond that point on toward the 'inside' surface of the earth, a reverse attraction geometrically increases in power, until the other one hundred and fifty miles of distance is traversed, which would bring us out on the 'inside' of the earth."

"Thus, if a hole were bored down through the earth's crust at London, Paris, New York, Chicago, or Los Angeles, a distance of three hundred miles, it would connect the two surfaces. While the inertia and momentum of a weight dropped in from the 'outside' surface would carry it far past the magnetic center, yet, before reaching the 'inside' surface of the earth it would gradually diminish in speed, after passing the halfway point, finally pause and immediately fall back toward the 'outside' surface, and continue thus to oscillate like the swinging of a pendulum with the power removed, until it would finally rest at the magnetic center, or at that particular point exactly one-half the distance between the 'outside' surface and the 'inside' surface of the earth." (THE SMOKY GOD, pp. 160-162)

Therefore, if we were to go over the 1,258 mile curve of the North Polar Opening, which is the semi circumference of a 800 mile estimate of the earth's shell, we would always be exerted toward the central sphere of gravity of earth's mass, which contrary to popular opinion is not primarily in the center of the earth but instead is actually a central *"sphere"* of gravity located between the inner and outer surfaces of the earth's shell.

As such the Arctic Ocean would no more empty into the *"hole"* than could the Australian Continent fall out into space.

Let us take Prytz's SECOND objection to the Hollow Earth Theory. He states, *"No unknown land masses exist in the area attributed to where the North Polar Opening theoretically exists."*

Prytz would have the reader believe that because no unknown lands are shown on north polar maps in high school texts that such lands do not exist!

But the observations of the polar explorers, on the other hand, do support the Hollow Earth Theory of lands inside the Polar Regions which have even a subtropical climate. Let us examine, therefore, the descriptions of polar explorers of what they observed in the polar region and then ask ourselves, does the evidence support orthodox Arctic theory or does it support the Hollow Earth Theory with polar openings leading to a land inside Our Hollow Earth?

Observations by north polar explorers indicate that there is indeed a land in the far north with a subtropical climate heated by a sister sun inside Our Hollow Earth. For example, explorers' reports of abundant animal and bird life in the summer time in the far north indicates a homeland in the north from which they extend in the summer further south and to which they are seen to migrate in the fall.

Explorer Hays observed abundant insect life in the far north. When he was in latitude 78 degrees, 17 minutes in early July he said,

> *"I secured a yellow-winged butterfly, and--who would believe it--a mosquito...ten moths, three spiders, two bees and two flies."* (THE OPEN POLAR SEA, p. 413)

Notice the element of surprise that many explorers expressed resulting from the discovery of conditions which they weren't expecting.

Explorer Greely, in his book, THREE YEARS OF ARCTIC SERVICE, in Grinnell Land in June of 1881, reports birds of an unknown species, butterflies, bumblebees, so many flies they couldn't sleep at night, and temperatures of 47 and 50 degrees at latitude 81 degrees 49 minutes north. He also found plenty willow to make fire, and much driftwood. (Chapter 26, Vol. I)

A Swedish expedition under Otto Torell, found near Trurenberg Bay in the Arctic Sea, trees floating with green buds on them and among them was found the seed of the tropical Entada Bean which measured 2.25 inches across. (Gardner, p. 253)

Explorer Sverdrup at 81 degrees north found so many hares that they named one inlet, Hare Fiord. Also nearly all expedition parties found enough game to keep their exploring parties well fed with meat. These included herds of musk-oxen and reindeer. (Gardner p. 254)

Captain Beechey saw so many birds on the west coast of Spitsbergen that sometimes a single shot killed thirty of them. (Gardner p. 254)

All explorers observed that not all animals migrate south to escape the cold Arctic winds in winter, but many instead go north. Where do they go? Greely, surprised at the tremendous amount of wildlife in a supposed frozen north wrote,

> *"Surely this presence of birds and flowers and beasts was a greeting on nature's part to our new home."*

Explorer Kane reported seeing several groups of Brent Geese, which is an American migratory bird, flying NORTHEAST in their wedge-shaped line of flight at 80 degrees 50' north at Cape Jackson, near Grinnell Land in late June 1854.

Explorer Greely makes this statement of the northward migration of bears,

> *"Lieutenant Lockwood, in May, 1882, noticed bear tracks (going NORTHEAST) on the north coast of Greenland, near Cape Benet in 83 degrees 3' N.,"* and commented, *"...I cannot understand why the bear ever leaves the rich hunting-field of the 'North Water' for the desolate shores of the northward."* (THREE YEARS, p. 366)

Greely also wrote about the Ross's Gull,

> *"...the observations of Murdoch at Point Barrow show that this bird, in thousands, passes over that point to the NORTHEAST in October, none of which were seen to return."* (THREE YEARS, p. 383)

Explorer Adolf Erick Nordenskiold, leader of a Swedish expedition, recorded in THE ARCTIC VOYAGE OF 1858-1878, that on May 23, they saw north of Amsterdam Island (by Spitsbergen),

> *"...great numbers of barnacle geese...flying towards the NORTHWEST, perhaps to some land more northerly than Spitsbergen. (There is no such land on our present-day maps) The existence of such a land,"* wrote Nordenskiold, *"is considered quite certain by the walrus-hunters, who state that at the most northerly point hitherto reached, such flocks of birds are seen steering their course in rapid flight yet farther toward the north."* (Gardner, p. 160)

Daines Barrington, in his book, ON THE POSSIBILITY OF APPROACHING THE NORTH POLE, wrote that observers in Spitsbergen have always noticed in spring, just before the hatching season, the wild ducks, geese, and other birds, fly in a northerly direction. There is also a heavy fall migration to the north.

In HEARNES JOURNAL, is told of observations around Hudson's Bay by Hearne of ten species of geese, particularly the snow goose, blue goose, brent goose, and horned wavy goose, lay their eggs and raise their young in some country which to Hearne was unknown. Explorers, Indians and Eskimos could never tell where these fowl bred and it was well known that they never migrated to the south.

Epes Sargent in his WONDERS OF THE ARCTIC WORLD tells that Franklin's second expedition saw large numbers of laughing geese migrating to the unknown north -- sure indication of land to the north. And this was observed on the north coast of Canada latitude 69 degrees 29" N., longitude 130 degrees 19 minutes W., on July 13. (Sargent, p. 163).

Newton in his ARCTIC MANUAL, wrote as follows concerning the migrations of the Knot,

"The knot...in the spring seeks our island (England) in immense flocks, and after remaining on the coast for about a fortnight, can be traced proceeding gradually northwards, until finally, it takes leave of us. It has been noticed in Iceland and Greenland, but not to stay; the summer there would be too rigorous for its liking, and it goes further and further north. Whither? Where does it build its nest and hatch its young? We lose all trace of it for some weeks. What becomes of it?"

"Toward the end of summer back it comes to us in larger flocks than before, and both old birds and young birds remain upon our coasts until November, or, in mild seasons even later. Then it wings its flight to the south, and luxuriates in blue skies and balmy airs until the following spring, then it resumes the order of its migration." (Gardner pp. 259-260)

Surely these migrations indicate a land further north than Greenland and Spitsbergen with an ideal climate for the breeding grounds of these migratory birds and animals.

Many explorers noticed a rise in temperature the farther north they went. For example, Nansen reported that a north wind in the winter is warmer than a south wind. On Jan. 18, 1894 at 79 degrees N. Lat., Nansen wrote,

"It is curious that there is almost always a rise of thermometer with these stronger winds...A south wind of less velocity generally lowers the temperature, and a moderate north wind RAISES it." (FARTHEST NORTH, Vol. I, p. 197) Two months later on March 4th, Nansen also wrote, *"It is curious that now the northerly winds bring cold*

and the southerly warmth. Earlier in the winter IT WAS JUST THE OPPOSITE."

This obviously indicates the existence of a warmer land toward the north from which the warm wind blows in the winter.

In Roald Amundsen's FIRST CROSSING OF THE POLAR SEA, by dirigible, May 12, 1926, this rise in temperature toward the pole was also recorded. Upon leaving Spitsbergen the temperature was minus 8 degrees centigrade. Then the temperature at the altitude of flight sank steadily from 5 degrees below freezing over King's Bay to 12 degrees below zero on 88 degrees north on the European side of the pole. FROM THIS PLACE IT BEGAN TO RISE SLOWLY. The temperature at the pole was 2 degrees below zero. That is an increase of 10 degrees! (FIRST CROSSING, p. 230)

In the soviet flight of Mikhail Gromov, of the Soviet Air Force, in an article titled, *"Across the North Pole to America,"* he recorded a similar increase of temperature at the pole. Flying above Franz Josef Land at 13,000 feet, the temperature was minus 16 degrees Centigrade. But at the pole at 8,850 feet the temperature was recorded at minus 8 degrees C--an 8 degree increase in temperature.

It must be understood that when explorers said that they had reached the pole, that this means that they had reached a point on the curvature of the earth in the Arctic or Antarctic where the angle of the sun over the horizon on their sextant readings indicated that they were farthest north for the north pole or farthest south for the south pole. Since it was difficult for explorers to measure distances in the Arctic or Antarctic directly, distances were measured by determining the latitude with the sextant. Explorers would travel a certain distance north, for example, take a reading with the sextant and based on that reading and the distance to the pole as shown on a map, they then calculated how many miles they had traveled.

A problem arises with locating the geographic pole with a sextant when taking into account the polar openings in the Arctic and Antarctic. My best estimate of the location of the North Polar Opening, for example, is that it is located on the Russian side of the pole at 85 N 130 E. This puts the geographic North Pole 102 miles within the perimeter of the polar opening, which is the location at which the earth begins to dip into the opening. Which means that a sextant determined pole will not actually be located at the geographic axis of the earth.

Today, however, with the advent of navigational satellites, geographic determination of location is determined with the GPS, the Global Positioning System. With hand-held devices, anyone can receive the satellite signals from the GPS to determine their latitude and longitude. With over flights of the poles and submarines criss-crossing the Arctic, one would rightly wonder why the polar openings haven't been openly discovered and published to the world. Certainly such a discovery must be world-shaking, mind shattering, and revolutionary to our sciences. Our physics books would have to be rewritten. Certainly the poles have been attained, which strongly suggests that any polar openings would have to be located to one side or other of the poles. Since this discovery has not been openly declared to the world, this means that the discovery of earth's polar openings is a World Top Secret.

Yet still, as strange it might seem, there are indications that the polar openings do exist!

Another evidence of the polar openings is found in the surprised comments of explorers upon finding the fog conditions at the poles. The May 1926, the Amundsen dirigible expedition noted that from Spitsbergen,

"For more than eleven hours we flew in brilliant sunshine. On 87 degrees latitude we met with fog, which, however, soon disappeared. Between 88 degrees and 89 degrees latitude we came into a new belt of fog. The fog lay, however, so low that we could fly over it by rising to 7,000 meters altitude."

From our theory, we obtain a ready answer concerning the origin of these belts of fog -- they result from the warm moist air currents issuing from the polar opening, which as they meet the lower cold air next to the ice, condense into fog.

Continuing, Malmgren wrote,

"At the Pole itself the fog thinned. The weather, as if for the occasion, at this longed-for spot on the earth's surface, can be described in a few words. The sky for the most part was covered with stratocumulus and altocumulus clouds. There was a complete cessation of wind. The temperature at about 300 meters altitude was 2 degrees below zero. From the pole we set our course towards Point Barrow. The journey from the Pole was, at the beginning, favored with good visibility, but between 86 and 85 degrees latitude we met with continuous fog." And notice was made of *"...the temperature in the layer of air nearest the ice was from 3-4 degrees lower than that higher up..."*

Commenting on this phenomenon, Malmgren, Amundsen's meteorologist on the expedition wrote,

"One of the problems that the expedition has brought to life refers to the polar fog. Why is it that over the monotonous plain which is formed by the polar sea there occur regions, close to one another, with and without fog, often without any changes in atmospheric temperature being observable? Are the lowermost air-layers so conservative that they can still, in the polar sea, retain memories from their more southerly existence? Or is the phenomenon due--which, however, appears incredible--to the variations in the heat development between the air and the underlying ice?" (FIRST CROSSING OF THE POLAR SEA, pp. 272, 280, 281)

Notice how Malmgren is puzzled by the temperature difference of the different layers of air suggesting currents from a more southerly or warmer climate.

In the book, POLAR AVIATION, by Lt. Col. C.V. Glines, USAF, is this comment about the temperature difference between the upper and lower air strata as a Douglas DC-3 transport descended in the first landing at the South Pole:

"Because of a polar phenomenon called inversion, the temperature dropped as the plane lost altitude." (p. 146)

You see, it is called a temperature inversion at the poles because normally in other parts of the world, air gets colder the higher up you go. Couldn't it be that higher layers of air are warmer at the poles because as warm air comes from the polar openings it rises above the colder, heavier air next to the ice?

Just where can heat and fog come from in the frozen polar *"ice-caps"*? Obviously, the answer is that they come from the warm, moist air currents emanating from the polar openings which are located farther and beyond the sextant located *"poles."*

Who then, are we to believe? Here are trained observers, explorers of the Arctic and Antarctic reporting what they actually saw. In the Arctic, they reported phenomena indicating there must be a warmer land towards the north from which all kinds of wildlife migrate from and return. Or are we to believe the theories of textbook writers who have never been there? Surely if we disbelieve the explorers' observations, we could go into the Arctic and see for ourselves that it gets warmer the farther north we go. We could observe for ourselves the wildlife migrating to and from the unknown northland.

However, not only does wildlife, and warm winds and fog come out of the north, but evidence points to the origin of icebergs in the far northland. The north polar sea is covered by ice which is frozen from fresh water and floats on a salty sea. The origin of so much fresh water ice that it covers thousands of square miles has been a long-time puzzle to scientists. The fact is that salty sea water does not freeze solid at temperatures found at the poles. If it did, the whole Arctic ocean would be frozen solid. The ice that covers the Arctic Ocean is salt-less.

Explorer Nansen noticed that the temperature of the water beneath the ice increased somewhat with depth, as also the temperature of the air above the ice, as measured from the crow's nest of his ship, was warmer than down next to the ice. He recorded that the icebergs in the Arctic Ocean are stratified and often contain driftwood, clay and rocks. This obviously indicates that these icebergs originate in rivers which have slowly frozen over causing the stratification layers of water freezing as it flowed over ice wedged between its banks, where rocks and clay were scraped off when the bergs eventually were pushed out to sea. However, there are not enough rivers or even glaciers around the polar sea to give origin to so much ice. So where do the icebergs come from?

A writer named William Reed, wrote in his book, PHANTOM OF THE POLES, in 1906 of his theory that these icebergs that fill the Arctic Ocean actually come from inside Our Hollow Earth. And this, in fact, is what was reported by an explorer who claimed he reached the land within the polar openings in 1829. On April 3 of that year, two Norwegian fishermen Olaf Jansen and his father, Jens Jansen left their home in Stockholm, Sweden on a voyage that took them past the Arctic ice in a direction northeast of the Franz Josef islands through leads in the ice-flows, and into the ice-free land on the other side of the north polar opening. There they were taken in by the people and lived among this advanced race for two years, then returned to the outside world by way of the south polar opening in 1831. Olaf later had his epic voyage published and can be obtained from out-of-print publishers by the title, THE SMOKY GOD. A writer named Willis George Emerson published it for him in 1908.

In his book, Olaf reports that,

"...about three-fourths of the 'inner' surface of the earth is land and about one-fourth water. There are numerous rivers of tremendous size, some flowing in a

northerly direction and others southerly. Some of these rivers are thirty miles in width, and it is out of these vast waterways, at the extreme northern and southern parts of the 'inside' surface of the earth, in regions where low temperatures are experienced that fresh water icebergs are formed. They are then pushed out to sea like huge tongues of ice, by the abnormal freshets of turbulent waters that twice every year, sweep everything before them." (THE SMOKY GOD, pp. 122, 123)

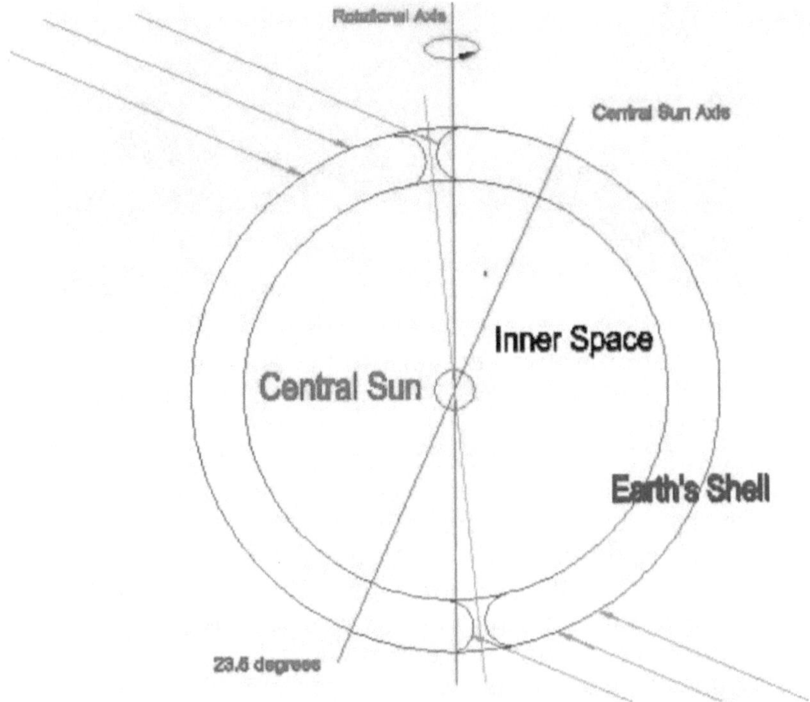

The Sun's Rays Strike the Polar Openings at Right Angles in Summer

Because of the earth's 23½ degree inclination to the plane of its orbit about the sun, the sun's rays, once each year, strike the polar lip at right angles, like at the equator, melting the ice loose at the mouths of the inner-earth rivers within the polar openings which then empty their fresh water icebergs into the Arctic and Antarctic oceans.

In connection with the origin of the icebergs is the origin of the remains of tropical wildlife which are found in the Arctic. Robert B. Cook, writing in the magazine KNOWLEDGE for 1884, tells of the remains of not only mammoths but of hairy rhinoceros, reindeer, hippopotamus, lion, and hyena, found in the northern glacial deposits and cannot explain why the supposedly extinct prehistoric mammoth is lying side by side of the remains of present-day wildlife. The truth of the matter is that all these animals were trapped in the frozen rivers in the interior and floated out in the icebergs many of which came to rest on the shores of Siberia and other northern coasts thereby depositing their trapped and preserved loads of frozen animals.

Discovery of the mammoth encased in ice.

In fact, a mammoth was actually found encased in an iceberg. In J.W. Buel's, THE WORLD'S WONDER'S, we read that in 1799, a fisherman of Tongoose, named Schumachoff discovered a tremendous elephant preserved in a huge block of ice clear as crystal along the banks of the river Lena that empties into the Arctic Ocean from Siberia. The flesh was cut off for dog meat and fed upon by wolves until as a skeleton it was removed to the St. Petersburg Museum of Natural History. Other fresh frozen mammoths were later discovered and scientific banquets featuring ancient foods including the supposedly ancient frozen mammoth were held. Scientists know that these frozen animals were frozen instantaneously, because when opening them up for the first time after their discovery, they have found undigested food in their stomachs, indicating that they had frozen so fast that the food that they had just eaten did not have time to digest.

Expedition member James Still and his wife pose in front of the
skeleton of a mammoth from Our Hollow Earth
in the Russian St. Petersburg museum.

In 1977, a baby mammoth was found in Siberia smashed by
the ice of the river it fell into in Our Hollow Earth where it froze
and later was pushed out to sea where it ended up on the
shores of our outer world.

The mammoth's true place of origin is from within Our
Hollow Earth. In Olaf Jansen's journeys inside Our Hollow Earth
he reported,

*"One day we saw a great herd of elephants. There
must have been five hundred of these thunderthroated
monsters, with their restlessly waving trunks. They were
tearing huge boughs from the trees and trampling smaller
growth into dust like so much hazel-brush. They would*

average over 100 feet in length and from 75 to 85 in height." (THE SMOKY GOD, p. 126)

Olaf further explains that from these vast herds many venture near the frozen river mouths in winter and fall into crevasses in the ice where they are instantly frozen and later when in the summer, our sun shines through the polar opening to thaw the ice loose, the rivers push the icebergs out to sea. The icebergs then gradually make their way to Arctic coasts of the outside world where vast graveyards of bone and even frozen animals have been discovered.

The claims of fisherman-explorer Olaf Jansen are truly fantastic. However, he was not the only explorer we have record of who attained the land of Our Hollow Earth and returned to tell of it. Another account is given to us by Dr. Nephi Cottam, (1883-1966) a Chiropractor of Los Angeles, California, originally from Salt Lake City, Utah, (who actually was my relative) in which he reported that one of his patients of Nordic descent told him the following account of his voyage into the land of Our Hollow Earth:

"I lived near the Arctic Circle in Norway. One summer my friend and I made up our minds to take a boat trip together, and go as far as we could into the North Country. So we put one months' food provisions in a small fishing boat, and with sail and also a good engine in our boat, we set to sea."

"At the end of one month we had traveled far into the north, beyond the Pole and into a strange new country. We were much astonished at the weather there. Warm, and at times at night it was almost too warm to sleep. Then we saw something so strange that we both were astonished.

Ahead of the warm open sea we were on was what looked like a great mountain. Into that mountain at a certain point the ocean seemed to be emptying. Mystified, we continued in that direction and found ourselves sailing into a vast canyon leading into the interior of the Earth. We kept sailing and then we saw what surprised us--a sun shining inside the earth!"

"The ocean that had carried us into the hollow interior of the Earth gradually became a river. This river led, as we came to realize later, all through the inner surface of the world from one end to the other. It can take you, if you follow it long enough, from the North Pole clear through to the South Pole."

"We saw that the inner surface of the earth was divided, as the other one is, into both land and water. There is plenty of sunshine and both animal and vegetable life abounds there. We sailed further and further into this fantastic country, fantastic because everything was huge in size as compared with things on the outside. Plants are big, trees gigantic and finally we came to GIANTS."

"They were dwelling in homes and towns, just as we do on the Earth's surface. And they used a type of electrical conveyance like a mono rail car, to transport people. It ran along the river's edge from town to town."

"Several of the inner earth inhabitants--huge giants-- detected our boat on the river, and were quite amazed. They were, however, quite friendly. We were invited to dine with them in their homes, and so my companion and I separated, he going with one giant to that giant's home and I going with another giant to his home."

"My gigantic friend brought me home to his family, and I was completely dismayed to see the huge size of all the objects in his home. The dinner table was colossal. A plate was put before me and filled with a portion of food so big it would have fed me abundantly an entire week. The giant offered me a cluster of grapes and each grape was as big as one of our peaches. I tasted one and found it far sweeter than any I had ever tasted 'outside.' In the interior of the Earth all the fruits and vegetables taste far better and more flavorsome than those we have on the outer surface of the Earth."

"We stayed with the giants for one year, enjoying their companionship as much as they enjoyed knowing us. We observed many strange and unusual things during our visit

with these remarkable people, and were continually amazed at their scientific progress and inventions. All of this time they were never unfriendly to us, and we were allowed to return to our own home in the same manner in which we came--in fact, they courteously offered their protection if we should need it for the return voyage." (RAINBOW CITY AND THE INNER EARTH PEOPLE, Michael X. Barton pp. 17, 18)

These evidences of wildlife in the north from whence the icebergs, warm winds, fog, and frozen mammoths come, and these histories of actual explorers reaching that land, help to establish the fact that there does exist great land masses within the north polar opening.

John M. Prytz, in his article, *"The Hollow Earth Hoax,"* states his third objection to the hollow earth theory: *"Consider the U.S. Atomic submarines which have traveled under the ice in crossing the Arctic Ocean and going under the pole could never have been possible (if the polar opening exists)."*

In the explorations of the Russians in the Arctic Ocean, it would seem that they already know about that land within Our Hollow Earth that Prytz says does not exist. From Scientific American, comes this revealing paragraph of the Russian discoveries:

"Exploration and research have shown that an enormous region of the earth's surface and correspondingly large realms of the UNKNOWN may be brought within the compass of human understanding in a very few years. The data thus far amassed by expeditions and ice stations fill more than 120 volumes; the list of books, monographs and articles that is emerging from THAT data already exceeds 600 titles..." (SCIENTIFIC AMERICAN, "The Arctic Ocean," by P. A. Gordienko, May, 1961)

Since the discovery of Our Hollow Earth by Admiral Richard E. Byrd in 1927, there has been an international cover-up of this, the WORLD'S TOP SECRET. And the publicizing of the atomic submarine's passing under the pole was part of that cover-up. If there has not been a cover-up, then just WHERE are all those 120 volumes proving there are "enormous regions of the earth's surface and correspondingly large realms of the UNKNOWN"? According to the textbooks, all large unknown regions of the earth's surface today are nonexistent! Especially in the Arctic Ocean, which supposedly has been crisscrossed thousands of times and completely mapped!

On February, 1927, Rear Admiral Richard E. Byrd of the United States Navy, before his seven-hour flight of 1,700 miles beyond the North Pole, said,

"I'd like to see that LAND BEYOND THE POLE. That area beyond the Pole is the center of the great UNKNOWN." (WORLD'S BEYOND THE POLES, F. Amadeo Gianinni)

Consider the testimony of the late Ray Palmer, of SEARCH and FLYING SAUCERS magazines, in which he testifies of Byrd's discovery of Our Hollow Earth. Palmer lived in Amherst, Wisconsin. He wrote in his magazine, that about three miles away is the hometown of the late Lloyd K. Grenlie who was a friend of his. Grenlie,

"...was the radioman on Admiral Byrd's expedition to the South Pole in 1926 and to both poles in 1929."

"It was emphatically denied that he made flights to BOTH poles in 1929." However, Palmer continued, *"That year a newsreel could be seen in America's theaters which described BOTH flights, and also showed newsreel photographs of the 'land beyond the pole (north) with its mountains, trees, rivers, and a large animal identified as a mammoth.'"*

"Today this newsreel apparently does not exist, although hundreds of my readers remember as I do, this movie short. Thus, I have it on my own personal viewing of this movie short, and from the radioman who went with Byrd to that land beyond the pole and SAW the things recorded on that film, that this unknown, unchartered, and presently denied land exists!" (FLYING SAUCERS, Sept. 1970)

According to our theory, the original geographic poles of the earth were located in space--in the center of the polar openings, several miles directly above a person standing on the polar lip. Since the earth was created in rotation, centrifugal force would have thrown matter outward from the center leaving a mass in the earth's core that later became the central sun. On the first day of creation when the earth's core was *"turned on"* in the process of lighting up a star, God said, *"Let there be Light."* At that time, the interior sun began to shine. Any matter further away from the central core would have been thrown outward away from the center by centrifugal force to form the earth's shell. This rotational centrifugal force would have resulted in a shell and a hollow interior with a core suspended in the center by the force of gravity acting on it from all directions. At the poles, centrifugal force would have formed the polar openings.

The Inner Earth Guide in ETIDORPHA explained that there is also a spiritual foundation to the earth,

> *"The earth forming principle consists of an invisible sphere of energy that, spinning through space, supports the space dust which collects on it, as dust on a bubble. By gradual accumulation of substance on that sphere a hollow ball has resulted, on the outer surface of which you have hither to dwelt. The crust of the earth is comparatively thin, not more than eight hundred miles in average thickness, and is held in position by the central sphere of energy* (center of gravity, or central sphere of gravity) *that now exists at a distance about seven hundred miles beneath the ocean level."* (ETIDORPHA, p. 193)

However, later, perhaps at the time of the world-wide cataclysm that produced Noah's flood, the earth has been tipped on its axis, perhaps even more than once, so that today, the polar openings are located off to the one side or other of the earth's axis. This would explain why the pole can be reached from some directions, but not another, such as the Soviet fliers that flew north from the Kara Sea and were lost when they flew into the polar opening. And yet Amundsen was able to fly over the pole in his dirigible when flying from Spitsbergen to Alaska.

There is a way to determine if someone has reached the pole or if he has only attained some point on the lip of the polar openings. At the geographic pole the sun should be the same distance above the horizon throughout any one Arctic or Antarctic day. Taking the Arctic as an example, if there is a big hole in the Earth located near the pole, and an explorer were approaching it from the south, the farthest north that person can go would be the rim of the polar opening.

If an explorer were located on the south rim of the north polar opening, he would be unable to go any further north, and the sun would appear to rise and fall throughout each day as it makes its apparent round of the Arctic sky indicating that he had not yet reached the polar axis of the earth. And yet, on the rim of the polar opening, a sextant reading would show that the explorer had reached the North Pole as indicated by the angle of the Sun above the horizon. Since at the exact polar axis of the earth, the sun should remain the same distance above the horizon all day long, and yet does not, but instead circles around up and down throughout the day, he will KNOW he has not arrived at the geographic North Pole, his sextant reading notwithstanding.

Explorers use an instrument called a sextant to determine if they have reached the pole. It is a navigation instrument that determines latitude north and south of the equator by determining the height the sun should be above the horizon for any specific latitude. At different times of the year, the degrees the sun would be above the horizon at the theoretical geographic pole are different. But on the summer solstice, the angle should be 23½ degrees, and if there are no polar openings, that angle should NEVER be greater than 23½ degrees, which is the maximum angle of the earth's axis to the ecliptic, the orbital plane of the earth about the sun. An angle of greater than 23½ degrees could be obtained only on the lip of a polar opening. As one advanced into the polar opening, the sun would appear to swing from low above the horizon to very high up to 90% -- such as happens at the equator.

If an explorer would stay at the pole during an entire season that the sun is above the horizon, he would notice that as the summer advances, the sun will rise higher and higher above the horizon. On any one day, it will circle around the sky at the same distance above the horizon if he is indeed located at the geographic polar axis of the earth. However, if the explorer discovers that the sun rises and falls throughout the day at angles greater than 23½ degrees, this would prove conclusively that he has not reached the geographic pole but instead has reached a point on the polar rim of great hole in the Earth.

This is what Norwegian explorer Olaf Jansen reported while he and his father were sailing their small fishing boat through the North Polar Opening. He noticed that, *"The sun was beating down slantingly, as if we were in a southern latitude, instead of in the far Northland. It was swinging around, its orbit ever visible and rising higher and higher each day…"* (THE SMOKY GOD, Loc. 288 at the end of Part Two).

For further discussion of the location and size of the polar openings see Table of Contents, Location and Size of the Polar Openings.

The fact that the magnetic pole does not coincide with the geographic pole is an evidence that our earth is hollow. Apparently, the earth has been tipped on its axis since creation leaving the interior sun in its original orientation. This caused the nonalignment of the earth's magnetic pole with its rotational axis. This is an evidence that the earth is hollow with a central sun which gives rise to the earth's geomagnetic field. The earth's shell rotating about the nearly stationary interior sun, both having electrical charges, positive for the inner sun and negative for the shell, is what produces the earth's geomagnetic field and causes the compass to point north. If the earth were full of matter throughout, the geomagnetic poles would coincide with the earth's rotational axis, if it could even have geomagnetic poles, which is doubtful since the earth would have to have an outer shell rotating about a stationary or near stationary core in order to produce a geomagnetic field. Instead, the magnetic poles revolve around the Arctic/Antarctic in magnetic orbits indicating that the earth is hollow and has been tipped from its original orientation with a core that is rotating at a slower rate than the shell. The core most likely would have retained its original orientation when the earth was tipped on its axis.

Raymond Bernard wrote concerning the revolving magnetic pole,

> "The first observation (of the magnetic declination) was made in London in 1580 and showed an easterly declination of 11 degrees. In 1814 the declination reached 24.3 degrees westerly maximum. This makes a difference of 35.3 degrees change in 235 years...The focal point, or the actual 'pinpoint' of the magnetic pole exists on only one portion of the circumference of that circle at a time, and moves progressively around the circle in a definite 'orbit'." (THE HOLLOW EARTH, pp. 57-58)

Here also is the reason why some polar explorers say their compass needle dips down in the far north and others say that it dips up, each depending on which side of the polar opening they are located. When the Russians reported that their compass needle pointed down for a thousand miles across the Arctic Ocean, they were on the side of the polar opening in which the magnetic pole is located. Olaf Jansen, on the other hand was on the opposite side of the polar opening in 1829, northeast of Franz Josef Land when he noticed his compass needle to point upwards across the polar opening to the other side where the magnetic pole was then located.

In the chapter of this book entitled, CHAPTER FIFTEEN A Proposal for an Expedition to Our Hollow Earth is shown how gyroscopes and radar readings can also be used to prove the existence of the polar openings.

Prytz' fourth objection states: *"The floor of the Arctic Ocean has been reasonably well charted and mapped -- no trace of any polar opening has been found."*

Without doubt, the Arctic has been mapped to a great degree of accuracy, but since all maps of the Arctic are provided by the CIA and other government agencies, no polar openings are taken into account. Certainly our textbooks contain no such maps.

One of Dr. Steven Greer's expert witnesses, Donna Hare, a NASA contractor, testified that a NASA technician told her that they are told to air brush out any UFO that appears on their photos from space, so my question is, would they not do the same to the location that UFOs come from – the polar openings?

While there does appear to be a cover-up of the discovery of the polar openings, in recent years, indications of the existence of the polar openings have begun to appear.

In December 1997, two NASA scientists, Dr. Rick Chappell and Dr. Barbara Giles published papers describing the discovery of polar fountains of energetic ions emitting from both polar regions with enough energy to light up the auroras and fill the Van Allen Radiation Belts, with no need for the source of this energy being our outer Sun's solar wind. In attempting to locate where this energy is coming out of the earth, I found two NASA Arctic images showing a hole in the ice pack, one dated February 25, 2011 from the Advanced Microwave Scanning Radiometer on NASA's Aqua Satellite, and the other dated September 16, 2012. Then in October 25, 2015, the US Naval Observatory published four images that showed that in this same exact location in the Arctic Ocean there was a lower ice concentration, a lower ice thickness, a lower sea surface salinity, and a higher sea surface temperature, all of which would occur if there is a polar opening into the earth at that location, which I have determined to be located centered at 85 N, 130 E.

Objection Six: *"Consider all the commercial airline flights that have been made between North America and Europe via the polar region. Not once has there been any observation of any polar opening, although the entire area has been crisscrossed."*

Prytz states that the polar airline flights have not seen any opening. Not so. My friend, Lord Ivars of Henwick, met an airline pilot that told him that he has seen the north polar opening many times while on transpolar flights. He said that all his airline pilot buddies know the polar opening exists, but said that if quoted, he will deny this confession because he values his job. He said that they don't try to fly across the opening because to do so would be like trying to fly into outer space. This airline pilot confirmed that the earth has the shape more like a donut, having a torus shape, with holes near the poles that lead into the interior of Our Hollow Earth.

In an article titled, "*MH370: What if it were to happen in the Arctic?*," written by Mia Bennett in Cryopolitics, of March, 2014 she shows a map she made from data from OpenFlights.org, using great circle distances of the current airline flights over the Arctic. You will notice on her map that no airline flies over the area where I have determined the North Polar Opening is located at 85 N Lat, 130 E Lon near Northland, Russia:

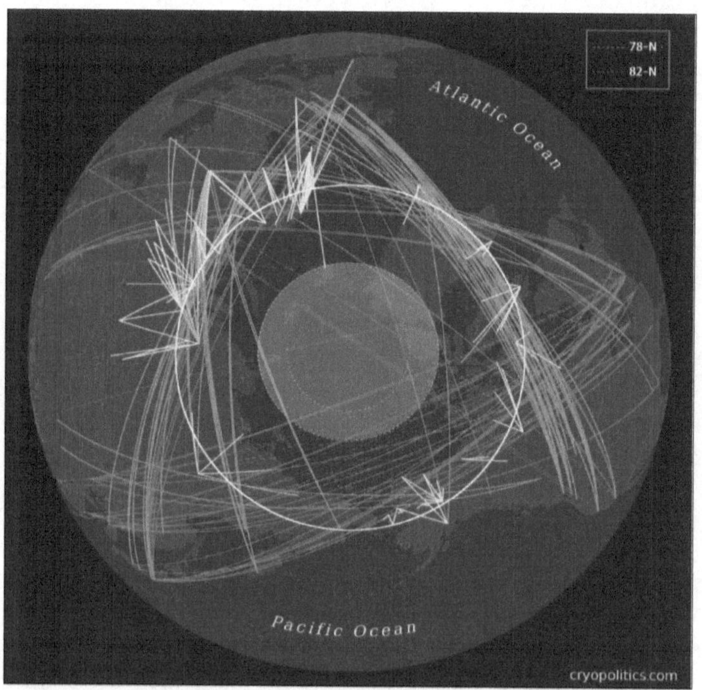

Objection Seven: *"Given that a North Polar opening exists, and that UFOs leave and enter there, why have they not been tracked by our D.E.W. line?"*

Undoubtedly they have been. But since the U.S. government considers UFOs and Our Hollow Earth top secret, any military personnel who dares divulge such information is severely dealt with. All such knowledge is considered too dangerous for common citizens to know by the International Illuminist Conspiracy which controls the U.S. military.

Objection Eight: *"There has never been any reported observations of polar openings by astronauts, in particular those with an Earth wide view such as Apollo 8, 10, and 11."*

Air Force Regulation 200-2 penalizes military employees with a $10,000 dollar fine and 10 years in prison if they publicize censored subjects such as UFOs. The astronauts undoubtedly were also under such strict regulations concerning their observations in space.

Objection Nine: *"No deep space satellite (even those in polar orbits) with photo coverage of the Earth has ever recorded a polar opening."*

There definitely is a dearth in polar opening photos from space. But that doesn't necessarily mean the polar openings don't exist. It could mean that NASA is covering them up by retouching any photographs of earth from space that might show a polar opening. There are NASA workers who have admitted to retouching space photos to eliminate anomalous UFOs (see Steven Greer's *Disclosure Project* DVD). Even one NASA scientist admitted that the polar openings exist, but that they retouch photographs of them to make them look like ice and snow.

I have included some photos of Earth taken by several satellites. The included Apollo 17 photo of Africa and Antarctica shows what may be an elliptical view of the South Polar Opening at the bottom of the picture.

NASA photo #72-HC-928 taken by Apollo 17

The elliptical area at the bottom of this photo shows the South Polar Opening on the Western Antarctic Continent.

NASA RadarSat Image of Antarctica

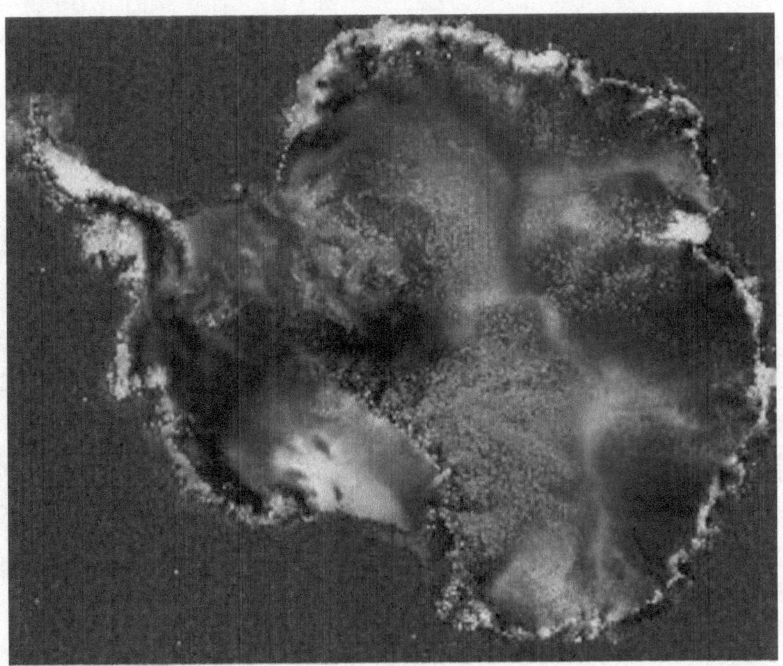

Taken in the Spring of 1997. Light areas reflect radar well, dark areas do not. There is, in fact, a dark circular area between the South Pole and the Weddell Sea that could be the South Polar Opening.

We have determined that ATS photo 67-HC-723 (below), which some hollow earth researchers thought might be showing the North Polar Opening -- is NOT. What appears at the top of the photo is the southern tip of Greenland AND an unusual cloud formation south of it that makes it look like it could be an opening.

NASA ATS photo #67-HC-723

The indentation at the top of this ATS photo has been determined to be a cloud formation on the southern tip of Greenland. It is NOT the North Polar Opening.

Here is the same image from the Earth Imaging website showing the location of the indentation in the clouds at the top of the ATS photo is really just the southern tip of Greenland:

This imagery of the ice cover in the Arctic from NASA seems to show the North Polar Opening, taken from: http://svs.gsfc.nasa.gov/vis/a000000/a003300/a003333/amsr e_sea_ice_640x480.mpg:

But rather than showing an opening into the earth, it actually just shows a hole in the satellite imagery.

The best imagery I have been able to find showing the location of the North Polar Opening is this one from the National Snow & Ice Data Center showing the extent of Arctic ice for September 16, 2012:

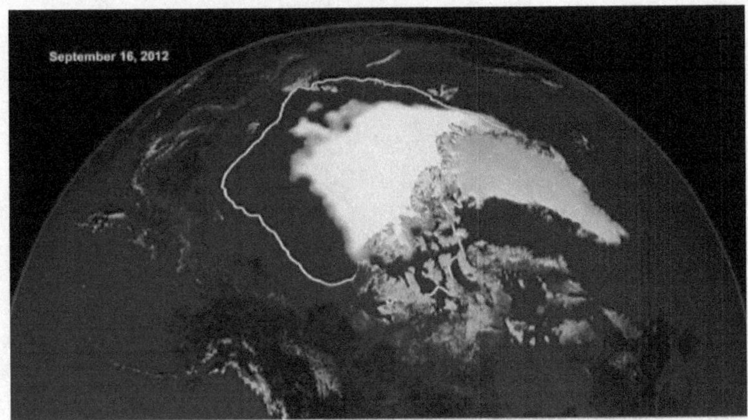

Notice the hole in the ice near Northland, Russia. This hole in the ice is the location of the North Polar Opening where warm air rises up out of the opening. Radiation from the Inner Sun also comes up out of this hole causing the auroras to light up during the Arctic winter months. Its location is 85 North Latitude, 130 East Longitude.

The same hole in the ice can be seen in this image from the year before from the Advanced Microwave Scanning Radiometer on NASA's Aqua Satellite of February 25, 2011:

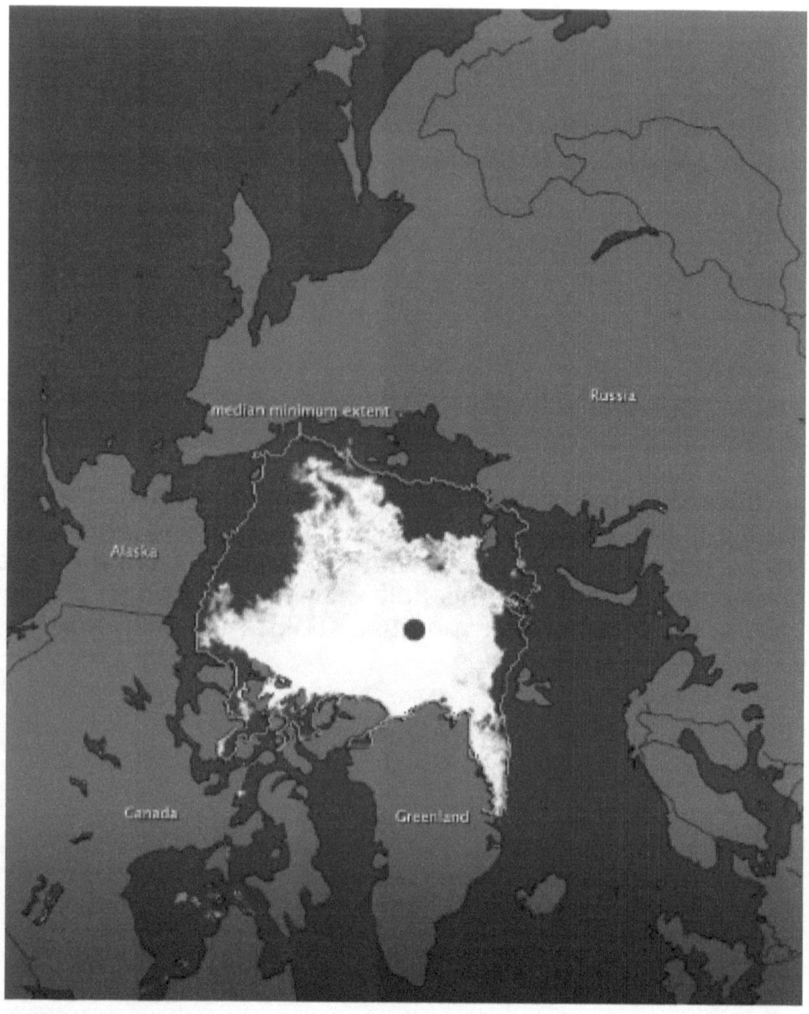

The black hole at the geographic North Pole is a hole in the satellite imagery. It is NOT the North Polar Opening. The North Polar Opening is the blue hole in the ice near Northland, Russia, which is an enduring feature of the Arctic Ice Pack as seen over time.

The South Polar Opening of Earth:

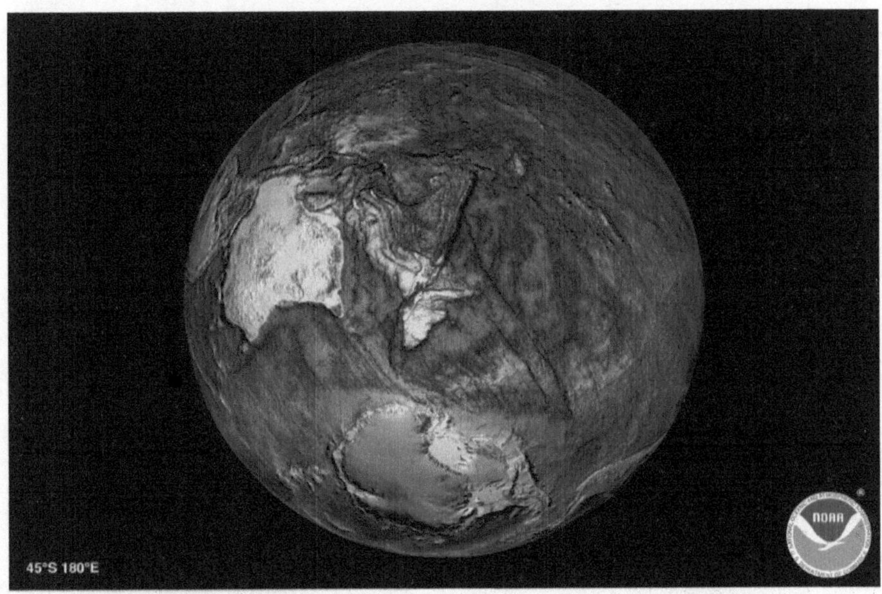

This NASA NOAA relief map of Antarctica shows the location of the South Polar Opening where the auroral radiation is emitting in the next photo is evidence our Earth is hollow.

Earth's Aurora Australis:

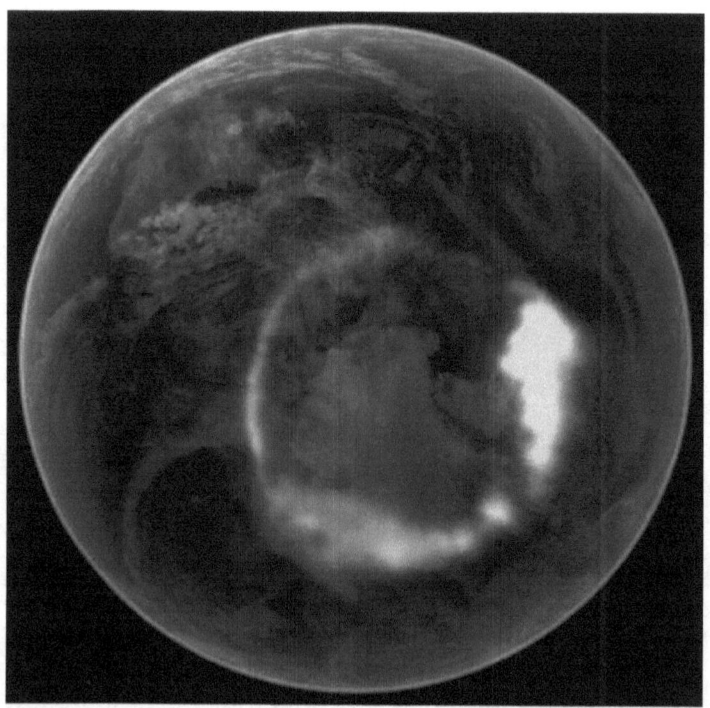

This NASA image of Earth's Aurora Australis shows the auroral radiation coming out of the Earth's South Polar Opening indicating the Earth is hollow.

Objection Ten: *"The South Pole and surrounding area (5,300,000) square miles as opposed to the Arctic, is not ocean, but dry land covered, however, with an ice cap many thousands of feet thick. This area has been extensively surveyed by air by foot in particular in preparation for and during the IGY. Of course, no southern polar opening was ever discovered."*

The scientists of the International Geophysical Year are controlled by the Illuminist Conspiracy. Therefore, we hear nothing of their discoveries of the land beyond the pole because it is considered World Top Secret.

The Polar openings were discovered. But U.S. and the International Illuminist Conspiracy intelligence clamped down on all further publication of that discovery.

Objection Eleven: *"No planet is naturally hollow-- regardless of which theory one believes in as to how a planet forms, the basic building block is that a planet grows from the center outward--therefore, no planet can be naturally hollow."*

On the contrary, since planets are formed in rotation, with a foundational spiritual bubble located at the central sphere of gravity, this causes the material to be thrown outward from the center of rotation by centrifugal force leaving a shell with space dust and rocks precipitated upon the central sphere of gravity creating a hollow interior with a central core. Additional space dust, rocks and gases accumulate on the outside portion of the central sphere of gravity augmenting the thickness of the earth's shell. Centrifugal force and gravity naturally makes all planets hollow. Many astronomical observations indicate that all the planets, moons and even the sun are hollow bodies.

In 1962, Dr. Gordon McDonald of NASA, published a report in the July issue of Astronautics, in which he stated that according to an analysis of the astronomical data, the Moon appears to be hollow:

"If the astronomical data are reduced, it is found that the data require that the interior of the Moon be less dense than the outer parts. Indeed, it would seem that the Moon is more like a hollow than a homogeneous sphere."
(ASTRONAUTICS, July 1962, pp. 14, 15)

MIT scientist, Dr. Sean C. Solomon admitted that the Moon might be hollow. He wrote,

"The Lunar Orbiter experiments vastly improved our knowledge of the Moon's gravitational field...indicating the frightening possibility that the Moon might be hollow."
(ABOVE TOP SECRET, by Jim Marrs, p. 227)

Furthermore, astronomers have observed gleams from the inner suns of Mars, Venus and Mercury shining from their polar openings. In fact, our own Aurora Borealis and Australis are caused by the electron and proton beam of highly charged radiation from the sun within Our Hollow Earth emanating through the polar openings, colliding with atmospheric atoms and thereby causing the beautiful *"Northern"* and *"Southern" Lights."* (See chapters Nine and Ten and for further details on the auroras.)

Objection Twelve: "The density of the Earth is 5.52 (on a scale of water equals 1) but the average density of the earth's crust is 2.7. Therefore, the interior of the earth must consist of a far greater density than the crust in order to get the overall average density of the earth. A hollow Earth would not be near compatible with these facts."

Here is where Prytz might have had something as proof of his side of the argument. Since surface rocks have a density of 2.7 times an equal volume of water, for the earth to have an overall density of 5.52, the interior would have to be in the range of 10 or greater. But would such densities preclude the earth from being hollow? Actually, they do not. Nevertheless, the discovery that all planets, moons, and stars are hollow bodies is bound to have a profound effect on gravity theory and how planets are formed. Perhaps this may even require a correction in the formula and gravitation theory. Let's review gravity theory as it is today and see where changes might be made and if present gravity theory is consistent with hollow planets.

The Gravity of Our Hollow Earth

Retired physicist Al Snyder wrote several books back in the 1970's disputing the claims of Newtonian scientists. In his NEWTON'S LAWS ARE FULL OF FLAWS, Snyder shows how illogical the gravitation formula attributed to Isaac Newton is. When Newton proposed in 1687 in his PRINCIPIA that *"there is a power of gravity pertaining to all bodies, proportional to the several quantities of matter which they obtain"* and that the *"force of gravity towards the several particles of any body is inversely as the square of the distances of places from the particles,"* he never actually stated the law of gravitation with its now-familiar equation containing the gravitational constant G.

In fact, Newton never stated that gravity is a pull. It could be a push. In his time, scientists believed that space was full of an etheric gas through which light propagated. If space is full of an etheric gas, it could be the source of that push. Gravity could consist of the ether of space flowing into all particles of matter. As the ether of space flows through matter, it would exert a pressure on that matter in the direction where most of the mass is located. Since most of the earth's mass is located inside it, the ether flowing towards the earth from space passes through us and keeps our feet firmly planted on the earth's surface.

Newton did not even say that the acceleration of gravity increases to the center of the earth, which scientists today claim. They also claim that anywhere inside a hollow sphere everything would be weightless. If such was the case, any people living inside the hollow earth would be floating around in zero gravity.

And there are other incongruencies in the Newtonian orthodox science, such as the equal gravisphere distances between bodies in space -- that place where a rocket after its initial burn in a trajectory coasting away from one body will begin to accelerate towards the other. For example, Newtonians place the equal gravisphere distance between the earth and the sun at 160,000 miles from Earth, and yet the moon is located at over 250,000 miles from earth. How then could the equal gravisphere distance between the Earth and the Sun be located BETWEEN the earth and the moon? If it, in fact, was, the sun's gravity would cause the moon to fall out of earth orbit towards the sun.

The same goes for the equal gravisphere distance between the Earth and the Moon. Prior to the Apollo missions which discovered by radar the distance of 54,828.7 nautical miles from the moon that the rocket stopped decelerating away from earth and started to accelerate towards the moon, Newtonians believed that the equal gravisphere distance between the earth and the moon was 1/81 of the distance from the moon to the earth, or about 3000 miles from the moon. That is why the first probes sent to the moon missed it altogether or crashed into the moon -- mission control was aiming at passing the equal gravisphere distance thought to be only 3000 miles from the moon instead of the 63,095.7 statute miles from the moon that it was later located at.

The calculation of tides by Newtonian scientists is even more incongruent. They even had to *"mickey mouse"* their Newtonian gravity formula to make it fit the observed tidal forces by cubing the distance between the earth and the moon instead of using the *"square of the distances"* as established by Isaac Newton in his PRINCIPIA. That is because, if you use the standard Newtonian gravitation formula of,

$$F = GmM/R^2$$

the sun by this formula exerts 99.5% of the gravity force on earth's tides and the moon only .05%. However, it has been known for millennia that the moon exerts the greater gravitation force on earth's tides, because the tides come up when the moon comes up, even when the sun is down. The moon comes up 50 minutes later every day and so also do the tides. We KNOW the moon exerts the greater gravity force on

the tides, yet the Newtonian gravity formula says it does
not. For further details and calculations see my gravity paper
in Contents, THE ORIGIN, CAUSE and CONTROL of GRAVITY --
FOUND!

There is a definite need to revise our physics and gravity
theories to resolve these and other incongruencies with the
current physics. The physics of hollow planets will need to be
included in the revision. Perhaps we can even achieve
realization of the illusive United Field Theory while we're about
it.

Since before the days of Albert Einstein, there have been
those who have noticed similarities between the different force
formulas. This has led to the belief in the possibility of a United
Field Theory in which perhaps one formula could describe all
the forces of nature. The underlying existence of an all
pervasive etheric substance that would contribute to all the
forces of nature would definitely be a start in the right direction
to the development of the United Field Theory.

The first step in developing the United Field Theory is to
realize that the ether that fills the immensity of space is a
reality. The Michelson-Morley light experiment performed in
1887 claimed a null value in the detection of the ether, but it
did not. The value was just so small that it was considered
within the range of error. The reason the detection was so low,
was the experiment was carried out in a heavy rock and brick
basement which blocked most of the ether flow. Edward Morley
continued extensive interferometer measurements with Dayton
Miller over a period of 25 years between 1902 and 1927 with
over 200,000 individual readings and determined conclusively
that an ether wind does indeed blow by the earth. See:
http://www.orgonelab.org/miller.htm

The next step in developing the United Field Theory would
be to find similarities between the forces and the formulas
describing those forces for a reconciliation of the formulas into
one United Field Theory formula that would apply to all the
forces of nature. For example, notice the similarity between
the above mentioned Newtonian gravitation formula and the
electrostatic formula,

$F = GmM/R^2$

$F = k\,Qq/d^2$

Both have constants; both have two bodies and both are
separated by a distance squared. The differences are that the
electrostatic bodies are quantified with charges of opposite sign
whereas the gravitation bodies are quantified by mass, and the

distance between electrostatic bodies is measured from their surfaces, but the gravitation bodies from their centers. If we were to unify these formulas into one, the differences would have to be resolved.

Let's start with the distances that separate them. Both are varied as to their distances squared. The only difference here is the starting point of the measurement one from the center, the other from the surface of the interacting bodies.

Remember that Newton stated in his PRINCIPIA that, the *"force of gravity towards the several particles of any body is inversely as the square of the distances of places from the particles."* Nothing is mentioned of measuring the distance from the center of the particles. Similar to the electrostatic forces between two bodies, the gravity force from the gravitational formula can only be applied to two bodies separated by a distance. It does not describe the gravity force within a body.

One of the definite flaws of Newtonian gravity theory is its assumption that the center of gravity is located in the center of the earth. Although gravity, which I define as an acceleration of the ether of space into matter, does flow into the central sun suspended in the hollow of the earth, it also flows towards the inner surface, thus allowing inner earth inhabitants to have their feet firmly planted on the inner surface instead of floating about as Newtonians maintain. The flaw in their theory of gravity is they assume that all gravity inside the earth gravitates towards the center of the planet. However, a closer reading of Newton's above statement on gravity indicates that the gravity force varies *"inversely as the square of the distances"* between all *"particles"* of matter in the earth.

Gravity accelerates towards the greatest concentration of matter. In a hollow planet, there are two concentrations of matter, the shell and the interior sun. Therefore, gravity will accelerate towards the shell from without and from within, AND towards the inner sun helping to keep it suspended in the hollow of the planet. This means that the center of gravity in a hollow planet's shell would be in its shell, not in the central sun. The guide in ETIDORPHA asserted that the center of gravity in the earth's 800 mile thick shell is 700 miles down from the outer surface. This indicates a greater concentration of denser matter towards the inner surface than towards the outer surface of the shell.

In fact, the center of gravity in the shell would actually be a sphere -- a central sphere of gravity where ether flowing in from the outside and inner surfaces meet. The resultant gravitational flow of the ether at this depth would be flowing in all directions with a resultant directional flow of zero. Thus a person located at the central sphere of gravity would float as if he were in space, only he would be surrounded by air and the matter of the earth's shell as if he were in a spaceship in space.

Newtonian theory maintains that the acceleration of gravity increases to the center of the planet resulting in tremendous pressures that create great heat that supposedly causes their *"outer core"* to be molten. However, the inner earth explorer that called himself I-AM-THE-MAN in ETIDORPHA, whom Hollow Earth researcher Bruce Walton discovered was a man named William Morgan, reported that as he and his guide descended through communicating caverns from an entrance in Kentucky, they gradually lost weight until at the central sphere of gravity they weighed zero. At a depth of around 200 miles from the outer surface he was able to walk with leaps and bounds with little surface gravity requiring almost no effort to move. At 700 miles from the outer surface they were floating in the air of the cavern.

Pressures increase for a short distance from either the outer or inner surface as the acceleration of gravity into the mass of the earth decreases, but then pressure decreases towards the central sphere of gravity. Resultant gravitational flow gradually cancels out and weight decreases to zero as the central sphere of gravity is approached. There is NO molten *"outer core."* The outer core is actually the hollow of the earth through which NO earthquake waves pass. See Chapter Eleven, *Earthquakes Prove Our Earth Is Hollow!* for further information on how earthquakes prove our earth is hollow.

Al Snyder pointed out another incongruency in the Newtonian gravitation formula. He did this by comparing two sets of magnets, one set 10 times more powerful than the first. Using the Newtonian formula, he showed that for the first set of magnets of power 1, separated by a distance of 1 unit,

$F = m * M / R^2$

$1 = 1 * 1 / 1^2$

But for the second set of magnets 10 times more powerful than the first, separated by a distance of that same 1 unit,

$100 = 10 * 10 / 1^2$

Newtonians would maintain that the second set of magnets is 100 times more powerful than the first set, instead of the actual 10 times more powerful that we KNOW they are. Therefore, Snyder concluded that in the Newtonian gravitation formula, F is actually squared,

$F^2 = m * M / R^2$

For the second set of magnets 10 times more powerful,

$10^2 = 10 * 10 / 1^2$

$F = 10$

Could this mean that the force we attribute to gravity is exerted by a much less quantity of matter than previously thought to be the case? And could this much less quantity of matter in a hollow earth exert the gravity force we observe the earth to have? Newtonians have presumed a much more massive and dense earth than a hollow planet would seem to have.

However, even if we assume that the Newtonian mass and density for the earth are correct, this does not preclude the earth being hollow. It could still be hollow even with a density of 5.5 gm/cc. Let's review how the mass and density of the earth are determined.

Newtonians assume, by Newton's Second Law, that the momentum of a small mass accelerating towards the earth near its surface is equal to the earth's gravitational force acting on that small mass:

$F = m \times a$ Momentum Formula (Newton's 2nd Law)

$F = GmM/R^2$ Newtonian Gravitation Formula

$m \times a = GmM/R^2$

Solving for **a**, the mass **m's** cancel out leaving,

$a = GM/R^2$

We can now solve for **M**, the mass of the Earth,

$M = a \times R^2 /G$

using the Newtonian Gravitational Constant and the acceleration of gravity at the surface of the Earth,

$980.665 \times 4.0678884 \times 10^{17} / 6.67259 \times 10^{-8}$

$= \mathbf{5.978541732 \times 10^{27}}$ **gm**   The Newtonian mass of the Earth

From the Density formula

$D = M/V$

we obtain the Newtonian density of the earth.

From the volume of a sphere formula,

$V = PiD^3/6$

the volume of the earth is 1.082×10^{27} cc.

The Newtonian density of the earth then is:

$5.978541732 \times 10^{27}$ gm $/ 1.082 \times 10^{27}$ cc

$= \mathbf{5.525 \ gm/cc}$

According to the Inner Earth Guide in <u>ETIDORPHA</u>, the average shell thickness of Our Hollow Earth is 800 miles, which makes the hollow interior of the earth consist of 50% of the earth's volume, with a density of 0 gm/cc.

The shell of the earth comprises the other 50% of the total volume of the earth and thus has an average density of **11.05 gm/cc**. So 0 + 11.05 gm/cc divided by 2 equals an average density of 5.525 gm/cc for the whole earth.

We do not know how many layers the shell of the earth consists of, but there are certainly more layers than two. At a minimum there should be three layers, one for the outer surface layer, another for the inner surface layer, and then a layer between them.

Surface rocks have an average density of 2.7 gm/cc. If we divide the 50% of the volume of the earth that the shell comprises into three layers, each layer would be 16.67% of the volume of the shell. The outer surface layer would have a density of 2.7 gm/cc. If the inner surface layer also has a density of 2.7 gm/cc, then the middle layer would have a density of **5.65 gm/cc**: 2.7 + 2.7 + 5.65 = 11.05 gm/cc. If the outer and inner layers are thinner in volume than 16.67%, then the middle layer would have a density greater than 5.65 grams per cubic centimeter.

This assumes that most of the earth's mass is located in its shell. As you can see, Newtonian physics requires an average shell density almost as dense as lead (11.3). And since surface rocks average 2.7 gm/cc in density, then the interior of the shell would have to be greater than the average density.

So we can say that a shell density of 11.05 gm/cc could be in the realm of possibility. After all, the earth DOES ring like a bell after a rather large earthquake. A bell is hollow and is made of metal, just as a hollow earth must be.

We might ask how much of the earth's mass would be contained by the interior sun? Actually, an interior sun of the estimated diameter of 600 miles would contain very little of the mass of the earth.

Evidence indicates that all stars are actually hollow crystal balls, instead of just balls of burning gas as Orthodox science claims. Therefore, assuming the interior sun has the density of glass, its mass would be only .01% of the mass of the Newtonian mass of the earth, as shown in the following calculations:

$$V = \text{pi } D^3 / 6 \quad \text{volume of a sphere formula}$$
$$\text{pi} * (600 \text{ mi} * 1.60934722 \text{ km} * 100,000 \text{ cm})^3 / 6$$

$= 4.714130881 \times 10^{23}$ cc   Volume of inner sun

Let's assume that the inner sun is also hollow and has a shell 10% of its diameter, or 60 miles. This would give the inner sun's hollow a volume of $2.413635011 \times 10^{23}$ cc. So the volume of its shell would be $2.30049587 \times 10^{23}$ cc multiplied by 2.6, the density of glass gives,

Mass = Volume * Density

$= 5.981289262 \times 10^{23}$ gm, Mass of inner sun

divided by mass of earth of $5.978541732 \times 10^{27}$ gm

$= .000100046 * 100 = .01\%$

By far, most of a hollow earth's mass is located in its shell, 99.99%.

Another possibility, you may say, is that the earth's shell is thicker than 800 miles which would give it a lower average shell density. This also, could be a possibility. Some method of determining the shell's thickness needs to be devised. This could be determined by entering the hollow of the earth through a polar opening and bouncing radar waves off the opposite side of the hollow interior to determine the diameter of the hollow interior.

In all, actually, I see nothing in the Newtonian mass and density of the earth that would completely exclude the earth from being hollow. Earthquake waves have been noticed to bend as they descend into the earth causing them to curve back up to the surface before hitting the discontinuity inside the earth that scientists claim is the outer core.   This indicates the earth does increase in density with depth which is consistent with an earth shell having a thickness of about 800 miles using the Newtonian mass of the earth.

In fact, if the earth is hollow and the Newtonian mass of the earth requiring an increased density with depth is correct, then that observed increase of density with depth would exclude their claim to a molten interior.   The discontinuity inside the earth could be the inner surface.   The reason the discontinuity of the inner surface is only 800 miles down instead of 1,800 as they claim, is because there must be a more rapid increase of density with depth than orthodox scientists estimate causing seismic waves to travel faster through the more dense depths. For further email discussions on gravity go here:

https://www.ourhollowearth.com/G-emails.htm

After applying Newtonian physics to the planets and assuming they are all hollow with shells having a thickness of 10% of their diameters, it turns out that all the planets including the sun would have solid surfaces (with one possible exception being Saturn with a shell density of 1.26 gm/cc which is closer to the density of water=1. However, if Saturn has a shell thickness of 5% of its diameter, its shell would also be solid).

With surface gravities close to that of earth and interior suns that create their planetary magnetic fields and which emit solar winds through polar openings to light up their auroras, it is even plausible that most planets contain inner atmospheric conditions similar to earth conducive to plant, animal and human life on their interiors. For a more detailed analysis of gravity, see my paper in the Contents Page, THE ORIGIN, CAUSE and CONTROL of GRAVITY -- FOUND!

Objection Thirteen: *"The earth's magnetic field could not be accounted for if the Earth were hollow, as it is the Earth's core acting like a dynamo which produces the magnetic field."*

On the contrary, a hollow earth with an inner sun would create a magnetic field easier than a solid earth with a solid core and a molten outer core of orthodox science's model of the earth. The rotation of the hollow earth's shell about the much slower rotating inner sun in the center of the earth's vacuum causes the electromagnetic field of the earth. (For more information on the earth's electromagnetic field see: CHAPTER ELEVEN EARTHQUAKES PROVE OUR EARTH IS HOLLOW! and CHAPTER TWELVE OUR HOLLOW EARTH AND THE PLATE TECTONIC SYSTEM.)

Objection Fourteen: *"The temperature of the Earth increases as the depth increases. This is known as the thermal gradient, and the value is 150 degrees Fahrenheit per mile. Thus after relatively short depths are reached, the temperature approaches the melting point of many rocks. If a race of people did live in the subterranean Earth, it would be just a mite hot for them."*

Professor Mohr of Bonn shed some true light on this subject. Marshall B. Gardner, in his book, <u>A JOURNEY TO THE EARTH'S INTERIOR</u> wrote,

> *"Every reader is acquainted with the fact, as reported by miners and other observers that the further one digs into the earth the hotter it gets. It was that idea that led people to believe that if they dug far enough they would come to a depth where it was so hot that everything would be in a molten condition. But that idea, too, must go, as being no longer in accordance with the evidence. Professor Mohr of Bonn has written a very important paper on*

thermometric investigations of a 4,000 feet boring at Speremberg who finds that while there is an increase of temperature, as we go down, the rate of that increase gets less and less all the time, so that soon it will be nil; that is to say there will no longer be any increase, and the point at which the heat would cease to increase would be about 13,550 feet." (Gardner, p. 357)

In this study, at a depth of 2.57 miles in the Speremberg pit, the temperature would stop rising if it were drilled to that depth, but in other areas of the Earth the rise in temperature could be different, such as in the cavern that William Morgan entered in Kentucky, where he reported no heat increase with depth.

Neither is the earth rigid enough to be solid all the way through. In Sir G. H. Darwin's book, THE TIDES AND KINDRED PHENOMENA OF THE SOLAR SYSTEM, he writes,

"The body of the earth, on which the oceans rest, cannot be absolutely rigid. No body is. It must be deformed more or less by the attractions of the Sun and Moon." So he shows how these gravitational interactions are calculated. He does this by measuring the fortnightly tide.

By fortnightly tide is meant *"...a minute inequality in the tide-height, having a period of about a fortnight, depending upon the inclination of the moon's orbit to the plane of the equator. Now the amount which the fortnightly oceanic tide would have if the Earth were absolutely rigid can be calculated."*

The results from these calculations show that the earth does yield to some extent under the force of the moon's gravity, and the yielding is not small enough to justify us in saying that the earth is practically rigid and it is not large enough to suggest that the earth has a liquid interior.

Gardner writes, *"Now, if the earth is not a solid, rigid body on the one hand or a shell-encrusted viscous or fluid body on the other hand--and as we have seen scientists can prove neither the one thing nor the other--there is left but one possibility--that the earth is hollow..."* (Gardner, pp. 342-50)

Concerning the liquid interior theory, Gardner writes:

"Of the old liquid interior people, it is not necessary to say very much. Their day is over. Scientists no longer put any credence in that notion--it is only in school books that it survives."

He quotes Grew in his, THE ROMANCE OF MODERN GEOLOGY, of the impossibility of a molten interior:

"For that would leave a molten ocean more than 7900 miles across any way in which it was measured: 7900 miles deep, 7900 miles broad, 7900 miles long if we take 8000 miles to be the diameter of the earth. We all know what great tides the sun and moon by their attractions raise in the earth's outer ocean of water. Think what tides they would raise in this inner ocean of molten rock and metal. The earth's crust would not be able to hold such tides in. The molten stuff would always be breaking through the flimsy thirty miles of outer solid rock as if it were eggshell. Twice a day there would be outbreaks of lava vast enough to submerge continents."

Since Gardner wrote his book in 1920, scientists have decided that the earth is partially solid and partially liquid -- since it can't be all solid nor practically all liquid. Today's theory of the earth's interior says the earth has a solid core surrounded by a liquid molten outer core. Then the mantle up to the crust is somewhat plastic with a solid crust at the surface.

But the fact that earthquakes occur down to 450 miles is evidence that the crust must extend down at least that far. And if it extends that far down, Prytz's thermal gradient is obviously wrong since earthquakes cannot occur in molten lava. And the reason earthquakes do not occur deeper than 450 miles is not because it is molten down there, but because the earth's shell has increased in density to such a degree that earthquakes cannot occur there.

Whenever a big earthquake occurs, the earth has been noticed to ring like a bell, with a fundamental period of 54 minutes. If the interior of the planet were full of molten lava, it could not ring like a bell. The liquid lava would absorb all the vibrations. (See in Contents, CHAPTER ELEVEN Earthquakes Prove Our Earth Is Hollow! for further details)

Objection Fifteen: *"Pressure also increases with depth. No cavity in the Earth can exist at a depth greater than 40 miles down due to the pressure of the overlying rocks."*

Pressure does increase with depth, but just like the temperature, the rate of the increase gets less with depth so that at a certain depth the rate of increase ceases and from that point on down the pressure decreases as the central sphere of gravity is approached. According to the inner earth guide in ETIDORPHA, at the depth of 700 miles the center of gravity in our earth's 800-mile thick shell is reached where everything is weightless and pressure is zero. Even as weight and pressure decrease as one descends into the earth, cavities get bigger and bigger. William Morgan who was taken into the Kentucky cavern by his guide as recorded in ETIDORPHA, finally came to a cavity 150 miles down with a length of 6000 miles in which lay a giant lake. First-hand observations certainly are more reliable than theories based upon the imagination.

In an article in livescience.com, scientists probing the Earth's interior have found evidence of a large reservoir of water equal to the volume of the Arctic Ocean beneath eastern Asia at a depth of about 620 miles. The finding was made by seismologist Michael Wysession and his graduate student Jesse Lawrence at Washington University in St. Louis, Missouri. (livescience.com, February 28, 2007 by Ker Than)

Objection Sixteen: *"Earthquake waves have been recorded traveling through the entire diameter of the Earth, many times, in many places. These earthquake waves can only travel through solids and liquids. If the earth were hollow, these quaking waves could not be detected."*

Earthquake waves traveling down into the earth's crust are known to bounce back to the surface. Scientists say they are bouncing off a hard core. They could also be bouncing off an inner surface that ends in air--the hollow interior. Every time there is a large earthquake, there is a large area on the opposite side of the earth starting at 103 degrees from the epicenter where no or very little earthquake waves reach.

Scientists claim this is caused by the molten outer core through which no S-type earthquake waves can pass. Weak P-type waves do reach into the shadow zone but even P-waves have a shadow zone on the opposite side of the earth from the epicenter of the earthquake. Since P-type waves CAN pass through liquid, establishment scientists claim the weak P-type wave reception in the shadow zone are actually passing through their molten outer core and solid inner core to reach the opposite side of the earth with some bending as they pass through different density layers. But P-waves could also bend around our hollow core to be received weakly in the shadow zone. Actually, the Shadow Zone IS the evidence of the hollow in the earth.

The earth is constantly trembling like a soap bubble, which is hollow. The cause of these ever present *"micro-earthquakes"* has never been explained by the solid-liquid earth theory but would be the natural expectation of a hollow globe. A liquid-solid interior would absorb these micro-earthquakes so they wouldn't even exist in a solid-liquid earth. The fact that whenever there is a large earthquake the earth vibrates like a bell is only more proof that our Earth is hollow. A bell is hollow. Since our earth rings like a bell with every large earthquake, the earth must be hollow! Apollo seismometers left on the moon found that the moon also rings like a bell when impacted, with a fundamental period of over three hours. This indicates the moon is also hollow. (See CHAPTER EIGHT and CHAPER ELEVEN for further details)

Objection Seventeen: *"If the Earth were hollow, it would cave in on itself due to pressure, weak points in the Earth's crust, meteor impacts, earthquakes, etc."*

Of course, this supposition that the hollow earth would cave in is based upon the false idea that gravity pulls everything to the center of the earth. Concerning this, Gardner replies,

"The answer to this is, that in gravitational pull it is not the geometrical position that counts. Center, in the geometrical sense of the word, does not apply. It is the mass that attracts. And if the great mass of the earth is in its thick shell, it is the mass of that shell that will attract, and not a mere geometrical point which is not in the shell at all, but 2900 miles away from it, as that is the approximate distance between the central sun and the inner surface of the earth. As a matter of fact, it is the equal distribution of the force of gravity all through the shell that keeps the sun suspended in the spot which is equidistant from every part of that shell. When we are on the outside of the shell it is the mass of the shell that attracts us to its surface. When we go over to the inside of the shell that same force will still keep our feet solidly planted on the inner side." (Gardner, p. 34)

A high concentration of metal in its purer state located in the central sphere of gravity -- the center of gravity -- gives the earth the rigidity of a steel ball. In fact, ENCYCLOPEDIA AMERICANA says:

"The very largest earthquakes cause the earth to vibrate like a BELL for several hours, with a fundamental period of vibration of 54 minutes."

Our earth is like a bell, which is HOLLOW!

Objection Eighteen: *"If the Earth were hollow, it would cause variations in the orbit of a satellite in particular around the openings wherever they might exist. If a satellite passes over an area where the mass is either higher or lower than usual, alterations in the orbit will occur. This is how the massive mascons on the Moon were pinpointed--through deviations from the normal of the orbiting lunar Orbiters. If any one of the thousands of pieces of Earth orbiting hardware passes over a hollow Earth opening, similar deviations would have to occur. None never have."*

The fact is that polar orbiting satellites are placed in orbit off to the side of the polar axis of the earth and do not go over the polar openings. If polar openings do exist, especially if the openings are not centered over the polar axis of the earth, then satellites would have to be placed in polar orbits farther away from the openings so that the lack of gravity over the openings do not mess up the orbits of the satellites.

If a satellite were placed in orbit over the polar openings at say a height of 100 miles and the openings are 125 miles in diameter, on the first pass the satellite would follow the earth's curvature through the opening and crash on the earth's interior. This apparently happened when the first satellites were put in into polar orbit in the 1950's.

The negative anomalies over the polar openings are so great that the U.S. has lost satellites over them. In the early 1950's when the U.S. was trying to put its first satellites into polar orbit, they kept losing them over the Arctic up near the pole until they decided to put them into orbit on either side of the polar opening. When they tried to send their satellites over the polar opening, several satellite cones were lost because they followed the earth's curvature into the earth's hollow interior where they crashed. (SECRET OF THE AGES, p. 130)

Ray Palmer wrote in 1959,

"Most recent evidence that there is something strange about the Poles of Earth comes in the launching of Polar orbit satellites. The first six of these rockets launched by the United States from the California coast were full of disappointments--and surprises. The first two, although perfect launchings, seemed to go wrong at the last minute, and although presumed to be in orbit, failed to show up on the first complete pass around the Earth. Technically speaking, they should have gone into orbit but they did not. Something happened, and the location of this something was the Polar area."

"The next two rockets fired did achieve orbits. This was done by 'elevating sights,' so to speak, and trying for a higher orbit, with a large degree of eccentricity, that is, a high point of orbit above the poles and a low point of orbit at equatorial areas. It was admitted that this eccentric orbit would produce a short-lived orbit, but it would also give the advantage of readings at widely varied heights above the Earth. Especially interesting was the readings expected above the Poles, because of the discovery of the radiation ring that surrounds the Earth like a huge doughnut, with openings at both Poles. (see Chapter Ten The Earth's Van Allen Radiation Belts Prove Our Earth Is Hollow!)"

"The next two satellites bore nose cones similar to those in which a future astronaut would be sent into orbit. In each one was a powerful radio transmitter, which was possible because the cone was the size of an automobile, and carried heavy batteries. Also included were powerful lights which could be illuminated at the proper time. The technique of releasing this cone from the satellite was to drop it by a radio-triggered device somewhere north of Alaska. Once dropped, the cone lost altitude and proceeded around the Earth for one more revolution on its orbit. Having come over the Pole it was then low enough (calculated the rocket men) to drop into the atmosphere over Hawaii, where a parachute would lower it slowly to the Earth's surface, and there huge planes awaited, rigged to "fish for" the descending cone, and take it into the plane before it dropped into the ocean and thus retrieve its important contents intact, without damage of crash landing."

"On both occasions the following happened: The powerful radio signals were not heard at all. The lights were not seen at all. Radar, with a range of at least 500 miles detected absolutely nothing. Each "pick-up" was a complete failure because there was nothing to pick up..."

"Each launching was perfect. Orbits finely determined as to exact distance, speed, etc. were achieved, and constantly tracked. Yet, when the final deed is done, and the cone is detached successfully according to monitoring devices signaling the detachment, everything goes wrong and the result is complete and inexplicable disappearance of the cone..."

"Can it be that the reason the descending cone does not come over the Pole on that last low pass is because the

Polar Area is mysterious in extent, not in the area calculated by the rocket men, and therefore not taken into consideration? Can it be that the nose cone fell to Earth inside that 'land of mystery' discovered by Admiral Byrd? Where else could they have gone? If the Earth at the Poles is as given on today's maps, could four successive 'low-level' launchings give the same inexplicable result -- unreasonable disappearance?" (LANDS BEYOND THE POLES by Ray Palmer, published by Gray Barker, pp. 13-14. See also DEEP BLACK by William E. Burrows, pp. 105-106.)

Today there are no satellites in polar orbit that go directly over the polar openings. Those in polar orbit all go to one side or the other of the polar openings. That is, all except two. There are two satellites that have been discovered, but instead of being in polar orbit, these are permanently stationed over the polar openings. These satellites belong to no known nation on earth. And they are different from our satellites. These two mysterious satellites consist of rock about 15 tons in size. (UFO REPORT Aug. 1977, p. 29)

It would seem that these two polar *"rock"* satellites belong to the nation inside Our Hollow Earth who know how to cause their satellites to compensate for the lack of gravity over the polar holes.

In summary of the scientific evidences of Our Hollow Earth, we maintain that:

1. Explorers have gone to that land beyond the poles and returned to report their discoveries. The prophet Enoch, together with his city were translated and taken by God physically into Heaven in 3,013 B.C. Several passages in the Book of Enoch (of the Lost Books of the Bible) indicate Enoch and his city were taken to the Inner Sun of Our Hollow Earth:

The Book of Enoch 76:6-7 (from LOST BOOKS OF THE BIBLE) says, *"Seven rivers I beheld upon earth, greater than all rivers, one of which takes its course from the west; into a great sea its water flows. Two come from the north to the sea, their waters flowing into the Erythraean sea, on the east, And with respect to the remaining four, they take their course IN THE CAVITY OF THE NORTH* (the four rivers that come out of the Garden of Eden), *two to their sea, the Erythraean sea, and two are poured into a great sea, where also it is said there is a desert* (the Arctic and Antarctic)."

The Book of Enoch 68.27 (from LOST BOOKS OF THE BIBLE) says, *"...and by this oath the ABYSS* (the hollow of the

71

earth) *has been made strong; nor is it removable from its station for ever and ever."*

The Book of Enoch 21:4-5 (from LOST BOOKS OF THE BIBLE) says, *"From thence I afterwards passed on to another terrific place; Where I beheld the operation of a GREAT FIRE BLAZING AND GLITTERING* (the Inner Sun), *in the midst of which there was a DIVISION* (between its day and night sides). *COLUMNS OF FIRE struggled together to the end of the ABYSS* (to light up the Auroras), *and deep was their descent..."*

The Book of Enoch 22:9-10 (from LOST BOOKS OF THE BIBLE) says, *"At that time therefore I inquired respecting him, and respecting the general judgment, saying, Why is one separated from another? He answered, THREE SEPARATIONS have been made between the spirits of the dead, and thus have the spirits of the righteous been separated. Namely, by a CHASM* (the hollow in the earth), *by WATER* (the oceans and shell of the earth), *and by LIGHT* (the Inner Sun) *above it."*

The Book of Enoch 22:9-10 (from LOST BOOKS OF THE BIBLE) says, *"In those days shall the earth deliver UP FROM HER WOMB* (the hollow in the earth), *and HELL deliver up from hers* (the earth's shell), *that which it has received; and destruction shall restore that which it owes."*

The Greek explorer Pytheas, between 330 BC and 320 BC, sailed several days north of Great Britain past the ice into a land where the sun did not set -- within the North Polar Opening.

Olaf Jansen sailed in his fishing boat with his father, Jens Jansen, to Our Hollow Earth reaching the Inner Continent through the North Polar Opening August 15, 1829. (THE SMOKY GOD, p. 89)

The fellow discovered by Hollow Earth researcher, Bruce Walton of Provo, Utah, to be William Morgan, who called himself "I-AM-THE-MAN," in his book, ETIDORPHA, wrote a book in about 1826 revealing the secrets of Masonry. Newspapers reported that he was killed by the Masons for writing his book revealing the secrets of Freemasonry and his body thrown into a river, but instead was actually taken by the Masons and sent on a journey with an Inner Earth Guide to Our Hollow Earth through communicating caverns beginning at an entrance in Kentucky. (see: https://etidorhpacontent.blogspot.com/)

Richard Evelyn Byrd, Admiral of the United States Navy flew his airplane through the North Polar Opening in February 1927, through the North and South Polar Openings in 1929, and through the South Polar Opening in February 1946 and 1947 as part of Operation High Jump. (SECRET OF THE AGES, p. 114, WORLDS BEYOND THE POLES, by F. Amadeo Giannini, and GENESIS FOR THE SPACE RACE, by John B. Leith, Chapter XI)

Captain Hubert Wilkins reached that land beyond the South Pole on December 12, 1929 (WORLDS BEYOND THE POLES, F. Amadeo Giannini)

Reinhold Schmidt was taken to Our Hollow Earth in a flying saucer from the Bakersfield quarry, Los Angeles, California in 1958 through the North Polar Opening (See in Contents, CHAPTER SIX The Origin of Flying Saucers -- FOUND! for more details.)

Karl Unger made it to the Hollow Earth in a German submarine at the end of World War II in 1943, and wrote a letter back to his friend in the United States delivered through a German colony in the Matto Grosso of Brazil which had found a way to the hollow earth through a cavern (see Contents Page, German Submarine U-209 Made It to Our Hollow Earth!).

Hank Krastman, a Los Angeles Attorney, while attending Northern Arizona University in his twenties in 1961 requested permission of the Hopi Indian elders Council of the Nine to visit Our Hollow Earth through their entrance near the Grand Canyon. He was granted that request and taken there by Kopavi, the great-great-grandson of Jacob Waltz. Jacob Waltz himself had been asked by some inner earth people who communicated to him through the Indian Elders of the Pima Indians to take them sacks of salt to their cavern city beneath the Superstition Mountains near Apache Junction, Arizona, for which they paid him in gold bricks, which he stored in the cavern. (see Contents Page, Interesting Hollow Earth Emails).

A Norwegian patient of Chiropractor, Dr. Nephi Livesay Cottam of Los Angeles, originally from Salt Lake City, Utah, visited the North Countries of Our Hollow Earth through the North Polar Opening with his friend in a sail boat that also had an engine, as related in Michael X. Barton's book, RAINBOW CITY AND THE INNER EARTH PEOPLE, page 17.

Retired Col. Billie F. Woodard was born in Our Hollow Earth as a hermaphrodite and delivered as a baby with his sister to our outer world by his parents to carry out a mission to make their world known and accepted by surface peoples. He was raised by a U.S. Air Force colonel. He was taken to Our Hollow Earth in a flying saucer at the age of 12. Later after graduating from High School, Woodard joined the military and after taking

Basic Training was assigned to Area 51 in Nevada where he went on several missions through Hollow Earth tunnel trains to Our Hollow Earth's capital City of Eden at their request to deliver a message to our government that the Hollow Earth nation does not want us using nuclear weapons (see Biography of Ret. Colonel Billie Faye Woodard in Contents Page).

2. The magnetic poles do not coincide with the geographic poles. The Earth's shell with a slower rotating interior sun gives rise to the geomagnetic field of the earth as the earth's shell rotates about the inner sun. Since the earth has been tipped on its axis in past geologic history, the interior sun still retains its original orientation and causes the geomagnetic poles to not coincide with the geographic axis of the earth.

3. Icebergs originate from rivers inside the polar openings which freeze over in winter and in summer when our outer sun shines into the polar openings, the icebergs break loose and are pushed out to sea filling the Arctic Ocean.

4. Mammoths, and other wild animals which live in the interior, such as the hairy rhinoceros, reindeer, hippopotamus, lion, giant deer, horse, and hyena, fall into inner earth rivers that empty into the Arctic Ocean within the North Polar Opening where they are frozen in winter and later in summer when the light of our outer sun shines in the opening and thaws out the river mouths and are then pushed out with the icebergs and end up on the shores of Siberia and other Arctic shores creating great bone yards where, sometimes, even fresh carcasses are found.

5. Numerous fish, such as, mackerel and herring; animals, such as, whale, seal, Arctic fox, reindeer and musk-oxen; birds, such as, knots, swans, snow geese, blue geese, brent geese, horned wavy geese, and ross's gull, migrate to and from the unknown north country each spring or fall to have their young or to escape the cold winter. (Gardner, Chapter 12) In the book by Jack Denton Scott, JOURNEY INTO SILENCE, page 107, he quoted the captain of the *Havella*, a Norway's ship, regarding the Ross Gull, *"Ross's Gull! We don't see them often. Most naturalists never see one. Mystery birds. They don't go south. Breed in eastern Siberia then fly north over the Polar Sea. No one knows where they are between October and June."* Michael Densley in his book, IN SEARCH OF ROSS'S GULL, said he found a colony of Ross's Gull on the northeastern coast of Siberia, and reported seeing several unknown strange species of birds among them.

6. A North Wind brings warmer weather. In fact, on the rim of the polar opening the sun strikes in the summer at right angles just as at the equator, raising the temperature substantially. Olaf Jansen reported that when they were in the far north on the polar rim, "The sun was beating down slantingly, as if we were in a southern latitude, instead of in the far Northland. It was swinging around, its orbit ever visible and rising higher and higher each day...The sun's rays, while striking us aslant, furnished tranquil warmth." (THE SMOKY GOD pp. 76, 83)

7. Eskimos say that their ancestors originated from a land in the north where the sun never sets. Marshall B. Gardner reports that "...in the efforts of these Eskimos to tell where they came from they would point to the north and describe a land of perpetual sunshine..." (A JOURNEY TO THE EARTH'S INTERIOR, p. 302)

8. In the very far north, beyond the pole, within the polar opening and past the ice exists an OPEN SEA to which few explorers have attained. At times the extent of this open sea has expanded down to the 80th parallel. In earlier times it seems to have extended farther south than today is the case making it easier for some explorers to reach that land beyond the pole.

Explorer Olaf Jansen and his father arrived at Franz Josef Land in late June, 1829, and finding an open lead in the ice, they followed it into the open sea within the polar opening and then into Our Hollow Earth. (THE SMOKY GOD, pp. 60, 61)

Several accounts of explorers reaching this open sea on the other side of the ice is found in a book by Dr. D. Barrington, THE POSSIBILITY OF APPROACHING THE NORTH POLE ASSERTED, published in 1818 in New York. Dr. Barrington writes that in 1751 a Captain MacCallam commanding a whaler, during a lull in the usual business of the voyage, thought he would make a dash for the North Pole. He reached a latitude of 83 degrees and found in front of him no further ice. In fact they had not seen a speck of ice for the last three degrees, he reported, but had to abandon his adventure as he did not wish to incur the displeasure of his owners.

Another voyage by a Dr. Dallie of Holland on a Dutch warship in supervision of the Greenland fisheries reached a latitude of 88 degrees and reported that the weather was warm, and the sea perfectly free from ice. Dallie pressed the captain to proceed but the captain felt he had already gone too far by having neglected his station.

Then a Mr. Stephens, sailing on another Dutch ship in 1754 was driven into latitude 84½ degrees north and reported that they did not find the cold excessive, used little more than common clothing, met with but little ice, and even less ice the farther northward they went.

9. The explorers find subtropical seeds, flowers, green plants and trees and much driftwood floating in the Arctic Ocean which could come from no other place than Our Hollow Earth through the North Polar Opening.

10. Often winds from the north carry so much pollen as to color the icebergs. The colored snow has been analyzed and the red, green and yellow have been found to contain vegetable matter, similar to the pollen of a plant. At different seasons of the year, the pollen has been observed to fall with different colors. Explorer Kane, in his first volume, page 44 wrote,

>*"We passed the Crimson Cliffs of Sir John Ross in the forenoon of August 5th. The patches of red snow, from which they derive their name, could be seen clearly at the distance of ten miles from the coast...All the gorges and ravines in which the snows had lodged were deeply tinted with it...for if the snowy surface were more diffused, as it is no doubt earlier in the season, crimson would be the prevailing color."* (ARCTIC EXPLORATIONS IN THE YEARS 1853-54-55)

The report of February 2, 2007, Multi-Colored Snow Falls over Three Siberian Regions by the Environmental News Network, confirms Olaf Jansen's report that pollen from the great flower fields from the Inner Continent at times blows out through the North Polar Opening and falls over the northern regions of the earth's exterior surface coloring the snow with different colors of pollen.

11. The Aurora Borealis and Aurora Australis are caused by the solar wind of the interior sun streaming through the polar openings following the earth's electromagnetic field lines and causing atoms in the atmosphere around the polar openings to light up and give off the beautiful Northern and Southern *"lights."* Scientists compare the Auroras to a television set but have no answer as to what takes the place of the cathode; the energy source of the auroras is not known. They admit it must be a solar wind, but the solar wind from our sun is deflected around the earth by the earth's electromagnetic field and is thus prevented from entering. It is here the Hollow Earth theory provides the perfect answer as to the source of the energy of the auroras, the cathode of the auroral television

tube. It is the sun within the hollow of our earth which emanates its high energy electrons and protons through the polar openings which when they hit the atmosphere around the polar openings cause the auroras to light up. Variations in the auroral lights at both poles have been observed to occur simultaneously indicating that the source of the auroras comes from a single location – the inner sun in the center of the earth. For more information on the auroras see CHAPTER NINE The Auroras Prove Our Earth Is Hollow!

12. The Voyager spacecraft and the Hubble space telescope have verified that most of the planets of our solar system have auroras. This has puzzled scientists because the solar wind from the sun is not strong enough to even cause earth's auroras much less the outer planets' auroras. The hollow planet theory provides the most logical solution: ALL bodies in space are hollow. Those having auroras and geomagnetic fields strongly indicate that they are not only hollow, but have interior suns and polar openings through which strong solar winds from those inner suns emit to light up their auroras as their interior solar winds impact on the atmosphere around their polar openings.

Rodney M. Cluff

Photo No.: STScI-PRC96-32 of Jupiter's Auroras by the Hubble
Space Telescope, Oct 17, 1996. The top picture shows Io, a
moon of Jupiter, interacting with Jupiter's electromagnetic field
and its auroras:

Another close-up of Jupiter's Auroras:

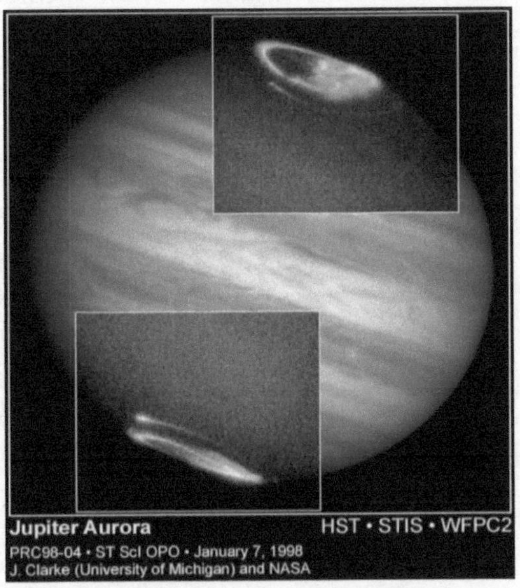

Jupiter Aurora — HST • STIS • WFPC2
PRC98-04 • ST ScI OPO • January 7, 1998
J. Clarke (University of Michigan) and NASA

Saturn's Aurora:

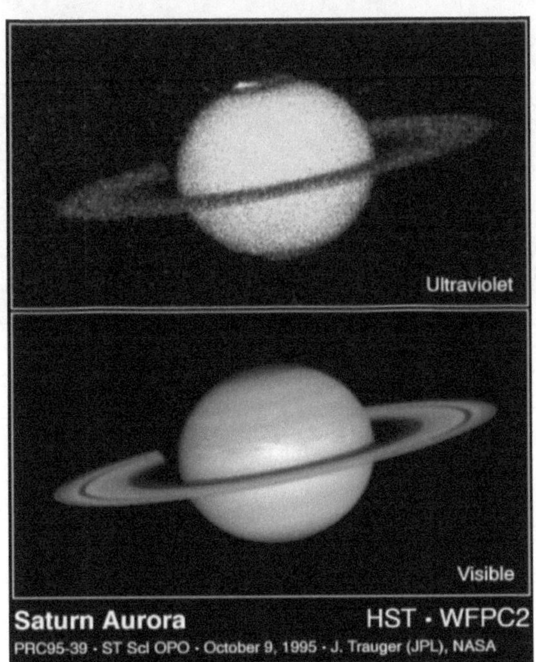

Saturn Aurora — HST • WFPC2
PRC95-39 • ST ScI OPO • October 9, 1995 • J. Trauger (JPL), NASA

13. Astronomer's observations of the polar lights on Mars, Venus and Mercury show them to be hollow with central suns shining through their polar openings.

14. Satellite images show polar openings not only on Earth, but also on Mercury, Venus, Mars, Jupiter, Saturn and Neptune.

Mar's North Polar Opening:

This Hubble Space Telescope image of Mars shows its North Polar Opening with clouds visible down within the depression.

Venus' South Polar Opening:

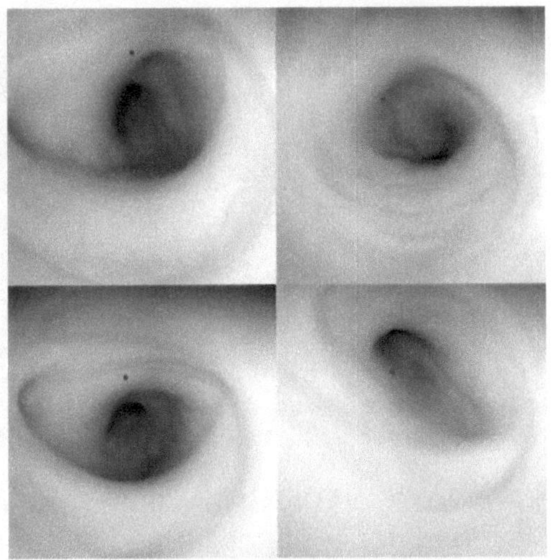

European Space Agency's (ESA) robotic Venus Express spacecraft took pictures of this vortex in the clouds at Venus' South Pole indicating that Venus has polar openings and is a hollow world.

Mercury's Polar Openings:

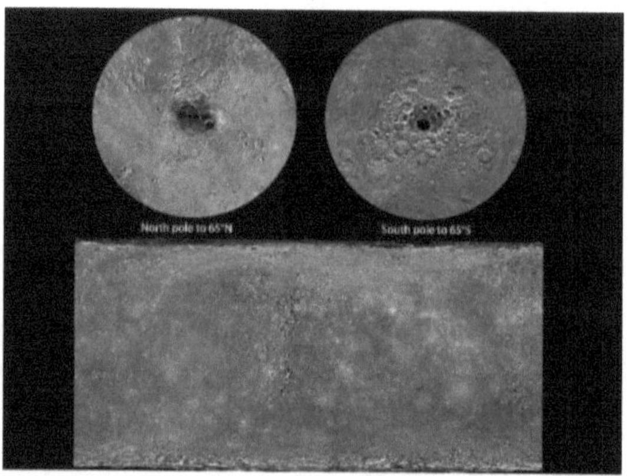

At the end of 2012, NASA's Messenger probe to Mercury recorded polar openings at its north and south poles. Mercury has been found to have a tenuous atmosphere. And since Mercury rotates only 1.5 times in its year, it's magnetic field can only be explained if it is hollow and has an interior sun that rotates faster than Mercury's shell.

Saturn's South Polar Opening:

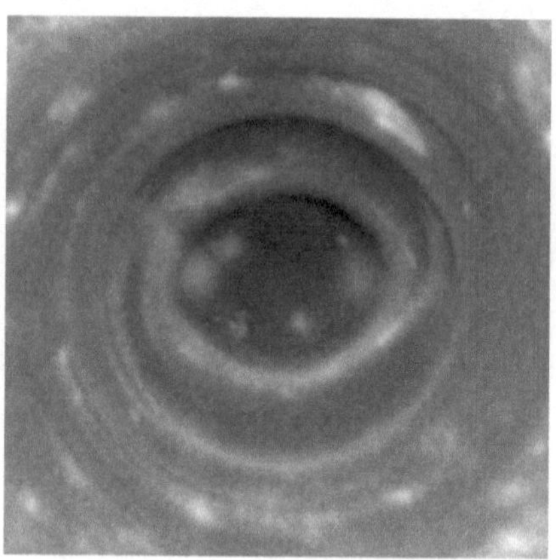

Filmed by NASA's Cassini spacecraft, the South Polar Opening of Saturn is considered a permanent feature that has a hurricane-like eye with winds over 325 miles per hour circling it.

Jupiter's South Polar Opening:

Neptune's North Polar Opening:

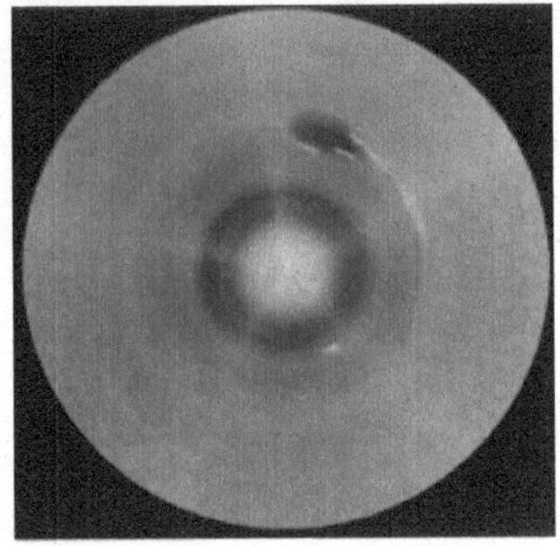

This image of Neptune's North Polar Opening shows its inner sun shining out through the polar opening

15. The Earth is not rigid enough to be solid all the way through, nor does it yield enough to tidal forces to have a liquid interior. As described in Sir G. H. Darwin's book, THE TIDES AND KINDRED PHENOMENA OF THE SOLAR SYSTEM, the results

from calculations of the gravitational interactions of the Sun and Moon on Earth's fortnightly tide show that the Earth does yield to some extent under the force of the Moon's gravity. The yielding, however, is not small enough to justify us in saying that the Earth is practically rigid and it is not large enough to suggest that the Earth has a liquid interior.

So if the Earth is not rigid enough to be solid all the way through and yet does not yield enough to the gravitational interactions of the Moon and Sun to have a liquid interior, then the Earth must be hollow.

After thoroughly investigating gravity theory, I have concluded that Newtonian physics does allow all the planets, the sun, moons, comets and asteroids to be hollow bodies. Assuming they are all hollow bodies with shells 10% of the planetary diameter, this would allow the planets and even the sun to have solid surfaces with shell densities within the range of known materials.

Nevertheless, in my study of gravity, I have come to the conclusion that gravity is not a mysterious pull, as orthodox science claims, but is a push of the ether of space as it accelerates into the nucleus of all atoms where it is being transformed into electromagnetism and electromagnetic radiation, and in fact gravity can be controlled to produce usable free energy. See in Contents, my paper, THE ORIGIN, CAUSE and CONTROL of GRAVITY -- FOUND!

16. The source of some long-delayed radio echoes is our hollow earth. Occasionally, radio waves bouncing off the ionosphere bounce down into the polar openings and echo around inside our hollow earth and then bounce back out of the polar openings to receivers on the outside of the earth. Scientists understand the radio echoes that travel around the earth in 1/7th of a second and the echoes of 2.6 seconds that bounce off the moon, but are befuddled as to the cause of those that take longer times of 30 seconds and even up to 5 minutes. These are the echoes that enter the Polar Openings and bounce around in the earth's interior before bouncing back out. (See HOLLOW PLANETS by Jan Lamprecht, pp. 316-317)

17. Mirages of land seen around the Arctic of land towards the north are actually doubly inverted mirages of land within the North Polar Opening reflected off warm air rising up out of our hollow earth. The Russians have seen what they call Sannikov land north of the New Siberian Islands. It was first sighted in 1811 by the Russian merchant and explorer Yakov Sannikov and later also sighted by Russian explorer Eduard

Toll. Dr. Frederick A. Cook on his trip to the North Pole in 1908 took a photo of what he called Bradley land at 84° 20′ N to 85° 11′ N on the 95° W meridian.

In his book, MY ATTAINMENT OF THE POLE, Dr. Cook wrote of his discovery of Bradley land,

"*On the western heavens, ever a blank mystery, cleared. Under it, to my surprise, lay a new land. I think I felt a thrill such as Columbus must have felt when the first green vision of America loomed before his eye. My promise to the good, trusty boys of nearness to land was unwittingly on my part made good, and the [Pg 244] delight of eyes opened to the earth's northernmost rocks dispelled all the physical torture of the long run of storms.*

As well as I could see, the land seemed an interrupted coast extending parallel to the line of march for about fifty miles, far to the west. It was snow covered, ice-sheeted and desolate. But it was real land with all the sense of security solid earth can offer. To us that meant much, for we had been adrift in a moving sea of ice, at the mercy of tormenting winds. Now came, of course, the immediate impelling desire to set foot upon it, but to do so I knew would have side-tracked us from our direct journey to the Polar goal. In any case, delay was jeopardous, and, moreover, our food supply did not permit our taking time to inspect the new land.

[Pg 245] This new land was never clearly seen. A low mist, seemingly from open water, hid the shore line. We saw the upper slopes only occasionally from our point of observation. There were two distinct land masses. The most southern cape of the southern mass bore west by south, but still further to the south there were vague indications of land. The most northern cape of the same mass bore west by north. Above it there was a [Pg 246] distinct break for 15 or 20 miles, and beyond the northern mass extended above the eighty-fifth parallel to the northwest. The entire coast was at this time placed on our charts as having a shore line along the one hundred and second meridian, approximately parallel to our line of travel.

At the time the indications suggested two distinct islands. Nevertheless, we saw so little of the land that we could not determine whether it consisted of islands or of a larger mainland. The lower coast resembled Heiberg Island, with mountains and high valleys. The upper coast I

estimated as being about one thousand feet high, flat, and covered with a thin sheet ice. Over the land I write "Bradley Land" in honor of John R. Bradley, whose generous help had made possible the important first stage of the expedition. The discovery of this land gave an electric impetus of driving vigor at just the right moment to counterbalance the effect of the preceding week of storm and trouble.

Although I gazed longingly and curiously at the land, to me the Pole was the pivot of ambition. My boys had not the same northward craze, but I told them to reach the land on our return might be possible. We [Pg 247] never saw it again. This new land made a convenient mile-post, for from this time on the days were counted to and from it.

A good noon sight fixed the point of observation to 84° 50′, longitude 95° 36″." (MY ATTAINMENT OF THE POLE by Frederick A. Cook, pp. 244-247)

This is the photo Dr. Cook took of what he called Bradley Land:

The land that Dr. Cook took a photo of and called Bradley Land, is a doubly inverted mirage of the Inner Continent within the North Polar Opening.

On June 24, 1906, from the peaks of Grant Land, Ellesmere Island, Admiral Peary sighted through his glasses *"the faint white summits of a distant land,"* and again on June 28, he *"could make out, apparently a little more distinctly, the snow-clad summits of the distant land in the northwest, above the horizon"* that he then named Crocker land. (NEAREST THE POLE by Admiral Robert E. Peary, pp. 202, 207)

American whaling Captain Keenan also sighted this mirage of land to the north from Harrison Bay, Alaska, which contributed to the conclusion in 1904 by Mr. R. A. Harris, of the United States Coast and Geodetic Survey that there exists land up near the pole. (NORTH POLE AND BRADLEY LAND, by Edwin Swift Balch, p. 1) Jan Lamprecht, in his book HOLLOW PLANETS, came to the conclusion that these mirages of land seen all around the Arctic are actually sightings of land within the North Polar Opening as a doubly inverted mirage. In the Arctic, a mirage occurs when warm air above the ice reflects land on the surface. The first inversion shows the land to appear as up-side down, while the second inversion above the first inversion is right-side up. Since the actual land within the polar opening was over the horizon, the first up-side down inversion was not visible, but the second above that WAS visible as up-right land on the horizon. (HOLLOW PLANETS, Jan Lamprecht, p. 441)

18. The earth appears flattened at the poles, which is caused in part by the existence there of openings into the interior of the planet.

19. The earth trembles like a soap bubble, which is hollow. And when a large earthquake strikes, the earth vibrates like a bell, which is also hollow. The earth's fundamental period of vibration is 54 minutes. This vibration of the Earth like a bell indicates that the earth must be a hollow sphere. Additionally, on the opposite side of the Earth from the epicenter of an earthquake starting at 103 degrees from the epicenter there is a shadow zone where no S type earthquake waves arrive because they cannot pass through the hollow of the Earth. The arrival of weak P type earthquake waves in the shadow zone could occur from repeated reflections from the inner surface of the shell and the outer surface. Reception of P-Waves past the shadow zone for P-Waves could occur because P-Waves are compression waves that, like sound, can travel around corners or obstacles like the hollow interior. Proof that no P-Waves pass through the hollow interior is the Air Force experiment carried out at Arizona State University where ultra-sonic beams were unable to pass through the hollow of the Earth.

20. The Van Allen Radiation Belts have holes at the polar extremities of the earth, coinciding with the earth's electromagnetic field there. Scientists are puzzled as to the source of the radiation of the belts, admitting that it comes from a solar wind, but what solar wind? The solar wind from our outer sun is deflected around the earth by the earth's electromagnetic field which prevents its entrance into earth's atmosphere. It is here that the Hollow Earth theory provides

the answer to that source of the solar wind that causes the auroras and the Van Allen Radiation Belts: It is the sun within the hollow of the earth emanating protons and electrons through the polar openings causing the atmosphere to light up with the auroras and thereafter becoming trapped in the electromagnetic field of the earth producing the Van Allen Radiation Belts.

21. While in the far north, explorer Nansen discovered that the north-south horizon became foreshortened while the east-west remained the same which would be true if a polar opening exists there. In fact, normal navigation instruments don't function in the Arctic/Antarctic as they do elsewhere because the polar openings are not taken into account. For example, a horizontal gyroscope will turn vertical as the polar opening is entered and the midpoint of the polar lip is reached. The magnetic compass will point up on one side of the opening and down on the other depending on which side one is located from the magnetic pole. In other places it may just spin.

22. No airline flies over the polar openings because this would risk flying into the openings.

23. No satellite can be put into a stable orbit directly over the polar openings without taking into account the lack of gravity there as result of no mass within the polar openings.

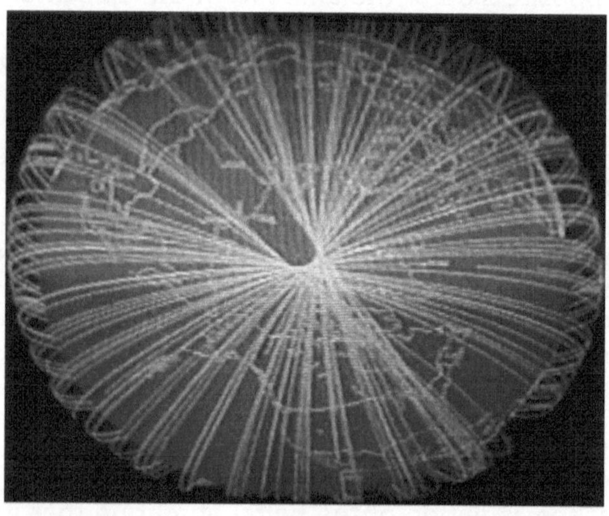

NORAD satellite paths show no orbiting satellites over the North Polar Opening. Notice North Land, Russia jutting out into the no satellite overflight zone.

<u>NASA's ICESAT</u> — Ice, Cloud and Land Elevation Satellite intentionally circumvents the polar regions of the planet in order not to go over the polar openings. This image of Antarctica shows the ICESAT orbital paths not passing over the area of the South Polar Opening.

24. X-ray, ultraviolet, and other pictures of the sun taken by Skylab revealed holes at the poles of the sun where gravity is greatly diminished because of the openings there into the hollow interior of that PLANET OF THE GODS. (See Chapter Eight)

25. Chernobyl radiated dust was found near the South Pole, but nowhere else in the southern hemisphere. This is because the air masses from either hemisphere do not mix. They are parted by a jet stream at the equator which keeps the two air masses separate. Since the Chernobyl nuclear power plant melt down occurred in the northern hemisphere in the Ukraine, the radiated dust from the meltdown only spread out over the northern hemisphere. Scientists were at a loss as how the radiated dust from the Chernobyl meltdown could reach the South Pole. Jan Lamprecht in his book, <u>HOLLOW PLANETS</u>, explained how it could get there: The dust was carried by prevailing winds up into the Arctic where it was sucked inside the earth through the North Polar Opening where it then traveled inside the earth to and out the South Polar Opening

and was deposited near the South Pole by Antarctic snows. (HOLLOW PLANETS, p. 501, 502)

26. Seismograph recordings placed on the Moon's surface by the Apollo missions indicate that the Moon is also hollow. When the Moon was hit by a large meteor, and another time by the spent lunar module, the seismographs recorded the Moon rang like a bell. A bell is hollow, so the Moon also must be hollow. (ABOVE TOP SECRET by Jim Marrs, p. 227)

27. Ozone holes at Earth's poles are created by the ozone free air emanating from the Earth's interior through the polar openings. These ozone holes in the atmosphere are greater in size in the seasons that greater quantities of air flow out of the polar openings.

Ozone Hole over Antarctica:

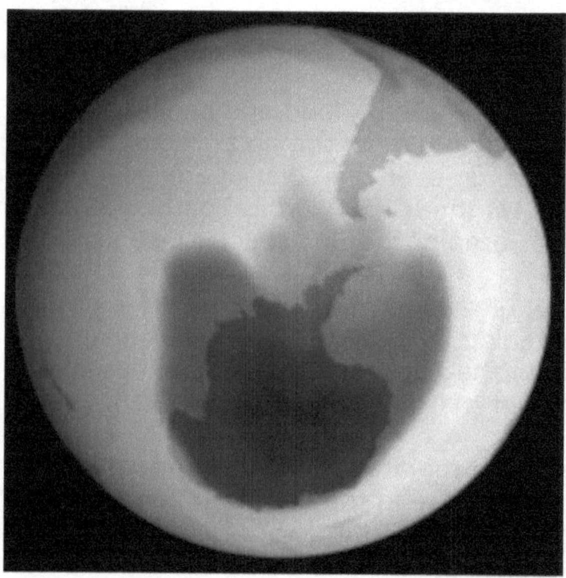

Ozone holes at earth's poles are created by the ozone free air emanating from the earth's interior through the polar openings.

Isn't it evident that there exists more scientific evidence that our earth IS hollow, and that all planets are hollow than can be dug up proving them to have combinations of solid-liquid interiors? Indeed, the evidence is almost overwhelming that Our Earth IS HOLLOW, yet it is not common knowledge because its discovery has been purposely kept hidden from the public by the powerful Illuminist Conspiracy. But the Conspiracy does not yet have a monopoly on truth even though it would like to. And anyone with a little study and investigation on his own may now be able to come to knowledge of the truth about the real structure of OUR HOLLOW EARTH!

CHAPTER THREE
The Garden of Eden -- Found!

William F. Warren in his scholarly work, <u>PARADISE FOUND, THE CRADLE OF THE HUMAN RACE AT THE NORTH POLE</u>, quotes from a translation by A.M. Sayce taken from a book called <u>RECORDS OF THE PAST</u>,

"We are told of a dwelling which 'the gods created for' the first human beings--a dwelling in which 'they became great' and 'increased in numbers,' and the location of which is described in words exactly corresponding to those of Iranian, Indian, Chinese, Eddaic and Aztecan literature; namely, 'in the center of the earth.'" (Warren, p. 240)

As we go back in history, perhaps a clue as to the whereabouts of the Biblical Garden of Eden wherein the human race had its beginnings can be obtained if we can trace the origins of Noah and his family. Ever since Noah's Ark settled on the mountains of Ararat in eastern Turkey and his descendants began once again to populate the earth and to name important landmarks, it has been supposed by students of Biblical history that the Garden of Eden must have been at the head of the river Euphrates in the Mesopotamian valley, as that river was named by Noah's early descendants. But surely the Mesopotamian valley was not the original home of Noah. In the 150 days that the Ark was afloat by the waters of the Great Flood, which the Bible maintains covered the entire surface of the planet, the ark would have moved a great distance being blown by the winds and driven by ocean currents.

Scientists are adamant that Noah's flood that covered the entire surface of the globe never occurred. However, all scientific evidence goes to prove otherwise. The fact that coal beds are found in the Arctic, Alaska and Spitsbergen, and in the Antarctic continent proves that these lands in times past grew luxuriant forests that later were violently covered with sand, mud and gravel by the turbulent waters of the great flood and thus turned into coal.

Immanuel Velikovsky, in his books, gives ample evidence that the Earth has been by-passed several times in past geological history by comets which have caused great catastrophes, changes in the earth's surface and shifting of the poles. Such a shift in the poles would be enough to inundate the globe with the subsequent melting of the polar icepacks. Enough water presently exists in these ice-packs to drown a large portion of the world even today according to Ian Cameron, who in his book, <u>ANTARCTICA, THE LAST CONTINENT</u>, writes,

> *"Seen from Space, the astronauts tell us, the most distinctive feature of our planet is the ice sheet of Antarctica, which 'radiates light like a great white lantern across the bottom of the world'. This ice sheet covers 5,500,000 square miles (an area greater than the United States and Central America combined); it averages more than 7,000 feet in thickness; it contains more than 90 per cent of the world's ice and snow, and if suddenly melted the oceans would rise to such height that every other person on earth would be DROWNED."* (Cameron, p. 12)

Modern revelation of God to the American prophet, Joseph Smith, infers that the original home of Noah was upon the American continent in a place called Adam-ondi-Ahman, Missouri, the land where Adam and his righteous descendants dwelt -- so in the five months that the Ark was upon the waters, the mighty storm and currents caused by the flood must have carried the Ark the 9,000 miles that lie between Missouri, the land of Adam-ondi-Ahman (<u>D&C 116</u>), and the Mountains of Ararat where the Ark came to rest in present day eastern Turkey.

Noah, carrying with him the writings of the prophets from Adam down to his time, had an account of the creation of the world and of the Garden of Eden, wherein was recorded that one of the rivers that flowed out of that garden was named the Euphrates. So in their new land, the descendants of Noah no doubt named one of the rivers, the Euphrates, but in no way must it be mistaken to be the actual river which flowed out of the Garden of Eden. For nowhere in Mesopotamia can be found a place from which flows four rivers and can thus be said to be the location of the original Garden of Eden.

And so scholars of the Bible have searched the world over in hopes of finding that garden from which flows four rivers wherein the human race had its start. In fact, peoples and explorers, ancient and modern, have embarked on expeditions in search of the Lost Garden of Eden. The 16th century Spanish scholar Bernardino de Sahagun recorded in his <u>HISTORY OF THE THINGS OF NEW SPAIN</u> that the original settlers of America, the ancestors of the American Indians, came in boats from the East in search of the terrestrial paradise. They settled in Central America near the highest mountains they could find, because they had with them an account that the earthly paradise is a very high mountain. (<u>ANCIENT AMERICA AND THE BOOK OF MORMON</u>, pp. 31, 156.) Wouldn't it, therefore, be of extreme interest to the Judeo-Christian world, if an explorer in actuality found the Garden of Eden that has been lost for thousands of years?

Perhaps a clue as to the location of the lost Garden of Eden can be taken from an ancient map published in 1569 by the foremost European mapmaker of the Renaissance era, Gerardus Mercator. Our maps to this day are based on his famous Mercator projection. In Mercator's map of the Arctic, centered on the North Pole, you can see inside the world a mountain where the Garden of Eden is located and from which flows four rivers to the four quarters of the inner continent.

Gerardus Mercator drew a map of Our Hollow Earth using the best reports he could obtain from explorers. His map shows the location of the Garden of Eden on the highest mountain plateau of the Inner Continent with the four rivers flowing out of the Garden of Eden towards the four main points of the compass. The River Hiddekel flows to the right and empties into the North Polar Opening near North Land, Russia.

Discovery of the Lost Garden of Eden

Since Christian explorers, among them Christopher Columbus, have sought earnestly the lost Garden of Eden, it is with interest that the actual discovery should fall to a non-Christian, in fact, a worshiper of the Norwegian gods, Odin and Thor who were their ancestors. This explorer, Olaf Jansen, actually a fisherman, has related in his book, THE SMOKY GOD, of his voyage with his father into the inside of the earth in their small sail boat via the North Polar Opening in 1829. They were taken in by the people that live there, taught their language and shown their country. Hence, their great discovery was brought to bear when after one year they were taken to the capital city of the inner earth which the inhabitants call, the City of Eden.

Olaf describes this city as a veritable garden in which,

"...all manner of fruits, vines, shrubs, trees, and flowers grow in riotous profusion. In this garden four rivers have their source in a mighty artesian fountain. They divide and flow in four directions. This place is called by the inhabitants, the 'navel of the earth' or the beginning, 'the cradle of the human race.' The names of the rivers are Euphrates, the Pison, the Gihon, and the Hiddekel." (THE SMOKY GOD, p. 114)

This is the narration of an explorer who was never a Christian. To his dying day, Olaf Jansen was the ardent worshiper of the Norwegian gods, Odin and Thor. His purpose was not to advance the reality of religion, but in recording his experience in the Garden of Eden, he was reporting exactly what he saw and what the people who live there told him. They explained to him that the garden he was taken to inside Our Hollow Earth was none other than the *"navel of the earth,"* and *"the cradle of the human race,"* and is called *"Eden."*

In the Book of Genesis we read about the Garden of Eden which Olaf Jansen discovered is located inside Our Hollow Earth:

"And the Lord God planted a garden eastward in Eden, and there he put the man whom he had formed. And out of the ground made the Lord to grow every tree that is pleasant to the sight, and good for food; the tree of life also in the midst of the garden, and the tree of knowledge of good and evil. And a river went out of Eden to water the garden; and from thence it was parted, and became four heads. The name of the first is Pison: that is it which

compasseth the whole land of Havilah where there is gold, And the gold of that land is good; there is bdellium and onyx stone. And the name of the second river is Gihon; the same is it that compasseth the whole land of Ethiopia, And the name of the Third river is Hiddekel; that is it which goeth toward the east of Assyria, And the fourth river is Euphrates." (GENESIS 2:8-14)

Notice may be given to the exactness of Olaf Jansen's description of the Garden of Eden to the account in the Bible, adding that the river which waters the garden and from which the other four rivers spring is a *"mighty artesian fountain."* This is supported by the dream of the Tree of Life given to the prophet Lehi as recorded in the Book of Mormon. He described a river that ran by the Tree of Life began as a fountain. What a beautiful sight it must have been for Olaf and his father to see this natural fountain in the midst of the most beautiful garden in the world!

The Garden of Eden in America?

It is the teaching of the Church of Jesus Christ of Latter-day Saints that the Garden of Eden was located in America. Brigham Young, the second prophet of the church, said,

"Joseph, the prophet told me that the Garden of Eden was in Jackson County, Missouri. When Adam was driven out, he went to the place we now call Adam-ondi-Ahman, in Daviess County, Missouri." (WILFORD WOODRUFF, by Cowley, p. 431)

The prophet Joseph Fielding Smith, a descendant of Joseph Smith's brother Hyrum wrote,

"In accord with the revelations given to the Prophet Joseph Smith, we teach that the Garden of Eden was on the American continent located where the City Zion, or the New Jerusalem will be built. When Adam and Eve were driven out of the Garden, they eventually dwelt at the place called Adam-ondi-Ahman, situated in what is now Daviess County, Missouri." (DOCTRINES OF SALVATION, Vol Three, p. 74)

Obviously, this statement originated with that of Brigham Young's statement as quoted above.

However, Joseph Fielding Smith does refer to some revelations of Jesus Christ to Joseph Smith on which he seems to base his statement. These are Section 57 and 84 of the Doctrine and Covenants, verses 1 to 3 in both sections. But upon reading these passages, I discovered that they refer to the place where the future New Jerusalem will be built and they do not mention nor infer anything about the Garden of Eden. To the knowledge of this author, Joseph Smith never recorded a revelation saying where the Garden of Eden is or was located.

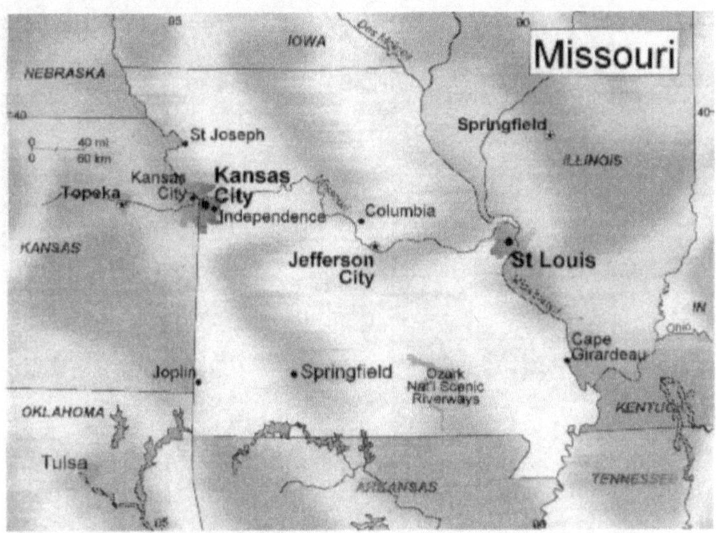

There is no indication that four rivers ever originated from Independence, Missouri to indicate it was the location of the original Garden of Eden.

Certainly, in looking at a map of Jackson County, Missouri, there is no evidence that four rivers ever emanated from Independence which could thereby be taken as the original location of the Garden of Eden. The reason is -- because the Garden of Eden is not located on the surface of the planet.

GLOBE SHOWING SECTION OF THE EARTH'S INTERIOR

The earth is hollow.   The poles so long sought are but phantoms.   There are openings at the northern and southern extremities.

William Reed's Conception of Our Hollow Earth

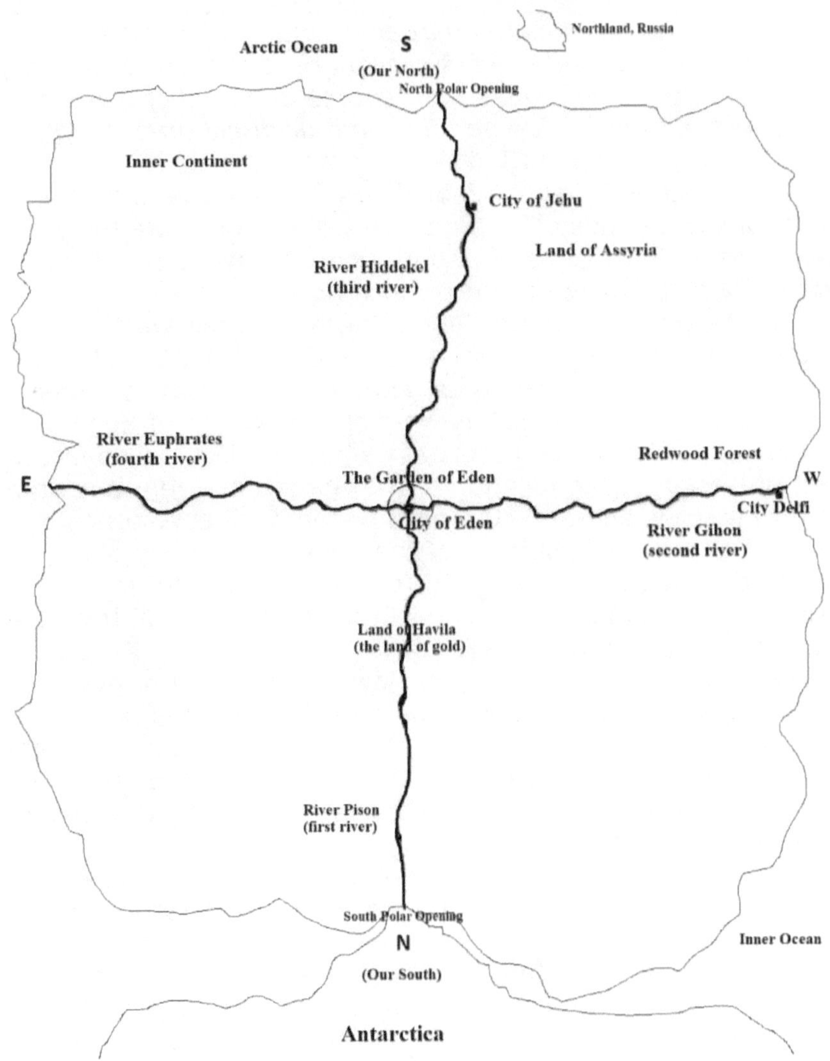

The Garden of Eden within Our Hollow Earth

In <u>THE SMOKY GOD</u>, Olaf Jansen gives a description of where the Garden of Eden is located. Inside the earth, Olaf Jansen says,

"The City of 'Eden' is located in what seems to be a beautiful valley, yet, in fact, it is on the loftiest mountain plateau of the Inner Continent, several thousand feet higher than any portion of the surrounding country. It is the most beautiful place I have ever beheld in all my

travels. In this elevated garden all manner of fruits, vines, shrubs, trees, and flowers grow in riotous profusion. In this garden four rivers have their source in a mighty artesian fountain. They divide and flow in four directions." (THE SMOKY GOD, pp. 113, 114)

The directions that the four rivers flow are the four main points of the compass. The first river, the Pison flows toward their north pole which is our south pole. It flows through the land of Havilah *"where there is much gold."*

The second river is the Gihon flowing to their west and empties into the Inner Ocean near the City of Delfi. Olaf Jansen reported that a Redwood Forest extends back from the coast from the City of Delfi which is west of the Garden of Eden. Curiously, our Redwood Forest in California, USA is also located west of the supposed location of the Garden of Eden, but which we now learn is actually 800 miles beneath Independence, Missouri on the inner surface of the planet. Whereas our Redwood Forest trees grow 300 feet tall and 30 feet in diameter, those in the Hollow Earth grow up to 1,000 feet tall and are 120 feet in diameter.

We know the Hiddekel River flows to their south towards our north pole because that is the river Olaf and his father sailed up soon after reaching the land of Our Hollow Earth after passing through the North Polar Opening. The Hiddekel River flows to the east of the land of Assyria -- a land by that name in Our Hollow Earth.

The fourth river, the Euphrates, then, flows to their east. Olaf Jansen reported that the points of the compass inside the earth are opposite ours.

Since Olaf and his father traveled far into the interior on a monorail train to reach the Garden of Eden upon the highest mountain plateau of all the Inner Continent, which the hollow earth people call the *"navel of the earth,"* and since out of the Garden of Eden flows the four rivers in four directions, we may conclude that the Garden of Eden must be located toward their southern center of the Inner Continent.

Olaf reported that the inside of our earth has only one continent and one ocean and the ocean is small in comparison with the land area. Olaf Jansen said that the ocean occupies one fourth of the total area inside the earth. This ocean also extends from the Arctic Ocean to the Antarctic, and it was upon this inner ocean that Olaf and his father navigated in their small sail boat when they entered the North Polar Opening. It was upon this ocean that they also navigated to come out the South

Polar Opening at the end of their two and half year stay in the land of Our Hollow Earth.

Therefore, the single continent inside the earth comprises three fourths of the inner surface area, and extends from one side of the inner ocean clear around the inner surface to the other side of that ocean. We can estimate the circumference of the interior from the 800-mile estimate of the earth's shell which gives us in rounded figures, 16,000 miles. The interior ocean being one-fourth of that figure is 4,000 miles across and the remaining 12,000 miles consists of continent. From this we can conclude that there is more land inside Our Hollow Earth than there is upon the outer surface of the earth.

As can be seen on William Reed's conception of the inner surface, the ocean inside the earth is located under Asia, the exact opposite of the earth from America. Thus the center of the Inner Continent would be under America. Since the Garden of Eden is located approximately in the southern center of the Inner Continent, then the Garden of Eden must be under North America. Therefore, when Joseph Fielding Smith wrote, *"We teach that the Garden of Eden was on the American Continent located where the City Zion, or the New Jerusalem will be built,"* he nevertheless came close to the location of the Garden of Eden, which according to Olaf Jansen could be located 800 miles BENEATH Independence, Missouri, the future location of the New Jerusalem.

One could ask why the Lord would be so elusive as to hide the Garden of Eden 800 miles beneath Missouri. But when we recall that the Lord let the people of Europe believe that the earth was flat, in order to preserve the Americas for the descendants of Lehi, the father of the American Indians, we may then understand why He has allowed the world of today believe that the earth is partially molten and partially solid inside to hide the Garden of Eden 800 miles beneath the surface of the earth of Missouri in order to preserve the land of Our Hollow Earth and its Garden of Eden as the inheritance for the people living inside the earth. They must be a chosen people of God to be given such a paradise to live in.

But unlike the ancient Americans who became wicked and their inheritance was allowed to be overrun by a more righteous European people, the people inside Our Hollow Earth are more righteous than we, and the Lord will thus continue to hide their land until the wicked are done away with on the surface world. Then will the Lord make known to the righteous, the land of the Garden of Eden. But until that revelation comes, the few who would believe may know for themselves the true location of the lost Garden of Eden.

Since Olaf Jansen found a city inside the earth built in a garden called "Eden" from which flows four rivers with the same names as given in the Bible, we may conclude that the human race had its beginning inside the earth. Adam and Eve, our first parents, had their first home in the Garden of Eden, which has now been FOUND to be located on the inside surface of Our Hollow Earth!

From THE FORGOTTEN BOOKS OF EDEN we read of the manner in which Adam and Eve arrived on the surface of the planet from the Garden of Eden within. In the book of ADAM AND EVE, we find that when Adam and Eve were expelled from the Garden for having partaken of the forbidden fruit, the Lord commanded them to go into the Cave of Treasures where they got lost and after wandering and praying for a long time, they emerged to the surface world -- their new home. (Chapter III:17, XII)

The sun that warmed them when they came out of the cave was not the same sun that gave them light in the Garden of Eden. Chapter XVI says that when Adam came out of the cave, he was afraid of the sun and its extraordinary heat because,

> "Inasmuch as while he was in the garden and heard the voice of God and the sound He made in the garden, and feared Him, Adam never saw the brilliant light of the sun, neither did the flaming heat thereof touch his body."

The sun that warmed the Garden of Eden was the sun inside our Hollow Earth, and is gentler than our hotter sun which warms the exterior of the planet. When Adam and Eve emerged from the cavern on the exterior of the planet, our outer sun warmed them with greater intensity than the interior sun did when they were in the Garden of Eden.

The Missouri-Kentucky area is noted for its extensive cavern system. The Mammoth Cave of Kentucky is so extensive that the end of it has not been found. Through a certain cavern in Kentucky, William Morgan, who called himself I-AM-THE-MAN in ETIDORPHA, reached the hollow interior of the earth with his inner earth guides when he was sent there by the Freemasons in 1827. Perhaps it was this or a similar cavern that Adam and Eve journeyed from their home within to their home without after being expelled from the Garden of Eden.

In the Book of Moses, Chapter 7, verse 48, we read,

"And it came to pass that Enoch looked upon the earth; and HE HEARD A VOICE FROM THE BOWELS THEREOF, saying: Wo, wo is me, THE MOTHER OF MEN; I am pained, I am weary, because of the wickedness of my children. When shall I rest, and be cleansed from the filthiness which is GONE FORTH OUT OF ME? When will my Creator sanctify me, that I may rest, and righteousness for a season abide UPON MY FACE?"

By passing through the cavern systems from the earth's *"womb"* inside the earth to come to the surface, Adam and Eve were *"born"* into this world by *"mother"* earth.

Perhaps you have heard that the American Indians believe that their ancestors came up out of the earth? I have been personally told this by a Navajo that installed a satellite internet dish on my farm for me in northeast Arizona.

When Jesus came to America soon after his resurrection, he first visited the Nephites and Lamanites in MesoAmerica who had survived the catastrophe that destroyed many of their cities during the three hours that Christ was on the cross in Jerusalem. Several days after His resurrection, He came to their temple in the Land of Bountiful, which I believe was the Temple of Kukulcan in Chichen Itzá that my wife and I visited in the Yucatan peninsula in 2013. (See The Book of Mormon, 3rd Nephi Chapter 11)

The sacred records of the North American Indians record the visit of Jesus to them after his visit to the Nephites and Lamanites in the Land Southward. As Jesus was teaching the people, in their capital city of Mentinah (what is today Manti, Utah) he said, *"After that he* (Adam) *had been cast out of the Terrestrial World* (from the Garden of Eden) *for a season, he built an altar and offered sacrifice unto the Lord."* (The Record of Ougou, verse 113, THE SACRED RECORDS OF THE NEMENHAH.)

So here we learn from our Savior that where the Garden of Eden is located is a Terrestrial World. The North American Indians said that when *"Adam and Eve emerged up out of the sheltered place where they had spent their spiritual infancy into the Telestial World they called it a lone and dreary world..."* (The Relation of Heavenly Father and Heavenly Mother, Chapter Three, verse 6, THE SACRED RECORDS OF THE NEMENHAH) We are living in that "lone and dreary world" called the Telestial World and the Terrestrial World is the inner surface of Our Hollow Earth where the Garden of Eden is located today surrounded by the City of Eden -- the capital city of the Hollow Earth nation.

The Norwegian fishermen Olaf Jansen and his father Jens in 1829 in their voyage beyond the "north wind" where they entered into Our Hollow Earth through the North Polar Opening discovered the lost Garden of the Eden on the highest mountain plateau of the Inner Continent. We now learn from the North American Indians that where the Garden of Eden is located is a Terrestrial World. And our outer world is a Telestial World. This is why the North American Indians tell us that their ancestors came from inside the earth. Their ancestors were Adam and Eve, and they give their genealogy back to Adam and Eve.

In 1960, a resident of Los Angeles, California, Larry Foreman, went camping out in the desert where he was befriended by people humans like us that came in a flying saucer. They wanted him to write a book, which he did, called, PASSPORT TO ETERNITY. Although the saucer men did not specifically say where they came from, many things they told Larry indicate to me that they came from Our Hollow Earth. Olaf Jansen said that the people he discovered in Our Hollow Earth speak a language known to us as Sanskrit. Sanskrit is a language spoken anciently in Iran where the Ten Tribes of Israel were carried into captivity in 721 B.C. by the Assyrians and where they lived for 34 years before escaping over the Caucasus Mountains into southern Ukraine north of the Black Sea.

The saucer men told Larry that their native language is Sanskrit, and that they are our relatives. They gave a message to Larry for all of us: Our planet is a prison planet and we were placed here for something that our ancestors did.

Our ancestors were Adam and Eve, who broke a commandment given them by the Creator in the Garden of Eden: Do not partake of the forbidden fruit. The punishment for braking that commandment was that it would cause them to become mortal beings, (but also changed their bodies so that they could have children -- us) and they were expelled from His presence from their home in the Garden of Eden to our surface world. We, as their descendants are all living out a prison sentence on this prison planet we call Earth. We are living a life of *"probation"* as the scriptures say, to see if we are willing to obey the Creator's commandments -- to see if we can be found worthy to return to the presence of our Creator.

The ancient prophet Lehi said, as recorded in the Book of Mormon,

"And after Adam and Eve had partaken of the forbidden fruit they were driven out of the garden of Eden, to till the earth. And they have brought forth children; yea, even the family of all the earth. And the days of the children of men were prolonged, according to the will of God, that they might repent while in the flesh; wherefore, their state became a state of probation, and their time was lengthened, according to the commandments which the Lord God gave unto the children of men. For he gave commandment that all men must repent; for he showed unto all men that they were lost, because of the transgression of their parents. And now, behold, if Adam had not transgressed he would not have fallen, but he would have remained in the garden of Eden. And all things which were created must have remained in the same state in which they were after they were created; and they must have remained forever, and had no end. And they would have had no children; wherefore they would have remained in a state of innocence, having no joy, for they knew no misery; doing no good, for they knew no sin." (2 Nephi 2:19-23)

But the good news is that our prison sentence is about up. The flying saucer people came to tell Larry that they have a project they call, *"Project Milana."* They told Larry that our prison sentence on this planet is about up (my estimate is about the year 2035 when I estimate the earth will be hit by a devastating solar flare caused by the passage of the same planet-sized comet that initiated Noah's flood). And that they are coming to take us all back home. That is, those of us that deserve to be taken back home, which I doubt would include murderers, liars, thieves, adulterers, and war mongers, in short, all who insist on breaking the Creator's laws.

The flying saucer people told Larry they can requisition enough space craft to evacuate the surface of the planet in ONE week! And take us home.

Our original home was the Garden of Eden. They say they come from the center of the Universe where there are two planets that share the same atmosphere -- which is a way of saying they are really from the Center of the Earth -- Our Hollow Earth, since we share our atmosphere with them through the polar openings. They told Larry that they live lives that we would consider *"heaven."*

Deuteronomy 30:4 calls the land of Our Hollow Earth, the *"outmost parts of heaven"* where the Lost Tribes of Israel were driven to. The prophet Isaiah wrote that in the last days God would *"...set up an ensign for the nations, and shall assemble the outcasts of Israel, and gather together the dispersed of Judah from the four corners of the earth."* (Isaiah 11:12) The four corners of the Earth is the Garden of Eden, where out of flows four rivers to the four cardinal points of the compass.

The flying saucer people told Larry that their capital city is on their southern hemisphere. Their southern hemisphere is under our northern hemisphere. Therefore, my conclusion is that the Garden of Eden is located 800 miles BENEATH Jackson County, Missouri. Their capital city is the City of Eden, built next to the original Garden of Eden, where Olaf Jansen found it in 1830 when he sailed there with his father *"beyond the north wind"* through the north polar opening, into the interior of our Hollow Earth where *"the chosen people"* live. They explained to Olaf that the garden he was taken to inside Our Hollow Earth was none other than the *"navel of the earth,"* and *"the cradle of the human race,"* and is called *"Eden"* and is located on the highest mountain plateau of the inner continent on their southern hemisphere. Since inside the earth the directions of the compass are reverse of what they are on our surface world, our south pole is their north pole, then their southern hemisphere is beneath our northern hemisphere, and their capital city is located beneath North America. The coordinates the saucer men gave Larry for their capital city were: South South West 1/4 9' 3" 10 SID, according to their coordinate system.

Our expedition member, James Still met Larry Foreman February 16, 1979, in Los Angeles, California, at which time Larry gave James a copy of his book, PASSPORT TO ETERNITY. After joining our Voyage to Our Hollow Earth expedition, James gave me a copy of Larry's book. In that book, the saucer men told Larry that *"the earth is not solid."* (p. 93) If the earth is not solid, then it must be hollow and the Garden of Eden is inside it. Ret. Col. Billie Woodard told me that while he was stationed at Area 51, he was taken in the tunnel trains beneath Area 51 to Our Hollow Earth where he met with leaders of the hollow earth nation in their capital city of Eden. (See the Contents page for Biography of Ret. Colonel Billie Faye Woodard.)

Could it be that the International Illuminist Conspiracy is hiding the discovery of our Hollow Earth because it hides the lost Garden of Eden, which discovery and publicity to the world would destroy their atheistic Theory of Organic Evolution which they use to destroy world morality and consolidate their power?

The day will soon come that the discovery of the Garden of Eden will be made known to the world. Then will faith increase in the earth, the Illuminist Conspiracy will lose its power over the minds of men, and God will be enthroned once more in the hearts of mankind because they will know that our first parents, Adam and Eve, did truly live in the Garden of Eden which will prove that the scriptures are true history as revealed from God, that God lives and loves us, his children.

CHAPTER FOUR
The Land of The Lost Ten Tribes of Israel -- Found!

One of the greatest discoveries of all time would be the actual location of the present descendants of the original *"lost"* Ten Tribes of Israel. The land where they are at present located and established as the richest, perhaps most populated, and without doubt the most powerful and most highly advanced nation of the earth has been hidden from the world since history began.

It was in the year 721 before the birth of Christ that the Israelite nation, then residing in Palestine and consisting of ten tribes, descendants of ten out of the 12 sons of the ancient prophet Jacob, were carried away captives into Assyria by the Assyrians. James E. Talmage writes in his ARTICLES OF FAITH,
> *"The people were led into Assyria and later disappeared so completely that they have been called the Lost Tribes. They seem to have departed from Assyria, and while we lack definite information as to their final destination and present location, there is abundant evidence that their journey was toward the north."*
(Talmage, p. 325)

Knowledge of their whereabouts has been given to certain persons throughout history, while the great majority of the people of our civilization ignore of such knowledge. Through his prophet Moses, the Lord admonished the Israelite nation to love the Lord their God and keep His commandments or else they would be scattered all over the earth. Still, he promised that he would not forget them, and in the latter days he would gather them once again to the land He gave unto their fathers. As history has shown, the House of Israel kept not the statutes given by Moses from the Lord their God, and so they were scattered all over the earth.

Two groups of Israelites came to America, the Mulekites, and the family of the prophet Lehi. The Mulekites were Jews, descendants of Judah, son of Jacob. When the Babylonians captured Jerusalem in 597 B.C. and the Jews were carried off to Babylon (today's modern Iraq), the palace guards escaped with King Zedekiah's son Mulek to America. They settled in Central America. Their descendants were the "king-men," spoken of in the Book of Mormon who where continuously trying to take over the government from the Nephites (Alma 51:5-8). They believed they had a right to the government because they were of the royal lineage of David.

The family of the prophet Lehi came to America before the Babylonian conquest of Jerusalem. They left Jerusalem in 600 B.C. and sailed to America and settled in Central America, further south than the Mulekites, but later joined the Mulekites. The family of Lehi divided into two groups when they arrived in America, the Nephites and the Lamanites, and were descendants of the Tribe of Manasseh of the house of Joseph, who was sold into Egypt by his brothers. These immigrants to America wrote the record known as The Book of Mormon -- a history of God's dealings with them in America. The American Indians are a remnant of their descendants.

The other larger group of Israelites was the Lost Ten Tribes, who had been carried away into captivity by the Assyrians in 721 B.C. Later they escaped and went north over the Caucasus Mountains to land north of the Black Sea. An indication where the Lord intended that the Lost Tribes would be scattered to is alluded to in verse 4 and 5 of Deuteronomy 30:

> *"If any of thine be driven out unto the OUTMOST PARTS OF HEAVEN, from thence will the Lord thy God gather thee, and from thence will he fetch thee: And the Lord thy God will bring thee into the land which thy fathers possessed, and thou shalt possess it; and he will do thee good, and multiply thee above thy fathers."*

To the prophet Jeremiah, the Lord gave a message to take to the Lost Ten Tribes. Said He,

> *Go and proclaim these words toward the north, and say, Return, thou backsliding Israel, saith the Lord; and I will not keep anger forever."*

> *"Only acknowledge thine iniquity, that thou has transgressed against the Lord thy God, and hast scattered thy ways to the strangers under every green tree, and ye have not obeyed my voice, saith the Lord."*

> *"Turn, O backsliding children, saith the Lord; for I am married unto you: and I will take you one of a city, and two of a family, and I will bring you to Zion."*

> *"And it shall come to pass, when ye be multiplied and increased in the land, in those days, saith the Lord, they shall say no more, The ark of the covenant of the Lord; neither shall it come to mind; neither shall they remember it; neither shall they visit it; neither shall that be done any more."*

"At that time they shall call Jerusalem the throne of the Lord; and all the nations shall be gathered unto it, to the name of the Lord, to Jerusalem; neither shall they walk any more after the imagination of their evil heart."

"In those days the house of Judah shall walk with the house of Israel, and they shall come together OUT OF THE LAND OF THE NORTH to the land that I have given for an inheritance unto your fathers." (Jeremiah 3:12-18)

The Land of the North from which in some future day the Ten Tribes will come out from and establish open communication with the rest of the world, is the land to which Esdras saw in a vision that the Ten Tribes went after escaping the Assyrians.

Esdras wrote,

"Those are the tribes which were carried away captives out of their own land in the time of Oseas (Hosea) the king, whom Shalmanezer, the king of the Assyrians, took captive, and crossed them beyond the river; so were they brought into another land. But they took counsel to themselves, that they would leave the multitude of the heathen, and go forth into a further country where never man dwelt, that they there might keep their statutes, which they never kept in their own land. And they entered in at the narrow passage of the river Euphrates. For the Most High then showed them signs, and stayed the springs of the flood till they were passed over. For through the country there was a great journey, even of a year and a half, and the same region is called Arsareth. Then dwelt they there until the latter time, and when they come forth again, the Most High shall hold still the springs of the river again, that they may go through." (2 Esdras 13, The Apocrypha)

Although the Ten Tribes of Israel were carried away into captivity by the Assyrians because of Israel's iniquity in worshiping heathen Gods, Jehovah had told Elijah before the captivity that there were still seven thousand righteous in Israel *"all the knees of which have not bowed unto Baal."* (1 Kings 19:18) These righteous were the sons of the prophets, (2 Kings 2:15) who were carried away by the Assyrians along with the rest of the Israelite Nation of the Ten Tribes. Later on it must have been one of these sons of the prophets who led their nation in an escape from the Assyrians into the North Country.

James E. Talmage, in his ARTICLES OF FAITH continues,

"The Lord's word through Jeremiah promises that the people shall be brought back 'from the land of the north,' and a similar declaration has been made through divine revelation in the present dispensation." (Talmage, p. 235)

The 10th Article of Faith of the Church of Jesus Christ of Latter-day Saints states,

"We believe in the literal gathering of Israel and in the restoration of the Ten Tribes; that Zion (the New Jerusalem) will be built upon the American continent; that Christ will reign personally upon the earth; and, that the earth will be renewed and receive its paradisiacal glory."

To one Latter-Day Saint, Benjamin F. Johnson, the prophet Joseph Smith gave an indication of where the Lost Ten Tribes are located.

In MY LIFE'S REVIEW, Benjamin F. Johnson relates, *"...that I was then really 'the bosom friend and companion of the Prophet Joseph.' ...Sometimes when at my house, I asked him questions relating to the past, present and future."* Some of those answers ended up in the Doctrine and Covenants of the church, Section 130.

Benjamin F. Johnson wrote in his journal, *"Other questions were asked* (of the Prophet Joseph Smith) *when Brother Clayton* (the prophet's secretary) *was not present* (to record it), *one of which I will relate: I asked where the nine and a half tribes of Israel were. 'Well,' said he, 'you remember the old caldron or potash kettle you used to boil maple sap in for sugar, don't you?' I said, 'Yes.' 'Well,' said he, 'they are in the north pole in a concave just the shape of that kettle. And John the Revelator is with them, preparing them for their return.'* (MY LIFE'S REVIEW, by Benjamin F. Johnson, pp. 88-89)

In the Doctrine and Covenants, Section 133:26-33, Joseph Smith recorded a revelation from Jesus Christ in 1831 concerning the return of the Lost Ten Tribes:

"And they who are in the north countries shall come in remembrance before the Lord; and their prophets shall hear his voice, and shall no longer stay themselves; and they shall smite the rocks, and the ice shall flow down at their presence."

"And an highway shall be cast up in the midst of the great deep."

"Their enemies shall become a prey unto them,"

"And in the barren deserts there shall come forth pools of living water; and the parched ground shall no longer be a thirsty land."

"And they shall bring forth their rich treasures unto the children of Ephraim, my servants."

"And the boundaries of the everlasting hills shall tremble at their presence."

"And there shall they fall down and be crowned with glory, even in Zion, by the hands of the servants of the Lord, even the children of Ephraim."

"And they shall be filled with songs of everlasting joy."

And They Who Are In the North Countries

Where are the North Countries? They must be further north than the ice barrier and icebergs in the Arctic Ocean because the scripture says that "the ice shall flow down at their presence" when the Lost Tribes come to Zion from the north.

Norwood Review of England, in its issue of May 12, 1884, summarized the surprised discovery of Arctic explorers of a warm country up near the pole,

"We do not admit that there is ice up to the Pole--once inside the great ice barrier, a new world breaks upon the explorer, the climate is mild like that of England, and afterward, balmy as the Greek Isles."

Of course, this is in the summer time, for in the winter the icebergs form. Referring to the origin of the icebergs, the Lord asked the prophet Job:

"Out of WHOSE WOMB came the ice?" (Job 38:29)

The answer is, out of the earth's womb. The polar openings do exist, and the Lord testifies that the icebergs come out of them. The North Countries of the Lost Ten Tribes must be located inside the earth's WOMB from whence the icebergs come.

From Olaf Jansen's account of his journey into the inside of Our Hollow Earth, the scriptural reference that the Ten Tribes live in whole COUNTRIES in the north can now be understood. The scripture does not make sense if one is to believe that the north is a frozen ice-cap, unfit for human habitation. But when one realizes that near the geographic North Pole there exists an opening that leads into the hollow interior of the earth where there is more land than upon the outer surface of the planet, then one can understand how millions of the Lost Tribes have lived in the *"North Countries"* for nearly 2,500 years.

In the observation of their compass which guided Olaf and his father to the inside of Our Hollow Earth, we find an explanation as to why the *"inside"* of the earth would be called the *"north countries"* in scripture. When the Israelites took their year and half journey through the north polar opening and reached the *"north countries"* inside Our Hollow Earth, had they been guided there with a compass, they would have noticed the same behavior of the compass that Olaf Jansen and his father noticed. They found that the compass continued to point toward that *"North Country"* even though after passing the polar opening they were indeed going south on the inside surface of the planet.

Olaf described this performance of their compass,

"My father and I commented between ourselves on the fact that the compass still pointed north, (to the *"north"* marking on the compass) *although we now knew that we had sailed over the curve or edge of the earth's aperture, and were far along southward on the 'inside' surface of the earth's crust, which, according to my father's estimate and my own, is about three hundred miles in thickness from the 'inside' to the 'outside' surface."* (THE SMOKY GOD, pp. 107-108)

The land of the North Countries, the land of the Ten Tribes of Israel, is truly there, and anyone who follows his compass north, if he travels directly toward the north geographic pole on a certain meridian in a straight line without deviating (130 E Longitude), will enter the North Polar Opening into the hollow of the earth and will observe, as Olaf Jansen did, that his compass still points "north" towards those North Countries where the Lost Tribes presently reside. Since the magnetic field lines of the earth come out of the earth at our South Magnetic Pole and go back inside the earth at the North Magnetic Pole, this causes the compass to point North on the outside surface of the planet and on the inside the planet it points to their North Pole which is our South Pole. The directions of the compass inside Our Hollow Earth are reverse of what they are on the outside surface of the earth.

The Scandinavians have a legend of a land of paradise in the far north, known as *"Ultima Thule."* Perhaps this is an indication that the Ten Tribes of Israel after escaping the Assyrian armies may have reached the North Country inside Our Hollow Earth on a path that took them through Germany, Denmark, and Norway and left some of their people behind in that journey.

The legend of Ultima Thule originated with the Greek explorer Pytheas from his travels between 330 BC and 320 BC, in which he sailed around Great Britain and even further north and reported that after he had sailed several days north past the ice, he arrived to a land where the sun did not set. Most likely, the Ultima Thule that Pytheas found was the inner continent of Our Hollow Earth accessed through the North Polar Opening during the summer months when there are leads of open water between the ice flows through which to navigate.

In their journey of a year and a half to the Hollow Earth North Countries, the Lost Tribes left their markings behind in Europe. One of the tribes is the Tribe of DAN. In Germany we find a river named Danube. Denmark could have been derived from his name. A portion of the tribes must have left the main group in this long journey and stayed behind in the countries through which they passed giving rise to the legends among the Norwegians, Swedish, Germans, Greeks and Romans of a paradise in the north where the "Chosen People" lived. The legends of the *"chosen"* ones of the Lost Ten Tribes who migrated into the North Country of Our Hollow Earth, concerning the paradise in the north remained with those who stayed behind in Europe thereby giving rise to the legend of *"Ultima Thule,"* the land of paradise in the far north.

Perhaps the people left behind in Europe were even revisited by the people in the North Country, from time to time, giving origin to the gods of the Norwegians, the Greeks, and the Romans.

"Plato and his contemporaries in ancient Greece were fervent believers in an underworld. He wrote, 'He is the god who sits in the center, on the navel of the earth; and he is the interpreter of religion to all mankind.'" (THE HIDDEN SECRETS OF THE HOLLOW EARTH, p. 164)

Olaf Jansen and his father were Norwegians with their home in Stockholm, Sweden. It was because of their legend of Ultima Thule that they decided to go on their voyage to that land. Olaf wrote,

"My father was an ardent believer in Odin and Thor, and had frequently told me they were gods who came from far beyond the 'north wind.' There was a tradition, my father explained, that still farther northward was a land more beautiful than any that mortal man had ever known, and that it was inhabited by the 'Chosen.'" (THE SMOKY GOD, pp. 62, 63)

There are scriptures from the Bible which refer to the polar openings which lead to the North Countries of Our Hollow Earth. Job wrote of the Lord that,

"He stretcheth out the north over the EMPTY PLACE, and hangeth the earth upon nothing." (Job 26:7, 9)

And Isaiah in writing about the ambitions of Satan referred to the polar opening in the north when he recorded,

"For thou hast said in thine heart, I will ascend into heaven, I will exalt my throne above the stars of God: I will sit also upon the mount of the congregation, IN THE SIDES OF THE NORTH; I will ascend above the heights of the clouds; I will be like the most High. Yet thou shalt be brought down to hell, to the SIDES OF THE PIT." (Isaiah 14:12-15)

The prophet Jeremiah and King David spoke of the Lord, that, "He causeth the vapours to ascend from the ends of the earth..." and ..."he bringeth the wind out of his treasuries." (Psalms 135:7, Jeremiah 10:13, Jeremiah 51:16)

The "ends of the earth" are the earth's polar openings at its northern and southern extremities and the "vapours that ascend" are warm winds, clouds and fogs that come out of the polar openings from an inner continent warmed by an inner sun beyond the ice where the Lost Tribes do dwell.

Olaf Jansen, in his journeys in Our Hollow Earth, reported that the inhabitants told him that the name of their God is JEHOVAH.

Now, Jehovah was THE God of the ancient Israelites and unique to that nation alone. The fact that the inhabitants of the Hollow Earth declare their God to be Jehovah proves that they are Israelites. And the fact that there are descendants of the Tribes of Ephraim, Dan and others of the Lost Tribes who stayed behind in Europe with a legend of a paradise in the North where the *"Chosen"* people went, gives reason to believe that the land which Olaf and his father, Jens Jansen, found beyond the North Wind must be the North Countries of the Lost Ten Tribes of Israel, which land is the inside surface of Our Hollow Earth -- the very *"outmost parts of heaven"*!

And Their Prophets Shall Hear His Voice

After a year spent in teaching Olaf and his father their language, the inhabitants of the hollow earth took them to their capital city located in the Garden of Eden where they were presented to their prophet-leader.

"The surprise of my father and myself was indescribable when, amid the regal magnificence of a spacious hall, we were finally brought before the Great High Priest, ruler over all the land. We were given an audience of over two hours with this great dignitary, who seemed kindly disposed and considerate. He showed himself eagerly interested, asking us numerous questions, and invariably regarding things about which his emissaries had failed to inquire." (THE SMOKY GOD, pp. 112-115)

The Lord Jesus Christ revealed to his prophet Joseph Smith that in the Last Days, *"... they who are in the north countries shall come in remembrance before the Lord; and their prophets shall hear his voice, and shall no longer stay themselves..."* From this we can conclude that the Ten Tribes have prophets who are *"staying themselves,"* hidden up in their *"north countries"* waiting until some future time when the Lord will command them to come forth. Olaf Jansen reported that their prophet is also their king. They called him, *"the Great High Priest, ruler over all the land."*

If their king literally is a *"High Priest,"* then the people of the hollow earth have the Melchizedek Priesthood, since High Priest is an office of that priesthood. D&C 107:91 says,

> *"And again, the duty of the President of the office of the High Priesthood is to preside over the whole church, and be like unto Moses--"*

The Great High Priest, the ruler over all the land of Our Hollow Earth therefore is the prophet of their church as well as their king. This would indicate that they are a very righteous people.

Regarding what Olaf reported that their God is Jehovah, Jehovah is the premortal name of Jesus Christ, therefore their church must be the Church of Jesus Christ.

Under the title, *"President of the Church,"* Elder Bruce R. McConkie writes in MORMON DOCTRINE,

> *"Upon the President of the Church the Almighty bestows the highest office and the greatest gifts that mortal man is capable of receiving. He is the earthly king of the kingdom of God, the supreme officer of the Church, the 'President of the High Priesthood of the Church...'"* (p. 532)

It would appear that the Tribes among whom Olaf went were a very righteous people living a spiritual-economic order similar to the United Order that was revealed by Jesus Christ to the prophet Joseph Smith. In this order, as explained in modern scripture, there are no poor among the people, which is the case among the people Olaf Jansen visited. They also live in small cities which is a pattern of the United Order. In this Order, the church government is also the government of the land. Such is so in Our Hollow Earth. The ruler over all the land is also the prophet of their church. He is their earthly king of the Kingdom of God.

This indicates that the entire population of Our Hollow Earth must be faithful members of the Church of Jesus Christ. For the United Order to function successfully, all citizens must be faithful members of the Church. If there were contentions and divisions among them, out of necessity the church would have to be separate from the state government as it is upon our surface world. The reason for this is because the laws of consecration and stewardship of the United Order must be complied with voluntarily, and if there were a percentage of the population who were not members or unwilling to live those voluntary laws, then a separate government based upon force would be necessary, as are the governments of our surface world which are set up to govern the more rebellious.

Perhaps the United Order was instituted among the Lost Ten Tribes at the time it was instituted among the Church at Jerusalem after Christ's resurrection and among the Nephites in America when he visited them at which time He said,

> *"But now I go unto the Father, and also to show myself unto the lost Tribes of Israel, for they are not lost unto the Father, for he knoweth whither he hath taken them."* (3 Nephi 17:4)

Even though the Ten Tribes have the Melchizedek Priesthood, and their ruler is a prophet of God, he doesn't have all the keys of the priesthood. It is evident that God has not given their prophet the keys of the sealing ordinances of the temple. The President of the Church of Jesus Christ of Latter-day Saints with headquarters in Salt Lake City, Utah,

> *"...is the one man on earth at a time who can both hold and exercise the keys of the kingdom in their fullness."* (MORMON DOCTRINE, p. 532)

These keys were given to the prophet Joseph Smith and Oliver Cowdery, his assistant in the Church after the dedication of the first temple of this, the Dispensation of the Fullness of Times (see Ephesians 1:10), in the temple at Kirtland, Ohio. They were in the temple on April 3, 1836 when Jesus Christ appeared to them, followed by Moses, Elias (Noah) and Elijah. Moses delivered to them the *"keys of the gathering of Israel from the four parts of the earth, and the leading of the ten tribes from the land of the north."* (D&C 110:11) Elias (this Elias was the prophet Noah, who held the keys of the dispensation of Abraham, see WHO AM I, by Alvin R. Dyer, Loc. 4777) *"committed the dispensation of the gospel of Abraham, saying that in us and our seed all generations after us should be blessed."* Elijah gave them the keys of the sealing of husbands and wives in eternal matrimony and the sealing of the children to their parents. This will unite families through all generations of time back to Adam and Eve so that all that will accept these ordinances performed in the temples of God can become part of the family of God.

Because the prophets of the Ten Tribes do not have the keys of these sealing ordinances of the temple, in the not too distant future they will lead their people out of the earth through the north polar opening, build a *"highway"* across the Arctic Ocean from the City of Eden, their capital, to the New Jerusalem in Jackson County, Missouri which will be the future capital of America for the political Kingdom of God, and there they will receive their endowments in the temples of God.

> *"And there they shall fall down and be crowned with glory even in Zion, by the hands of the servants of the Lord, even the children of Ephraim."* (D&C 133:32)

The Book of Mormon prophet Ether, the last of the Jaredite prophets wrote about the New Jerusalem that will be built up in America, which the Lost Ten Tribes will help build together with the Latter-day Saints and the indigenous Indians of America who are descendants of Judah and Manasseh of the House of Israel. In finishing his summary of the record of the Jaredites, the prophet Moroni commented on Ether's teachings on the New Jerusalem thus:

"And now I, Moroni, proceed to finish my record concerning the destruction of the people of whom I have been writing.

For behold, they rejected all the words of Ether; for he truly told them of all things, from the beginning of man; and that after the waters had receded from off the face of this land it became a choice land above all other lands, a chosen land of the Lord; wherefore the Lord would have that all men should serve him who dwell upon the face thereof; And that it was the place of the New Jerusalem, which should come down out of heaven (speaking of the return of the City of Enoch), and the holy sanctuary of the Lord.

Behold, Ether saw the days of Christ, and he spake concerning a New Jerusalem upon this land (the future capital of America). And he spake also concerning the house of Israel, and the Jerusalem from whence Lehi should come—after it should be destroyed it should be built up again, a holy city unto the Lord; wherefore, it could not be a new Jerusalem for it had been in a time of old; but it should be built up again, and become a holy city of the Lord; and it should be built unto the house of Israel — And that a New Jerusalem should be built up upon this land (of America), unto the remnant of the seed of Joseph, for which things there has been a type.

For as Joseph brought his father down into the land of Egypt, even so he died there; wherefore, the Lord brought a remnant of the seed of Joseph out of the land of Jerusalem, that he might be merciful unto the seed of Joseph that they should perish not, even as he was merciful unto the father of Joseph that he should perish not. Wherefore, the remnant of the house of Joseph shall be built upon this land; and it shall be a land of their inheritance; and they shall build up a holy city unto the Lord, like unto the Jerusalem of old; and they shall no more be confounded, until the end come when the earth shall pass away.

And there shall be a new heaven and a new earth; and they shall be like unto the old save the old have passed away, and all things have become new (speaking of the end of the earth's temporal existence at the end of the millennium, its death and resurrection as the celestial abode of the righteous). *And then cometh the New Jerusalem* (the holy city being now built within our outer hollow sun, which will be brought to earth after the earth's death and resurrection); *and blessed are they who dwell therein, for it is they whose garments are white through the blood of the Lamb; and they are they who are numbered among the remnant of the seed of Joseph, who were of the house of Israel. And then also cometh the Jerusalem of old; and the inhabitants thereof, blessed are they, for they have been washed in the blood of the Lamb; and they are they who were scattered and gathered in from the four quarters of the earth, and from the NORTH COUNTRIES, and are partakers of the fulfilling of the covenant which God made with their father, Abraham."* (Ether 13:1-11)

And They Shall Bring Forth Their Rich Treasures

The people inside the earth are fantastically rich in precious stones and metals. In Olaf Jansen's book, THE SMOKY GOD, he describes the throne room of the Great High Priest,

"The immense room in which we were received seemed to be finished in solid slabs of gold thickly studded with jewels of amazing brilliancy." (p. 113)

On page 100 he says, *"They wore knee-breeches and stockings of fine texture, while their feet were encased in sandals adorned with gold buckles. We early discovered that gold was one of the most common metals known, and that it was used extensively in decoration."*

On page 105, Olaf remarked, *"I never saw such a display of gold. It was everywhere. The door-casings were inlaid and the tables were veneered with sheetings of gold. Domes of the public buildings were of gold. It was used most generously in the finishings of the great temples of music."*

An abundance of their treasures will be brought with them when they come to the New Jerusalem as verse 30, Section 133 of the D&C indicates,

"And they shall bring forth their rich treasures unto the children of Ephraim, my servants."

And will undoubtedly be used to help build the temples in the New Jerusalem where the Ten Tribes will come to receive their endowments.

Another scripture refers to the source of the Ten Tribe's riches. As Olaf Jansen says in his book, the Garden of Eden is located inside the earth and the river Pison is one of the four rivers that flow out of the garden. In Genesis we read that,

"The name of the first is Pison: that is it which compasseth the whole land of Havilah where there is gold. And the gold of that land is good; there is bdellium and the onyx stone." (Genesis 3:11,12)

With reference to the building of the New Jerusalem and the twenty-four temples to be built there, Joseph Smith said that, *"The Ten Tribes of Israel will help you build it."* (Prophecy recorded by Edwin Rushton and Theodore Turley, PROPHECY-- KEY TO THE FUTURE, by Duane S. Crowther, p. 117)

Reason, perhaps, to show why the ten tribes will need to build their own temple at the New Jerusalem temple complex is Olaf Jansen's description of their gigantic stature. After crossing the Arctic Ocean in their journey through the north polar opening in their small boat, Olaf and his father came to land whereupon they found a river. They went up the river and met a ship coming down and were invited aboard. Olaf wrote,

"If my father and I were curiously observed by the ship's occupants, this strange race of giants offered us an equal amount of wonderment. There was not a single man aboard who would not have measured fully twelve feet in height. They all wore full beards, not particularly long, but seemingly short-cropped. They had mild and beautiful faces, exceedingly fair, with ruddy complexions. The hair and beard of some were black, others sandy, and still others yellow. The captain, as we designated the dignitary in command of the great vessel, was fully a head taller

than any of his companions. The women averaged from ten to eleven feet in height. Their features were especially regular and refined, while their complexion was of a most delicate tint heightened by a healthful glow."

"Both men and women seemed to possess that particular ease of manner which we deem a sign of good breeding, and, notwithstanding their huge statures, there was nothing about them suggesting awkwardness. As I was a lad in only my nineteenth year, I was doubtless looked upon as a true Tom Thumb. My father's six feet three did not lift the top of his head above the waist line of these people." (THE SMOKY GOD, pages 98-100)

Some have wondered how it is possible that the inner earth inhabitants could possibly be so large in stature, and have asked, "Isn't that highly unlikely that they are giants, especially if they are the Lost Ten Tribes of Israel and emigrated to the Hollow Earth from our exterior world where the people are much smaller?"

Olaf Jansen gave the answer to this when he reported that because of the ideal climate and environment in the Hollow Earth, as well as having 2/5^{ths} of our surface gravity, everything grows much larger than it does on the exterior of the planet where we live in much harsher climates, environments and more gravity.

But even on our exterior world, there have been reports of giant peoples and animals having lived here in ages past. William F. Warren reported in his, PARADISE FOUND, THE CRADLE OF THE HUMAN RACE AT THE NORTH POLE that skeletons of people have been found in Italy and Palestine with heights up to 35 feet tall. Fossils and skeletons of very large animals have also been found, giant tortoises up to 20 feet long, giant lions, deer, mammoths, and, of course, dinosaur bones with heights of up to 75 feet tall.

There is evidence that the people who lived on our exterior world before the Flood of Noah, were very large in stature. For example, the human footprints in sand stone discovered along the Puluxy River in Texas measured 16 inches long, 9 inches wide and had a stride of 6 feet -- surely a giant of a man. Pres. Spencer W. Kimball, a previous President of the Church of Jesus Christ of Latter-day Saints, in his book, THE MIRACLE OF FORGIVENESS, relates an incident in which David Patten was riding in the woods of Tennessee in 1835 on his mule when he suddenly noticed someone walking beside him. He looked to his side and saw a large *"Bigfoot"* entity that wore no clothes, was covered with hair, and had a very dark skin. The entity spoke to David and said that he was Cain, the son of Adam. And explained that the Lord had cursed him and would not let him die because the Lord had condemned him to be a *"fugitive and a vagabond in the earth"* for having conspired with Satan to kill his brother Abel. Cain was large in stature. His head was even with David's shoulder even as he was riding on his mule, so Cain must have been at least 8-10 feet tall.

In the Bible is recorded that when the Tribes of Israel entered into their Promised Land after having wandered 40 years in the wilderness for having refused to invade Palestine, they destroyed a race of people who possessed the land who were giants. It was because the Israelites had been afraid to attack the giants in the first place that they had refused to invade Palestine the first time. It wasn't until the fearful ones had died off during their 40 years of wandering that they were successful in the invasion of Palestine in redeeming the land that been given them by God as promised to their ancestors Abraham, Isaac and Jacob.

From Milton R. Hunter's book, ANCIENT AMERICA AND THE BOOK OF MORMON, we learn from the Works of Ixtlilxochitl, an Aztec prince (1568-1648) that the original settlers of America came from the Tower of Babel in submarine barges, and that they were giant in stature. The Book of Mormon contains a history of them and calls them the Jaredites after one of their original leaders.

My brother attended a State Fair in Phoenix, Arizona many years ago and related to me how he paid to view a skeleton of a giant woman 12 feet long that was on display, apparently discovered in some cavern tomb.

So, *"giants in the earth"* (Genesis 6:4, Moses 8:18) is not such of an unusual thing, even though it may seem so to us of this modern age where most of us don't lift our heads above 6 feet.

And An Highway Shall Be Cast Up

Verse 27 of the Doctrine and Covenants 133, says,
> *"And an highway shall be cast up in the midst of the great deep."*

This highway will be used to carry the people who inhabit the inside surface of the earth to the New Jerusalem to receive their endowments in the temples of God. The highway undoubtedly will be used also to establish a link between the government of the Kingdom of God on the inside of Our Hollow Earth and the government of the Kingdom of God on the outside as it is expanded to the outside of the earth. This highway will pass from the continent inside the earth over the Arctic Ocean or the "great deep" to the North American continent to the New Jerusalem which will be built at Independence, Jackson County, Missouri.

This highway that will be cast *"...up in the midst of the great deep"* shall be called the *"Way of Holiness."*
Isaiah 35:8 says,

> *"And an highway shall be there, and a way and it shall be called the Way of Holiness; the unclean shall not pass over it, but it shall be for those: the wayfaring man, though fools, shall not err therein."*

Perhaps this highway will be a monorail train, which is the type of inter-city transport Olaf reports the inhabitants of the inside of the earth use. He wrote,

> *"We were taken overland to the city of 'Eden,' in a conveyance different from anything we have in Europe or America* (written in 1908). *This vehicle was doubtless some electrical contrivance. It was noiseless, and ran on a single iron rail in perfect balance. The trip was made at a high rate of speed. We were carried up hills and down dales across valleys and again along steep mountains, without any apparent attempt having been made to level the earth."* (THE SMOKY GOD, pp. 110-111)

That the mode of transportation over the highway that will be cast up could be a monorail train, might be why in verses 9 and 10 of Isaiah 35, it says,

> *"...no lion shall be there, nor any ravenous beast shall go up thereon; it shall not be found there; but the redeemed shall walk there: And the ransomed of the Lord shall return, and come to Zion with songs and everlasting joy upon their heads: they shall obtain joy and gladness, and sorrow and sighing shall flee away."*

Olaf Jansen reported that the people of Our Hollow Earth were *"learned to a remarkable degree in their arts and sciences, especially geometry and astronomy."* (THE SMOKY GOD, Part Four) This was confirmed by Larry Foreman, who reported in his book, PASSPORT TO ETERNITY that the Flying Saucermen he met out in the desert outside Los Angeles were especially skilled in medical technology. They told Larry that they will come to *"take us all home,"* which *"home"* I believe is the Hollow Earth where our first parents were first placed on this planet in the Garden of Eden. Once our sentence is up on this planet, we will all be taken home where they will fix all our ills. If we have lost an arm, or leg, or any other part of our bodies, they will take a cell from our body and *"grow"* us another body part. And since it is a cell taken from our own body, our bodies won't reject the new body part when it is installed.

Vilhjalmur Stefansson, in his book, MY LIFE WITH THE ESKIMO, reported that in a conversation with the Eskimos he lived with on Victoria Island in northern Canada, he told them how our surgeons can transplant body organs from one man into the body of another. One Eskimo listening to the conversation said that he had a friend who suffered from backaches until one of their great medicine-men fixed his back. He recounted that while the patient slept, the medicine-man removed his entire spinal column and replaced it with a complete new set of vertebrae and there was not so much as a scratch on the patient's skin to show that the exchange had been made. Stefansson had to admit that such skills were beyond what our surgeons could do, so the Eskimos concluded that in point of skill our doctors were not the equal of theirs. (MY LIFE WITH THE ESKIMO, p. 118) Most likely, the Eskimos' *"medicine men"* are the highly skilled doctors within the Hollow Earth, and the Eskimos know how and where to go to get their medical needs taken care of by them.

A Musical People

There is much reference in the scriptures to the fact that the people of the Ten Tribes of Israel are a very musical people. In verse 9 of Isaiah 35, it says that when the Ten Tribes come to Zion they will *"...come to Zion with songs of everlasting joy upon their heads."* In the D&C it also says, *"And they shall be filled with songs of everlasting joy."* (D&C 133:33)

Of the people among whom Olaf Jansen and his father lived for two years inside Our Hollow Earth, Olaf says,

"The people are exceedingly musical, and learned to a remarkable degree in their arts and sciences, especially geometry and astronomy. Their cities are equipped with vast palaces of music, where not infrequently as many as twenty-five thousand lusty voices of this giant race swell forth in mighty choruses of the most sublime symphonies."

"The children are not supposed to attend institutions of learning before they are twenty years old. Then their school life begins and continues for thirty years, ten of which are uniformly devoted by both sexes to the study of music." (THE SMOKY GOD, pp. 121, 122)

The Mysteries of God Unfolded

Notwithstanding these evidences of Our Hollow Earth and its inhabitants being the Lost Ten Tribes of Israel, many will say, *"But aren't the mysteries of God prohibited to us? Isn't it somehow wrong to know where the 'lost' Ten Tribes are? And if they have now been found, why hasn't the prophet declared it openly?"*

The answer is that unto the wicked the mysteries of God are hidden. The truth is hid from those unwilling to accept the truth and live by it. The Lord will not command his prophet to reveal it openly until there are enough righteous people who will accept it. But unto the righteous few, God gives as much as they are willing to search for. Christ says,

"Seek not for riches, but for wisdom, and behold, the mysteries of God shall be unfolded unto you, and then shall you be made rich. Behold, he that hath eternal life is rich." (D&C 6:7)

If the eyes of the people were uncovered, and were able to see the riches of the world "inside," they would covet that land for themselves. For this reason, the Lord hid the Americas from Europe so many years before Columbus. As father Lehi said,

"And behold, it is wisdom that this land should be kept as yet from the knowledge of other nations; for behold, many nations would overrun the land, that there would be no place for an inheritance." (2 Nephi 1:8)

The day came when Lehi's descendants became wicked and God permitted other nations to overrun their inheritance. But the people inside the earth are still a just and righteous people, living the United Order with prophet-leaders to guide them. God will continue to protect their land from other nations. Part of that protection is a disbelief among the masses, and the teaching of accepted science that the earth is liquid-solid, and that any account saying that the earth is hollow is a story and a myth.

A belief does not make a truth. Even as God hid the Americas, by allowing the people of Europe believe the earth was flat, even so He can hide the land of the Ten Tribes by allowing the people of the surface world to believe that the earth is solid or molten inside. However, unto those who would like to know, He says,

> *"And if thou wilt inquire, thou shalt know mysteries which are great and marvelous; therefore thou shalt exercise thy gift, that thou mayest find out mysteries, that thou mayest bring many to the knowledge of the truth, yea, convince them of the error of their ways."* (D&C 6:11)

Frequently, the real truth of a matter has been a mystery all along and is only discovered by a diligent searcher after truth.

Christ, in his visit to the Americas after his resurrection told the Nephites that a person can know where the Ten Tribes are located by the inspiration of the Holy Ghost. He said to them,

> *"And I command you that ye shall write these sayings after I am gone, that if it so be that my people at Jerusalem, they who have seen me and been with me in my ministry, do not ask the Father in my name, that they may receive a knowledge of you by the Holy Ghost, AND ALSO OF THE OTHER TRIBES WHOM THEY KNOW NOT OF..."* (3 Nephi 16:4)

So here we see that it IS possible to ask God and receive an answer as to the true location of the Ten Tribes if an effort is exerted to ask.

Keeping the commandments of God is an indispensable requisite in obtaining the truth.

> *"And no man receiveth a fullness unless he keepeth his commandments. He that keepeth his commandments receiveth truth and light, until he is glorified in truth and knoweth all things."* (D&C 93:27, 28)

If the righteous enquire, they may know by the power and inspiration of the Holy Ghost that the Ten Tribes live within Our Hollow Earth!

The scriptures do give clues as to the Ten Tribe's present location. Perhaps the most revealing passage is in the Doctrine and Covenants, Section 84:99-102, which is part of a song the Lord's people will sing in the Millennium. It reads:

"The Lord hath gathered all things in one."

"The Lord hath brought down Zion from above." (This refers to the City of Enoch which was taken to heaven 3,013 years before the birth of Christ and which will return to the earth's surface at the beginning of the Millennium. See Chapter FOURTEEN, The City of Enoch – FOUND!)

"THE LORD HATH BROUGHT UP ZION FROM BENEATH."
(This refers to the present location of the Lost Ten Tribes. They are "beneath" our feet within the hollow of the earth and will come up from beneath to receive their endowments in the temples of the New Jerusalem and expand their political Kingdom of God to the surface world.)

"The earth hath travailed and brought forth her strength;" (When the Ten Tribe nation expands their Kingdom to the surface world, *"mother"* earth will *"travail"* because of the close passage of a comet, and give forth her strength: The powerful Ten Tribe nation will at last come out and help overcome the Satanic Illuminist Conspiracy with the help of their FLYING SAUCERS and help preach the gospel of Jesus Christ to the world by the power of the Holy Melchizedek priesthood.)

"And truth is established in her bowels;" (Truth is established in the earth's bowels because people live there who have the truth and live by it. During the Millennium there will be three world capitals, Jerusalem in Palestine, the New Jerusalem in Missouri, and the City of Eden inside the "bowels" of the earth from whence the word of the Lord will issue forth to all the world.)

In fact, the scriptures refer directly to a people who live within the earth's crust in giant cavern cities, and to the Ten Tribes who live within Our Hollow Earth. In Section 88, verse 104, it is written:

"And this shall be the sound of his trump, saying to all people, both IN HEAVEN and IN EARTH, and that are UNDER THE EARTH—for every ear shall hear it, and every knee shall bow, and every tongue shall confess, while they hear the sound of the trump, saying: Fear God, and give glory to him who sitteth upon the throne, forever and ever; for the hour of his judgment is come."

In our search for the truth, it is imperative that we search the scriptures. The test of the truthfulness of any scientific theory is that it must be supported by the revealed word of God which we receive through His inspired servants, the prophets. All truth is revealed by God and whether obtained by the scientific method or by direct revelation, the truth cannot contradict itself. Therefore with the testing rod of the scriptures, the Lord has given us the commandment to search out the truth in religion, astronomy, geography, geology, history, current events, prophecy, sociology, and government. The glory of God is intelligence, (D&C 93:36) and would be ours also if we are successful in obtaining true knowledge and use it correctly.

Christ has said,

"Teach ye diligently and my grace shall attend you, that you may be instructed more perfectly in theory, in principle, in doctrine, in the law of the gospel, in all things that pertain unto the kingdom of God, that are expedient for you to understand; Of things both IN HEAVEN and IN THE EARTH, and UNDER THE EARTH...and also a knowledge of countries and of kingdoms." (D&C 88:77-79)

If we would go *"under the earth's"* crust about 800 miles, we would learn about a country and a kingdom where the Ten Tribes of Israel dwell.

CHAPTER FIVE
Paradise -- Found!

The existence of a place called *"paradise"* or "heaven" to where the spirits of all people go at death is mentioned in the writings of the prophets of God.

Generally, it is the desire of all Christians, as also the people of other religions, who strive hard to live the commandments of God, to go to Paradise or Heaven when they die. To John, the Lord Jesus Christ said,

"To him that overcometh will I give to eat of the tree of life, which is in the midst of the paradise of God." (Revelation 2:7)

Paradise is a welcome place to go after death and is considered a resting place from the cares of the world. The ancient American prophet Moroni, in concluding the history of his destroyed nation said,

"And now I bid unto all, farewell. I soon go to rest in the paradise of God, until my spirit and body shall again reunite, and I am brought forth triumphant through the air, to meet you before the pleasing bar of the great Jehovah, the Eternal Judge of both quick and dead. Amen." (Moroni 10:34)

The spirit world, wherein Paradise, the place of rest after death, is located, had its beginnings before the world was created physically. The Lord told Moses,

"For I, the Lord God, created all things, of which I have spoken, spiritually, before they were naturally upon the face of the earth." (Moses 3:5)

Thus the fact was revealed to Moses that there were two creations which brought our world into being: First, the creation of the spirit world, and Second, the creation of the physical world.

In understanding the origins of our physical and spirit world, the word *"creation"* must not be misunderstood. Joseph Smith, the great American prophet of the 19th century, in his King Follet sermon said,

"Now, the word create came from the word Baurau, which does not mean to create out of nothing; it means to organize; the same as a man would organize materials and build a ship." (TEACHINGS OF THE PROPHET JOSEPH SMITH, pp. 350-352)

When God created the earth, he organized it from pre-existing materials. God revealed to Joseph Smith in 1833 the great truth that, *"The elements are eternal,"* a fact, which the scientists of the twentieth century confirmed, to which they stated: Matter cannot be destroyed. It can be changed from one state to another or even to energy but is never destroyed.

Joseph Smith went on to state further that,

"There is no such thing as immaterial matter. All spirit is matter, but it is more fine or pure, and can only be discerned by purer eyes. We cannot see it; but when our bodies are purified we shall see that it is all matter." (D&C 131:7, 8)

Clairvoyants are persons who can see the spirit world about us. There are actually persons who are born with this ability. One such person was the author of THE BOY WHO SAW TRUE, published in 1953 in London by Neville Spearman. The book is the diary of a boy, beginning at age 5, who wrote about experiences in his life which resulted from the fact that he could see the auras which surround people. He could also see disembodied spirits and communicate with them, yet failed to realize for a long while that other people were not similarly gifted. For example, his Uncle Willard had lived with his family before he died. But Uncle Willard's spirit continued living with them after the death of his physical body. Several times Uncle Willard was sitting in his father's big chair and his father unable to see spirits as his son could, would come home and sit on top of Uncle Willard. Seeing this, the boy would protest and say, *"Father don't sit on Uncle Willard!"*

In consequence of this and similar experiences this young boy was misunderstood and suffered many indignities from his parents and others who could not understand him for what he could see and hear.

The ancient prophet, Enoch, who lived before the Flood of Noah, was a Clairvoyant. God gave him this gift of being able to see the spirit world around us when he was called to be His prophet. (Moses 6:35, 36)

Experiences such as these help to establish the reality of the existence of the spirit world around us and its inhabitants who are the disembodied spirits of those who have died, as well as the evil spirits of Lucifer and his devils who never had bodies of flesh and bone but were cast upon this earth from their rebellion in the pre-existent world in heaven. (Revelation 12:7-17)

Jude spoke of the rebellious ones who kept not their first estate in heaven and so were cast down to Hell here to this planet,

> *"And the angels which kept not their first estate, but left their own habitation, he hath reserved in everlasting chains under darkness unto the judgment of the great day."* (Jude 6)

An experiment performed by Sir Richard Crooks, supports Joseph Smith's statement that spirit is matter when he,

> *"...actually weighed a dying man, bed and all, and found that the scale indicated a loss of approximately three ounces at the instant of death."*

Ray Palmer, in an article in his SEARCH MAGAZINE after stating the foregoing experiment concludes,

> *"If this is true, then we have a spirit upon which the laws of gravity still function (if we say gravity is the attraction of matter). We also have a material spirit."*

(SEARCH Magazine, *"Heaven is Solid,"* p. 18, Spring, 1977)

Joseph Smith asserted that not only are there two types of matter, one finer than the other, but that every living thing has a spirit body in the likeness of its physical body. Its location is within the physical body.

> *"That which is spiritual being in the likeness of that which is temporal; and that which is temporal in the likeness of that which is spiritual; the spirit of man in the likeness of his person, as also the spirit of the beast, and every other creature which God has created."* (D&C 77:2)

Joseph Fielding Smith, the great nephew of Joseph Smith, stated that, *"This earth is a living body."* (DOCTRINES OF SALVATION, Vol. I p. 72)

In the Doctrine & Covenants, the earth is also spoken of as being a living entity,

> *"And again, verily, I say unto you, the earth abideth the law of a celestial kingdom, for it filleth the measure of its creation, and transgresseth not the law. Wherefore, it shall be sanctified; yea, notwithstanding it shall DIE, it shall be quickened again, and shall abide the power by which it is quickened, and the righteous shall inherit it."* (D&C 88:25, 26)

Yes, our earth is a living body. And just as our physical human bodies have spirit bodies in the same form and likeness of our physical bodies, so does the earth have a spirit body in the same shape and form as the physical world. The earth's spirit body is the spirit world -- the habitation of the spirits of all who die.

The Mormon Apostle, Bruce R. McConkie, in his book, MORMON DOCTRINE wrote,

> *"The spirit that enters the body at birth leaves it at death, and immediately finds itself in the spirit world. That world is upon this earth."* (MORMON DOCTRINE, p. 68)

In a study of the scriptures we find that,

> *"The Spirit World is divided into two parts: PARADISE which is the abode of the righteous, and HELL which is the abode of the wicked."* (MORMON DOCTRINE, "Spirit World" p. 68)

This division of the spirit world into two distinct locations is clearly evident in the writings of the ancient American prophet, Jacob, the brother of Nephi. In explaining the resurrection of all men from the grave, he said,

> *"And this death of which I have spoken, which is the spiritual death, shall deliver up its dead; which spiritual death is hell; wherefore, death and HELL must deliver up their dead, and hell must deliver up its captive spirits, and the grave must deliver up its captive bodies, and the bodies and the spirits of men will be restored one to the other; and it is by the power of the resurrection of the Holy One of Israel."*

> *"O how great the plan of our God! For on the other hand, the PARADISE of God must deliver up the spirits of the righteous, and the grave deliver up the body of the righteous; and the spirit and the body is restored to itself again, and all men become incorruptible, and immortal, and they are living souls, having a perfect knowledge like unto us in the flesh, save it be that our knowledge shall be perfect."* (2 Nephi 9:12, 13)

Paradise and Hell can deliver up their captive spirits only if they have a distinct location in which they hold the spirits of the dead captive.

Since the spirit world has the same shape and form of our physical world, there must exist physical counterparts to Paradise and Hell. These counterparts must be separate places in the physical world also as they are separate places in the spirit world.

The separation between Paradise and Hell in the spirit world is called a *"great gulf"* in the scriptures. In the book of Luke in the New Testament, we find that

"...Abraham told the rich man in hell that between him and Lazarus (who was in Paradise) there was a great gulf fixed so that none could go from paradise to hell or from hell to paradise." (MORMON DOCTRINE p. 682)
Abraham told the rich man in Hell,

"And beside all this, between us and you there is a GREAT GULF FIXED: so that they which would pass from hence to you cannot; neither can they pass to us, that would come from thence." (Luke 16:26)

This great gulf separating Paradise and Hell must also be a gulf in the physical world. The word *"gulf"* means a *"wide separation."* And the characteristics of this gulf is that it prevents the spirits in Paradise and Hell from traveling back and forth, which they could not do until Christ came and gave the righteous in Paradise the power to cross that gulf into Hell to preach His gospel.

Now, let us look at our earth and try to find the physical location of Paradise and Hell. Since the spirit world is in the likeness of the physical world, and since Hell and Paradise are locations in the spirit world separated from each other by a great gulf and yet are a part of this earth, we must also find counterparts of Paradise and Hell in the physical world where two physical locations are separated by a great Gulf of space.

It can be assumed that the physical location of Hell is on the earth's surface, atmosphere and within the crust because Satan and his angels in Hell are here with us to tempt us. If they were separate from us, the devils could not tempt us. Therefore, the earth's shell must be the physical location of Hell.

However, it is evident that even though the devils can come up to the surface to tempt us, their home, or place of abode is within the earth's shell. Therefore Hell can be termed as being *"down"* in the earth's crust. Christ said of those who go to Hell at death,

"These are they who are liars, and sorcerers, and adulterers, and whoremongers, and whosoever loves and makes a lie...These are they who ARE CAST DOWN TO HELL..." (D&C 76:103-106)

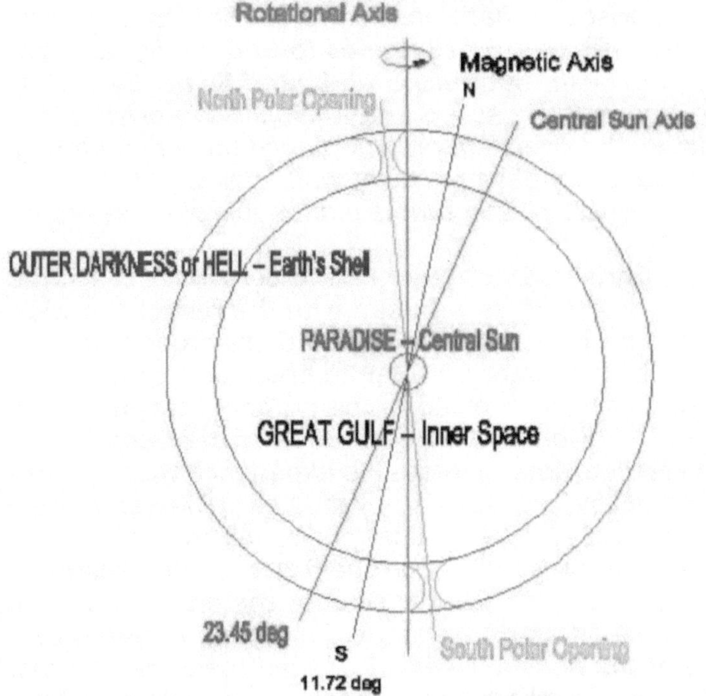

Our Hollow Earth and it's Spirit World Counterparts

Paradise – A Sun Inside Our Hollow Earth

If Hell is in the earth's crust, then where is the physical location of Paradise? Surprisingly, the scriptures indicate that Paradise is also *"down"* inside the earth, actually located in a sun within the hollow of our earth. And polar explorers claim to have seen this sun shining through the polar openings from the hollow interior of our earth.

We can get an indication as to the location of Paradise from the writings of the prophets. They state that at the death of Christ, while his body lay in the tomb, his spirit passed into the spirit world. Christ's statement to the thief on the cross, *"Today shalt thou be with me in paradise,"* (Luke 23:43) indicates that the place in the spirit world where Christ's spirit went to was Paradise.

A statement by Peter indicates that Christ's mission to the spirit world was to preach the gospel to the wicked spirits. He wrote,

"For Christ also hath once suffered for sins, the just for the unjust, that he might bring us to God, being put to death in the flesh, but quickened by the Spirit: By which also he went and preached unto the spirits in prison; Which sometime were disobedient, when once the longsuffering of God waited in the days of Noah, while the ark was a preparing, wherein few, that is eight souls were saved by water." (I Peter 3:18-20)

However, Christ did not go to Hell to personally teach the thief who died on the cross with him with the rest of the wicked spirits. Instead he went to Paradise and there organized missionaries which he sent into Hell to preach his gospel to the wicked dead. This was revealed to the prophet Joseph Fielding Smith Sr. in his *"Vision of the Redemption of the Dead."*

Joseph Fielding Smith relates his experience thus,

"On the third of October, in the year nineteen hundred and eighteen, I sat in my room pondering over the scriptures and reflecting upon the great atoning sacrifice that was made by the Son of God for the redemption of the world...While I was thus engaged, my mind reverted to the writings of the apostle Peter...I opened the Bible and read the third and fourth chapters of the first epistle of Peter, and as I read I was greatly impressed more than I had ever been before...As I pondered over these things which are written, the eyes of my understanding were opened, and the Spirit of the Lord rested upon me, and I saw the hosts of the dead, both small and great."

"And THERE WERE GATHERED TOGETHER IN ONE PLACE an innumerable company of the spirits of the just, who had been faithful in the testimony of Jesus while they lived in mortality...They were assembled awaiting the advent of the Son of God into the spirit world to declare their redemption from the bands of death...While this vast multitude waited and conversed, rejoicing in the hour of their deliverance from the chains of death, the Son of God appeared, declaring liberty to the captives who had been faithful...But unto the wicked he did not go...I perceived that the Lord went not in person among the wicked and the disobedient who had rejected the truth, to teach them; but behold, from among the righteous he organized his forces and appointed messengers, clothed with power and authority, and commissioned them to go forth and carry the light of the gospel to them that were in darkness, even

to all the spirits of men. And thus was the gospel preached to the dead." (D&C 138:1-30)

Here, Joseph F. Smith relates that the righteous spirits in Paradise are "gathered together in one place," and considered themselves in prison. In the vision, Joseph F. Smith saw that, *"...the Son of God appeared, declaring liberty to the captives who had been faithful."* Even though the spirits of the righteous were in Paradise or Heaven, they could not be delivered from the chains of death until Christ came to resurrect them.

In the writings of Paul to the Ephesians we find that the place or location of Paradise must be inside the earth. Paul wrote of Christ saying,

> *"Wherefore he saith, When he ascended up on high* (referring to the ascension into heaven after his resurrection)...*Now that he ascended, what is it but that he also DESCENDED FIRST INTO THE LOWER PARTS OF THE EARTH?..."* (Ephesians 4:8,9)

The period of time when Christ *"descended first into the lower parts of the earth,"* was while his body was dead. This is clarified in Matthew 12:40. Here Christ says,

> *"For as Jonas was three days and three nights in the whale's belly; so shall the Son of Man be three days and three nights IN THE HEART OF THE EARTH."*

Since the *"heart of the earth"* can be considered the *"center"* of the earth, Paradise must be located in the center of the earth, or as Paul expressed it, in *"the lower parts of the earth."*

However, since both Hell and Paradise are located *"down"* in the earth, somewhere inside the earth the two must be separated by a *"great gulf."* Since the spirit world has its counterparts in the physical world, *"that which is temporal in the likeness of that which is spiritual,"* then Paradise and Hell must be two separate physical locations inside our earth. Since paradise is in the HEART or center of the earth, Hell would then be the earth's shell which surrounds the center or Paradise.

A counterpart in the physical world of the great gulf in the spirit world which separates Paradise and Hell must also exist. Such a gulf or separation in the physical world could only consist of a stretch of space separating the physical center-paradise and the physical hell in the shell of the earth.

This arrangement of the physical counterparts to Paradise and Hell describes a hollow earth in which the shell or Hell extends down a few hundred miles whereupon the earth's shell or Hell ends and extending across the center of the earth would be pure space--a great Gulf. And suspended in the center of this great hollow in the earth by gravity, electrostatics and electromagnetism would be a physical mass – the location of Paradise, the Heaven of this Earth!

Such is a description of a hollow earth. Proponents of the Hollow Earth theory base their conclusions on the observations of polar explorers who claim that instead of finding only polar ice-caps, they find that the Polar Regions contain openings into the hollow interior of our planet. And what do they see suspended in the center of the earth where Paradise would be located? They see a SUN!

In their journey into the far north explorers Olaf and Jens Jansen caught sight of the interior sun. Records Olaf Jansen,

"One day about this time, (38 days sailing northeast of Franz Josef Land about August 1, 1829) my father startled me by calling my attention to a novel sight far in front of us, almost at the horizon. 'It is a mock sun,' exclaimed my father. 'I have read of them; it is called a reflection or mirage. It will soon pass away.'"

"But this dull-red, false sun, as we supposed it to be, did not pass away for several hours; and while we were unconscious of its emitting any rays of light, still there was no time thereafter when we could not sweep the horizon in front and locate the illumination of the so-called false sun,

during a period of at least twelve hours out of every twenty-four."

"Clouds and mists would at times almost, but never entirely, hide its location. Gradually it seemed to climb higher in the horizon of the uncertain purply sky as we advanced."

"It could hardly be said to resemble the sun, except in its circular shape, and when not obscured by clouds or the ocean mists, it had a hazy-red, bronzed appearance, which would change to a white light like a luminous cloud, as if reflecting some greater light beyond."

"We finally agreed in our discussion of this smoky furnace-colored sun, that, whatever the cause of the phenomenon, it was not a reflection of our sun, but a planet of some sort--a reality." (THE SMOKY GOD, pp. 85-87)

Olaf Jansen further described this sun as they observed it throughout their two and half year stay in the land of Our Hollow Earth:

"The great luminous cloud or ball of dull-red fire, fiery-red in the mornings and evenings, and during the day giving off a beautiful white light, 'The Smoky God,'--is seemingly suspended in the center of the great vacuum 'within' the earth, and held to its place by the immutable law of gravitation..."

"The base of this electrical cloud or central luminary, the seat of the gods, is dark and non-transparent, save for innumerable small openings, seemingly in the bottom of the great support or altar of the Deity, upon which 'The Smoky God' rests; and, the lights shining through these many openings twinkle at night in all their splendor, and seem to be stars, as natural as the stars we saw shining when in our home in Stockholm excepting that they appear larger. 'The Smoky God,' therefore, with each daily revolution of the earth, appears to come up in the east and go down in the west, the same as does our sun on the external surface. In reality, the people 'within' believe that 'The Smoky God' is the throne of their JEHOVAH, and is stationary. The effect of night fand day is, therefore, produced by the earth's daily rotation." (THE SMOKY GOD, pp. 108-110)

Amazingly, Olaf's description of the interior sun as a *"great luminous cloud"* and the *"throne"* of Jehovah is very similar to a conversation between Jehovah and the prophet Job when He said,

> *"Hearken unto this, O Job: stand still, and consider the wondrous works of God. Doest thou know when God disposed them, AND CAUSED THE LIGHT OF HIS CLOUD TO SHINE?"* (Job 37:14, 15)

In another place, Job tells his friends of the North Polar Opening and of Jehovah's throne in the shining cloud:

> *"He stretcheth out the north over the EMPTY PLACE, and hangeth the earth upon nothing...He holdeth back the FACE OF HIS THRONE, and spreadeth his CLOUD upon it."* (Job 26:7, 9)

Further support of this hollow earth theory location of Paradise is an ancient American prophet's advice to his son showing the importance of this life in preparing for what we will receive after death. Alma said,

> *"Now, concerning the state of the soul between death and the resurrection--Behold, it has been made known unto me by an angel, that the spirits of ALL MEN, as soon as they are departed from this mortal body, yea, the spirits of all men, WHETHER THEY BE GOOD OR EVIL, are taken HOME to that God who gave them life."*

> *"And then shall it come to pass, that the spirits of those who are righteous are received into a state of happiness, which is called PARADISE, a state of rest, a state of peace, where they shall rest from all their troubles and from all care, and sorrow."*

> *"And then shall it come to pass, that the spirits of the wicked, yea, who are evil--for behold, they have no part nor portion of the Spirit of the Lord; for behold, they chose evil works rather than good; therefore the spirit of the devil did enter into them, and take possession of their house-- and these shall be cast out into OUTER DARKNESS; there shall be weeping, and wailing, and gnashing of teeth, and this because of their own iniquity, being led captive by the will of the devil."*

> *"Now this is the state of the souls of the wicked, yea, in darkness, and a state of awful, fearful looking for the fiery indignation of the wrath of God upon them; thus they remain in this state, as well as the righteous in paradise, until the time of their resurrection."* (Alma 40:11-14)

In our departure to the spirit world at death, our spirits are *"taken home to that God which gave us life."* That home is Paradise, the Heaven of this earth. Christ told the thief while both were on the cross, *"Today shalt thou be with me in paradise."* (Luke 23:43) Alma corroborates this with the statement that, *"the spirits of ALL men, whether they be good or evil, are taken home to that God who gave them life."*

That God which gave us life is Christ. Jesus Christ is the Judge of all the Earth. As recorded by John, Jesus said,

> *"For the Father judgeth no man, but hath committed all judgment unto the Son."* (John 5:22)

Alma infers that the reason we are taken home to that God which gave us life is to be judged by Him to see if we will be assigned to Paradise or Hell while we await our resurrection. Therefore, when the spirits of all men, both good and bad, are taken home to be judged of Christ, they are taken to Paradise. This explains why the people who live inside Our Hollow Earth told Olaf Jansen that the sun inside the earth is the throne of Jehovah. There is where we are all taken when we die to be judged.

Jehovah is just the premortal name for Jesus Christ. This fact was well established in a vision manifested to the American prophet Joseph Smith and his counselor in the Church, Oliver Cowdery, wherein Christ appeared to them in the temple at Kirtland, Ohio, April 3, 1836:

> *"The veil was taken from our minds, and the eyes of our understanding were opened. We saw the Lord standing upon the breastwork of the pulpit, before us; and under his feet was a paved work of pure gold, in color like amber. His eyes were as a flame of fire; the hair of his head was white like the pure snow; his countenance shone above the brightness of the sun; and his voice was as the sound of the rushing of great waters, even the voice of JEHOVAH, saying: 'I am the first and the last; I am he who liveth, I AM HE WHO WAS SLAIN; I am your advocate with the Father.'"* (DOCTRINE AND COVENANTS, 110:1-4)

After we are taken to Paradise to be judged of Jehovah-Christ, Alma continues to explain that, *"The spirits of those who are righteous are received into a state of happiness which is called paradise..."* The spirits of the just are received into Paradise to stay once they are there, but not so with the wicked spirits. Once they are judged, and seen Heaven, they are then cast out. Alma says of the spirits of the wicked that *"these shall be cast out* (of paradise) *into outer darkness,"* into hell where *"...they remain in this state, as well as the righteous in paradise until the time of their resurrection."* Of course, faith in our Savior, Jesus Christ, and repentance can be a wicked spirit's passport to Paradise which is the heaven of this earth. (Alma 40:13-14)

Now, if the sun inside the earth is the physical location of Paradise, then the place the wicked spirits are *"cast out"* of Paradise *"into outer darkness"* could only refer to the shell of the earth. The earth's shell would be the *"outer darkness"* because Hell's proper location is inside the earth's shell where naturally it is dark because the light of neither sun nor the stars penetrate. The earth's shell as the location of Hell could be understood as being *"outer,"* away from the center sun or location of Paradise.

Paradise is also a place of flaming fire as our earth's inner sun could be considered to be. This was revealed by the prophet Joseph Smith, who wrote,

> *"The spirits of the just are exalted to a greater and more glorious work; hence they are blessed in their departure to the world of spirits. ENVELOPED IN FLAMING FIRE, THEY ARE NOT FAR FROM US, and know and understand our thoughts, feelings, and motions, and are often pained therewith."* (TEACHINGS, p. 326)

In the Book of Mormon, we find in the Vision of the Tree of Life of the prophet Lehi and the interpretation of that vision or dream given to his son Nephi, supporting evidence for the location of Paradise, Hell and the Great Gulf that separates them in the Spirit World of this Earth as we have determined in this chapter.

In the dream, Father Lehi found himself in a dark and dreary wilderness. He then saw a man dressed in a white robe who asked him to follow. As he followed, he then found himself in a dark and dreary waste. And after traveling many hours, he began to pray unto the Lord to have mercy on him. After he prayed, he then saw a large and spacious field in which he found a tree, *"whose fruit was desirable to make one happy."* He ate of the fruit of the tree which *"filled his soul with exceedingly great joy."*

Lehi then wanted his family to partake of the fruit of the Tree of Life. Looking about to see if he could see his family, Lehi then noticed a fountain (which is the artesian fountain in the Garden of Eden inside Our Hollow Earth) that was at the head of a river of water that passed by the tree. He could see the head of the river not far away, and there was his family not knowing which way to go. He called and beckoned them to come. They came and partook of the fruit also, all except two of his sons who would not come.

It was then that he noticed a Rod of Iron extending alongside the river and it led to the tree where he stood. He also noticed a strait and narrow path which came along by the Rod of Iron and that led to the tree also. The path also extended the other direction past the head of the fountain at the head of the river to a large and spacious field that seemed to be the world (our outer surface world).

Then he saw numberless concourses of people pressing forward to obtain the path which led to the tree, but as they would commence in the path there arose a mist of darkness so that many lost their way and were lost. Others pressed forward and caught hold of the Rod of Iron and made it through the mists of darkness clinging to the Rod of Iron until they made it to the tree and partook of the fruit of the tree. But many after partaking of the fruit then cast their eyes about as if they were ashamed. It was then that he noticed on the other side of the river a great and spacious building which stood high in the air above the earth. The building was filled with people in fine dress in the attitude of mocking and pointing their fingers at those who had come to partake of the fruit. These who had tasted then were ashamed because of those who scoffed and so fell away into forbidden paths and were lost. Others who partook of the fruit did not pay any attention to the scoffers and were happy in Paradise.

Lehi's son, Nephi, wanting to know the meaning of the dream or vision his father had received, went and prayed to the Lord. He was caught away to a high mountain where the Spirit of the Lord showed him the future of the earth and how the dream related to the future. It is in the interpretation given to Nephi that we find the parallels describing the components of the Spirit World of this Earth.

To Nephi it was revealed that the fruit of the tree represents the LOVE OF GOD, which *"is the most desirable above all things."* The Love of God was personified in a vision he saw of the birth and ministry of the Savior. He saw that the Rod of Iron was the Word of God which led to the Fountain of Living Waters and to the Tree of Life which tree and waters also represented the Love of God.

Nephi saw in his vision how the Lord called his twelve apostles and set up the Church of Jesus Christ. He saw the crucifixion of the Lord and the multitudes of the earth gathered together to fight the apostles of the Lamb. These multitudes were in a great and spacious building that was *"the pride of the world."* He saw the formation of a *"great and abominable church, which slayeth the saints of God."* He saw a fountain of filthy water and river which depths are the *"depths of hell,"* and was given to understand that the mists of darkness are the temptations of the devil *"which blindeth the eyes, and hardeneth the hearts of the children of men, and leadeth them away into broad roads, that they perish and are lost."* He saw that the large and spacious building is the *"vain imaginations and the pride of the children of men. And a great and terrible Gulf divideth them"* from the Tree of Life in Paradise.

Nephi saw in his vision the discovery of America by Columbus, and how the Spirit of God *"wrought upon the man"* to discover America. He saw the wars of Independence and how the power of God would deliver the peoples of *America "out of the hands of all other nations."* He saw the people carrying a *"book"* -- the Bible -- *"which contains the covenants of the Lord, which he hath made unto the house of Israel."* But he saw that the Bible had been altered by the great and abominable church which took away from the book *"many parts which are plain and most precious; and also many covenants of the Lord have they taken away."* And because of the *"things taken away out of the gospel of the Lamb, an exceedingly great many do stumble, yea, insomuch that Satan hath great power over them."*

Nephi saw that the Gentile peoples who came to America would not be permitted by God to *"utterly destroy the mixture of thy seed, which are among thy brethren"* of the American Indians. Nephi was permitted by the Lord to see how the record he and his descendants would write would be *"hid up, to come forth unto the Gentiles,"* in the last days as The Book of Mormon through the prophet Joseph Smith to restore the gospel of the Lamb of God and His church, the Church of Jesus Christ again upon the earth. He saw how the Book of Mormon would *"establish the truth of the"* Bible, and *"shall make known the plain and precious things which have been taken away from them, and shall make known to all kindreds, tongues, and people, that the Lamb of God is the Son of the Eternal Father, and the Savior of the World; and that all men must come unto him, or they cannot be saved."*

The Spirit of the Lord made known unto Nephi that all peoples who do not belong to the church of the Lamb of God, *"belongeth to that great church, which is the mother of abominations; and she is the whore of all the earth"* which has *"dominion over all the earth."* The numbers of the church of the Lamb were *"few"* in comparison, but were also found upon all the face of the earth. And *"I, Nephi, beheld the power of the Lamb of God, that it descended upon the saints of the church of the Lamb...and they were armed with righteousness and with the power of God in great glory. And...I beheld that the wrath of God was poured out upon that great and abominable church, insomuch that there were wars and rumors of wars among all the nations and kindreds of the earth."* Nephi then saw that John, the apostle of the Lamb would write the rest of the future history (the Book of Revelation in the Bible). Nephi did see the fall of that great and abominable church, and the *"fall thereof was exceedingly great."*

Now, let's analyze this vision with our hollow earth theory in mind.

Nephi's vision, or the recording of it, was more complete than that of his Father, Lehi. In it he saw that there is a place like unto Paradise, *"and the brightness thereof was like unto the brightness of a flaming fire,"* (1 Nephi 15:30) wherein is located a Fountain of Living Waters and the Tree of Life. He saw the world where numberless concourses of people were pressing forward toward either the Rod of Iron and its Strait and Narrow Path, or toward the large and spacious building. In the world, Nephi also noticed a fountain of *"filthy waters"* and river of water or *"many waters"* which *"was a representation of that awful hell, which the angel said unto me was prepared for the wicked."* Between Hell and Paradise was *"an awful gulf which separated the wicked from the tree of life, and also from the saints of God."* (1 Nephi 15:28)

Nephi saw that the way to make it to Paradise was by taking hold of the Rod of Iron which is the Word of God. The Word of God is the Bible, the Book of Mormon and the revealed Word of God given to the living prophets of God, and inspiration from the Holy Ghost to one's own soul. Those that follow the Strait and Narrow Path take hold of the Rod of Iron which is the Word of God and partake of the fruit of the Tree of Life which is the Love of God and become members of the Church of Jesus Christ. Those that do not, belong to the Church of the Devil by default. Some who partake of the fruit or become members of the true church of God, feel ashamed because of the pride and mocking of the world and fall away. Those who are faithful, some are murdered by the Church of Satan, but in the end all the faithful receive of the fullness of the joy of the Lord. By passing out of Hell, through the Gulf, the righteous enter into the Paradise of God to await a glorious resurrection in peace and happiness.

Thus, we see that the scriptures describe a Hollow Earth when describing our Earth and its Spirit World, which Spirit World the scriptures and modern prophets say is a part of and within the physical Earth thus making Earth a Living Being that obeys God and serves as a prison for us while we are on probation before the Lord.

The Book of Mormon, like the Bible, describes our Earth and its Spirit World as divided into two places, Paradise and Hell, which are separated by a great "Gulf" of space. When compared to the Hollow Earth theory, Paradise, or Heaven is the Inner Sun, a place of "flaming fire," located in the center or "heart" of the Earth. Hell is the shell of the planet, which is "down" from our point of view and "outer darkness" when compared to the central Inner Sun Paradise, and is separated from the central Inner Sun Paradise by the great "Gulf" or hollow of the Earth.

The Bible, the Book of Mormon, as well as the living prophets of God, further clarify that our Earth is a Prison Planet where we are all on probation to see if we are willing to obey the commands of God, which if we do, because of the Atonement of our Savior, Jesus Christ, who paid the penalty of our breaking the commandments of God, which penalty caused Him to sweat blood in the Garden of Gethsemane and die on the Cross of Calvary, we can then be allowed, on conditions of repentance, to pass out of the darkness of Hell through the shell of the planet and through the great hollow "Gulf" of the Earth into the Inner Sun Heaven or Paradise located in the Center or "Heart" of the Earth that is a place of "flaming fire" to await a glorious resurrection.

CHAPTER SIX
The Origin of Flying Saucers -- Found!

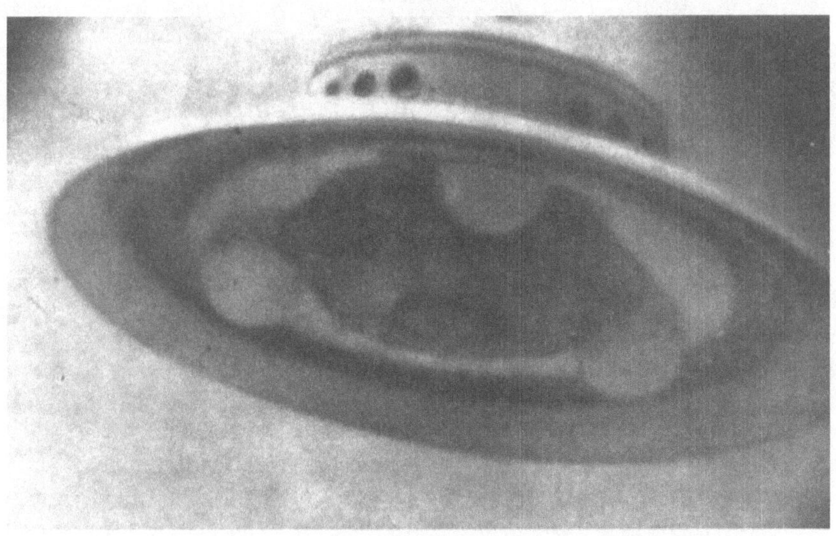

Adamski Venusian Flying Saucer

A Venusian flying saucer or 'Scout Ship' photographed at 9:10 am December 13, 1952 at Palomar Gardens, California, by George Adamski through his six-inch telescope. About 35 feet in diameter, this little space craft was made of a translucent metal. Notice the portholes and spherical landing gear. Not shown in this picture was a lens or light on top of the cabin dome. Above the portholes appeared to be some form of a power coil around the base of the dome.

An integral part of the World's Top Secret is the origin of the Unidentified Flying Objects, abbreviated as UFO's and commonly known as flying saucers. That flying saucers originate inside Our Hollow Earth was the decisive conclusion of the world's top UFO researcher, the late Ray Palmer of Amherst, Wisconsin.

In the December 1959 issue of his FLYING SAUCERS magazine, Palmer wrote:

"Flying Saucers magazine has amassed a file of evidence which its editors consider unassailable, to prove that the flying saucers are native to the planet Earth: That the governments of more than one nation know this to be a

fact; that a concerted effort is being made to learn all about them, and to explore their NATIVE LAND; that the facts already known are considered so important that they are the WORLD'S TOP SECRET; that the danger is so great that to offer public proof is to risk widespread panic; that public knowledge would bring public demand for action, which would topple governments both helpless and unwilling to comply; that the inherent nature of the flying saucers and their origination area is completely disruptive to political and economic status-quo."

Aime Michel, in his book, FLYING SAUCERS AND THE STRAIGHT-LINE MYSTERY, published in 1958, cited extensive observations proving that most of the flight patterns of flying saucers are in a north-south direction, which would indicate that their origin is polar, coming from the polar openings.

The frequency of observations of UFOs during a normal week might indicate that they originate from a Christian nation, since they seem to work on a seven-day week pattern with weekends off. Computer Analyst David R. Saunders comments,

"There is a particular deficiency of (UFO) sightings on Saturday with Friday and Sunday down from the rest of the week. Monday through Thursday are essentially equal."
(UFO REPORT, Dec. 1976, p. 20)

UFOs use an anti-gravity field which envelopes them to enhance their maneuverability. Thus the effect of gravity and inertia do not have effect upon them which allows them to make right angle turns at high speeds, fly up as fast as they can forward, and allows them to travel under water at these same high velocities. With conventional science, such feats are considered impossible.

Although Olaf Jansen in his journeys to Our Hollow Earth did not mention that they at that time had UFO craft, they could easily have constructed such craft. He reported that they had an anti-gravity system, which they used on their ships and monorail train systems to facilitate their operation. Olaf wrote concerning their anti-gravity system,

"We were taken overland to the city of 'Eden,' in a conveyance different from anything we have in Europe or America. This vehicle was doubtless some electrical contrivance. It was noiseless, (UFOs are reported to be noiseless also) *and ran on a single iron rail in perfect balance. The trip was made at a very high rate of speed. We were carried up hills and down dales, across valleys and again along the sides of steep mountains, without any*

*apparent attempt having been made to level the earth as
we do for railroad tracks. The car seats were huge yet
comfortable affairs, and very high above the floor of the
car. On the top of each car were high geared flywheels
lying on their sides, which were so automatically adjusted
that, as the speed of the car increased, the high speed of
these flywheels geometrically increased. Jules Galdea
explained to us that these revolving fanlike wheels on top
of the cars destroyed atmospheric pressure, or what is
generally understood by the term gravitation, and with this
force thus destroyed or rendered nugatory the car is as
safe from falling to one side or the other from the single
rail track as if it were in a vacuum: the flywheels in their
rapid revolutions destroying effectually the so-called power
of gravitation, or the force of atmospheric pressure or
whatever potent influence it may be that causes all
unsupported things to fall downward to the earth's surface
or to the nearest point of resistance."* (THE SMOKY GOD,
pp. 110-112)

Their Enemies Shall Become A Prey Unto Them

In Section 133 of the Doctrine and Covenants we learn that
the Ten Tribes have enemies. At that future date when the Ten
Tribes build their highway to the New Jerusalem from the North
Country, the scripture says, *"Their enemies shall become a prey
unto them."* (verse 28) It would seem that their enemies are on
the outer surface of the planet because, as the scripture says,
they will win over their enemies as a beast devours his prey
when they come to Zion.

It is the conclusion of this author that the enemy of the
Lost Tribes of Israel is the International Illuminist-Communist
Conspiracy and their Jesuit controllers in the Vatican, whose
goal it is to take over the whole world. In their global conquest,
they will not be content until they have in their hands the most
rich, and most powerful country of the world -- that world
INSIDE. But such an enemy as this Conspiracy is really the
enemy of the peoples of the WHOLE world, including those
inside the Earth.

More specifically to the people of the United States did the
ancient American prophet Moroni speak his warning of this
enemy of all peoples:

"Wherefore," he wrote, *"O ye Gentiles, it is wisdom in
God that these things should be shown unto you, that*

thereby ye may repent of your sins, and suffer not that these murderous combinations shall get above you, which are built up to get power and gain (the goal of the Super-Rich of the U.S. and Europe is to obtain complete possession and control of the whole world) *-- and the work, yea, even the work of destruction come upon you* (from ten to twenty five percent of the population of a nation is slain when taken over by Communism), *to your overthrow and destruction if ye shall suffer these things to be."*

"Wherefore, the Lord commandeth you, when ye shall see these things come among you that ye shall awake to a sense of your awful situation, because of this secret combination which shall be among you; or wo be unto it, because of the blood of them who have been slain (over 50,000,000 in the Soviet Union, 70,000,000 in Red China, etc. etc.); *for they cry from the dust for vengeance upon it* (The Communist Conspiracy), *and also upon those who buildeth it up* (The Order of the Illuminati whose members are the Super-Rich International Bankers of the United States and Europe and the Jesuits that control them). *FOR IT COMETH TO PASS THAT WHOSO BUILDETH IT UP SEEKETH TO OVERTHROW THE FREEDOM OF ALL LANDS, NATIONS, AND COUNTRIES: AND IT BRINGETH TO PASS THE DESTRUCTION OF ALL PEOPLE, FOR IT IS BUILT UP BY THE DEVIL, WHO IS THE FATHER OF ALL LIES..."* (Ether 8:23-25)

In Gary Allen's book, NONE DARE CALL IT CONSPIRACY, is the proof that the International Illuminist-Communist Conspiracy is that secret combination which the Book of Mormon says *"...seeketh to overthrow the freedom of all lands, nations, and countries..."* And it plainly shows that those *"who buildeth it up"* are the Super-Rich Bankers of America and Europe who financed the Communist revolution in Russia and China. And since they came in control of the U.S. government in the early 20th century, they have used the power and resources of the U.S. government and its people to place communist and socialist governments in power all around the world.

The Mormon Apostle, Elder Ezra Taft Benson, speaking at General Conference of the Church of Jesus Christ of Latter-day Saints, April 6, 1972, said, *"There is no conspiracy theory in the Book of Mormon -- it is a conspiracy fact. And along this line I would highly recommend to you the book, 'None Dare Call It Conspiracy,' by Gary Allen."* (Discourse titled, "CIVIC

<u>STANDARDS FOR THE FAITHFUL SAINTS</u>") The recommendation to read Gary Allen's book was taken out of his printed speech because the lives of church members in Communist countries are valued. Elder Benson afterwards stated that he would gladly send his original speech to anyone who requests it to him personally. Such a copy I have in my files. (Elder Benson later became President of the church and passed away in 1994.)

Upon the reverse side of the book, <u>NONE DARE CALL IT CONSPIRACY</u>, is this statement of the former Secretary of Agriculture, Ezra Taft Benson,

> *"I wish that every citizen of every country in the free world and every slave behind the Iron Curtain might read this book."*

Since the goal of the International Illuminist-Communist Conspiracy and their Jesuit controllers in the Vatican is to take over the whole world and take away our freedom, which is so dear to us citizens and slaves, surely we must honestly begin to consider them our enemies. This Conspiracy represents such a powerful force in the world today that it is only logical that the Hollow Earth Nation would also consider them as enemies.

Ever since Admiral Richard E. Byrd's great geographical discovery of that land beyond the polar openings, the U.S. government and all the governments around the world hold this discovery as WORLD TOP SECRET. F. Amadeo Giannini, in his book, <u>WORLDS BEYOND THE POLES</u>, speaks of the censorship the U.S. government placed on his syndicated news releases of Admiral Byrd's 1927 flight of...

> *"...1,700 miles beyond the North Pole in which the Admiral reported by radio that he saw below him, not ice and snow, but land areas consisting of mountains, forest, green vegetation, lakes and rivers, and in the underbrush saw a strange animal resembling the mammoth found frozen in Arctic ice."*

Giannini comments,

> *"These accounts described Byrd's 1,700 mile flight of seven hours over land and fresh water lakes BEYOND the assumptive North Pole 'end' of the Earth. And the dispatches were intensified until a strict censorship was imposed from Washington."*

Flying Saucers From Inside the Earth

In Alec Maclellan's book, THE HOLLOW EARTH ENIGMA, is an account he found of a man of German descent who became interested in the flying saucer phenomena. Reinhold Schmidt was interviewed by reporter Charles Longcroft of the Los Angeles Examiner who wrote,

"This was the first time I have ever been face to face with someone who claims to have contacted space men or to have been inside a saucer ... My impression is that the man has definitely seen something and is not making the whole story up as a publicity stunt."

Reinhold Schmidt (1897-1974) related to *UFO Report* that at age 61 he was contemplating on the subject of flying saucers after reading Frank Scully's book, BEHIND THE FLYING SAUCERS (1950) when on August 14, 1958 he had what he thought was a communication telling him to drive to a quarry in Bakersfield, California. After sitting around several hours, he saw a silvery circular craft come down from the sky. Access to it was by sliding doors and a ramp lowered to the ground.

A figure appeared in the doorway and flashed a ray on him that momentarily paralyzed him, apparently to synchronize his body's electromagnetic field to that of the craft. Others appeared and escorted him into the craft. They also drove his car up the ramp so as not to leave it in the quarry. They then took off and flew north towards Alaska and up over the Polar Regions.

The crew consisted of four men and two women. Reinhold described them as tall, with noble features dressed in gray, one-piece, skin-tight suits. The women were especially beautiful. They appeared in all respects as humans as we are. They spoke between themselves in what Reinhold recognized as *"high German"* which had been taught him by his parents. However, throughout the entire trip, he was addressed in perfect English.

The craft appeared to be transparent which allowed him to look out in all directions except where obscured by machinery, control panels, couches, chairs or small tables. The latter did not seem to be attached to the floor, but never moved with all the craft's maneuvering. He was able to see out the transparent walls of the craft throughout the journey up to the Arctic Ocean, where he said, *"We seemed to go under the*

Arctic Ocean and enter a huge hole." They then passed over strange earth landscapes, but never landed.

Reinhold stated that his *"hosts"* never told him exactly where they came from, although he became convinced that their homeland was somewhere in the region of the pole. He could see that they were highly advanced in technology, but seemed to be a rather peaceful people. He said that if their mission had a purpose, from what he gathered, it was to observe mankind and to keep us from destroying the planet.

Schmidt's flying saucer trip lasted for five days. He remembers seeing a land which was lit by a glowing sun rather different from our Sun, and twice had the impression of crossing a large curve of ocean where the horizon dipped and fell and then righted itself. This would be a perfect description of having passed into and out of a polar opening. So this author believes Schmidt was actually taken to the hollow earth via the North Polar Opening. On August 18, 1958, Reinhold was returned by his flying saucer friends to the Bakersfield quarry with his Buick. It was then he noticed the paint on his car had turned luminous.

Just the fact that the governments of the world consider the Hollow Earth, its inhabitants and their flying saucers as the WORLD TOP SECRET indicates that flying saucers come from a highly advanced race of this earth. Certainly it must be something sinister that the Conspiracy is trying to hide from the people. Knowledge of the good people inside our earth surely could do us no harm.

But suppose that the people inside Our Hollow Earth knew something which if they were able to communicate it to the people of the surface world that knowledge would endanger the hold the International Illuminist Conspiracy has upon the world? Almost unbelievable to most people is the fact that a Conspiracy is in control of the very government of the United States 60-80%. THE ENEMY IS WITHIN OUR BORDERS TODAY! And slowly but surely they are taking over our freedoms, destroying our national moral character, our economy, our religions, our health, and changing our government. And they will not be satisfied until they have complete control over our minds and bodies.

The Conspiracy knows that if they would allow the people of the United States to establish contact and communication with the people inside the earth, that the knowledge the people of Our Hollow Earth have of the Conspiracy which they obtain from their ever vigilant flying saucers and agents among us, were given to the people of the United States, it would topple the Conspiracy's hold upon us and the world. The conspiracy can live only in secrecy. The Conspiracy's leaders shield themselves from the people with our ignorance of them, which is the only obstacle preventing us from overthrowing the Conspiracy. They maintain our ignorance of them through their control of the banks, the media and educational institutions, many of which they own. If the people were to be educated with the TRUTH, then WE, THE PEOPLE could get the Conspiracy's leaders and supporters out of the government.

Once the people of the United States cleared the Conspiracy out of the U.S. government then Communism and Socialism would fail the world around since it is the U.S. government which has supported and even financed Communist governments and revolutions around the world at the command of the international bankers. Without U.S. support in money, food and technology, Communism and Socialism, the world's parasite, would fail.

The Illuminati with their international bankers are in the process of destroying the United States with a death spiral of debt. It's a Ponzi scheme of the highest order in which Congress has mortgaged the whole country to the international bankers with a debt they cannot pay, and if they do nothing to correct this debt spiral, will bankrupt our country. The bankers will then try to repossess our country with foreign nations they have bankrolled. Once more we find ourselves fighting to preserve our freedom and our country, as our ancestors did during the Revolutionary War of Independence, and the Civil War between the States.

Nevertheless, when the Lost Ten Tribes come out from inside the earth from the City of Eden to the New Jerusalem, which will be the future capital city of America, the Conspiracy will be overthrown, as the scripture says, *"Their enemies shall become a prey unto them."* The Illuminati will be powerless to resist them.

Ever since Admiral Byrd discovered that land beyond the polar openings, the nations controlled by the Conspiracy have been operating a secret offensive against the Israelite Nation inside the earth. Many nations have set up bases in Antarctica from which they are exploring the land beyond the south polar opening in preparation for the coming *"space wars"* between their flying saucers and our military craft.

In August 1945, the United States dropped atomic bombs on Hiroshima and Nagasaki, Japan. Two years later on February

15, 1947, Admiral Byrd of the U.S. Navy made his historic flight to the inside of the earth via the South Polar Opening. Four months later on June 24, 1947, Kenneth Arnold made his historic sightings of Flying Saucers over the Cascade Mountains of the state of Washington in the first modern intensive wave of UFOs. The Conspiracy had knowingly violated the airspace of the greatest nation of the earth and they responded in defense with their own military aircraft over our airspace.

When Arnold sighted the Flying Saucers, they were traveling SOUTH. The Strategic Air Command has scrambled many fighter-bombers since in attempts to intercept Flying Saucers coming from the polar areas.

In Dr. Steven M. Greer's book, DISCLOSURE, is recorded expert witness testimony he has obtained from ex-military and government personnel in the United States describing how flying saucers have kept a continual vigilance of our nuclear weapons sites and even have on occasion shut them down to show our military their capability to prevent nuclear war. Nuclear tipped missiles have been intercepted and destroyed in flight by UFO's including a nuclear warhead in route to the moon.

It is the testimony of many witnesses contacted by flying saucer occupants that they are concerned with the nations of the earth's use of atomic weapons. Obviously, they are concerned for the welfare and security of their nation. The United States, Russia and China have most of their atomic missiles aimed at each other over the Arctic. In a twist of the game of war, the U.S., Russia and China could turn allies against the Hollow Earth Nation and attack them through the north polar opening. The Hollow Earth Nation therefore, is doing everything they can to convince the Conspiracy not to use nuclear weapons.

CIGAR-SHAPED SPACE CRAFT RELEASING FLYING SAUCERS
The last of a rapid series of four telescopic pictures taken by George
Adamski on 5 March 1951. In the first picture only one saucer is
visible. In each successive picture more saucers have left the
mother-ship until in this exposure, six are visible.

Nevertheless, when the Ten Tribes come from inside the earth the scriptures maintain that *"their enemies shall become a prey unto them."* Flying Saucers, the military aircraft of the Ten Tribes, indicate how much more powerful the Hollow Earth nation is than the Conspiracy. Their craft are built with a technology far in advance than our own. Their flying saucers are so versatile that anyone of them can outperform all the functions of our submarines, ships, airplanes, tanks and spacecraft. They even have aircraft carriers in the air which we do not have.

"These include Air Force sightings of UFOs in the sky, under the surface of the sea, and entering and leaving the water, and the active participation of an underwater UFO in a U.S. Navy maneuver off the east coast of Puerto Rico in 1963, during which it was checked at underwater speeds of up to 200 knots and tracked to a depth of 27,000 feet." (THE BERMUDA TRIANGLE, Doubleday, 1974)

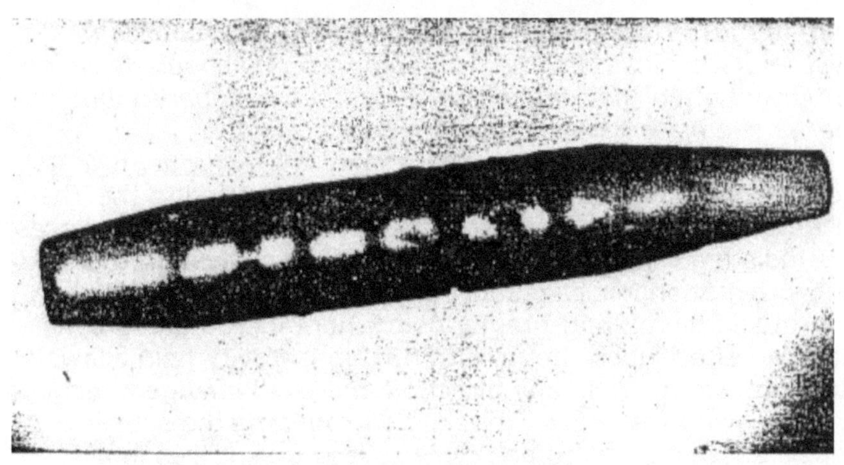

CIGAR-SHAPED SPACE CRAFT
A similar giant carrier ship appeared again at 7:58 a.m. 1 May 1952. It
was of a shining silvery appearance and almost identical to the one that
brought the visiting 'Scout Ship' six months later on 20 November. This
telescopic photo was taken as it hovered over a mountain peak some
thirty miles distant.

UFOs fly circles around our fastest aircraft and fly at speeds
impossible for us. During the summer of 1952 at Terre Haute,
Indiana, three CAA control-tower operators at the airport saw a
saucer streak across the sky and estimated its speed at 42,000
miles an hour! One of our aircraft would burn up in seconds
flying at that speed in the atmosphere. UFOs are continually
seen to make immediate right angle turns while in flight, which
is impossible for our planes to accomplish.

*"On June 28, 1947, at Maxwell AFB Alabama, two pilots
and two intelligence officers of the Air Force saw a star-like
object zigzagging with bursts of speed and making a 90-
degree turn. Even more startling than sharp turns,
perhaps, are those UFOs that reverse their course without
the slightest change in speed, like a Yo-Yo suddenly being
pulled back."*

UFOs are continually seen to fly straight up at blinding
velocities.

*"On July 10, 1947, in Southern New Mexico, an
astronomer saw an elliptical UFO hover and wobble, then
make a 'remarkably sudden ascent estimated at 600 to 900
mph."* (WHAT WE REALLY KNOW ABOUT FLYING SAUCERS,
Otto Binder, pp. 33, 34)

This fantastic maneuverability of the UFO can be
accomplished because *"the owners of the saucers have been*

able to master the physics of gravitation. The propulsion system used must in some way apply what is popularly called antigravity...For not only has gravity been conquered, but inertia seems to have been conquered also."

Dr. Raymond Bernard, in his book, THE HOLLOW EARTH goes on to explain on pages 183-4 that according to the Theory of Relativity of Albert Einstein, gravitational and inertial forces are indistinguishable and equal, and that any system that is anti-gravitational is also anti-inertial. Thus, flying saucers operate within an anti-gravity field where inertia and gravity have no effect upon their craft. This anti-gravity field allows their craft to move in any direction and even change directions while moving with ease. The field surrounding the saucer craft repels anything that comes close to it -- thus it flies in a vacuum. As such, it can fly through water or air equally well.

As we began the exploration of outer space, more and more astronauts have admitted they have seen UFO's in earth's atmosphere, out in space and on the moon. In fact, Astronaut Neil Armstrong has stated that we were warned off the moon by extraterrestrials and their craft. Most likely this is why we haven't returned to try to set up a base or moon city there, as previous space planners had envisioned of using the moon as a launching pad to the other planets. Instead, we have turned our energies towards building a space station closer to earth.   Why? One reason is because the moon is already in control of extraterrestrials.

One of the early contactees, Howard Menger, describes in his book, THE HIGH BRIDGE INCIDENT: THE STORY BEHIND THE STORY, how he witnessed this first hand, back in the '50's when he was befriended by Venusian visitors to earth over a period of many years and was at one point taken by them in one of their flying saucers on a visit to a domed extraterrestrial base on the far side of the moon. There on the Moon, he said, he met people from all the planets in our solar system.

Dr. Greer has provided separate expert witness testimony in his book, DISCLOSURE that verifies that our Apollo flights around the moon took pictures of this alien base on the far side of the moon. Menger discovered that the moon has a tenuous atmosphere and that potatoes grown in lunar soil have a greater percentage of protein than potatoes grown on Earth. Menger even claims that from what he learned from these extraterrestrials and their craft, that he was able to build a small flying saucer craft himself. He even helped the government in a secret project at Colorado Springs, Colorado to build a working flying saucer.

More and more evidence has surfaced to indicate that the U.S. government and other governments are heavily involved in cracking the secret of the antigravity-inertia propulsion system of flying saucers. It is even rumored that Saucers have been captured and intensely studied to discover the secrets to their propulsion system and method of construction. Scientists keep surfacing to report that they have been contacted by the military to analyze these craft.

Obviously, our military is hard at work trying to match their technology in order to fight them more effectively. Dr. Greer also has provided expert witness testimony in his book, DISCLOSURE, that our military has successfully back engineered alien flying saucer technology sufficiently to build what is called *"alien reproduction vehicles"* capable of interstellar travel at the speed of light and more.

Nevertheless, it is still agreed that alien flying saucer technology is far superior to anything we have. It is manifestly evident that they also are not hostile, but appear to be on reconnaissance missions only to get information of enemy territory. However, the world is enemy territory only to the extent that it is controlled by the International Conspiracy. Certainly the people of our Conspiracy-controlled nations are not considered enemies by the Hollow Earth Nation.

Otto Binder, in his book, lists the evidence that the pilots of flying saucers are not hostile, which would be true if they were performing purely reconnaissance flights for their national defense. Binder notes that,

"A) To date, no death can be directly attributed to a UFO as a DELIBERATE action. B) The saucers have gone out of their way to observe us without interfering with our affairs or causing disturbances. C) Though fired upon by Air Force planes, they have never been known in any documented report to have fired back. D) The majority of 'little men' seen outside of landed saucers have been gathering soil, stones, plants, and perhaps insects, quite like scientists gathering the oddities of a new world."

In his 1997 book, THE DAY AFTER ROSWELL, Retired Col. Philip J. Corso confirmed that autopsies of captured aliens reveals that the so called little saucer men are actually androids. They have two brains, one a human-type brain and the other an interfaced integrated circuit computer brain. They have no sexual organs, no digestive organs, no vocal cords and communicate telepathically. In fact, their bodies are an integral part of their craft and cannot operate without these androids.

"E) At any rate, being vastly superior to us in flying technology, they could have easily invaded earth years ago, whereas for 20 years they have made no concerted hostile move. Hostilities have not yet broken open between the saucer-craft and our aircraft. Or at least only one-way 'hostility' has been displayed--by our jets. The saucerians simply disdain shooting back, apparently 'invulnerable' to our weapons." (WHAT WE REALLY KNOW ABOUT FLYING SAUCERS, pp. 148, 150)

If anyone is hostile, it is our military which is controlled mainly by the Conspiracy. Otto Binder's listings of where the most frequent U.S. sightings of UFOs occur shows that they ARE on reconnaissance missions of enemy territory:

"1. The U.S. atomic energy installations, particularly Los Alamos. 2. U.S. Air Force bases throughout the country. 3. Naval and Marine bases (even those around the world). 4. White Sands missile range in New Mexico. 5. Aircraft plants wherever the industry is most heavily concentrated. 6. Most of America's major cities. UFOlogists find it significant that the city most 'surveyed' was Washington, D.C., the capital of our country, what with some 67 UFOs buzzing over the city in 1952. We can surmise that the Russians' key bases, installations, cities, and such have also been minutely 'scouted' by a kind of 'Project Earth Reconnaissance' of the UFOs." (Otto Binder, pp. 146, 147)

In addition to a basic reconnaissance of the nations under the Conspiracy's control, it is evident that since 1947 when Admiral Bryd of the U.S. Navy flew through the South Polar Opening into the land of the Ten Tribes of Israel, that their nation set their military Flying Saucer craft into operation to counter the attempts of the International Jesuit-Illuminist-Communist Conspiracy to penetrate their land. UFO investigators do not know the purpose of the flying saucer reconnaissance of the nations of the world, but their analysis of UFO operations clearly shows that it is a MILITARY reconnaissance and it must be against the Conspiracy which is International. Members of this Conspiracy even call themselves "Internationals."

John Keel, in his article, *"Are Extraterrestrials Blackmailing Earth?"* describes how the flying saucers are staging a subtle but effective counter-offensive against the Conspiracy designed not to cause war but to prevent it. Keel emphasizes the INCREASE in intensity of UFO operations around the world since the last half of the 20th century. In addition, I would note that this increase in UFO activity directly corresponds to the increase in Communist-Socialist takeovers and the spreading control of the International Illuminist Conspiracy in the nations of the world. This all points to a vital concern the flying saucers are taking in the progress the Conspiracy is making in subjugating the world under Communist and Socialist rule. Therefore, their defense and offense has assumed a position of a Blackmail.

The Hollow Earth nation's Blackmail is two-pronged. It consists of a power-show, and what we may call an uncover-up threat. In view of the Conspiracy's relentless drive to conquer the world including the Hollow Earth nation of Israelites, a star wars confrontation of the flying saucers against the leaders of the Conspiracy will undoubtedly come to pass. Nevertheless, it will be a one-sided war, because the military under the Conspiracy's control will be powerless to oppose the flying saucers.

In addition to a power-show blackmail of the International Illuminist Conspiracy in which the Ten Tribes threaten destruction with their powerful flying saucers and their light-rays in defense of their country, is the threat of an *"uncover-up"* of the secret Conspiracy. Because of the fantastic worldwide reconnoitering of the International Jesuit-Illuminist-Communist Conspiracy by the Ten Tribes, they know all about the Conspiracy. And since the Conspiracy can operate successfully only in secrecy, an uncover-up of the leaders of the Conspiracy and their treasonous activities would bring about an unquenchable demand by the people for their overthrow. The UFO pilots' ability to appear on TV and interrupt radio programs indicates their ability to communicate this knowledge of the Conspiracy's insiders and their treasonous actions to the people.

Perhaps this threat of an uncover-up blackmail against the Conspiracy has not been used yet because to uncover the Conspiracy would force the Conspiracy to uncover their secret knowledge of Our Hollow Earth and in a last grand effort for survival try to influence world opinion that the Ten Tribes are our enemies and that we must wage nuclear war against them.

Nonetheless, the Hollow Earth Nation is not afraid of the peoples of outer earth, nor their leaders. They are on a mission for the Creator to keep the Illuminati from blowing up the planet. And their message to the people of the world is that of peace, love and harmony.

We find a sort of balance of power and secrets between the Conspiracy and the Hollow Earth nation. But of course we would have no *"cold war"* between the International Illuminist Conspiracy and the Hollow Earth Nation if the Conspiracy was not active in its goal of world conquest and control.

It wasn't until the Conspiracy became interested in the Hollow Earth after its discovery by Admiral Byrd in his expeditions of 1927, 1929, and 1947 into the hollow interior of our planet through the polar openings that the Hollow Earth Nation began sending its flying saucers out to the surface world in defensive action against the Conspiracy.

With the breakup of the Soviet Union in 1992, many believe that communism has also failed in its drive for world conquest. They are deceived. A book called, NEW LIES FOR OLD, by Anatoliy Golitsyn, a top KGB agent who defected to the west in the December 1961, contains the ongoing Plan the communists have for world conquest. They work for the International Bankers. YOU need to know that Plan in order to safeguard your freedom!

Before defecting to the United States, Anatoliy committed to memory the long range goals of the Soviet Empire. After spending a couple of decades trying in vain to convince our government leaders of the threat and real Plan for world conquest by the communist world, in 1984 he decided to take his warning to the people.

In his 1984 book, Anatoliy described how the Soviets and Chinese would play war games to make the West believe they were enemies. Then he described how world communism was financed by western International Bankers and governments by claiming each time that they were becoming *"democratic,"* and how through their propaganda machine, they made it appear that those who financed them were really their enemies. They fomented the lie that Communism and Capitalism are mortal enemies when, in fact, they are bedfellows. It is through decades of financing by the Capitalists of the West that Communism has survived at all. Their economic system is so inefficient that it couldn't possibly survive on its own. The reason is Communism is a parasitic system -- it can only live off the life blood of others. The Illuminati is also a parasitic system. Through their control of the world's banking system, they prey upon the people through economic booms and busts which allows them to repossess real property for worthless money created out of just numbers on their computers.

Anatoliy discloses that the Plan for World Conquest by the communist empire was that when they were at the pinnacle of their power, they would suddenly feign weakness to get the West to disarm before their final intense strike and conquest of the West. By feigning a breakup of the communist empire and the staging of democratic elections, the West was led to believe they now have no enemies. Therefore, the former communist, now deceptively turned democratic, countries are being admitted into the NATO partnership. And of course, the greatest reason for feigning a breakup of the Soviet Union was to get the West to disarm. After all, if you no longer have an enemy, why spend so much on defense?

The proposed breakup of the Soviet Union (which happened in 1991) was to be highly publicized and things like the unifying of the German Republic with communist Germany (happened in 1992) and the bringing down of the Berlin Wall (happened November 9, 1989) -- were to all be a part of that strategy -- including the overthrow of satellite communist governments by massive popular movements of the people. Remember, this was a long-time strategy Anatoliy committed to memory before defecting to the United States in 1961. His book was published in 1984. That was about 8 years before the breakup of the Soviet Union and the unifying of the Germanys actually came about.

In 2001, Dr. Steven Greer held a press conference in Washington, D.C. of the Disclosure Project with the help of about 15 expert witnesses who had first-hand knowledge of contacts with flying saucers. One, Carol Rosen, was an executive with Fairchild Industries, who became a close friend of Wernher Von Braun, founder of NASA. Von Braun was privy to the plans of the Illuminati and International Bankers and told her that the cold war between Russia and the United States was a big lie fomented by the International Bankers, who finance both sides of every war. He told her before it ever happened, that the cold war would be followed by a War on Terror, which would be followed by the weaponization of space with the excuse to fight asteroids.

That was to be followed by a staged Star Wars in which the Illuminati would stage an invasion from space to be fought with reverse engineered flying saucers from real flying saucers our military has knocked out of the sky with high-powered radar, and with biological engineering on kidnapped children they are creating "programmed life forms" to pilot their reverse engineered flying saucers. With these they would stage a fake invasion by "aliens" from space – all for the purpose of unifying the world behind their one world government which would be a military dictatorship to make all mankind the abject slaves of the Super-Rich.

But the Hollow Earth Nation's flying saucers are preventing the Illuminati from weaponizing space. And they have shown that they can stop our use of nuclear weapons. One military satellite was blown up by a small six-sided transparent ball that whips around the earth keeping tabs on what the military puts up in space, causing one military officer to comment that our earth is under quarantine by the extraterrestrials.

In view of the fact that traitors in our country are working night and day in delivering the world into the control of the International Jesuit-Illuminist-Communist Conspiracy, and lastly, we in America will be attacked by world forces in an attempt to enslave us to repossess our country for all the debt they have gotten us into when our government goes bankrupt, it is important that freedom loving people do all in our power to prevent this headlong rush into slavery.

I have written Congress to try to get them to implement my Asset Based Receipt Monetary System which would solve the problem of our national debt, and change our monetary system from the current system where money is created only as debt, to an asset-based receipt monetary system where all fractional reserve banking is made illegal. Our national bank would be 100% owned by the people of the United States instead of allowing private banks to control our money as is the case with the present Federal Reserve system. All new money created by my Asset Based Receipt Monetary system would be given to the people as a receipt for voluntary taxes paid. An Asset Based Receipt Monetary System would give the people a stable monetary system with no debt, no inflation nor deflation, and will more than double the income of all taxpayers every ten years. See:

https://www.ourhollowearth.com/saveourcountrynow.htm

And since the military unacknowledged black projects are actively engaged in discovering the secrets of flying saucer technology for the benefit of their banker bosses, which they hope to use to enslave us, We, The People, must urgently get involved in the discovery and development of this fantastic technology. Technology is power, and those that have it first will have the advantage in the war between freedom and slavery. With this in mind, more and more scientists and inventors are going public in an attempt to get this advanced technology out to the people so that the One Worlders can't enslave us.

In a radio talk show in Utah in the summer of 1995, the radio announcer interviewed an American scientist named Stan Deyo, regarding his involvement in the development of flying saucer technology. He had graduated from a US military Academy and later was working in Texas when he was approached by the inventor of the Hydrogen bomb, Edward Teller. Stan had been dabbling in antigravity propulsion when he was tapped on the shoulder and asked if he would be interested in going to Australia and continue his work on antigravity technology. He accepted and after working with them for several years, he discovered that they were heavily monitored and controlled by certain Internationalist groups that are involved in developing a world government with intentions of subjugating the world. It was then that Stan decided to part ways with them. His main disagreement with them was that they did not want to make this technology available to the peoples of the world. They wanted to use it to control the peoples of the world.

Another self-made scientist in England, by the name of John R. R. Searl, developed flying saucer and free energy technology that he discovered in the 1940's. In the book, ANTIGRAVITY: THE DREAM MADE REALITY, THE STORY OF JOHN R. R. SEARL, by John A. Thomas, Jr., an electrical engineer from the state of New York, Prof. Searl's technology and inventions have been documented. Also available are ten books on the Law of the Squares by Prof. John R. R. Searl.

John R. R. Searl (May 2, 1932 - December 4, 2018) was born May 2, 1932 at the Downs, Newbury Road, Wantage, England. At 4 years of age, John's father abandoned his mother and children and from her inability to support the family, the government placed the children in foster homes where John was often mistreated. From age 4.5 to 10, John had 2 dreams twice a year from which he developed his Law of the Squares -- the basis of the technology that he developed producing free-energy electric generators and flying saucer-type craft.

From his revolutionary work on free-energy technology and devices, John R. R. Searl was awarded an honorary degree as Professor of Mathematical Structures of Creation and Energy, was listed in the International Registry of Who's Who in the World and had many years of study at schools and universities coupled with the practical experience in the design and construction of electric motors, generators, medicine, navigation, electronics and computers.

John's first Searl Effect Generator (SEG) was assembled when he was 15. The most notable effect produced by the generator was that it would suddenly lose its gravity and become airborne. From what John learned over the years from his SEG, he and his friends built about 40 flying saucers. John also powered his home for 30 years with one of his SEG generators.

After discovering that his SEG generators can become airborne, John and his friends gave it a body and it became a flying saucer craft. The first SEG's John built would become airborne and would shoot off into space and were lost. After he resolved the problem of flight control and after many years of setbacks, Searl finally found a company in southern California that is working once more to get his technology out to the world.

John R. R. Seal's Flying Saucer Craft

In 1968, they were within three months of having a completed manned craft in the space race to the moon, when John was poisoned and while he was in the hospital, his home was broken into, his walls were hacked where he was hiding his SEG generator and it was stolen. The field in the countryside he was using to build his flying saucer was sold and his saucer was sold for junk. His wife also disappeared.

Because of the opposition to his revolutionary free-energy technology, John R. R. Searl decided to make some of this technology available to the world in a series of books, The Law of the Squares. He hoped to get the price of his SEG generators down to about 300 Pounds so that the average homeowner can buy one. It would produce all the free electricity a homeowner can use. It is silent and only about 16 inches in diameter. In fact, it can fit inside a wall of your house. It improves the electric wiring in your home instead of deteriorating it and injects negative ions into the air of your home to clean the air and gives energy to the inhabitants so they don't ever get sick. Besides generators, and flying saucer-type space craft, there are endless applications for his technology, including building homes that would be fire proof, earthquake proof, tsunami proof, EMP proof, and even solar flare proof using his SEG technology to protect them.

After attempting for many years to recover his technology and find someone who would support his work, John Searl found a position as technical advisor to SEG Magnetics in Spring Valley, California until his death on December 4, 2018 due to Pneumonia and other health issues at age 86. SEG Magnetics is first attempting to recreate his SEG generator technology, and then eventually his Inverse-G craft technology. See: Research

and Development Center | SEG Magnetics, Inc.:
https://segmagnetics.com/

Another inventor that I met at the 1991 Flying Saucer Conference in Phoenix, Arizona was Howard Menger (February 17, 1922 to February 25, 2009). A UFO contactee, Howard Menger, described in his book, THE HIGH BRIDGE INCIDENT: THE STORY BEHIND THE STORY, how he was befriended by extraterrestrials from Venus at a very young age as he was playing with his brother in a forest on his father's farm. From his life-long association with these Venusians and flights on their craft, he came to the conclusion that gravity is a push from space and can be countered by electrostatic forces. Using this knowledge he constructed a three-foot-in-diameter model flying saucer in 1951. He flew it by remote radio control from the ground. However, it flew out of range and he lost it. Later FBI agents visited him with parts of his crashed saucer and expressed interest in the propulsion system.

From this contact, in 1961 the Pentagon set up a high tech laboratory installation near Colorado Springs where Howard Menger helped the government and participating big industry build a full size flying saucer craft which Menger test flew successfully. For this the government promised him a tax-free $1,500 check each month for the rest of his life. However, after one year the checks quit coming. He checked with some associates that worked on the project, and found that the project had gone underground into the deep Black Projects. Subsequently, Howard began working on building a full-size flying saucer in his garage. He also published an updated version of his book, THE HIGH BRIDGE INCIDENT: THE STORY BEHIND THE STORY, which describes the years of contacts he made with the human aliens from Venus.

These pioneers have contributed to our knowledge of gravity that will help us achieve electrogravitic propulsion. It is just a matter of putting the details together for a flying saucer to be actually constructed. This craft will be a silent electrogravitic propelled saucer shaped craft capable of underwater, air and space travel.

The advantages of this saucer craft will be its unlimited range, record breaking speed, the elimination of costly depletable fuels, and its ability to fly underwater, in the air and into space. This would make it more difficult for enemy missiles to hinder a mission. With this craft, missions could easily be undertaken to the moon and other planets. Perhaps the surfaces of the other planets in our solar system are uninhabitable on their surfaces, however, in all probability, their interiors are gardens of Eden, and inhabited by other civilizations of God's children, perhaps even closely related to our race of humans.

Saucer technology is based on the discoveries of several scientists, Townsend Brown, Horace C. Dudley, Joseph Newman, Al Snyder, and inventors John R. R. Searl of England and Howard Menger of Florida, in combination with additional knowledge concerning the Ether of space as taught by 19th Century scientists.

These inventors and scientists believe that by empowering the people, the Conspiracy will find it very difficult, if not impossible, to enslave us. Toward this end, I have decided to make available to the public my research in a report on THE ORIGIN, CAUSE and CONTROL of GRAVITY -- FOUND! (See Contents Page)

In that report, I disclose what I have been able to find regarding gravity control and flying saucer technology. You will read in that paper how I recalculated the surface gravities of all the planets showing that they are all within 60% of earth's surface gravity which means that people could possibly live within the hollow interiors of those planets.

In fact, there is evidence that we are not alone in our solar system -- or the universe, if only we don't cover up our eyes when the evidence presents itself.

The worlds are and were created by Jesus Christ for the purpose of populating them with *"sons and daughters unto God."* In a revelation given to the prophet Joseph Smith, a precious portion of the Bible has been restored which had been taken away. In it, God revealed to Moses that *"worlds without number have I created, and I also created them for mine own purpose; and by the Son I created them, which is mine Only Begotten."* (Moses 1:33, the Pearl of Great Price) A little while later God told Moses his purpose for creating worlds and populating them with his children, when He said, *"For behold, this is my work and my glory—to bring to pass the immortality and eternal life of man."* (Moses 1:39, the Pearl of Great Price)

In another vision, in 1832, Joseph Smith and Sidney Rigdon testified,

> *"For we saw him, even on the right hand of God; and we heard a voice bearing record that he is the Only Begotten of the Father -- That by him, and through him, and of him, the worlds are and were created, and the*

*inhabitants thereof are begotten sons and daughters unto God." (*D&C 76:23, 24*)*

In a small book called, MY FRIEND FROM BEYOND EARTH, Dr. Frank E. Stranges (October 6, 1927 to November 17, 2008), a minister of religion from Los Angeles, California was invited by a friend to the Pentagon in 1959 where he met a visitor from Venus. The visitor had landed his flying saucer craft near a close-by town and was picked up by the police. The visitor requested to be taken to the Pentagon where he could speak with those in authority. He had been there several weeks when Dr. Frank Stranges was introduced to him. He said his name was Val Thor. He looked in every way like us humans, except he had no fingerprints. He spoke perfect English and said he was head of the Council of the Twelve on Venus and commander of a Starship 14 miles long, 7 miles wide and 3.5 miles tall. Val later let Dr. Stranges know that his people live INSIDE the planet Venus and that Venus is hollow like our Earth is hollow.

When Dr. Stranges met Val Thor, he was wearing normal earth clothes. He explained that he had given his flying saucer uniform to the people at the Pentagon to run tests on it. He showed the garment to Dr. Stranges. It was a one-piece garment that glittered as he brought it toward the sunlight streaming through the window. The tests run on it determined the garment to be indestructible. It was heated to temperatures above the melting point of steel without damaging it, acid rolled off it like water off the back of a duck, and a diamond-point drill bit overheated and snapped when brought in contact with the garment.

He stated his purpose in coming to earth was to help mankind return to the Lord. He always spoke with a smile on his face, but said that God was displeased with the fact that mankind was farther away from Him than ever before, but that there still was a chance for mankind to find salvation if they would look for it in the right place. He had been on earth for three years, offering advice to the leaders of our country, but so far only a few would even listen to him. When asked what he thought of Jesus, Val Thor replied,

"I know that Jesus is the alpha and omega of yours and everyone's faith. He today has assumed His rightful position as the ruler of the universe and is preparing a place and a time for all who are called by His name to ascend far above the clouds to where His Power and Authority shall never again be disputed."

Dr. Stranges then recorded,

"As he spoke these wonderful words, my own heart burned within me and tears filled my eyes...When I asked him did they need the Bible to guide them, he stated, turning to me again: 'Why would we need the book when we are still walking in unbroken fellowship and harmony with the author?'

"I asked him if there is life on other planets. His reply was, "There is life on many other planets of which people on earth know nothing. There are more solar systems for which man has not even given God credit. There are many beings that have never transgressed the perfect laws of God. Man does not possess the right to condemn the whole of God's creation because he himself has broken the perfect laws of God through disobedience."

Subsequently, on a couple of occasions, Dr. Stranges was invited by Val Thor on board his flying saucer craft. He was continually amazed at the technology and personal powers of Val and his people.

Dr. Frank Stranges also wrote a book on the Hollow Earth and another book on THE STRANGER AT THE PENTAGON.

In an interesting series of books, another author Charles Hall wrote about some extraterrestrials that live underground in some bunkers that the United States military built for them on Range Four of the Nellis Air Force Base, northeast of Las Vegas, Nevada.

Charles Hall was a military meteorologist stationed on Range Three. While there, he witnessed the aliens arriving regularly from space, a hanger door on a mountain would open and the spacecraft would go in. On many occasions, children from the alien base would come down to his weather shack to play and look through his theodolite and watch him launch his weather balloons. Charles Hall called these aliens, "the Tall Whites" because although they grow very slowly, they live very old life spans up to 800 years or more and never stop growing. So the old Tall Whites can get up to nine feet tall.

In his books, named MILLENNIAL HOSPITALITY, six books in all, Hall described the interactions of the Tall Whites with people on the base, but the military required that the Tall Whites go disguised as humans whenever they would go into Las Vegas. The Tall Whites negotiated a deal with the United States military to allow Charles Hall the freedom to come and go as he pleased and report what he learned about them. They liked him because he wasn't so deathly afraid of them as other previous meteorologists that would only stay an hour or so at the Range Three weather shack upon their horror of learning there were aliens in those hills.

These Tall White aliens had very keen sensory abilities. When they played at the Las Vegas casinos, they could always win because with their keen eyesight, they could see the fingerprints on the cards of their adversaries in the games. Their native language sounded like dog barks, but with their advanced abilities they could speak English perfectly, even to imitate Hall's friends on the base. He had to listen very carefully when the phone at the weather shack rang, because by what was said he could tell it was Range Four Harry, a Tall White that Hall interacted with often imitating one of Hall's friends on the phone. Another of the Tall Whites that Hall often interacted with was a female called "The Teacher." She and other Tall White mothers would come down to his weather shack with their children so they could play, and told Hall about her interactions with our founding father, George Washington. She told Hall that the Tall White mothers love their children more than human mothers love their children because they take care of their children better than human mothers do.

Although, the Tall Whites never told him where they came from, his intuition was that they came from the solar system of the star Arcturus. This alien base was like a stop off location for spacecraft of the Tall Whites. He would often see their spacecraft bring tourists from their solar system. One tourist spaceship stopped at his weather shack and the tourists would off board and walk around his weather shack. When they got ready to go, they were leaving an old Tall White man, who had sat down and leaned against his weather shack and gone to sleep. Hall had to run to the spaceship and yell to them that they were leaving one of their party.

Charles Hall said the Tall Whites had antigravity suits, and pencil-like weapons they always carried. One of the female Tall Whites had an assignment to come with her father to the weather shack and dance with Hall, but became nervous and shot Hall with her pencil-like weapon. It caused his thyroid to overheat and start to bleed in his throat. Usually, when a human would be shot with this weapon, they would take off running afraid of the Tall Whites, and bleed to death internally. Charles Hall didn't do this. He knew he had to get the bleeding to stop, so he got on the ground and made a poultice with dirt and blood and shoved it in his throat to stop the bleeding. The Military shortly arrived to chew out the aliens for almost killing

Charles Hall. The Tall Whites apologized and when Hall got well, the father brought her daughter again to pass the dance test. By passing the dance test with Charles Hall, she would be allowed to go to meetings with our government and contractor industry partners to negotiate transfer of some of their technology to us in return for something the aliens wanted.

Charles Hall wrote his books as fiction, but later told audiences that the events he described in his books actually happened to him on the Nellis Air Force Base in Nevada, that he changed the names and places of the persons in his books to protect their identities.

In conclusion, I could say that:

1) Admiral Byrd reported that the inner earth inhabitants have flying saucers and that most saucers seen around the world most likely originate from within Our Hollow Earth, but that

2) Earth most definitely has been visited by spacecraft and extraterrestrials from other planets and solar systems, and that

3) The US military black unacknowledged projects have successfully downed many extraterrestrial flying saucers and have reversed engineered them and are using them to explore the Solar System and our Galaxy in a secret astronaut program separate from NASA and are the ones behind the cattle mutilations and most of the abductions in hopes of turning the general population against aliens so that they can stage an invasion from space with Alien Reproduction Vehicles piloted by Programmed Life Forms they have created from abducted children in underground bases in a last-ditch effort to unite the world behind their One World Government project of the United Nations that would be a dictatorship of most ghastly proportions, and

4) if We, the People would get behind our inspired inventors and scientists who have already discovered flying saucer technology, we could be right now flying around in flying saucer craft ourselves with free energy generators powering all our homes and vehicles.

IT IS STILL NOT TOO LATE, if we will ACT to save our world for FREEDOM, our God, our Religion, our Property, our Families, and ultimately our Happiness!

CHAPTER SEVEN
"And They Shall Hunt Them...
Out of the Holes of the Rocks"

A comprehension of Our Hollow Earth would not be complete without the knowledge of the great civilizations that live within the Earth's shell. A greater population could possibly live within the crust of this planet than upon either the outside or inside surfaces. However, many people who perhaps could accept the truth that our earth is hollow and inhabited by the lost Ten Tribes of Israel, the thought that millions of people are living in caverns within the earth's crust is wholly preposterous. The evolutionists have taught for so long that cavemen were the most unintelligent man-apes, that to believe that whole civilizations of highly advanced and intelligent people live in giant caverns within the earth's crust is hard to accept. But the truth must come to light sooner or later.

In the last days before the second coming of Our Lord and Savior Jesus Christ, when the Ten Tribes are to come out of the north to the New and Old Jerusalem, the Lord will choose 144,000 missionaries, 12,000 from each of the twelve tribes of Israel to take the gospel to the whole world for the last time before our Savior's Second Coming to save those found worthy from the burning. In a revelation given to the prophet Joseph Smith, March 1832, the answer is given to the question:

"What are we to understand by the sealing of the one hundred and forty-four thousand, out of all the tribes of Israel--twelve thousand out of every tribe?

Answer: We are to understand that those who are sealed are high priests, ordained unto the holy order of God, to administer the everlasting gospel; for they are they who are ordained out of every nation, kindred, tongue, and people, by the angels to whom is given power over the nations of the earth, to bring as many as will come to the church of the Firstborn." (D&C 77:11)

At that time the gospel will be taught to those civilizations living within the earth's crust.

That was the word of the Lord to Jeremiah many centuries ago:

"Therefore, behold, the days come, saith the Lord, that it shall no more be said, The Lord liveth, that brought up the children of Israel out of the land of Egypt;"

"But, the Lord liveth, that brought up the children of Israel from the land of the north, and from all the lands whither he had driven them; and I will bring them again into their land that I gave unto their fathers."
"Behold, I will send for many fishers, saith the Lord, and they shall fish them; and after will I send for many hunters, and they shall hunt them from every mountain, and from every hill, and out of the HOLES OF THE ROCKS." (Jeremiah 16:14-16)

This scripture says that the gospel will be preached to people who live in *"the holes of the rocks."* It is a direct reference to the cavern people. Surely then, they do exist. The false teaching of the Conspiracy opinion-molders that the earth is overpopulated certainly does not take into account the 800 miles of the earth's shell wherein people do live.

The Lord has said that there is enough space for all and even to spare.

"I, the Lord, stretched out the heavens, and built the earth, my very handiwork; and all things therein are mine. And it is my purpose to provide for my saints, for all things are mine...For the earth is full, and there is enough and to spare..." (D&C 104:14-17)

It is the testimony of a few explorers who have been privileged to visit those civilizations who live in giant cavities within the earth's crust that there is enough land and to spare within the earth's shell in which to populate God's children.

Jules Verne's work of fiction, JOURNEY TO THE CENTER OF THE EARTH, must have been taken from accounts of actual explorers who had been within the earth's crust. Compared to other accounts of actual explorers having been there, Jules' descriptions are highly similar. However, Verne's explorers did not go to the center of the earth as the book title infers, but to a giant cavern 75 miles beneath the Atlantic Ocean.

So similar is Verne's story to ETIDORPHA by John Uri Lloyd that it is very probable that he based his adventure upon one similar to that of Lloyd's character, I-AM-THE-MAN, as he preferred to call himself. Bruce Walton, author of GUIDE TO INNER EARTH, claims I-AM-THE-MAN was a man named William Morgan who had joined a Masonic lodge and subsequently published a book of the secret rituals of the Masons. For that the Masons simulated his murder, but actually kidnapped him and condemned him to a life-time journey to the Hollow Earth through a certain cave near the Cumberland River in Kentucky.

Jules Verne wrote his book in 1864, and I-AM-THE-MAN's adventure began on August 12, 1826, 38 years earlier. ETIDORPHA was published in 1895 by John Uri Lloyd, who was a friend of Llewellyn Drury to whom I-AM-THE-MAN delivered the manuscript of his story. As I-AM-THE-MAN explained in his story, he was taken by fellow members of a secret society to which he had joined. They entered the earth's crust by way of a cavern into which a creek was emptying near the Cumberland River in Kentucky.

Jules Verne's explorers found their way inside the earth's crust by way of a volcanic crater in Iceland. They obtained the clue to find the opening into the cavern world from some writings they had discovered written by a previous explorer.

In Lloyd's account of I-AM-THE-MAN's journeys, his guide took him 150 miles under the Atlantic Ocean where they crossed a lake 6000 miles long in a small boat. Jules Verne's explorers discovered a big lake 75 miles under the Atlantic Ocean which they crossed on a raft.

Both described plant life within the earth's crust as being giant mushrooms. Wrote I-AM-THE-MAN, *"There could be no doubt that I was in a forest of colossal fungi..."* (ETIDORPHA OR THE END OF EARTH, p. 106) and made the interesting discovery that they were not only edible but tasted like strawberries, pineapple and other delicious fruits. Such fantastic amounts of food could feed millions and so deliciously!

Both Jules Verne's account and Lloyd's said that at one end of the lake was the origin of a volcano which erupts periodically on an island of Italy. Both agree that volcanoes are caused by the reaction of water with explosive minerals such as potassium and sodium which ignite in the presence of air and water. Magnesium, which is the eighth-most-abundant element in Earth's crust by mass reacts violently with water, and so can be one of the principle ingredients of volcanic lava.

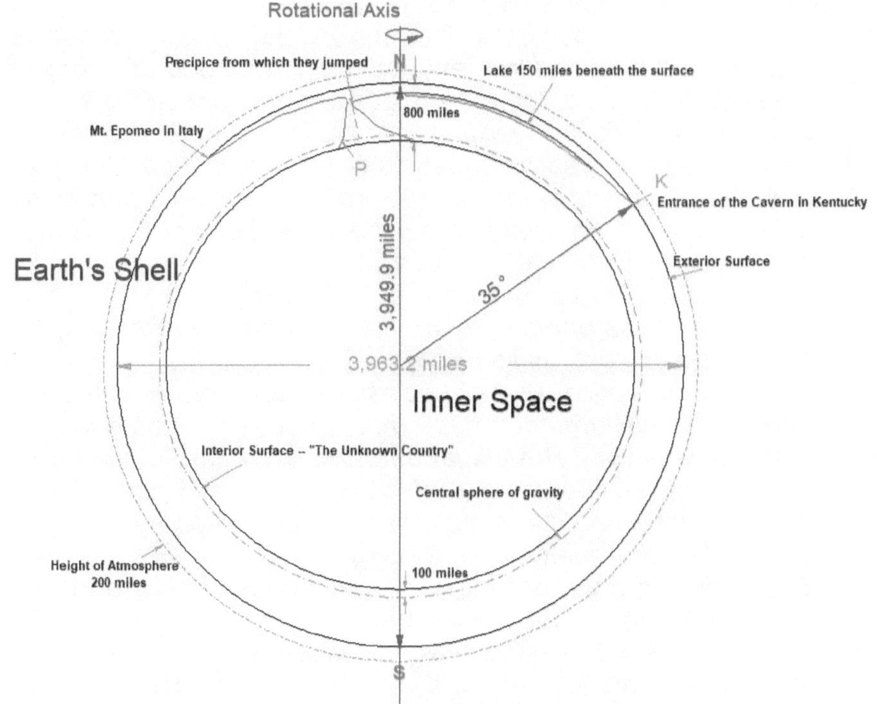

Description of journey from K (Kentucky) to P - The End of Earth

A significant difference, however, between the two accounts is that Jules Verne never infers that the earth is a hollow sphere. I-AM-THE-MAN on the other hand was taken to the inner surface of Our Hollow Earth by his guides through communicating caverns that reached from the outside surface of our planet to the inner. In fact, one of his guides gives us the thickness of the earth's crust at 800 miles from the outside to the inside surface with the center of gravity 700 miles down-- closer to the inside surface than to the outside surface. I-AM-THE-MAN gives us a diagram of Our Hollow Earth with the cavern entrance into which he was taken, but neither author mentions nor infers of the existence of the polar openings, nor of the inner sun.

Both Verne and Lloyd say there is light inside the earth's crust. Verne, perhaps unable to believe actual explorer's accounts of that light, or getting his account mixed up with other explorer's accounts of the Hollow interior with its sun, resorted to a powerful electric light over the lake while I-AM-

THE-MAN's guide explained to him the light that they discovered within the earth's crust thus:

> *"I will only say that this luminous appearance about us is produced by a natural law, whereby the flood of energy, invisible to man, a something clothed now under the name of darkness, after streaming into the crust substance of the earth, is at this depth* (about 10 miles), *revivified, and then is made apparent to mortal eye, to be modified again as it emerges from the opposite earth's crust but not annihilated."* (ETIDORPHA, p. 101)

I-AM-THE-MAN explained this light further,

> *"There was apparently no central point of radiation; the light was such as to pervade and exist in the surrounding space, somewhat as the vapor of phosphorus spreads a self-luminous haze throughout the bubble into which it is blown. The visual agent surrounding us had a permanent, self-existing luminosity, and was a pervading bright, unreachable essence that without an obvious origin, diffused itself equally in all directions."* (ETIDORPHA p. 74)

Such lighting must permit untold millions to live within the 800 mile thickness of Our Hollow Earth's crust.

Although I-AM-THE-MAN was not taken to any city of the cavern people in his journey to the hollow of the earth, he apparently got glimpses of them. He wrote,

> *"From time to time I experienced strains of melody, such as never before had I conceived, seemingly choruses of angels were singing in and to my very soul. From empty space about me, from out of the crevices beyond and behind me, from the depths of my spirit within me, came these strains in notes clear and distinct, but yet indescribable. Did I fancy, or was it real? I will not pretend to say. Flowers and structures beautiful, insects gorgeous and inexplicable were spread before me. Figures and forms I cannot attempt to indicate in word descriptions, ever and anon surrounded, accompanied, and passed me by...Sometimes I begged to be permitted to stop and live forever 'mid those heavenly charms, but with as a firm a hand as when helping me through the chambers of mire, ooze, and creeping reptiles, my guide drew me onward."* (ETIDORPHA, p. 268)

In the UNDER-PEOPLE, Eric Norman comments on these sounds of choirs singing within the earth's crust:

> *"The choral singing of men and women is frequently heard in certain parts of the world. Occultists claim these*

*'celestial choirs' are an indication of a tunnel leading to the
subterranean tunnels of the Under-People."*
He goes on to quote from an article in the SEARCH
magazine in which Will Carson and Jeannie Joy told of a couple
who were exploring in the Casa Diablo region north of Bishop,
California when they discovered a circular hole in the earth. The
hole was approximately nine feet in diameter and the couple
impulsively decided to explore the unusual formation. The hole
turned into a sloping tunnel and, armed with a flashlight, the
couple reported walking through the horizontal corridor that...
"...could only have been carved by human hands."

*"At the end of the short passage, they discovered a
huge door of solid rock. They attempted to open the door,
but it did not yield. After their return to the surface, the
wife turned to her husband and remarked: 'Do you know,
while I stood down there I heard music--the strangest,
most weird music I've ever heard. But it seemed to come
from everywhere at once, or inside my own head. I guess it
was my imagination.'"*

*"Her husband turned pale. 'My ---, I thought it was my
imagination. I heard it too -- like music from another
world!'"* (THE UNDER-PEOPLE, pp. 148-189)
In THE HIDDEN SECRETS OF THE HOLLOW EARTH, Warren
Smith gives an account of his acquaintance with Robert Maxwell
II, author of LEMURIA--FACT OR FICTION?, who was camping
out in the woods of Siskiyou County, California, one night near
Mount Shasta where he was befriended by a resident of a
Lemurian colony and later taken to see their city deep within
Mount Shasta in a giant cavern.
In his interview with Maxwell, Warren Smith asked:
"'So we have to take your story on faith,' I said.

*"'Not quite,' answered Maxwell. 'There are ways of
proving the city is there.'*
"'Have you been there?'

*"Maxwell nodded. 'I was led through the tunnel and
Mokla took me to the inner opening that overlooked the
cavern. I could look down and see the city. I was not
allowed to enter. Outsiders are forbidden beyond the inner
opening of the tunnel.'*

*"'Why jeopardize their security by showing you the
entrance?' I asked.*

"'I was taken to the outside entrance when blindfolded,' Maxwell said. "I was blindfolded after my visit to the inner city.'

"'What did the city look like?'

"Maxwell thought for a moment. 'The architecture was beautiful,' he said. 'Looking down from our vantage point when the tunnel opened into the cavern, I saw temples and spires like those you might see in a picture of some fabled Biblical city.'

"'What about lighting?'

"'They have a bright light in the top of the cavern that is an eternal light.'

"I asked, 'What about power sources?'

"'They lead a simple life and don't require power.'

"'What about the eternal light,' I asked.

"'I was told that the light came from Lemuria before it broke up,' (in the Pacific Ocean) said Maxwell. 'It doesn't need an outside power source. The energy comes from magnets; Lemuria knew the secret of getting power from a magnetic source. The light will burn forever unless someone dismantles it.'

"...I asked Maxwell about the alleged Lemurian flying saucers."

"'Mokla told me they also operate on a magnetic principle,' he went on. 'Mokla claims their saucers are left over from their ancestors who flew them from Lemuria to Mt. Shasta. He also said that they don't have all the flying saucers. There are others who have them, although he refused to elaborate on that statement.'"

Maxwell later brought Smith a Lemurian coin with hieroglyphics on it which he said Mokla gave him as proof to Smith that the city indeed exists. Smith since has taken the coin to many coin dealers none of which have been able to identify the coin. The people who live in the cavern of Mt. Shasta were described by Maxwell as being about five feet tall, and dressed in loose-fitting gray robes with a hood thrown back over the shoulders. (Warren Smith, pp. 33-37)

There are however, various races that live within the earth's crust. Jules Verne, who could have based his story on the experience of actual explorers, had his adventurers get a glimpse of a giant man 12 feet tall tending a herd of mammoths in his cavern world. Similarly actual explorers have encountered evidences of a race of giants living within the earth's shell.

"And They Shall Hunt Them...
Out of the Holes of the Rocks"

In <u>LOST MINES AND HIDDEN TREASURE</u>, the author Leland Lovelace tells of two prospectors who discovered a series of caves in the mountains of southwestern Nevada. Within the giant caverns they discovered furniture pieces of enormous size as if they had been constructed for giants. Dishes of gold and other precious metals also were found in the caves.

Lovelace also tells of a prospector named J.C. Brown who, in 1904, claimed to have discovered a tunnel cut into the slopes of the Cascade Mountains of California. He followed the enormous tunnel through solid rock and came into a large, cavern-like room lined with tempered copper. Gold shields and other artifacts hung upon the walls. Strange drawing, undecipherable hieroglyphics, and the skeletons of giant humans were discovered in other rooms. (<u>THE UNDER-PEOPLE</u>, p. 147)

Also in THE UNDER-PEOPLE, Eric Norman quotes the story of a woman, Mrs. Margaret *"Maggie"* Rogers who was taken into the cavern cities of the *"Nephli"* civilization who live in giant caverns near Mexico City. In the three years she lived with the Nephli people, she was shown the way of life of this race of giants, who are highly advanced in technology; who have mastered space travel and claim to have colonies on other planets, and their religion in which they claim to know and communicate directly with God.

Mrs. Rogers insists her story is true. She writes,

"I have been cajoled, tempted, even threatened, in an effort to make me tell what I know. It is futile...This is my story, a vindication of my friends, the Nephli, and a Tribute to Tamil (their God)."

An interesting assignment was given to Maggie before she was brought again to the surface: She was told to seek out those who had Nephli blood and inform them of their great heritage and fellow race in the caverns below. To Maggie they said,

"You will remember everything. You will say nothing, though, until the time is ripe. Then you will tell just what we tell you to say. The truth. From that truth you will tell, you will find five of the undiluted blood of the Nephli, many who have a strain of Nephli mixed with surface who will eventually remember, or who will dream and in dreams be shown their heritage."

Coming to a knowledge of their heritage will be a great occasion. As Maggie described,

181

"The next morning, or I should say, the end of sleep-time, my friends took me to the room called Tamion, There I saw the three new soon-to-be residents of Nephli-land. There were two women and one man. The man looked like a German and the two women like Mexicans. Judging from the expressions on their faces, they were very happy about the whole thing. We only stayed for a moment inside, long enough to see them lie down in front of a tall stone. At first glance, the stone seemed to be a shaft of granite, but then I could see that a soft rosy glow made it nearly transparent."

"Sixteen hours later we went back and those three, who had entered old, wrinkled, gray and worn, came forth young, beautiful and strong. They were forthwith taken to another room, the enlarging room. I would say it was two hours they stayed there, and although I am not by nature a curious person, I was all agog with excitement, for I wished to be assured it was true and that I would some time be able to do the same."

"When they came out they were as large as Arsi and Mira." (about 12 feet tall) (THE UNDER-PEOPLE, pp. 58, 59)

Another civilization was discovered in the caverns under Europe by Sir Bulwer Lytton in the 19th Century. From an opening in a mine shaft he found his way into the cavern civilization of the Vril-ya where he was taken in and lived for one year. Their civilization, consisting of (at that time) a million and a half small cities of about 50,000 inhabitants, is both technologically and spiritually highly advanced. They have discovered a power they call VRIL, which according to Lytton, they command by the means of a small tube or staff which every citizen of their nation constantly carries with him.

Writes Lytton,

"It can destroy like the flash of lightning; yet differently applied, it can replenish or invigorate life, heal, and preserve, and on it they chiefly rely for the cure of disease, or rather for enabling the physical organization to cure itself. By this agency they rend way through the most solid substances, and open valleys for culture through the rocks of their subterranean wilderness. From it they extract the light which supplies their lamps, (since they live close to the surface in the dark zone) finding it steadier, softer, and healthier than the other inflammable materials they

had formerly used." (VRIL, THE POWER OF THE COMING RACE, p. 55)

Their civilization claims to be antediluvian in which they entered the caverns of the earth's crust in order to escape the inundation of their surface world before the final big flood which covered the earth's entire surface at the time of Noah. Following them into the earth's cavities were the antediluvian reptiles and creatures which on the surface have become extinct but which Lloyd, in ETIDORPHA, also reported still live in the caverns of the earth's shell.

The Vril-ya civilization is highly advanced in science and transportation. Nevertheless, they had no knowledge of our surface world other than the ancient traditions of their forefathers. They concerned themselves exclusively to their way of life. The city which Lytton found was in a giant cavern. Surrounding the city of 12,000 families was agriculture land, a lake and tunnels communicating between cities in caverns miles away. At all times the caverns and tunnels were lighted with lamps powered by Vril.

As Lytton described them, they lived in…

"…communities of moderate size. The tribe amongst which I had fallen was limited to 12,000 families. Each tribe occupied a territory sufficient for all its wants, and at stated periods the surplus population departed to seek a realm of its own…What we call crime was utterly unknown to the Vril-ya; and there were no courts of criminal justice. The rare instances of civil disputes were referred for arbitration to friends chosen by either party, or decided by the Council of Sages…But though there were no laws, such as we call laws, no race above ground is so law-observing. Obedience to the rule adopted by the community has become as much an instinct as if it were implanted by nature."

"Poverty among the Ana is as unknown as crime…none had become absolutely poor…If they did so, it was always in their power to migrate, or at the worst to apply, without shame and with certainty of aid, to the rich; for all the members of the community considered themselves as brothers of one affectionate and united family." (VRIL, pp. 56-60)

Lytton's description of this civilization compares closely to the spiritual-economic order set up by Joseph Smith by commandment of Jesus Christ in 1834. Called the United Order, its purpose was to,

"...provide for my saints, because all things are mine. But it must needs be done in mine own way, and, behold, this is the manner in which I, the Lord, have decreed to provide for my saints, that the poor be exalted, in the which the rich are made low... Therefore, if anyone partakes of the abundance I have created, and does not impart a portion to the poor and the needy, in conformity to the law of the gospel, from hell will he lift his eyes with the wicked, being in torment." (D&C 104:15, 18)

And it will be remembered, that each city of the United Order, as envisioned by Joseph Smith, was to be a small city of 20,000 to 50,000 people surrounded by their farms and factories much as the Vril-ya civilization functions.

The Vril-ya civilization is, in fact, a religious and righteous people. Continues Lytton,

"...divorces and polygamy are extremely rare, and the marriage state now seems singularly happy and serene among this astonishing people...It will be observed that in the relationships of the sexes I have spoken only of marriage, for such is the moral perfection to which this community has attained, that any illicit connection is as little possible amongst them as it would be to a couple of linnets during the time they agreed to live in pairs." (VRIL, pp. 72-73)

Of their religion Lytton says,

"This people have a religion...it has these strange peculiarities; firstly, they all believe in the creed they profess; secondly, that they all practice the precepts which the creed inculcates. They unite in worship of the one divine Creator and Sustainer of the Universe...they offer their devotions both in private and public...The Vril-ya unite in a conviction of a future state, more felicitous and more perfect than the present." (VRIL, pp. 89, 90)

That the earth's crust is literally full of inhabitants is the conclusion given us by Lytton. He wrote,

"And according to all the accounts I received, vast tracts immeasurably deeper beneath the surface, and in which one might have thought only salamanders could exist, were inhabited by innumerable races organized like ourselves."

As one of the Vril-ya citizens told Lytton,

"Wherever the All-Good builds," said she, *"there, be sure He places inhabitants. He loves not empty dwellings."* (VRIL, p. 75)

In a similar manner did Isaiah write,

"For thus saith the Lord that created the heavens; God himself that formed the earth and made it: he hath established it, he created it not in vain, HE FORMED IT TO BE INHABITED..." (Isaiah 45:18)

And other scriptures would seem to indicate that people do indeed live within the confines of the earth's crust -- a world of caverns.

"This shall be the sound of his trump, saying to ALL PEOPLE, both in HEAVEN (the paradise inner-sun) *and IN EARTH,* (the cavern people) *and THAT ARE UNDER THE EARTH...*(the hollow interior)" (D&C 88:104)

And there are other civilizations of cavern people.

So when the Conspiracy teaches that the earth is overpopulated and no room for more, we should listen to God. *"He loves not empty dwellings."* He formed the earth *"...to be inhabited."* There truly is space *"...enough and to spare"* within the confines of the earth's shell. And when the Ten Tribes come down out of the North Countries, we will come to a knowledge of those great expanses within the earth's shell wherein literally millions of God's children dwell. God will send His fishers and he will fish them out of the *"holes in the rocks"* so that we may come to know them and enjoy their brotherhood in the gospel of Jesus Christ. Then will the Kingdom of God truly fill the entire earth.

CHAPTER EIGHT
The Celestial Destiny of Our Hollow Earth

The great galaxies and stars which come to our view with the help of modern telescopes have truly shown us the immensity and beauty of God's great handiwork. Christ, the creator of heaven and earth said,

> *"Behold, all these are kingdoms, and any man who hath seen any or the least of these hath seen God moving in his majesty and power."* (D&C 88:47)

An understanding concerning our own earth's creation might therefore be enlightened by a look at the stars and planets in the heavens.   Perhaps the most obvious of God's creations which illustrates the hollow nature of planets are the planetary nebulae.

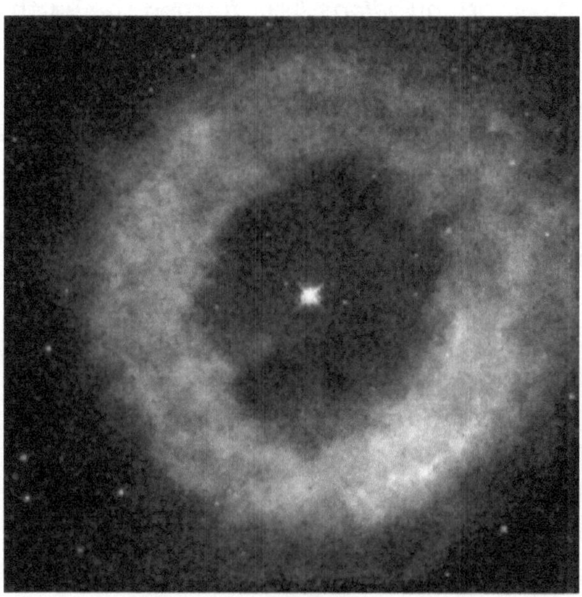

Planetary Nebula NGC 6369 taken by the Hubble Space Telescope

Marshall B. Gardner, in his JOURNEY TO THE EARTH'S INTERIOR, accumulated much astronomical evidence showing that all planets are hollow creations. Gardner quoted from H.D. Curtis of the Astronomical Society of the Pacific in an article in SCIENTIFIC AMERICAN on October 14, 1916:

"Fifty of these nebulae have been studied photographically with the Crosly reflector, using different lengths of exposure in order to bring out the structural details of the bright central portions as well as of the fainter, outlaying parts. MOST PLANETARY NEBULAE SHOW A MORE OR LESS REGULAR RING OR SHELL STRUCTURE, GENERALLY WITH A CENTRAL STAR."

Even as planetary nebulae are hollow with central stars, all planets are also hollow with central suns. In the creation process, planets are formed in rotation. Matter is placed into orbit around a central sun forming an outer shell and a hollow interior. And since the rotation in the polar regions throws matter away from the axis of rotation, polar openings into the interior are developed from the very beginning of the planetary formation.

So when God said in the beginning, *"Come, let us go down, for there is matter unorganized, from which we can build an earth whereupon these may dwell,"* they, the Gods, caused matter to go into rotation around a central sun bringing about the formation of Our Hollow Earth.

In forming the earth, physical matter was controlled and governed by spirit matter. Therefore, a hollow spirit world with a central paradise-sun was organized from spirit matter upon which the physical matter was superimposed. So explained the guide to I-AM-THE-MAN in Lloyd's ETIDORPHA:

"Matter has no strength, matter obeys spirit, and spirit dominates all things material. Energy in some form holds particles of matter together, and energy in other forms loosens them...The spirit that pervades all material things gives to them form and existence. Take from your earth its vital spirit, the energy that subjects matter, and your so-called adamantine rocks would disintegrate, and sift as dust into the interstices of space." (ETIDORPHA, p. 253)

This, in fact, will occur at the end of the earth's temporal existence when it will die. With its spirit taken out, the earth will disintegrate and then be brought forth in its final resurrected status by Jesus Christ.

In a revelation to the prophet Joseph Smith, in 1832, concerning this earth, Christ said, *"...wherefore, it shall be sanctified; yea, notwithstanding IT SHALL DIE, IT SHALL BE QUICKENED AGAIN, and shall abide the power by which it is quickened, and the righteous shall inherit it."* (D&C 88:26)

It is the earth's spirit which holds it together. And upon which the physical earth was founded.

The guide in ETIDORPHA, again well explained,

"The earth forming principle consists of an invisible sphere of energy that, spinning through space, supports the space dust which collects on it, as dust on a bubble. By gradual accumulation of substance on that sphere a hollow ball has resulted, on the outer surface of which you have hither to dwelt. The crust of the earth is comparatively thin, not more than eight hundred miles in average thickness, and is held in position by the central sphere of energy (center of gravity, or central sphere of gravity) *that now exists at a distance about seven hundred miles beneath the ocean level."* (ETIDORPHA, p. 193)

In the Genesis account of creation we find a perfect fit for the hollow earth theory.

The Lord spoke to Moses saying,

"In the beginning, God created the heaven and the earth."

Notice that *"heaven"* here is singular. Our Hollow Earth does have a heaven. It is a central sun inside the hollow of the earth. It is also the physical location of Paradise in the spirit world of this earth.

"And the earth was without form and void."

The earth was made out of space dust, rocks, and gases.

"And darkness was upon the face of the deep."

The space dust, rock and gases were brought together as a hollow earth in rotation about a central mass thereby shutting out star light.

"And the Spirit of God moved upon the face of the waters."

Water was placed on the planet and the oceans were made to cover the entire surface of the globe.

"And God said, Let there be light; and there was light."

The central mass was turned on, and became the interior sun producing light.

"And God saw the light, that it was good; and God divided the light from the darkness. And God called the light Day, and the darkness he called Night. And the evening and the morning were the FIRST DAY."

The interior sun was given the required brightness and heat to make life conditions on the earth's interior surface possible. He then divided the interior sun. As Olaf Jansen reported, the inner sun on one side emanates white light for day and the other side is a reddish-brown for night.

"And God said, Let there be a firmament in the midst of the waters, and let it divide the waters from the waters. And God made the firmament and divided the waters which were under the firmament from the waters which were above the firmament; and it was so. And God called the firmament Heaven. And the evening and the morning were the SECOND DAY."

In Hebrew, firmament meant *"expanse."* The hollow, inside our earth is an *"expanse,"* or firmament. The waters above and below the firmament referred to the oceans covering the entire inner surface. Therefore, the firmament, or expanse was literally in the *"midst of the waters."* And this firmament God called *"Heaven."*

Without an understanding of the Hollow Earth, it is impossible to understand how the Lord divided the waters *"above"* the Heaven from the waters *"below"* the Heaven. This fits the Hollow Earth perfectly since the ocean on the opposite side of the hollow interior of the earth is directly *"above"* a person standing on the other opposite side, and suspended in the hollow expanse of the earth is the interior sun or *"Heaven"* of this earth.

Creation scientists have speculated as to how there could exist waters *"above"* the firmament in the creation account, which *"firmament"* they have interpreted to mean *"sky."* They have surmised that a pre-flood ice canopy surrounded the earth several miles above the earth's surface and that at the time of Noah's flood that it was perhaps broken by asteroid hits and that the ice then fell into the atmosphere vaporizing and condensing into the rain that caused the great flood. Their theory, however, by their own admission has many problems which they have been unable to resolve. For example, what of all the asteroid impacts whose craters have been found all over the world and that most certainly existed before the flood? Any of these asteroids or meteors impacting the Earth would have broken up their ice canopy before the flood of Noah. The Hollow Earth theory gives the source of the waters of Noah's flood as the Inner Ocean sloshing out through the polar openings and pouring out on to the surface from underground lakes caused by upheavals of the close passage of a planet sized comet.

The Hollow Earth theory, on the other hand, fits the *"waters above"* the firmament much better. The interior sun and hollow of the earth on the first and second days of creation certainly was in the *"midst"* of the waters and there were waters above it and below it and towards the sides of it on the inner surface of the planet.

In Genesis 1:9,

"And God said, Let the waters under the heaven be gathered into one place, and let the dry land appear: and it was so."

In the scriptural accounts, on the Third day of the earth's creation, God caused land to come up out of the ocean. It is only with a knowledge of Our Hollow Earth that a person can understand how there could be a continent stretched out *"above the waters"* as spoken of in a song of the Book of Psalms, 136:6,

"To him that stretched out the earth above the waters: for his mercy endureth for ever."

Olaf Jansen reported that inside Our Hollow Earth there is one continent that extends three-fourths the distance around the interior surface with one-fourth being ocean. This scripture from the Psalms then would make sense to a person on a ship on the inner earth ocean because he could look up and see the inner continent stretching 6,400 miles above his head on the opposite inner surface of the planet. The inner continent would literally appear as stretched out *"earth above the waters"* of the inner ocean.

Not only would the inner continent be above the waters, but the inner continent would also be above the heaven.

Since the heaven was in the midst of the waters, with waters above the firmament and waters below the firmament, when the waters were gathered into one place UNDER the heaven or firmament, then the dry land that came up out of the waters had to have come out of the waters located ABOVE the firmament or heaven. Such could only occur in a hollow earth.

"And God said, Let the waters under the heaven be gathered together into one place, and let the dry land appear; and it was so. And God called the dry land Earth; and the gathering together of the waters called he Seas; and God saw that it was good."

Again *"the heaven"* is singular, referring to the hollow interior, expanse, or firmament that includes the Inner Sun.

Dry land was caused to appear, leaving one ocean and one continent. The interior, reported Olaf Jansen, has one ocean and one continent.

"And God said, Let the earth bring forth grass, the herb yielding seed, and the fruit tree yielding fruit after his kind, whose seed is in itself, upon the earth; and it was so."

"And the evening and the morning were the THIRD day."

Since the earth was created within the time reckoning of the Lord, whose home planet *"Kolob"* has a day lasting 1,000 earth years (Abraham Chapter 3, 2 Peter 3:8), it wasn't until the fourth day, or fourth thousand years that God put the earth into orbit around our sun.

This then, strongly indicates that all plant life on this earth began in the earth's hollow interior, on the THIRD day, where the interior sun had now been shining for 2000 years, the first two days of the creation period, building up a soil suited for good plant growth.

Olaf Jansen reported that the primeval Garden of Eden is located on the continent inside Our Hollow Earth. William F. Warren, although he was unaware of the polar openings and the hollow nature of the earth, devotes many chapters of his book PARADISE FOUND, OR THE CRADLE OF THE HUMAN RACE AT THE NORTH POLE, quoting scientific evidence that all plant and animal life originated from the north ... which possibly could have migrated from the interior through the polar openings.

"And God said, Let there be lights in the firmament of the heaven to divide the day from the night; and let them be for signs, and for seasons, and for days, and years; And let them be for lights in the firmament of the heaven to give light upon the earth; and it was so.

Olaf Jansen described how the Inner Sun is divided between its day and night sides, the day side giving off a white light and the dark side giving off a dark reddish-brown light, and in the center of the night side of the Inner Sun, Olaf described a large circular area that is opaque with holes letting white light through that shine like stars at night.

"And God made two great lights; the greater light to rule the day, and the lesser light to rule the night; he made the stars also. And God set them in the firmament of the heaven to give light upon the earth, And to rule over the day and over the night, and to divide the light from the darkness; and God saw that it was good. And the evening and the morning were the FOURTH day."

Now it was time to create a favorable environment on the earth's external surface to which the plant life in the interior could spread. The earth was then placed into orbit around our sun and the moon was placed into orbit around the earth, thereby creating days, years and seasons.

OBVIOUSLY, THE SUN THAT LIT UP ON THE FIRST DAY OF CREATION IS DIFFERENT FROM THE SUN THAT WAS SET IN THE HEAVEN ON THE FOURTH DAY.

Remember that it was on the third day that plant life was placed on the earth, but not until the fourth day (fourth thousand years) that the earth was placed in orbit about the Sun. And yes, the scripture would seem to indicate that there are TWO heavens in which the TWO suns are set. Verse one of Chapter 2 says, *"Thus the HEAVENS and the earth were finished..."*--one heaven in the planet's interior and one on the exterior, heaven being defined as *"space"* or *"expanse."*

The scriptural accounts have the Lord place plant life on the earth on the THIRD day with the Sun, Moon and stars placed in the sky on the FOURTH day.

Without an understanding of the hollow nature of our planet, it is impossible to understand how plant life could be planted on the THIRD day before the Sun, Moon and stars are placed in the sky on the FOURTH day as all three scriptural accounts of the creation relate (Genesis of the Bible, the Book of Moises, and the Book of Abraham). But when one understands that on the first day of creation the Lord lit up a sun on the interior of the Earth, *"Let there be light; and there was light,"* then it would be possible for Him to plant vegetation on the THIRD day on the Earth's interior continent where the interior sun had now been shining for two thousand years, two of the Lord's days.

With an understanding of the Hollow Earth, the Sun placed in the sky on the FOURTH day obviously was different than the Light that lit up on the first day of the Earth's creation.

The scriptural accounts fit the Hollow Earth viewpoint because they describe the creation as God sees it. And thus we see that the scriptural accounts of the earth's creation contribute to the evidence that our Earth is Hollow!

Where was the earth prior to being placed into orbit around our Sun? The Hollow Planet theory would suggest that perhaps planets are *"born"* from the *"wombs"* of larger planets.

Immanuel Velikovsky wrote that,

"The Greek authors described the birth of Athene (which became the planet Venus), *saying she sprang from the head of Jupiter."* (WORLDS IN COLLISION, p. 175)

If the planet Jupiter is hollow as our theory maintains that ALL planets are hollow, then the *"head"* or north polar area would be the logical place for a planet to be *"born"* or ejected --

out of its polar opening. Such an ejection could be effected by manipulating the electromagnetic fields of the two planets.

A perplexing discovery was made by Apollo astronauts when they brought moon rocks back to earth to be analyzed. Prior to the Apollo moon shots, scientists had assumed that the moon was made out of the same materials as the earth. Yet the rocks brought back from the moon were found to consist of materials very different from those found on earth, containing very dense rare earth minerals. Testing also determined that the moon is much older than the earth.

A prominent LDS geologist, Eric N. Skousen, (son of the famous LDS author, Cleon Skousen) believes this indicates that the moon may have been here first. According to his theory, the moon originally possessed earth's orbit about the sun BEFORE the earth was placed here, and that the earth was created somewhere else INSIDE another planet in an actual planetary birth process.   (See, pp. 92-96, EARTH, IN THE BEGINNING, by Dr. Eric N. Skousen)   Dr. Skousen quotes one of the earlier LDS church leaders, Heber C. Kimball, as asking, *"Where did the earth come from?"* To which he answered his own question: *"From its parent earths."* (November 8, 1857, JOURNAL OF DISCOURSES, 6:36)

Dr. Skousen proposes that the earth could have been ejected out of the planet where it was built by the Gods, and then was guided into our solar system and placed into the moon's orbit about the sun in close enough proximity that it caused the moon to go into permanent orbit about the earth.

This would also explain the paradox creation scientists have been faced with regarding star light. Since creation scientists believe the earth and the stars were created in six 24-hour earth days, they are faced with the dilemma of how to explain how the light of the stars light years away suddenly became apparent on the fourth day of creation. The light from the stars would take many years to arrive on earth, so how could the Lord *"place"* the stars in the sky on the fourth day of creation? This is simply resolved when we realize that the light from the stars was already here and just became apparent on the fourth day of creation when the earth was placed in orbit about our outer Sun after being expelled from its birth place from within another larger planet.

And if planets are born of larger planets, perhaps at least the spirit world of this earth was born of Kolob which is the star-planet of God -- since the time reckoning of the earth's creation period was computed according to that of Kolob.

The Lord declared to Abraham who wrote,

"And I Abraham, had the Urim and Thummim, which the Lord my God had given unto me, in Ur of the Chaldees; And I saw the stars, that they were very great, and that one of them was NEAREST UNTO THE THRONE OF GOD; and there were many great ones which were near unto it; And the Lord said unto me, by the Urim and Thummim, that Kolob was after the manner of the Lord, according to its times and seasons in the revolutions thereof; that one revolution was a day unto the Lord, after his manner of reckoning, it being one thousand years according to the time appointed unto that whereon thou standest. This is the reckoning of the Lord's time, according to the reckoning of Kolob." (ABRAHAM, The Pearl of Great Price, 3:1-3)

Scientists say that the center of our galaxy The Milky Way, is in the constellation of Sagittarius. That there are so many stars towards the center that it cannot be seen within the human visibility range of light. But there is no doubt that the largest stars would be in the center, as Abraham saw them. In the explanation of Facsimile No. 2 in the book of Abraham, Figure 2 says,

"Stands next to Kolob, called by the Egyptians, Oliblish, which is the next grand governing creation near to the celestial or the place where God resides...Fig 4...the time of Oliblish, which is equal with Kolob in its revolution and in its measuring of time."

Therefore, as revealed through Abraham, at the center of our Milky Way Galaxy which is in *"the same order* (in the same galaxy) *as that* (Earth) *upon which thou standest,"* are two stars of gigantic size, in rotation about a common center of gravity -- a double star system with the rest of the galaxy rotating about them. Both Kolob and Oliblish are the same size and take 1,000 of our earth years to complete one revolution or day. Oliblish *"...is the next grand governing creation near to the celestial or the place where God resides..."* and Kolob is the *"...nearest unto the throne of God."* Therefore, I conclude that the throne of God is the interior sun INSIDE Kolob and Kolob is HOLLOW as are all stars and planets. In fact, the stars ARE planets -- the planets of the Gods.

Since a hollow planet or star essentially consists of matter in orbit about its central sun, perhaps we could use Kepler's 3rd Law of Planetary Motion to calculate the size of Kolob. Using Kepler's 3rd Law of Planetary Motion, we can use the following formula to calculate orbital radii,

R=VxT
where
R = a planet's distance from the sun in Astronomical Units (AU). One AU equals 93,000,000 miles, the distance from the earth to the sun.

T = the time for a planet to orbit the sun in earth years

V = velocity is 1 divided by the square root of R.

Knowing this formula and the time it takes for one revolution of Kolob,

100 AU = 1/10 earth velocity X 1000 years

From this, we can deduce that the distance to Kolob's outer surface from the center of its inner sun, *"...the celestial or the place where God resides..."* would be 100 astronomical units, or 9.3 billion miles. That is 2.5 times the distance from the sun to the planet Pluto. The diameter of Kolob and its companion double star then, would be about 18.6 billion miles -- plenty of space could be found within its hollow interior for the creation of planets.

Image of Mars taken by the Hubble Space Telescope, showing the North Polar Opening where the polar rim is clearly visible with a few clouds down within the depression:

Other worlds are also hollow creations.

Marshall B. Gardner revealed evidence that Mars, Venus and Mercury are also hollow worlds with polar openings through which astronomers have seen glimmers of their central suns.

Gardner quotes Professor Percival Lowell in his book, MARS:

"Meanwhile an interesting phenomenon occurred in the cap (polar cap) on June 7 (1894). On that morning at about a quarter to six, (or, more precisely, on June 8, 1 hour, 17 minutes, G.M.T.), as I was watching the planet, I saw suddenly two points like stars flash out in the midst of the polar cap. Dazzlingly bright upon the duller white background of the snow, (actually clouds) these stars shone for a few moments and then slowly disappeared. The seeing at the time was very good." (MARS, p. 86)

Director O.M. Mitchell of the Cincinnati and Dudley Observatories, made a similar observation of Mar's central sun shining through its polar opening. He wrote in his book, <u>A CONCISE ELEMENTARY TREATISE ON THE SUN, PLANETS, SATELLITES, AND COMETS</u>, the following:

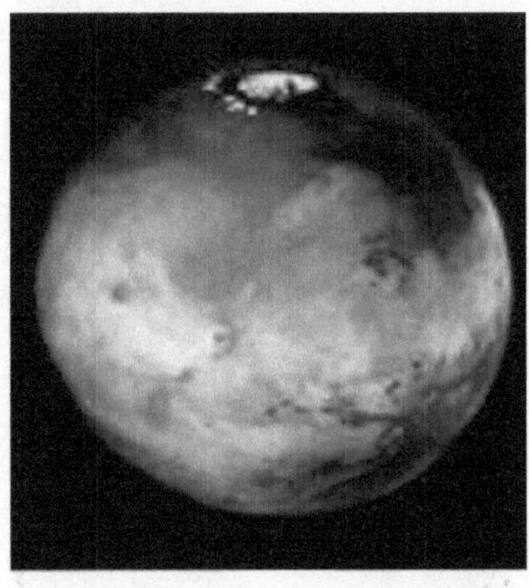

"On the evening of the 30th of August (1845), I observed, for the first time, a small bright spot, nearly or quite round, projecting out of the lower side of the polar spot. In the early part of the evening the small bright spot seemed to be partly buried in the large one...After the lapse of an hour or more, my attention was again directed to the planet, when I was astonished to find a manifest change in the position of the small bright spot...(caused by a shift in the cloud position within the polar opening). *In the course of a few days the small spot gradually faded from sight and was not seen at any subsequent observation."* (The planet's orbit changed the angle of observation.)

In his book, THE GREATEST CHALLENGE: THE INCREDIBLE ADVENTURE AND SPLENDID DESTINY OF MAN IN EXPLORING SPACE, Martin Caidin wrote,

"Both American and Russian Astronomers in recent years have observed a series of very bright flashes, lasting about five minutes, and followed by mushroom-shaped clouds..." emanating from Mar's polar area.

The clouds which come out of Mar's polar opening at times pile high above the planet's surface and the interior sun shining on them appear to project beyond the planet's surface. English Astronomer, J. Norman Lockyer in 1892 reported:

"The snow-zone was at time so bright that, like the crescent of the young moon, it appeared to project beyond the planet's limb. This effect of irradiation was frequently visible; on one occasion the snow spot was observed to shine like a nebulous star..." (Gardner p. 85)

Obviously, this astronomer didn't know how to interpret his observation of Mar's interior sun shining through the polar opening, but called it a snow spot that shone like a star.

A French astronomer, Trouvelet also observed lights coming from Venus' polar opening which he interpreted as reflections of polar ice, which scientists have since proven to be impossible for Venus' hot atmospheric temperature. Wrote Trouvelet:

> *"Their surface is irregular, and seems like a confused mass of LUMINOUS POINTS separated by comparatively sombre intervening spaces* (clouds). *This surface is undoubtedly very broken, and resembles that of a mountainous district studded with numerous peaks, or our polar regions with numerous ice needles BRILLIANTLY REFLECTING THE SUNSHINE."* (Gardner p. 95)

Astronomer Richard A. Proctor also reported his observation of the planet Mercury's inner sun shining through its polar opening:

> *"It has been supposed that a certain bright spot seen in the black disc of Mercury when the planet is in transit, indicates some sort of illumination either of the surface of the planet or in its atmosphere...the bright spot supposed to belong to Mercury has been seen when the strongest darkening-glasses have been employed. But there can be no manner of doubt that the bright spot is an optical phenomenon only."* (Gardner, pp. 96, 97)

As Mercury passes in front of the sun, in its dark, round disc is the bright spot-light of its interior sun shining at us through its polar opening. Scientists pass off this bright spot supposing it to be only an optical illusion since they don't believe in the hollow planets theory.

On page 22 of the March-April 1995 issue of FINAL FRONTIER, we find these comments on the planet Mercury:

> *"Temperatures that climb as high as 800 degrees Fahrenheit ... Researchers at the California Institute of Technology in Pasadena have identified what they believe is a water ice cap more than 180 miles in diameter on Mercury's North Pole. The researchers saw a bright area at the north pole...'We were amazed.'"*

Obviously, the researchers were not seeing a polar ice cap at 800 degrees Fahrenheit, but instead they were actually observing Mercury's north polar opening from which its interior sun is shining.

Evidence confirms that even our earth's moon is hollow.

Commenting on the different specific gravities of our moon and the earth, Marshall B. Gardner wrote in 1920,

"Only on our theory that both the moon and earth are hollow can this difference be explained." (Gardner, p. 376)

Gardner's prediction that our moon is also hollow has been confirmed by the Apollo shots to the moon. But even before the astronauts were sent to the moon, a NASA scientist published a report in the July issue of Astronautics of 1962 saying that studies of the moon indicated that that planet is hollow. He wrote:

"If the astronomical data are reduced, it is found that the data require that the interior of the Moon be less dense than the outer parts. Indeed, it would seem that the Moon is more like a hollow than a homogenous sphere." (ASTRONAUTICS, July 1962, pp. 14-15)

Since it is the accepted doctrine of scientists that no planet is hollow, Dr. Gordon McDonald went on to say in his article that his conclusion must somehow be wrong.

However, McDonald's conclusion that our moon is hollow was reinforced by further experiments on the moon. Apollo 12 astronauts set up very sensitive seismometers in November, 1969, on the Moon's Sea of Storms and NASA then guided various stages of the Apollo booster rockets into collision courses with the moon. The vibrations set up by the impact, as recorded by the seismometers, were very similar to that of a bell. At first the vibrations were large. Then they tapered off and finally died off after 3 hours, indicating a high metallic content in the moon's shell.

Also, with the seismometers set up at different locations on the moon's surface, by the Apollo 14 and 15 missions, they were able to record the vibrations of subsequent bombardments of the moon's surface as traveling down into the moon's crust. They were recorded as traveling down 15 miles into the crust whereupon they picked up speed and traveled at the speed they would travel through metal another 45 miles down at which point they bounced back to the surface, indicating that they had reached the inside surface of a 60-mile thick shell. (OUR SPACESHIP MOON, pp. 99-103)

Clementine Probe Composite Photo of the Moon's South Pole:

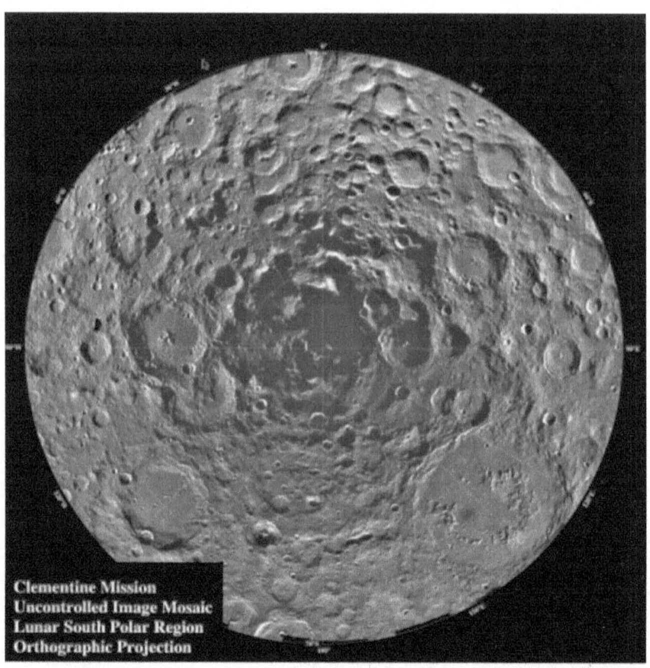

Could the dark area at the moon's South Pole be a polar opening instead of just ice as NASA claims?

The Apollo missions to the moon proved that the moon's shell contains a much greater percentage of metals than does the earth. Since the moon is hollow, as is the earth, and its metallic content is greater than earth's, the specific gravity of the moon should be greater than earth's. Newtonian physics gives the earth a density of 5.5 and the moon 3.3. And yet the high metallic content of the moon rocks brought back from the moon and the results from the seismic tests showing the moon to "ring" like a bell with a period of over three hours, whereas the earth "rings" with a period of 54 minutes, this may indicate that the moon's shell is much denser than the earth's.

If we consider that our Hollow Earth has a shell thickness of 10% of the earth's 8,000 mile diameter for a shell thickness of 800 miles and a shell volume of 50% of the Earth's total volume, the density of the earth's shell would be 11.05 gm/cc. Since NASA discovered that the moon has a shell thickness of only 60 miles, which is only 2.78% of the moon's diameter, this gives the moon's shell a density of 21.256 gm/cc, which is much denser than earth's shell. This gives the Moon a greater planetary density of (0 gm/cc + 21.256 gm/cc)/2 = **10.628 gm/cc** than the Earth (0 gm/cc + 11.05 gm/cc)/2 = **5.5 gm/cc**. Thus the hollow nature of the moon gives it a shell density consistent with the seismic tests carried out by NASA astronauts and the moon rocks brought back to earth.

If these observations are correct, the hollow interior of the moon could be 2,040 miles across. And if the moon contains, in its interior, an atmosphere one-sixth the size of earth's 26-mile high atmosphere, taking into account the moon's gravity of one-sixth that of the earth's -- its mass is .01232 of the earth's mass -- that would leave 2,031 miles of pure space within the moon's hollow interior wherein a small sun could be suspended to give light and life. There could exist there, as well as in the interior worlds of all the planets, civilizations of God's children living in ecologically balanced environments. In fact, evidence of the possibility of such life existing in the moon's interior is the occasional observation by the instruments left on the moon of escaping clouds of water vapor from certain points on the moon's surface. (OUR SPACE SHIP MOON, pp. 105-107)

To look into space and see the many worlds in just our own solar system, it is hard to believe we are alone, that they are all dead worlds just because we cannot detect any life on their surfaces. However, to discover that they are all hollow and habitable within is consistent with the testimony of the prophet Joseph Smith concerning Jesus Christ and his creations, when he wrote:

"And now, after the many testimonies which have been given of him this is the testimony, last of all, which we give of him: That he lives!"

"For we saw him, even on the right hand of God; and we heard the voice bearing record that he is the Only Begotten of the Father--"

"That by him, and through him, and of him, THE WORLDS ARE AND WERE CREATED, AND THE INHABITANTS THEREOF ARE BEGOTTEN SONS AND DAUGHTERS UNTO GOD." (D&C 76:22-24)

Nevertheless, we do not need to suppose that we have to go to the moon or the other planets to come to a knowledge of their civilizations. The Israelite nation of the Ten Tribes living within our own earth, as also the Nephli civilization living in the cavern cities under America have already established contact, communication and even shuttle service to the planets and civilizations of our solar system.

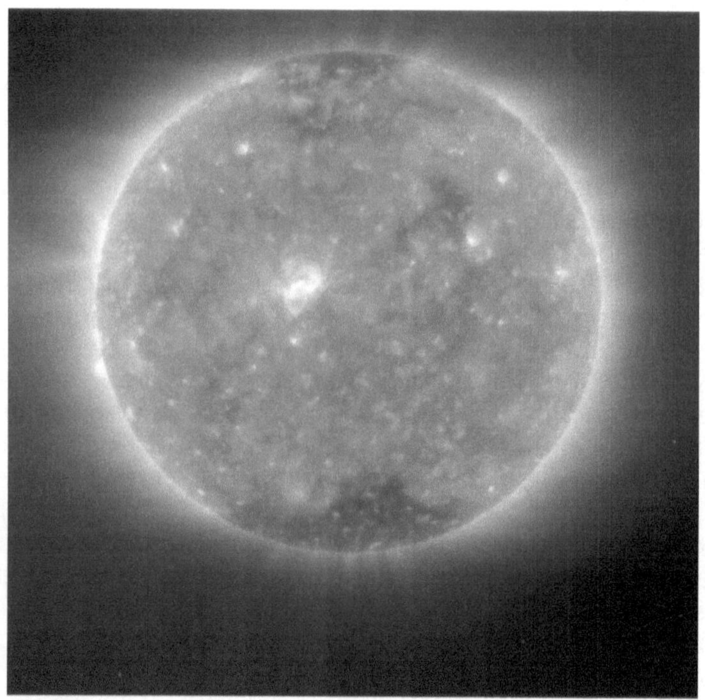

X-ray and Ultraviolet images of the Sun show polar coronal holes as enduring features

But, there is one civilization to which they cannot establish a shuttle service. That civilization which lives inside our HOLLOW SUN has certain requirements for entrance. In fact, the prophets have declared that God has prepared this earth to become a celestialized body. As a sun it will shine forth in its resurrection as the stars in the heavens. We read from DOCTRINES OF SALVATION, by Joseph F. Smith, of the Celestial Destiny of our earth:

"The earth will be cleansed again. It was once baptized in water. When Christ comes, it will be baptized with fire and the power of the Holy Ghost. At the end of the world, the earth will die; it will be dissolved, pass away, and then it will be renewed, or raised with a resurrection. It will receive its resurrection to become a celestial body, so that they of the celestial order may possess it forever and ever. Then it will shine forth as the sun and take its place among

the worlds that are redeemed. When this time comes the terrestrial inhabitants (those living terrestrial laws) *will also be taken away and be consigned to another sphere suited to their condition. Then the words of the Savior will be fulfilled, for the meek shall inherit the earth."*

The Meek are those living celestial laws -- which includes baptism in the true church of Jesus Christ and subsequent righteous living throughout this life. Faith, repentance, baptism by an authorized servant of Jesus Christ and reception of the Gift of the Holy Ghost, and continued living *"by every word that proceedeth out of the mouth of God"* (Matt 4:4) are the requirements for entrance into the Celestial World.

Then President Smith, of the Church of Jesus Christ of Latter-day-Saints, continued:

"It is my opinion that the great stars that we see, including our sun, are celestial worlds; at least worlds that have passed on to their exaltation or other final resurrected status. This is in conflict, of course, with the teachings of scientific men, who declare that the sun is losing energy and gradually cooling off and will eventually be a dead world. I do not believe the Lord has any such thing in his plan. The Lord lives in 'everlasting burnings' we are informed. President Brigham Young has said that this earth when it is celestialized will shine like the sun, and why not?" (DOCTRINES OF SALVATION pp. 88, 89)

The phrase *"everlasting burnings"* referred to is located in Isaiah 33:14-16, which reads:

"The sinners in Zion are afraid; fearfulness hath surprised the hypocrites. Who among us shall DWELL WITH THE DEVOURING FIRE? WHO AMONG US SHALL DWELL WITH EVERLASTING BURNINGS?: He that walketh righteously, and speaketh uprightly; he that despiseth the gain of oppressions, that stayeth his hand from holding of bribes, that stoppeth his ears from hearing of blood, and shutteth his eyes from seeing evil: He shall dwell on high..."

The prophet Joseph Smith taught,

"The angels do not reside on a planet like this earth; But they reside in the presence of God, on a globe like a SEA OF GLASS AND FIRE, where all things for their glory are manifest, past, present, and future, and are continually before the Lord."

"The place where God resides is a great Urim and Thummim."

"This earth, in its sanctified and immortal state, will be made like unto CRYSTAL and will be a Urim and Thummim to the inhabitants who dwell thereon, whereby all things pertaining to an inferior kingdom, or all kingdoms of a lower order, will be manifest to those who dwell on it; and this earth will be Christ's." (D&C 130:6-9)

To those of us who aspire to a more glorious future beyond the grave, Apostle Orson Pratt wrote,

"Who, but the most abandoned, does not desire to be counted worthy to associate with those higher orders of beings who have been redeemed, exalted, and glorified together with the worlds they inhabit, ages before the foundations of our earth were laid? O man, remember the future destiny and the glory of the earth, and secure thine everlasting inheritance upon the same, that when it shall be glorious, thou shalt be glorious also." (MILLENNIAL STAR, Vol. 12, p.72)

The prophet Daniel asserted that celestial beings shine with the same brightness as the heaven or star in which they live when he wrote,

"And they that be wise shall shine as the brightness of the firmament; and they that turn many to righteousness as the stars for ever and ever." (Daniel 12:3)

Therefore, we may conclude that the stars, including our Sun, were once earths populated by the children of God, as we are, and that their earths have since been resurrected as celestial worlds together with those inhabitants worthy to become Gods who are resurrected with celestial bodies, and that this earth when celestialized will become a SUN.

Astronomers report celestial events that occur at a rate of about ten per year called Novae and are considered the birth of new stars. Novae always occur in binary systems, which indicates that the new star perhaps was a planet orbiting a star before lighting up as a Nova. Perhaps Novae are planets that die and then are resurrected to their final celestial status, but still continue to orbit their primary star after being resurrected. At the moment of resurrection, the new star emits a bright flash of light that gradually diminishes over the next few days and weeks to its regular steady magnitude of light emission.

In the book of Revelation, the last book of the Bible, we read of the death and resurrection of this earth and of the City of God, the New Jerusalem, descending out of heaven to occupy its position suspended in the hollow of the resurrected Earth.

In _Revelation Chapter 20:11_, John the Beloved speaks of the death of the earth, _"And I saw a great white throne and him that sat on it, from whose face the earth and the heaven fled away; and there was found no place for them."_ Notice the special mention of _"the earth and THE heaven."_ This _"heaven"_ is a direct reference to the interior sun of our Hollow Earth.

Then in Chapter _21:1-3_ is described the resurrection of the Earth and the Earth's Heaven,

"And I saw A NEW HEAVEN and A NEW EARTH: for the first heaven and the first earth were passed away; and there was no more sea. And I John saw the holy city, new Jerusalem, coming down from God out of heaven, prepared as a bride for her husband. And I heard a great voice out of heaven saying, Behold, the tabernacle of God is with men, and he will dwell with them, and they shall be his people, and God himself shall be with them, and be their God."

A description of the Celestial City, the New Jerusalem, is given with exact dimensions in verses _10-27 of Chapter 20_, and _Chapter 22:1-5_. John was carried away in the spirit to a great and high mountain where the Lord showed him...

"...that great city, the holy Jerusalem, descending out of heaven from God, having the glory of God: and her light was like unto a stone most precious, even like a jasper stone, clear as crystal; and had a wall great and high, and had twelve gates...on the east three gates; on the north three gates; on the south three gates; and on the west three gates. And the wall of the city had twelve foundations, and in them the names of the twelve apostles of the Lamb. And he that talked with me had a golden reed to measure the city, and the gates thereof, and the wall thereof. And the city lieth foursquare, and the length is as large as the breadth: and he measured the city with the reed, twelve thousand furlongs. The length and the breadth and the height of it are equal. And he measured the wall thereof, an hundred and forty and four cubits, according to the measure of a man, that is, of the angel."

The dictionary says a furlong is 220 yards. John, the Apostle said the city was 12,000 furlongs cube. This turns out to be a city that is 1500 miles tall, 1500 miles long and 1500 miles wide -- 1500 miles cube. More than likely, however, the shape of the city would be a four-sided pyramid, the cap stone being the dwelling place of God. Perhaps the ancient Egyptian, Chinese and Central American pyramids, even the pyramids of the whole world, are modeled after this future New Jerusalem in the sky. The walls of this city are 144 cubits thick. The dictionary definition of a cubit is 18-22 inches. An average would be 20 inches. This gives us a thickness for the walls of the New Jerusalem of 1/2 mile thick.

A city 1500 miles high on the surface of this planet would surely stick out like a sore thumb. The 99.9% of Earth's atmosphere today only extends up to about 26 miles with the last traces at 600 miles. This Celestial City, if placed on the surface of the planet, would extend 1,000 miles above the atmosphere into space! From the Hollow Earth perspective, the ideal place for the Lord to place his gigantic Holy City would be suspended in the hollow of the celestialized Earth. If the resurrected, celestialized Earth has the same dimensions as the Earth is now, the Holy City suspended in the hollow interior would not come within 2,450 miles of the inner surface of the Earth now become a sun.

The description of this fabulous city is further detailed by John in the Book of Revelation. He continued recording that,

> *"...the building of the wall of it was of jasper: and the city was pure gold, like unto clear glass. And the foundations of the wall of the city were garnished with all manner of precious stones...And the twelve gates were twelve pearls; every several gate was of one pearl: and the street of the city was pure gold, as it was transparent glass. And I saw no temple therein: for the Lord God Almighty and the Lamb are the temple of it. And the city had no need of the sun, neither of the moon, to shine in it: for the glory of God did lighten it and the Lamb is the light thereof."* (Revelation 21:18-23)

Of us who wish to enter therein John wrote,

> *"And there shall in no wise enter into it any thing that defileth, neither whatsoever worketh abomination, or maketh a lie: but they which are written in the Lamb's book of life."* (Revelation 21:27)

Concerning those who will be privileged to live on our earth when it becomes a sun and they become Gods and angels, Jesus Christ revealed to the prophet Joseph Smith, July 12, 1843:

> *"And again, verily I say unto you, if a man marry a wife by my word, which is my law, and by the new and everlasting covenant, and it is sealed unto them by the Holy Spirit of Promise, by him who is anointed, unto whom I have appointed this power and the keys of the priesthood; and it shall be said unto them--Ye shall come*

forth in the first resurrection; and shall inherit thrones, kingdoms, principalities, and powers, dominions, all heights and depths--then it shall be written in the Lamb's Book of Life, that he shall commit no murder whereby to shed innocent blood, and if ye abide in my covenant, and commit no murder whereby to shed innocent blood, it shall be done unto them in all things whatsoever my servant hath put upon them, in time, and through all eternity; and shall be of full force when they are out of the world; and they shall pass by the angels, and the gods, which are set there, to their exaltation and glory in all things, as hath been sealed upon their heads, which glory shall be a fullness and a continuation of the seeds (children) *forever and ever."*

"Then shall they be GODS, because they have no end; therefore shall they be from everlasting to everlasting, because they continue; then shall they be above all because all things are subject unto them. Then shall they be GODS, because they have all power, and the angels are subject unto them."

"Verily, verily, I say unto you, except ye abide my law ye cannot attain to this glory."

"For strait is the gate, and narrow is the way that leadeth unto the exaltation and continuation of the lives (posterity), *and few there be that find it, because ye receive me not in the world neither do ye know me."*

"But if ye receive me in the world, then shall ye know me, and shall receive your exaltation; that where I am ye shall be also." (D&C 132:19-23)

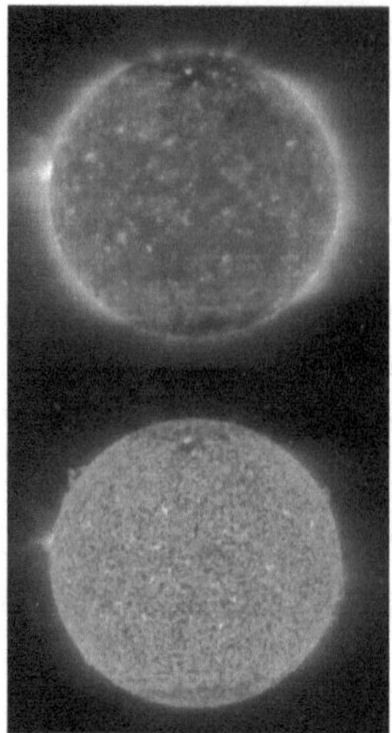

Where God now lives, is in a sun, because the stars are celestial worlds. And hollow worlds at that. X-ray pictures of the Sun show *"holes"* at the Sun's poles where no or very little x-rays are emitted. Nine months of Skylab observation of the Sun revealed coronal holes as enduring features at both poles as recorded from ultraviolet and x-ray emission indicating that the Sun has polar openings that lead into its hollow interior.

The Sun also has an electromagnetic field caused by the rotation of its crystal shell about the Sun's core suspended within its hollow interior. The Sun's electromagnetic field rotates once every 22 earth years, which indicates that the Sun's core rotates at this rate. And since the Sun's shell rotates once every 27.3 earth days, the faster rotation of the shell about the slower rotating core gives rise to the Sun's electromagnetic field.

If the Sun were an entirely gaseous planet, it could not produce an electromagnetic field. From an article in the Scientific American comes this significant confession:

"In 1934, Thomas G. Cowling of the University of Leeds in England proved that single, symmetrical fluid flows cannot generate magnetic fields...Astronomers cannot explain the galaxies or solar magnetic fields." (*Mystery of the Missing Dynamo*, Scientific American, p. 24, January 1995)

Because astronomers believe stars are gaseous and planets have liquid interiors they cannot explain the observed magnetic fields in suns or planets. Scientists teach that the Sun and stars and the giant outer planets of our solar system are composed entirely of gases.

However, if the Sun has a shell that is perhaps 10% of its diameter, I have calculated that its shell would be solid having a density of 2.86 gm/cc, which is a little denser than glass. As such, the Sun could be a giant crystal globe.

Confirming that the Sun has a solid surface is the scientific work of Michael Mozina. The evidence he presents on his website at:

http://www.TheSurfaceOfTheSun.com (now defunct) from solar observing satellite imagery is incontrovertible. For example, on Michael Mozina's website is a video taken by a solar observing satellite of a sunquake on the surface of the Sun showing a tsunami wave going out from the epicenter clearly passing over a mountain on the surface of the Sun, showing that the Sun has a solid surface covered by some type of liquid Mozina thinks is silicon.

The significance of this is that if the Sun has a solid surface, it MUST be hollow, since it does not have enough mass to be solid all the way through.

Also shown on Michael Mozina's website are running difference images of the Sun's surface captured by the SOHO NASA satellite using the 195A filter sensitive to iron ion emissions coming off the surface of the Sun showing permanent surface features such as mountains and valleys on the surface of the Sun that can be seen to rotate all at the same rate from the poles to the equator. In fact, the surface of the sun is relatively cool compared to its upper atmosphere allowing the Sun to have a solid surface.

NASA's SERTS program used spectroscopy to determine that the Sun has several layers. Mozina reasons that those layers most likely would stack up above the sun's surface by specific gravity, the lighter elements at the top, such as hydrogen, then below that helium, then neon, which produces the light we see with our eyes as it reacts to electric arcs rising from the surface of the Sun, then a layer of liquid silicon, and below that the solid surface of the Sun containing calcium, iron, magnesium, manganese, chromium, aluminum, sulfur and nickel. See *The Surface of the Sun* by Michael Mozina.

Our Sun IS HOLLOW!, and our Hollow Earth will also continue to be hollow when it becomes celestialized. In Joseph Smith's vision of the Celestial Kingdom, he records:

"The heavens were opened upon us, and I beheld the celestial kingdom of God, and the glory thereof, whether in

the body or out I cannot tell. I SAW THE TRANSCENDENT BEAUTY OF THE GATE THROUGH WHICH THE HEIRS OF THAT KINGDOM WILL ENTER, WHICH WAS LIKE UNTO CIRCLING FLAMES OF FIRE; also the blazing throne of God; whereon was seated the Father and the Son. I saw the beautiful streets of that kingdom, which had the appearance of being paved with gold." (D&C 137:1-4)

The gate Joseph Smith saw which leads into the celestial world *"which was like unto circling flames of fire"* is a good description of a star's polar opening where the scriptures say the angels and Gods are set to receive heirs to the Celestial Kingdom into its interior where the city of God is with streets having *"the appearance of being paved with gold."*

In 1933, Phoebe Marie Holmes published a book, MY VISIT TO THE SUN, of her visit to the Celestial City of God in our outer Sun!

Phoebe describes how she was taken in the Spirit by angels to the *"heart of the Sun"* where the New Jerusalem is being built by Jesus Christ, and all the resurrected holy prophets and Saints of God. Phoebe reported that the Sun is hollow and that the city inside our Hollow Sun is the New Jerusalem, as described by the Apostle John in the Book of Revelation Chapter 21. Phoebe reported that the New Jerusalem is a giant terraced "mountain" with a square bottom -- having a pyramid shape.

Christ, in his Sermon on the Mount, said that the *"meek"* shall inherit the earth, and indeed they will! A mansion is being built there right now for each of us, in the New Jerusalem, by our good actions here on earth. Phoebe was taken by the angels of God to visit her unfinished mansion, where she found her husband, who had already passed on. She then was brought back to earth to finish her life's work.

May we live the laws of God as revealed by his living prophets so as to be able to pass through the gate and live in the city of God, and as Gods, inherit for us and our children that celestial world that OUR HOLLOW EARTH shall surely become.

CHAPTER NINE
The Auroras Prove Our Earth Is Hollow!

The auroras are a beautiful display of light in the atmosphere above the Arctic and Antarctic regions. In the northern hemisphere it is called the Aurora Borealis, or the northern lights, and in the southern hemisphere, the Aurora Australis, or southern lights. The light of the aurora is caused by the glow of atoms in the thin upper atmosphere as they are hit by fast-moving electrons and protons.

The aurora appears as an oval curtain surrounding the polar regions of the earth. The radius of the auroral oval varies with the intensity of the solar wind. When there is little solar activity, the oval shrinks poleward, but is yearly observable only in the Arctic or Antarctic nighttime. During periods of great solar activity, about two days after an intense solar flare, the radius of the auroral oval increases southward opposite from the sun's position. (Figure 9-1)

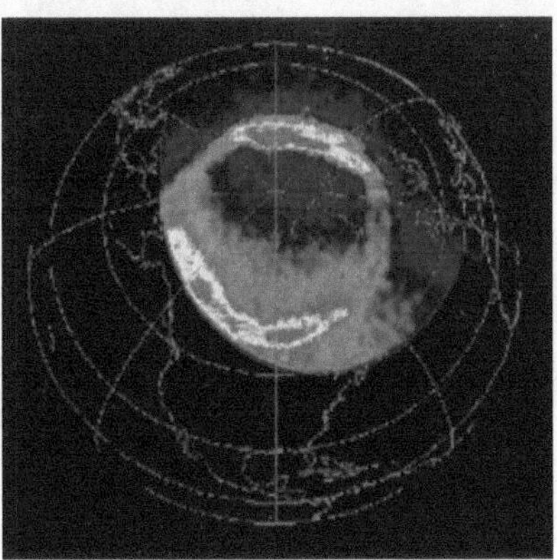

Figure 9-1. Earth's Northern Auroral Oval from the Polar Ultra Violet Imager. (The day side is on the upper portion of the picture.)

Since the aurora conforms to the shape of the earth's electromagnetic field, the solar wind from our outer sun deforms that field and hence the auroral oval also. Although the solar wind from our outer sun cannot enter the electromagnetic field of the earth but is deflected around it, the solar wind, consisting mainly of protons and electrons and traveling at about 880 kilometers per second does cause the auroral oval to elongate toward the equator on the opposite side of the earth from the Sun's position by stretching the earth's electromagnetic field. The auroral oval, however, rarely extends further south than 60 degrees latitude.

The appearance of the auroral curtain is generally greenish-white in color. The lower border of the auroral curtain is located at a height of about 60 miles, extending as high as the upper limits of the atmosphere. The auroral curtains are only a few hundred meters thick and when the auroral oval is observed from any one place, close-up, it appears in "arcs" -- curving curtains of light.

Figure 9-2. Aurora Borealis. Multiple auroral arcs photographed in Alaska.

When least active, the auroral curtains or arcs are quite motionless. When more active, the aurora develops up to five curtains or arcs (Figure 9-2) and when more active still, the arcs become wavy or folded. Also, there are active bands with ray structure where rays of the electrons and protons can be seen traveling up with a speed of about 300 feet per second. When a rayed band is seen from relatively nearby, the rays appear to emerge from a small region, resulting in a fan-shaped form called a corona. When the rays are not clearly seen, this broken-up form looks like a group of cumulus clouds called auroral patches.

Another important form of the aurora is the veil. The veil is an intensive dark red glow over a large region of the sky and is the type of aurora seen the farthest away.

The different colors of the aurora are caused by intense sheets of energetic electrons that emit from the polar region into the atmosphere where they collide with molecular nitrogen, molecular oxygen and atomic oxygen particles. The most common light of the aurora, the greenish-yellow color is emitted from excited oxygen atoms hit by low energy electrons. The dark red crimson color is emitted by oxygen molecules (O_2) when hit by more energetic electrons.

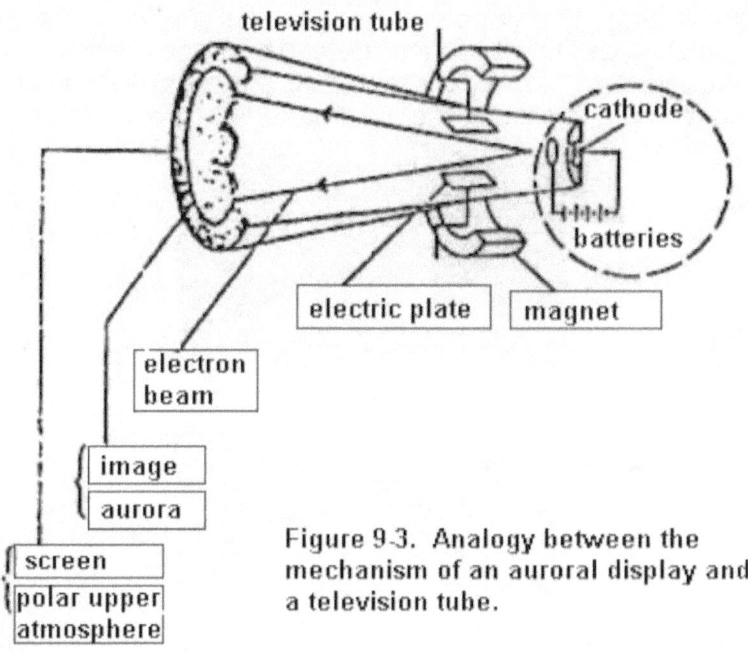

Figure 9-3. Analogy between the mechanism of an auroral display and a television tube.

Scientists compare the aurora to a gigantic television tube. (Figure 9-3) The polar upper atmosphere corresponds to the screen on a television tube that has a diameter of about 2,500 miles. The aurora corresponds to an image on the screen (atmosphere) of the television tube. A television screen is coated by a fluorescent material that emits light when it is hit by an electron beam. When an electron beam hits the polar upper atmosphere, it produces an aurora. In a television tube, the electron beam is acted upon (modulated) by a pair of electric plates and an electromagnet that create a moving image on the screen. The movement of the aurora during an auroral storm is also caused by the modulation of the electron beam that causes the aurora to light up.

Several fundamental questions still puzzle scientists concerning the comparison of the aurora to a television tube. It is unclear what processes play the roles of the tube's cathode (the energy source), or what processes play the roles of the electroplate and electromagnet during an auroral substorm. It is not known how the energy carried by the solar wind is transformed into the energy of the aurora. The problem has been even more mystifying because although it has been firmly established that the auroras are caused by high-powered electron beams, the source of these beams cannot be the solar wind which is deflected around the earth by the earth's electromagnetic field. Where then, are the electron beams coming from which cause the auroras to light up?

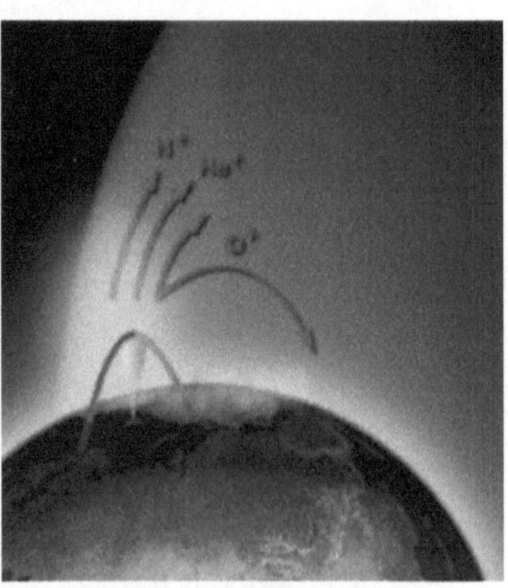

In a NASA article titled, "*Polar Fountains Fill Magnetosphere with Ions,*" a NASA scientist admits that evidence from polar satellites indicates that energetic ions accelerating *"upward"* from the poles cause the auroras to light up and continue out into space to fill the earth's magnetosphere.

The accelerating mechanism is not understood unless the Hollow Earth is taken into consideration. When so done, it becomes obvious that this polar fountain of ions is a solar wind coming from the inner sun within Our Hollow Earth shooting out through the polar openings.

Here is where the Hollow Earth Theory explains the unfathomable source of the auroral energy. The comparison of the television tube fits the hollow earth theory origin of the auroral lights perfectly. (Figure 9-4) The energy source of the electron beam (the cathode) is the sun inside the earth. The electromagnet is the polar openings, and the electroplates that variate the auroral display are winds and electrical storms on the polar lip which cause the electron beam to modulate or move about. A variating solar wind from sun spots on our outer Sun also modulates the electron beam by causing the electromagnetic field of the earth, and with it the auroral oval to elongate away from the direction of our outer Sun by putting a pressure on the electromagnetic field of the Earth.

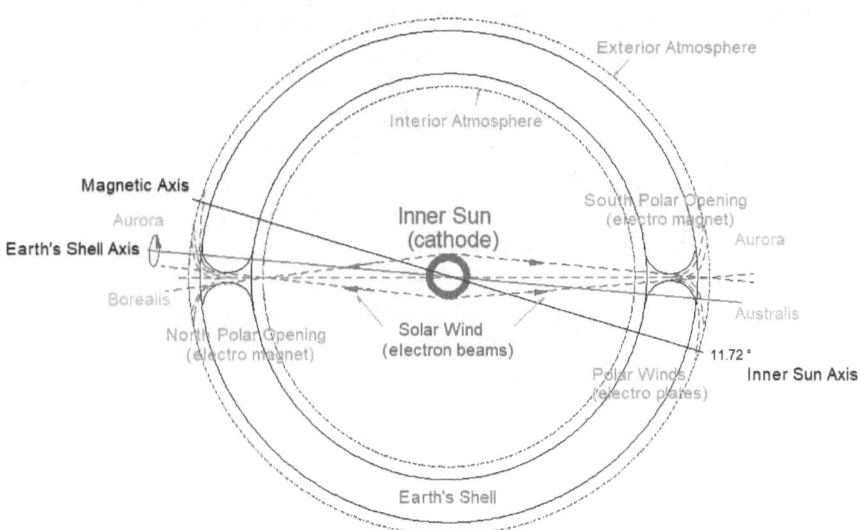

Figure 9-4. Comparison of the Hollow Earth origin of the aurora to a television tube -- the cathode is the sun inside Our Hollow Earth, the source of the electron beam that produces the auroras.

The auroral oval is shaped into the form of an oval when the solar wind from the inner sun comes out the polar openings and follows the electromagnetic field lines of the earth. The auroral oval is also deformed by the solar wind of our outer Sun as it pushes on the earth's electromagnetic field away from the direction of our outer Sun. The high velocity electrons and protons emitting through the polar openings from the sun within the earth assume the shape of the earth's electromagnetic field which has holes at the poles. This is the reason the aurora oval curtain produces no auroral light in the center of the oval -- because the aurora conforms to the earth's electromagnetic field which has holes at the poles.

After passing through the atmosphere around the polar openings the electron beam emanating from the sun inside Our Hollow Earth continues to follow the earth's electromagnetic field lines from the North Polar Opening south and from the South Polar Opening north. The radiation becomes trapped thereafter in the electromagnetic field of the earth above the equator and results in what is known as the Van Allen Radiation Belts.

Professor Neil Davis, a geophysicist at the University of Alaska in Fairbanks, found evidence that indicates that the auroral energy comes from inside the earth. In his book, THE AURORA WATCHER'S HANDBOOK, Professor Davis lays out the current orthodox thinking on the Auroras.

Current orthodox science claims the auroral lights are caused entirely by our outer Sun solar wind. Yet they admit they do not know how the Sun's solar wind gets inside the earth's magnetic field to cause the auroras to light up, and since the outer Sun's solar wind does not have enough energy to light up the auroras, they also admit they do not know how the Sun's solar wind can increase in energy enough to light up the auroras.

There are three flaws in the orthodox science's theory that the outer Sun's solar wind causes the auroral lights.

1.   The energy from the Sun's solar wind is not powerful enough to light up the auroras.

2.   The Sun's solar wind cannot get inside the earth's magnetic field, but instead is deflected around the earth's magnetic field.

3.   The solar wind that is causing the auroral lights travels UP from the earth, NOT down from space as is required by their theory.

I have been to Alaska and observed the Aurora Borealis myself and personally observed the auroral energy causing the auroral lights travels UP toward space from the earth.

The experiment conducted by Professor Davis that proves the auroral energy comes from inside the earth was carried out in 18 paired flights to both polar regions between 1967 and 1971, as described in Jan Lamprecht's book, <u>HOLLOW PLANETS</u>, pages 300-304. They used two long-range military versions of the Boeing 707 equipped with sensitive cameras and synchronized clocks. One aircraft took off from Anchorage, Alaska flying north, and the other from Christchurch, New Zealand flying south. What they discovered was that pulsations in the auroras occurred exactly at the same time in the southern aurora as in the northern aurora. Professor Davis concluded that, "Whatever the cause, it would seem that the triggering of the pulses must occur in the equatorial plane."

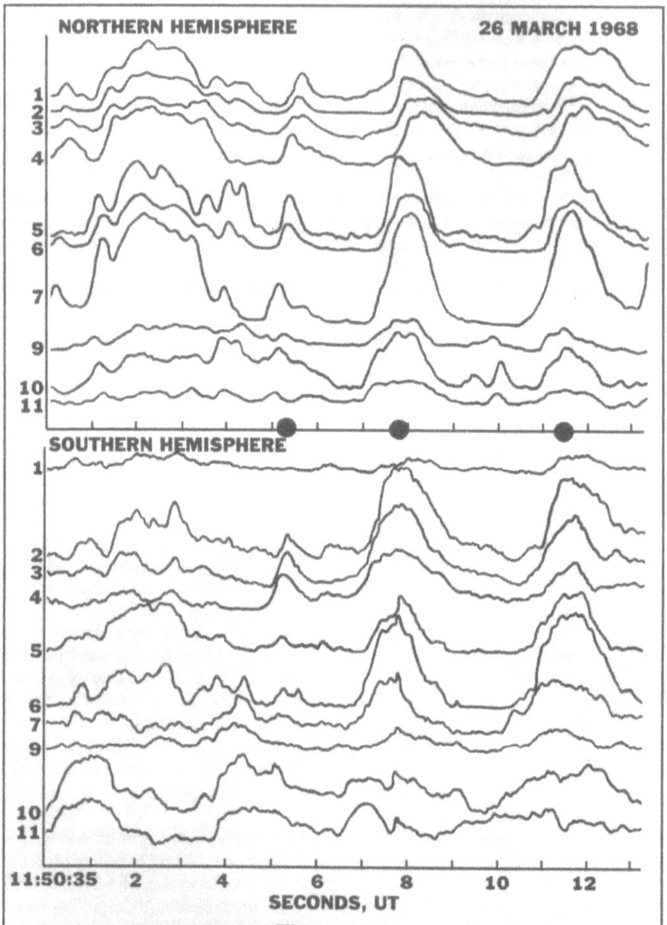

Figure 10.8

Photometer tracings of television recordings of pulsating aurora which shows exactly sychronous pulsations in the two hemispheres. (Each tracing line represents a particular portion of the television image.) Note the prominent sychronous pulses at right, marked by black circles at center. Credit: *The Aurora Watcher's Handbook*, page 123.

Diagram of Variations in the Auroral Lights at both poles show that they occur simultaneously.

Therefore, my conclusion from this evidence is that the Inner Sun of the Earth is on the equatorial plane, and is the source of the solar wind that lights up the auroras at both poles simultaneously as the solar wind from the Inner Sun comes out of the Hollow Earth through the polar openings, which then follows the electromagnetic field lines of the earth and at the height of 40-60 miles the air is rarified enough so that the highly energetic particles in the Inner Earth Solar Wind can knock photons out of the magnetic field of the atoms of oxygen in the air to create the auroral lights.

Storms and winds on the lip of the polar openings also variate the electron beam coming from the earth's interior sun through the polar openings. Polar explorers invariably noticed increased meteorological disturbance from the Polar Regions after a brilliant auroral display. Explorer Hall on page 300 wrote,

"At 6:30 the usual haziness of the sky after the occurrence of these (auroras) was noticed." On page 83 he wrote also, *"A smart breeze from the north was blowing nearly the whole night. This seemed to add to the briskness of the merry dancers as they crossed the heavens to and fro."*

Bernacchi wrote on page 130,

"But what was the greatest interest in the observation of the aurora was the connection which appeared to exist between it and approaching atmospheric disturbances. A strong gale (at Cape Adare, Antarctica) from E.S.E and S.E. was almost invariably preceded by a most brilliant and rapid auroral display. This was not a mere coincidence, but a fact repeatedly observed." (THE PHANTOM OF THE POLES, Chapter Seven)

The fact that these winds from the polar areas which cause the auroral displays to variate are warm is further evidence that the north wind originates in the earth's interior where the interior sun warms them.

Concerning this warm wind, explorer Peary writes on pages 214 and 215 of his work, NEAREST THE POLE:

"I expected to hear later of our February foehn in other parts of Greenland, and I was not disappointed. Lieutenant Ryder was living for nine months at Scoresby Sound, on the coast of East Greenland, while we were at McCormick Bay. He was about four hundred and fifty geographical miles south of us. The maximum temperatures he recorded occurred in February and May. He says (Petermanns Mittheilungen, XI, 1892, p. 256) *that these high temperatures were due to severe foehn storms, one of which, in February suddenly, raised the thermometer to 50 F, 8 degrees higher than my instruments had recorded."*

Foehn storms are warm winded storms which come out of the north Arctic in winter.

Explorer Fridtjof Nansen at about 79 degrees north latitude, on January 18, wrote of this warmer north-wind,

"It is curious that there is almost always a rise of the thermometer with these stronger winds; today it rose to 13° F below zero (-25° C). A south wind of less velocity generally lowers the temperature, and a moderate NORTH WIND RAISES IT." (FARTHEST NORTH, p. 373, Vol. I)

In the winter of 2018, a warm air intrusion into the Arctic brought the temperature above freezing and lasted for more than a week.

The weather station at Cape Morris Jesup on the northern coast of Greenland reported the temperature on February 25, 2018 was above freezing for about 24 hours. This is very unusual at this time of the year since there is no sunlight 24 hours a day at this location. The scientists analyzing this warm air intrusion, speculated that the warm air came into the Arctic from the Atlantic, but the Arctic Map showing this anomaly shows the warm air coming from the Russian side of the pole -- exactly where I have estimated the North Polar Opening is located.

Where could these observed warm north winds come from in the middle of the winter? Surely this is evidence of an opening into the earth that leads to a land warmed by an inner sun.

If this is not evidence enough, surely scientists are hard put to explain where Arctic animals and birds migrate to when they are observed in the Arctic regions to go north in the winter instead of south.

In his book, JOURNEY INTO SILENCE, Jack Denton Scott commented on Ross's Gull that,

"We don't see them often. Most naturalists never see one. (They are) mystery birds. They don't go south. (They) breed in eastern Siberia then fly north over the Polar Sea. No one knows where they are between October and June." (JOURNEY INTO SILENCE, 1976 p. 107)

Wally Herbert reported in his book, ACROSS THE TOP OF THE WORLD, that on his dog sled crossing of the *"pole,"* he found polar bears nearly up the *"pole."* He considered the Hollow Earth theory as so much *"crap."* And equally derided the eskimo belief of a *"hole"* in the Arctic Ocean. Perhaps the polar bears Herbert saw up near the pole knew something he didn't, as do many birds which migrate to the unknown north.

Driftwood that washes up on Arctic coasts drifts out from rivers that empty into the Arctic Ocean within the North Polar Opening of Our Hollow Earth.

In Daines Barrington's book, ON THE POSSIBILITY OF APPROACHING THE NORTH POLE, we read that driftwood is driven on the north coast of Iceland which could come from no other quarter than the north, but that among other fresh pieces whole trees were found which yet had their buds on them something which would have been absolutely impossible if this wood had drifted long distances from southern climes. It is obvious that a very few months in salt water would kill the buds, but here were trees which had evidently been growing only a short time before. He further tells us that observers in Spitsbergen have always noticed that in spring, just before the hatching season, the wild ducks, geese, and other birds, fly in a northerly direction. There is also a heavy fall migration to the north.

The land from which this drift wood, warm winds, even fogs, and to which animals and birds migrate to is the interior of Our Hollow Earth where a sun gives its rays to a perfect climate for all living things.

That sun has been seen by Arctic explorers.

Olaf Jansen and his fisherman father, who attained the land of Our Hollow Earth in a voyage in their small fishing boat through the North Polar Opening, tells of their first sighting of the interior sun. It was the first part of August 1829. They had been sailing about 15 days northeast of Franz Josef Land when Olaf records:

"One day about this time, my father startled me by calling my attention to a novel sight far in front of us,

almost at the horizon. 'It is a mock sun,' exclaimed my father. 'I have read of them; it is called a reflection or mirage. It will soon pass away.'"

"But this dull-red, false sun, as we supposed it to be, did not pass away for several hours; and while we were unconscious of its emitting any rays of light, still there was no time thereafter when we could not sweep the horizon in front and locate the illumination of the so-called false sun, during a period of at least twelve hours out of every twenty-four."

"Clouds and mists would at times almost, but never entirely, hide its location. Gradually it seemed to climb higher in the horizon of the uncertain purply sky as we advanced."

"It could hardly be said to resemble the sun, except in its circular shape, and when not obscured by clouds or the ocean mists, it had a hazy-red, bronzed appearance, which would change to a white light like a luminous cloud, as if reflecting some greater light beyond."

*"We finally agreed in our discussion of this smoky furnace-colored sun, that, whatever the cause of the phenomenon, it was not a reflection of our sun, but a planet of some sort -- a reality." (*THE SMOKY GOD*, pp. 85-7)*

Olaf and his father Jens Jansen were on the edge of the polar opening in their sailboat when they first saw the interior sun over the horizon directly to the north. As they continued over the lip of the polar opening into the hollow interior of the earth, the interior sun appeared to rise higher in the sky. They came to land a few days later and found a river. As they were going up it, they met a ship containing inhabitants of Our Hollow Earth. They were taken in by this race of intelligent and highly advanced people and lived with them for two and half years after which they made their return to our surface world by way of the Antarctic polar opening.

In his book, Olaf Jansen gives a few descriptions of the interior sun which should give scientists who ignore the creation-origin of our earth something to think about. As Olaf and his father boarded the ship and the inhabitants took them further inside the earth's hollow interior, Olaf writes,

"In the meantime, we had lost sight of the sun's rays, but we found a radiance 'within' emanating from the dull-red sun which had already attracted our attention, now giving out a white light seemingly from a cloudbank far

away in front of us. It dispensed a greater light, I should say, than two full moons on the clearest night."

"In twelve hours this cloud of whiteness would pass out of sight as if eclipsed, and the twelve hours following corresponded with our night. We early learned that these strange people were worshipers of this great cloud of light. It was 'The Smoky God' of the 'Inner World.'" (THE SMOKY GOD, pp. 102, 103)

He further writes,

"The great luminous cloud or ball of dull-red fire -- fiery-red in the mornings and evenings, and during the day giving off a beautiful white light, 'The Smoky God,'--is seemingly suspended in the center of the great vacuum 'within' the earth, and held to its place by the immutable law of gravitation..."

"The base of this electrical cloud or central luminary, the seat of the gods, is dark and non-transparent, save for innumerable small openings, seemingly in the bottom of the great support or altar of the Deity, upon which 'The Smoky God' rests; and, the lights shining through these many openings twinkle at night in all their splendor, and seem to be stars, as natural as the stars we saw shining when in our home at Stockholm, excepting that they appear larger. 'The Smoky God,' therefore, with each daily revolution of the earth, appears to come up in the east and go down in the west, the same as does our sun on the external surface. In reality, the people 'within' believe that 'The Smoky God' is the throne of their Jehovah, and is stationary. The effect of night and day is, therefore, produced by the earth's daily rotation." (THE SMOKY GOD, pp. 108-110)

The apparent size of our outer sun is 1/2 degree. The moon is only slightly larger in apparent size and thus can eclipse the sun. On the other hand, if the interior sun is 600 miles in diameter and is 3,000 miles from the inner surface of the planet, its apparent size would be 11.5 degrees, viewing it from the interior surface -- 23 times bigger than the apparent size of our outer sun or moon. If you took a 5 inch disk you would need to move it 25 inches from your eyes to give you the apparent size of the inner sun. So the inner sun would appear to fill more of the sky than does our outer sun. As such, the bright side of the inner sun would appear to come out in the morning on the east side of the inner sun's disk, move across the face of the disk and disappear on the west side of the disk of the inner sun in the evening (analysis of apparent size of the Inner Sun was provided by Scott Macklin.)

The interior sun imparts a warmth perfectly suited to the growth of plant, animal and human life. There are no seasons inside the earth except near the polar openings where the light and heat of our sun varies the weather conditions throughout the year. These seasons near the North Polar Opening cause different colors of flowers to grow at different seasons and the pollen from these flowers color the icebergs in the Arctic Ocean.

Olaf Jansen writes of this phenomena:

"The valleys of this inner Atlantis Continent, bordering the upper waters of the farthest north are in season covered with the most magnificent and luxuriant flowers. Not hundreds and thousands, but millions, of acres, from which the pollen or blossoms are carried far away in almost every direction by the earth's spiral gyrations and the agitation of the wind resulting therefrom, and it is these blossoms or pollen from the vast floral meadows 'within' that produce the colored snows of the Arctic regions that have so mystified the northern explorers." (THE SMOKY GOD, p. 163)

La Chambre, in an account of Andree's balloon expedition, on page 144, says,

"On the isle of Amsterdam the snow is tinted with red for a considerable distance, and the savants are collecting it to examine it microscopically. It presents, in fact, certain peculiarities; it is thought that it contains very small plants. Scoresby, the famous whaler, had already remarked this."

Explorer Kane noticed that at different times of the year the pollen-colored snow had different colors, or hues. This probably results from the different flowers that blossom at certain times of the seasons near the polar opening.

The interior sun contains a small temperature difference between the light and darker side which helps to cause slow wind patterns and the daily mist-rains which make such remarkable plant growth in the interior world. Olaf describes the interior winds as consisting of breezes that come up once each day.

After reaching the inner continent from their long voyage of 53 days across the Arctic ocean into the polar opening, Olaf wrote:

"We sailed for three days along the shoreline, then came to the mouth of a fjord or river of immense size. It seemed more like a great bay, and into this we turned our

fishing-craft, the direction being slightly northeast of south. By the assistance of a fretful wind that came to our aid about twelve hours out of every twenty-four, we continued to make our way inland, into what afterward proved to be a mighty river, and which we learned was called by the inhabitants, Hiddekel." (THE SMOKY GOD, pp. 91, 92)

The temperature difference on the night side of the interior sun also causes it to rain in the hollow earth. Olaf writes,

"There is a hazy mist that goes up from the land each evening, and it invariably rains once every twenty-four hours. This great moisture and the invigorating electrical light and warmth account perhaps for the luxuriant vegetation, while the highly charged electrical air and the evenness of climatic conditions may have much to do with the giant growth and longevity of all animal life." (THE SMOKY GOD, p. 128)

The slight temperature variation of the interior sun between day and night sides causes a breeze to blow in the day-time and a drizzle type rain to fall in the evening. These climatic conditions cause every living thing to grow to their perfection. All life forms grow to giant proportions compared to life on the earth's exterior surface.

It is known that red light causes plants to grow taller than usual. Perhaps the light from the night side of the interior sun, which Olaf described as having *"a hazy-red, bronzed appearance"* is a contributing factor in the giant growth of plant, animal and human life in the interior world, as well as abundantly mineralized soils with favorable temperatures and moisture. Compared to our exterior world whose antediluvian fertile soils have been washed into the sea by the waters of the great flood of Noah and our atmosphere bathed in harmful ultraviolet light from our exterior sun, in contrast the interior world contains an ideal environment for abundant life and longevity.

Here are a few descriptions of life in the interior as given by Olaf Jansen:

"There was not a single man aboard (the pleasure ship 'Naz') who would not have measured fully twelve feet in height. They all wore full beards, not particularly long, but seemingly short-cropped. They had mild and beautiful faces, exceedingly fair, with ruddy complexions. The hair and beard of some were black, others sandy, and still others yellow. The captain, as we designated the dignitary in command of the great vessel, was fully a head taller

than any of his companions. The women averaged from ten to eleven feet in height. Their features were especially regular and refined, while their complexion was of a most delicate tint heightened by a healthful glow."

"Both men and women seemed to possess that particular ease of manner which we deem a sign of good breeding, and, notwithstanding their huge statures, there was nothing about them suggesting awkwardness. As I was a lad in only my nineteenth year, I was doubtless looked upon as a true Tom Thumb. My father's six feet three did not lift the top of his head above the waist line of these people." (THE SMOKY GOD, pp. 98-100)

"We learned that the males do not marry before they are from seventy-five to one hundred years old, and that the age at which women enter wedlock is only a little less, and that both men and women frequently live to be from six to eight hundred years old, and in some instances much older." (THE SMOKY GOD, p. 118)

"Vegetation grew in lavish exuberance, and fruit of all kinds possessed the most delicate flavor. Clusters of grapes four and five feet in length, each grape as large as an orange, and apples larger than a man's head typified the wonderful growth of all things on the 'inside' of the earth. The great redwood trees of California would be considered mere underbrush compared with the giant forest trees extending for miles and miles in all directions." (THE SMOKY GOD, pp. 105-106)

"In our travels we came to a forest of gigantic trees, near the city of Delfi. Had the Bible said there were trees towering over three hundred feet in height, and more than thirty feet in diameter, growing in the Garden of Eden, the Ingersolls, the Tom Paines and Voltaires would doubtless have pronounced the statement a myth. Yet this is the description of the California SEQUOIA GIGANTEA; but these California giants pale into insignificance when compared with the forest Goliaths found in the 'within' continent, where abound mighty trees from eight hundred to one thousand feet in height, and from one hundred to one hundred and twenty feet in diameter; countless in numbers and forming forests extending hundreds of miles back from the sea." (THE SMOKY GOD, pp. 120-1)

"Whether inland among the mountains, or along the seashore, we found bird life prolific. When they spread their great wings some of the birds appeared to measure thirty

feet from tip to tip. They are of great variety and many colors. We were permitted to climb up on the edge of a rock and examine a nest of eggs. There were five in the nest, each of which was at least two feet in length and fifteen inches in diameter." (THE SMOKY GOD, pp. 124-5)

On an island near the South Polar Opening, Olaf reported seeing penguins,

"After breakfast we started out on an inland tour of discovery, but had not gone far when we sighted some birds which we recognized at once as belonging to the penguin family. They are flightless birds, but excellent swimmers and tremendous in size, with white breast, short wings, black head, and long peaked bills. They stand fully nine feet high. They looked at us with little surprise, and presently waddled, rather than walked, toward the water, and swam away in a northerly direction." (THE SMOKY GOD, pp. 134-5)

However, Olaf did see birds like those on our surface world. He wrote,

"We saw innumerable specimens of bird-life no larger than those encountered in the forests of Europe or America. It is well known that during the last few years whole species of birds have quit the earth. A writer in a recent article on this subject says: 'Is it not possible that these disappearing bird species quit their habitation without, and find an asylum in the 'within world'?'" (THE SMOKY GOD, pp. 123-4)

Not only is the human, plant and bird life huge in dimensions but also animal life. Olaf writes,

"After we had been in the city of Hectea about a week, Professor Galdea took us to an inlet, where we saw thousands of tortoises along the sandy shore. I hesitate to state the size of these great creatures. They were from twenty-five to thirty feet in length, from fifteen to twenty feet in width and fully seven feet in height. When one of them projected its head it had the appearance of some hideous sea monster."

"The strange conditions 'within' are favorable not only for vast meadows of luxuriant grasses, forests of giant trees, and all manner of vegetable life, but wonderful animal life as well."

"One day we saw a great herd of elephants. There must have been five hundred of these thunder throated monsters, with their restlessly waving trunks. They were

tearing huge boughs from the trees and trampling smaller growth into dust like so much hazel-brush. They would average over 100 feet in length and from 75 to 85 in height."

"It seemed, as I gazed upon this wonderful herd of giant elephants, that I was again living in the public library at Stockholm, where I had spent much time studying the wonders of the Miocene age. I was filled with mute astonishment, and my father was speechless with awe. He held my arm with a protecting grip, as if fearful harm would overtake us. We were two atoms in this great forest, and, fortunately, unobserved by this vast herd of elephants as they drifted on and away, following a leader as does a herd of sheep. They browsed from growing herbage which they encountered as they traveled, and now and again shook the firmament with their deep bellowing." (THE SMOKY GOD, pp. 126-7)

The person who Olaf obtained to edit his book and publish it, Willis George Emerson, wrote concerning those mammoths thought to be prehistoric found on the northern coasts of Siberia and Alaska,

"On the northern boundaries of Alaska, and still more frequently on the Siberian coast, are found boneyards containing tusks of ivory in quantities so great as to suggest the burying-places of antiquity. From Olaf Jansen's account, they have come from the great prolific animal life that abounds in the fields and forests and on the banks of numerous rivers of the Inner World. The materials were caught in the ocean currents, or were carried on ice-floes, and have accumulated like driftwood on the Siberian coast. This has been going on for ages, and hence these mysterious boneyards." (THE SMOKY GOD, pp. 44-5)

On this subject, William F. Warren, in his book, PARADISE FOUND, OR THE CRADLE OF THE HUMAN RACE AT THE NORTH POLE, wrote:

"The arctic rocks tell of a lost Atlantis more wonderful than Plato's. The fossil ivory beds of Siberia excel everything of the kind in the world. From the days of Pliny, at least, they have constantly been undergoing exploitation, and still they are the chief headquarters of supply. The remains of mammoths are so abundant that, as Gratacap says, 'the northern islands of Siberia seem built up of crowded bones.' Another scientific writer, speaking of the islands of New Siberia, northward of the

mouth of the River Lena, uses this language: 'Large quantities of ivory are dug out of the ground every year. Indeed, some of the islands are believed to be nothing but an accumulation of drift-timber and the bodies of mammoths and other antediluvian animals frozen together.' From this we may infer that, during the years that have elapsed since the Russian conquest of Siberia, useful tusks from more than twenty thousand mammoths have been collected."

In conclusion, we may say that if scientists would consider the true source of these mammoth remains, driftwood, warm arctic winds and even the auroras, they would discover a land in the north which Arctic explorer Fridtjof Nansen declared *"must be a Canaan, flowing with milk and honey."* They would also discover that our Earth IS Hollow with polar openings and contains an interior SUN.

CHAPTER TEN
The Earth's Van Allen Radiation Belts
Prove Our Earth Is Hollow!

Conforming to the true form of Our Hollow Earth are the Van Allen Radiation Belts. Even as the earth has holes near its poles, so do the radiation belts. These belts, discovered in May, 1958, by J.A. Van Allen, are two doughnut-shaped zones of charged particles trapped in the magnetic field of the Earth. The belts are most intense over the Equator and are effectively absent above the poles.

Scientists note that there is clearly a connection between the auroras and the Van Allen Radiation Belts, but the details remain uncertain. Direct measurement show the intensity of trapped particles in the outer belt to increase at the time of prominent auroras. Therefore, they conclude, it seems that some common source produces both the auroras and the outer belt. But they do not know what that source is.

The Hollow Earth theory clearly shows that the source of both the auroras and the Van Allen Radiation Belts is the sun suspended within the hollow interior of the earth. High velocity electrons and protons emit from the earth's polar openings from the sun within containing energies of 10,000 to 100,000 electron volts and with intensities of up to one million million (10^{12}) particles per square centimeter per second. These high energy electrons and protons strike the atmosphere above the Arctic and Antarctic in an oval shape at about 100 kilometers elevation where they excite air atoms to emit light causing the auroras.

After the energetic electrons pass through the atmosphere around the polar openings causing the auroras, they continue to follow the earth's magnetic field lines from the North Polar Opening, south, and from the South Polar Opening, north. The electron and proton particles when following a magnetic field line revolve around it as they advance describing paths having the form of a helix (corkscrew).

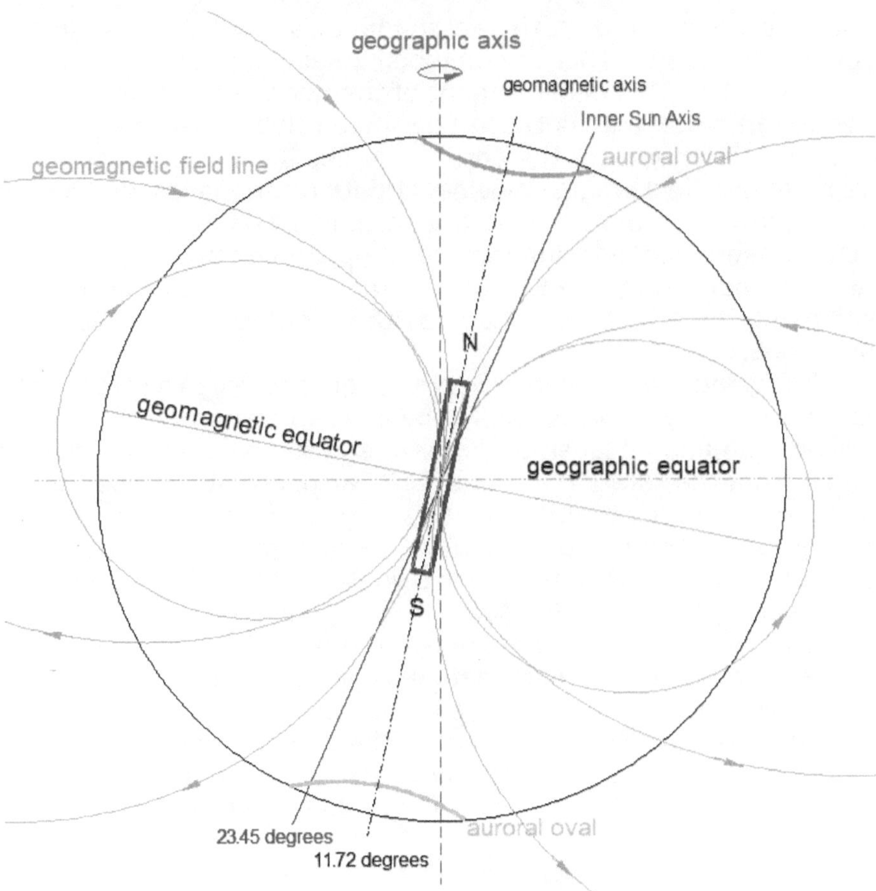

Figure 10-1. The earth's geomagnetic field does not conform to the rotational axis of the earth as does a common dynamo because the inner sun is oriented on a different rotational axis than the earth's rotational axis.

The Van Allen Radiation Belts are charged particles (electrons and protons) that conform to the earth's geomagnetic field. This field has holes at the poles but does not conform exactly to the physical structure of the earth at those points. (Figure 10-1) The earth's geomagnetic field has the form approximately that of a dipole with the north magnetic pole displaced 793 miles from the earth's geographical axis toward latitude 78.5 degrees north, longitude 69.1 degrees west (in 1965), and was tilted 11.5 degrees relative to the geographic axis.

231

Scientists have no idea why earth's geomagnetic field does not coincide with the earth's rotational axis, as is the case in a common dynamo. This can only be explained by the Hollow Earth Theory. This displacement of the geomagnetic axis from the geographical axis is caused by the earth's hollow nature and its interior sun. If the earth were like a common dynamo, its magnetic field would coincide with its rotational axis. The displacement occurs because the earth has TWO electromagnetic fields superimposed upon each other. One field is produced by the SUN within the hollow interior of the earth and the other is produced by the rotating shell of the hollow earth.

The geomagnetic axis revolves about the earth's rotational axis very slowly. This is caused by a slow rotation of the interior sun and because the interior sun's rotational axis is offset from the rotational axis of the Earth's shell. It also appears that the interior sun's axis of rotation is wobbling giving rise to one complete orbit of the geomagnetic poles about the polar axis of the earth's shell taking perhaps about 794 years.

The magnetic poles are known to move. The rate of movement of the North Magnetic Pole in 1965 was about 8 miles per year, but in recent years has accelerated as it goes over the Arctic. NOAA reports that the most recent survey determined that the North Magnetic Pole is moving north-northwest at 34 miles per year and in 2020 was located at 164.036° W, 86.502° N (https://www.ngdc.noaa.gov/geomag/GeomagneticPoles.shtml)

It appears that this movement may be caused by a slow rotation of the interior sun inside Our Hollow Earth that coincides with an axial wobble, making one complete rotation and wobble in about 794 years. This I have calculated assuming that in the creation period, God placed the Inner Sun in rotation at the same rate as God's planet Kolob in the center of our Milky Way Galaxy, which He revealed to Abraham, was rotating at the rate of 1,000 earth years to one day of the Lord. (Abraham 3:4, 2 Peter 3:8) As part of this calculation, I have taken the length of the Earth's original year of 290 days as depicted on the calendar of the ancient ruins of Tiahuanaco in the Andes Mountains of Bolivia. If Tiahuanaco's was earth's original calendar, then 1,000 290-day years should equal the present 365¼-day years times the rotation rate of the Inner Sun. The calculation for this is $1000*290 = 365.25x$, $x =$

793.98 years for the rotation rate of the Inner Sun and its magnetic field.

Scientists attempt to explain the auroras as being caused by radiation from the Van Allen Radiation Belts. But this is unlikely because it is known that there is not enough energy in the belts to sustain the always constant auroras. Scientists should consider the evidence that the source of the radiation comes from the poles and not the equator.

VAN ALLEN RADIATION BELTS

Van Allen radiation consists of high-energy electrons and protons in a doughnut-shaped ring around the earth. The electron radiation is most intense in regions of darker shading. A belt of proton radiation overlaps the electron radiation. Its peak intensity is indicated by the dashed lines:

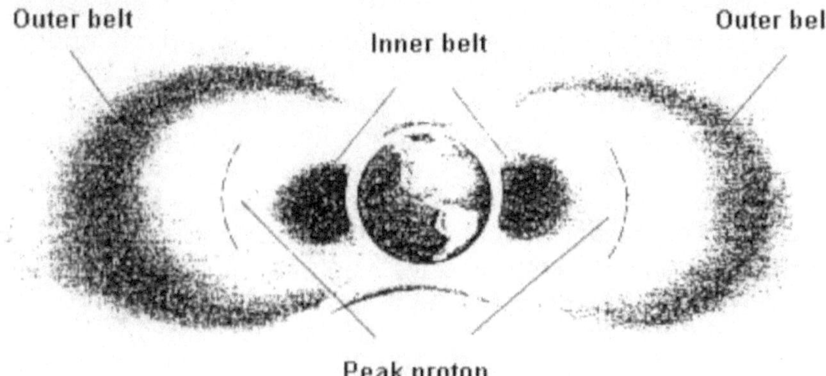

Figure 10-2. The Van Allen Radiation Belts have holes at the poles conforming to the electromagnetic field of the earth and form belts above the equator.

After leaving the polar atmosphere, the electrons and protons emanating from the earth's interior sun through the polar openings, divide into two layers or belts of electrons and two proton belts. (Figure 10-2) The belts are in the following order, outward from the earth's equator: electron, proton, electron and proton belts. Particle energies are highest closest to the earth where the magnetic force lines are strongest. Scientists try to explain this by saying that the particles come from the solar wind of our outer sun and upon approaching the earth are accelerated toward the earth by the geomagnetic field. But in their conjectures, scientists contradict themselves.

First of all, the energetic particles are observed to emit from the Polar Regions with sufficient energies to cause the atmosphere to light up the auroras. Then the particles with ever decreasing energies stream outward from the earth's surface and away from the poles following the earth's electromagnetic field lines.

Secondly, our outer sun's solar wind could not cause the auroras nor the radiation belts because they are much lower energetically. The solar wind is composed of protons with energies of about 1,000 electron volts and electrons with about 10 electron volts compared to the source of the auroras with electron volts of 10,000 to 100,000.

Actually the source of the auroras and the radiation belts is closer than our outer sun. That source is the sun within our earth's hollow interior.

Further evidence that our outer sun's comparatively energy deficient solar wind is not the cause of the auroras and the radiation belts is the fact that as the solar wind reaches the earth's proximity, the earth's magnetic field DEFLECTS the solar wind around the earth. (Figure 10-3) A boundary is formed when the deformation of the geomagnetic field just balances the pressure of the solar wind. Now, if the solar wind is deflected around the earth, and it is, how is it supposed to get into the earth's polar region and the geomagnetic field and increase the energies of its particles 10 to 1000 times?

Protons with energies up to 10,000 electron-volts are found in the magnetosphere but not in the free solar wind. Electrons with energies exceeding 40,000 electron-volts are also found in this region, while the solar wind electrons exhibit levels of only about 10 electron-volts. Scientists say that it seems highly likely that these low energy particles from the solar wind *"somehow"* get inside the magnetosphere where they *"somehow"* pick up energy, but the *"possible"* modes of entry are not understood. (ENCYCLOPEDIA BRITANNICA)

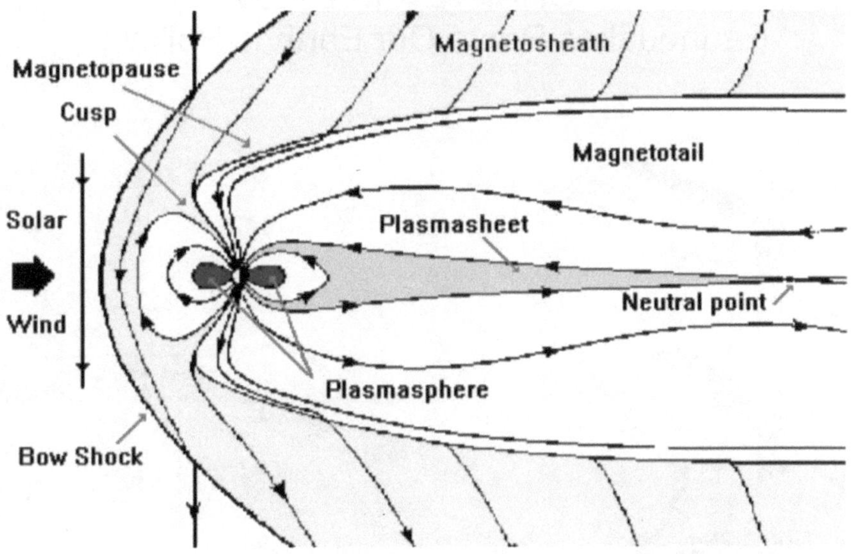

Figure 10-3. The earth's magnetosphere deflects the Sun's solar wind around the earth, and thus cannot be the source of energy in the Van Allen Radiation Belts or the Auroras.

It's time that scientists recognize that where the highest energy concentrations are located is the source of the auroras and Van Allen Radiation Belts. That source is within the polar areas. The sun inside Our Hollow Earth emanating its high energy electrons and protons through the earth's polar openings is the source of the solar wind that lights up the auroras and fills the Van Allen Radiation Belts, NOT the solar wind from our outer sun.

CHAPTER ELEVEN
Earthquakes Prove Our Earth Is Hollow!

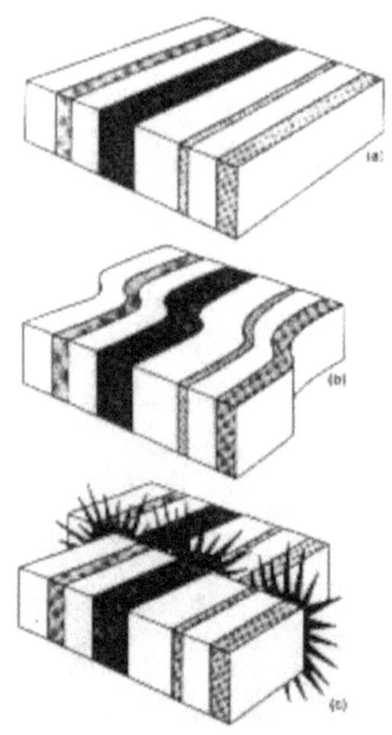

Diagram showing the relationship between the focus and the epicenter of an earthquake. The focus is the point of initial movement on the fault. Seismic waves radiate out from this point. The epicenter is a point on the earth's surface directly above the focus.

Figure 11-1. The origin of earthquakes. Strain is built up in earth's tectonic plates by tidal forces, which are the gravitational interactions of the moon, sun and planets acting on the earth until the plates rupture. Additionally, the expansion of the earth as it "grows" in size will cause plates to rupture. Energy is then released and seismic waves move out from the point of rupture.

Earthquakes are waves generated in the earth's interior and are caused by ruptures of rocks that are strained beyond their elastic limits. (Figure 11-1) Similar to a dry stick that is bent until it snaps, the rock layers at an earthquake site have energy stored up as they are bent by the gradual expansion of the earth as it "grows" and by the gravitational forces of the sun, moon and planets acting on the earth much as the moon's gravitational interaction with the earth produces the tides in the ocean. When the elastic limit of the rocks is reached by the bending, the stored energy is released by the rupture causing the fractured ends to vibrate and send out earthquake waves.

There are three major types of seismic (earthquake) waves. Primary P waves are compression waves similar to sound waves. P-waves cause the matter through which they pass to vibrate back and forth in the direction the waves travel. These are the fastest seismic waves. Secondary S waves travel about 1/2 as fast as P-waves and cause matter to vibrate from one side and to the other at right angles to the direction the waves travel. The slowest surface waves have the form of water waves. (Figure 11-2)

Fence line prior to seismic disturbance.

Motion produced by a P wave. Particles are compressed and then expanded in the line of wave progression.

Motion produced by an S wave. Particles move back and forth at right angles to wave progress.

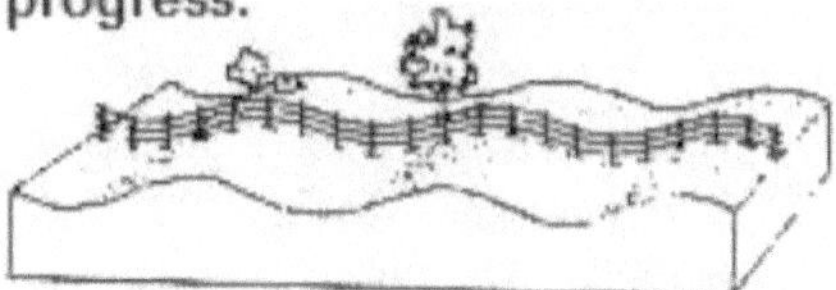

Motion produced by a surface wave. Particles move in a circular path at the surface and diminish with depth.

Figure 11-2. Motion produced by the various types of seismic waves. Each type of seismic wave produces a characteristic motion that can be illustrated by the distortions they produce in a straight fence line.

A significant difference between P and S waves is the fact that P compression waves can travel through any substance while S shear waves cannot travel through gases or liquids.

Seismic waves are important in the study of the earth's interior because next to volcanoes, they are virtually the only means that scientists have to determine the internal structure of the earth. Since they are waves, seismic waves obey the laws governing waves. That is, like light and sound waves, seismic waves move in straight lines through a homogeneous body, but when encountering a boundary between different substances, they are both reflected and diffracted (bent).

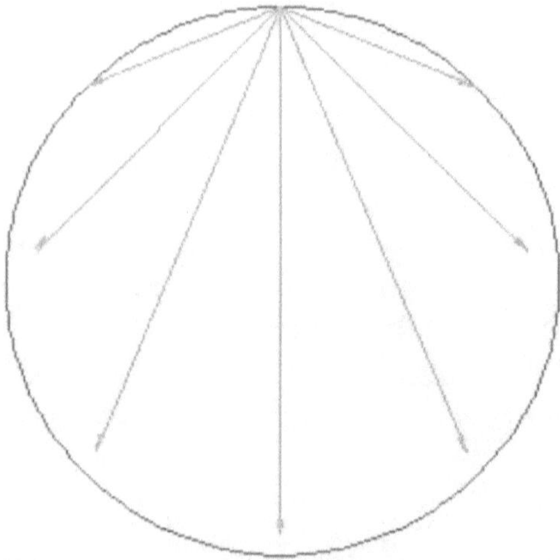

Figure 11-3. If there were a uniform density through all the earth, seismic waves would follow straight paths through the earth.

If the earth were a homogeneous sphere, seismic waves would travel in straight lines at a continuous velocity throughout the body, as illustrated by Figure 11-3. However, actual measurements of earthquake wave velocities show that seismic waves arrive at points progressively farther away from the earthquake epicenter faster than if they were traveling at uniform speed. (The epicenter is the point on the earth's surface directly above the focal point. The hypocenter, or focal point, is where the earthquake is calculated to have actually occurred.) Since seismic waves travel faster through denser material, it is concluded that as the waves travel down into the earth's interior they encounter progressively more dense materials. If the material toward the earth's interior becomes gradually more dense, then it would be expected that seismic waves would gradually be diffracted and caused to curve with the resulting path of Figure 11-4.

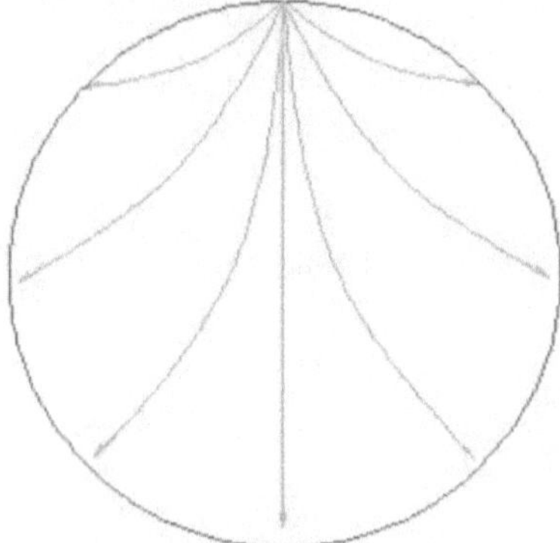

Figure 11-4. Paths followed by seismic waves in a planet where wave velocities increase with depth because of increasing density. The change in density would cause the waves to be refracted, so the wave would follow curved paths.

Figure 11-5. The shadow zone produced by an earthquake in Japan. The shadow zone in which no direct P waves are received is a band on the earth's surface.

A major discovery was made in 1906 when it was observed that whenever a major earthquake occurred there was a region on the opposite side of the planet between 103 degrees and 143 degrees from the earthquake focus that no seismic waves were detected except in comparatively very weak forms. This zone, called the shadow zone (Figure 11-5) is the basis upon which scientists build their theory of the core structure of the earth.

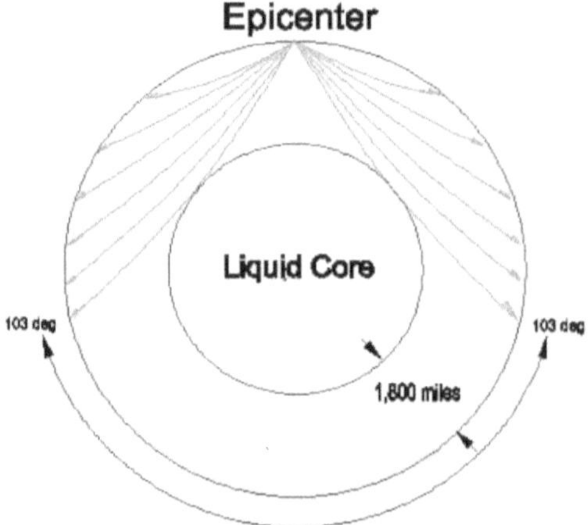

Figure 11-6. S-waves are received from an earthquake's focus to 103 degrees, but there is a large area on the opposite side of the planet from the epicenter where S-waves are not received. It is called the Shadow Zone. Scientists postulate that since S-waves cannot travel through liquid that the outer core must be molten liquid magma.

But neither would S-waves travel through a hollow core, so we maintain that the core is hollow, and that there is no liquid magma outer core.

First of all, S seismic shear waves do not even travel through the core and are not received at all past the 103 degree point (Figure 11-6). Since S-waves do not travel through liquids, scientists conclude that the outer core boundary is a liquid. The distance to this boundary was calculated in 1914 by Gutenberg, a German seismologist, to be 2,900 kilometers (about 1,800 miles). Since P compression waves can travel through liquids, the P-waves received past the 143 degree point are thus justified as traveling through the liquid outer core. However, since P-waves are not received in the shadow zone they are considered to refract as they hit the core as Figure 11-7 shows.

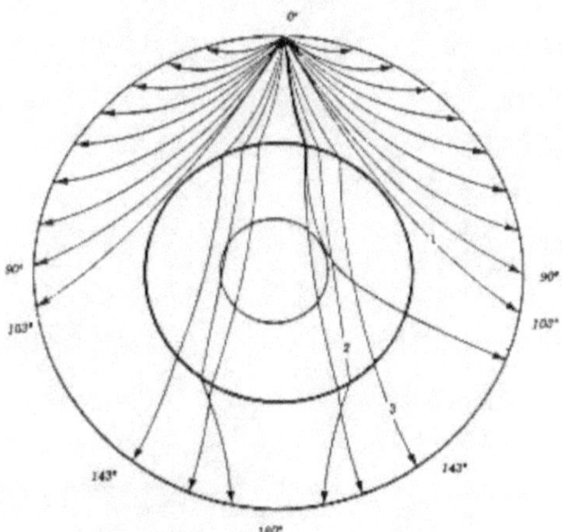

Figure 11-7. Orthodox science interprets the P-wave shadow zone by postulating that the earth has a central core through which the P waves travel relatively slowly. Ray 1 would just miss the core and would be received at a station located at 103 degrees from the earthquake focus. Steeper rays, such as ray 2, would encounter the boundary of the core and would be refracted. Ray 2 would travel through the core then refracted again at the core's boundary to be received at a station 180 degrees from the focus. Steeper rays would do the same until ray 3, which emerges at the surface at 145 degrees from the focus. Rays steeper than ray 3 would be bent so severely by the core they would not be received in the shadow zone.

   The hollow earth interpretation would reduce the distance to the core, and declares that the core is hollow through which neither P nor S seismic waves pass. The shadow zone is THE evidence of the hollow in the earth.

Figure 11-7 is a cross section through the earth. The true nature of the shadow zone is shown in Figure 11-5. Recent studies in the shadow zone showing a reception of weak P-waves there give scientists a belief in a solid inner core from which the P-waves that penetrate the outer core are deflected. (Figure 11-8)

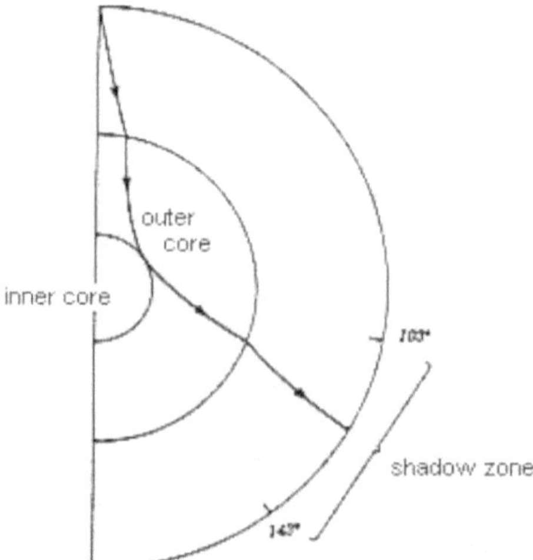

Figure 11-8. Orthodox science explains the weak P wave reception in the shadow zone suggests to them the existence of a solid inner core.

However, in a hollow earth, weak P wave reception in the shadow zone could result from multiple reflections between the inner surface and the outer surface causing them to be weak by the time they reach the shadow zone.

It has also been observed that P-waves which reach the opposite side of the earth beyond the shadow zone from the earthquake focus get there faster because they supposedly travel through a solid inner core than waves which travel more slowly through the supposedly liquid outer core.

However, this could also happen if these P-waves are traveling through a denser layer of the hollow earth's shell below the 450 mile depth where no earthquakes occur.

A solid inner core surrounded by a liquid outer core also gives scientists an explanation of the source of the earth's strong electromagnetic field. The reasoning is that the solid inner core maintains a more stationary position as the earth's outer parts rotate around it with the liquid outer core as a buffer zone. These dynamo-like counter-movements of the core and outer mantle-crust of the earth would then create the electromagnetic field of the earth. The problem with this orthodox science explanation of the Earth's magnetic field is that it would cause the magnetic poles to coincide with the rotational geographic axis of the Earth, which they do not.

The Hollow Earth Theory has no problem creating the earth's strong electromagnetic field. It is caused by the rotation of the Earth's shell about the very slowly rotating Inner

Sun that has a different inclination of its axis of rotation than the rotational axis of Earth's shell.

There exist other boundaries in the earth's outer layers as measured by seismic wave velocities as indicated by Figure 11-9. Geologists, considering that perhaps meteorites are fragments of an exploded planet have noticed that there are two general types of meteorites. One type is composed mostly of silicate minerals and the other of nickel and iron. By analogy they then conclude that the earth's mantle consists of silicate minerals and the core of iron and nickel. This they say is supported by evidence from density, seismic and magnetic studies also.

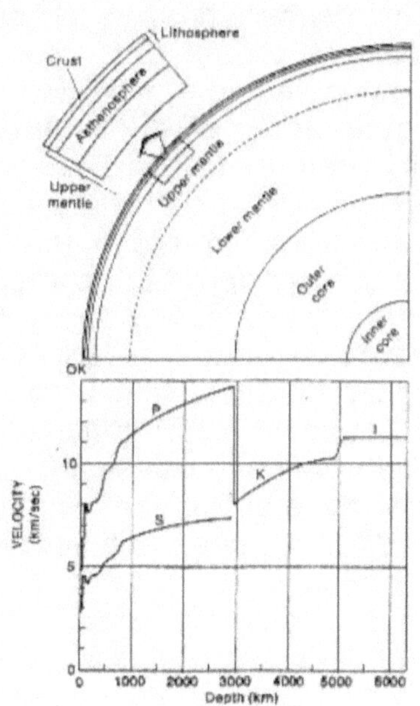

Figure 11-9. With this diagram, orthodox science interprets the internal structure of the earth as deduced from variations in seismic wave velocities with depth. They surmise that velocities of both the P and S waves increase to a depth of about 3,000 km (1,800 miles), where both change abruptly. The S wave disappears and does not travel through the outer and inner core, and the velocity of P waves decreases drastically. This is the most striking discontinuity in the earth and is considered to be the boundary between the core and the mantel. Another discontinuity occurs at a depth of 5,000 km, indicating an inner core. A low-velocity layer at depths from 100 to 400 km is called the asthenosphere.

The hollow earth interpretation would reduce the distance to the outer core to 800 miles caused by a higher increase in density than orthodox science estimates. The outer core discontinuity would be the inner surface. The asthenosphere discontinuity is the depth at which a more dense earth foundation exists consisting of metallic content with looser less dense rocks lying above it. Neither P nor S waves pass through the hollow of the earth. P waves bend around the hollow as well as through multiple reflections between the outer surface and the inner surface allowing weak P wave reception beyond the shadow zone. S waves cannot pass through the core because the earth's core is hollow.

The movement of continental plates in the crust and lithosphere are based on the idea that the asthenosphere is plastic. This plasticity is assumed to be caused by an increase in temperature with depth which brings the material in this region close to the melting point. And even though earthquakes occur down to a depth of 750 km (450 mi) the apparent contradiction between plasticity and shear ability is explained thus: Below the asthenosphere the rock is somewhat plastic but will fracture under sudden stress.

Now obviously, scientists do a marvelous job trying to explain observable data on our earth's interior. However, their conclusions, in reality are only just ONE theory of the earth's interior. It IS just a theory. No scientist has been down there to see if temperature increases with depth and creates a liquid outer core while at a greater depth the same increase in temperature creates a solid core. The farthest scientists have been able to penetrate down into the earth's interior is the Russian Kola Superdeep Borehole which reached a depth of 40,230 feet, or 7.619 miles in 1989 (Wikipedia.org, 2016). Yet, hollow earth explorers have traveled from the outer surface to the inner surface of the earth and find no molten interior and discovered that the earth's hollow interior is about 50% of the earth's total volume.

Illustrating the incongruity of orthodox science interior earth theory is a scientist's own admission that,

> *"We don't know too much about the behavior of materials at temperatures and pressure which would occur in this region of the earth, so the core is probably not like any liquid we encounter on the surface. Still, the fact that shear waves are not propagated through the core indicates that it is also not much like any solid we know about on the surface either."* (PHYSICAL SCIENCE 100, by Merrill, Hamblin and Thorne, Burgess Publishing Co., Minneapolis, 1978, pp. 11-13)

Now, if the core is admittedly not like any liquid or solid of which we know, could it perhaps not be better explained as being a gas or even empty space? After all, S waves will not travel through a gas or vacuum either. And there is nothing which scientists can conjure up to indicate why P waves cannot travel AROUND the core instead of through it. P-waves have that characteristic property of being able to bend around corners or obstacles like sound waves.

Let us consider the core to be an obstacle consisting of empty space as shown in Figure 11-10. In this case, S shear waves would not pass through this hollow core. P compression waves would be refracted (bent) around it allowing them to reach the opposite side of the earth. Also P waves traveling through the lower more dense portion of the mantle would carry the waves to the opposite side of the earth faster than those traveling through the upper, less dense crust.

The fact that neither P nor S waves are received in the shadow zone is an indication that NEITHER type of wave travels through the core. The shadow zone is THE evidence of the existence of a core. If some experiment could be performed which would prove that P waves do not travel through the core, then the core could not be a solid, liquid or gas, but pure space. It would be a hollow within our earth.

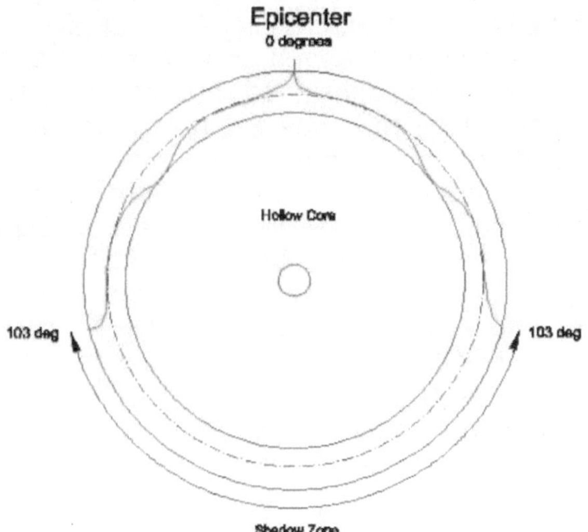

Figure 11-10. S-wave paths in a hollow earth arrive back on the surface at 103 degrees from the Epicenter. From the Epicenter, an S-wave curves more with depth because of increasing density. At the discontinuity at about 450 miles, all earthquakes cease, and S-waves at that point begin to curve in toward the inner surface as density decreases. At the inner surface, the S waves skim the inner surface and then follow the same pattern back out to the outer surface. No S waves pass through the hollow core and so none are received in the shadow zone past 103 degrees from the epicenter. The shadow zone is THE evidence for the hollow core.

Such an experiment was carried out by the U.S. Air Force. In 1977, this author had a friend attending the Arizona State University in Tempe who participated in an experiment sponsored by the Air Force. With an ultrasonic beam (which are compression waves similar to P-waves) and with recording devices, the experimenters aimed their ultrasonic beam straight into the earth with the purpose of trying to send a beam of sound straight through the earth. But they were unable to. No beam would go through the earth, but would bounce off the core or deflect around it. They made a map of the beams deflecting around the core and the combined paths described a sphere corresponding to the core boundaries.

My friend did not understand the significance of this experiment even though he participated in it. He figured that the ultrasonic beam was bouncing off something real hard down there. But sound waves travel faster through material the denser it is. For example, sound waves travel through steel about 15 times faster than it travels through air. The denser the material, the faster sound is going to travel through it. Therefore, the ultrasonic beam, which is a high frequency sound beam, would have changed directions at the core boundary but would have continued through the core if it were solid or liquid. Moreover, if it was a denser material, the beam would have continued through it at an even greater velocity. But it didn't. No beam was able to go straight through the earth to the other side but all were either bent around the core or reflected off it. My friend couldn't remember at what depth the reflection was occurring but said it was in the hundreds of miles. When I asked him to ask his professor if I could see the study, he returned and informed me that it was confidential material belonging to the U.S. Air Force.

What does all this mean? Many who have been in a high school physics class will remember the vacuum experiment with sound. A bell is started ringing in a sealed bottle with a vacuum pump attached. At first, the sound of the bell ringing is loud and clear but as the pump begins to pump out the air, less sound is transmitted and the ringing gets dimmer and dimmer until even though the ball can be seen hitting the bell, the sound cannot be heard through the vacuum. Sound does not travel through a vacuum. If the inside of our earth has a core of pure space, as Our Hollow Earth Theory states, then we have the reason why the experiments with the ultrasonic beam could not send their beam through the earth.

One cannot but wonder why such a fantastic discovery by the Air Force has not been shared with the scientific world. Actually, the conspiratorial forces controlling the U.S. government have known of the existence of Our Hollow Earth since 1920 when Marshall B. Gardner published his book, A JOURNEY TO THE EARTH'S INTERIOR, and subsequently Admiral Richard E. Byrd under the command of the U.S. Navy began the first explorations into the hollow of the earth in 1927. One can be sure that the government task forces charged with exploration of Our Hollow Earth know much more about the subject than does this author.

But what is there but obstinance to prevent some University or scientist from repeating this Air Force experiment? Surely the results can only be revolutionary for interior earth theories.

As far as the electromagnetic field of the earth is concerned, the Hollow Earth Theory doesn't need a liquid outer core and a solid inner core to explain it. Consider an atom. The fact that a piece of iron can be magnetized by cooling it from the molten state in the presence of a strong magnetic field indicates that atoms possess electromagnetic fields. In the process of making a magnet, atoms are aligned with their electromagnetic fields in the same direction thereby producing north and south poles on the combined atoms composing the metal piece. The reason a piece of iron is not naturally magnetized before the magnetization process, is because of the random orientation of the atoms.

Atoms are in fact, hollow, consisting of an electron shell orbiting about a proton-neutron nucleus with a magnetic field that consists of gyroscopic particles spinning at the speed of light being ejected out the south polar opening of the atom and traveling between layers of orbiting electrons and returning into the inside of the atom through its north polar opening in a never ending cycle. A rather large percentage of the diameter of an atom consists of empty space between the electron shell and the inner nucleus.

Even as atoms are mostly empty space, so are planets. If scientists would consider what the atom of the solar system and galaxy is, they would become aware of the universal hollow nature of the cosmos. The atoms of the solar system and galaxy are the planets. Although scientists are right in comparing the solar system to atomic structure, that is, the sun being the nucleus and the planets the electrons with the sun having a magnetic field, yet they need to discover the truth that planets are also built upon the same pattern. Even as atoms are mostly space, so are planets.

ALL planets are hollow creations. The shell in orbit about the interior sun compares to the electron shell and the interior sun is the proton-neutron nucleus. In fact, the earth's shell is negative in charge, and the interior sun is positive in charge. And since the interior sun rotates much slower than the shell, the revolving earth shell about the inner sun acts as a dynamo creating an electromagnetic force field around the earth.

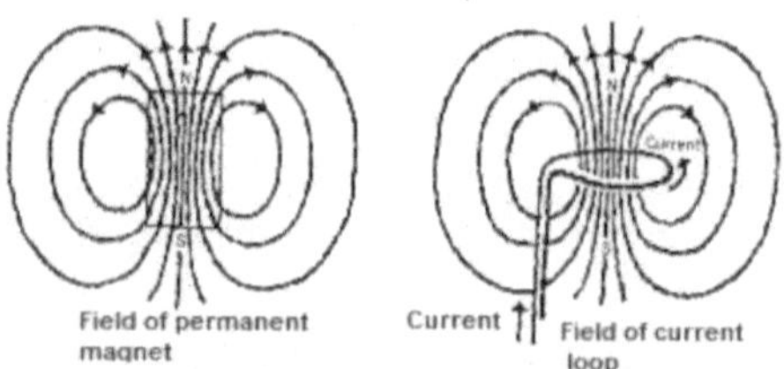

Figure 11-11. The electromagnetic field of a current in a wire loop or permanent magnet corresponds to the field lines in a hollow earth. The north and south directions of flux flow of the magnet and wire loop are the same as given by the points of the compass within the earth, which directions are opposite to those on the outer surface.

Figure 11-11 shows that the electromagnetic field of a current in a wire loop has the same configuration of field lines as the hollow interior of our earth.

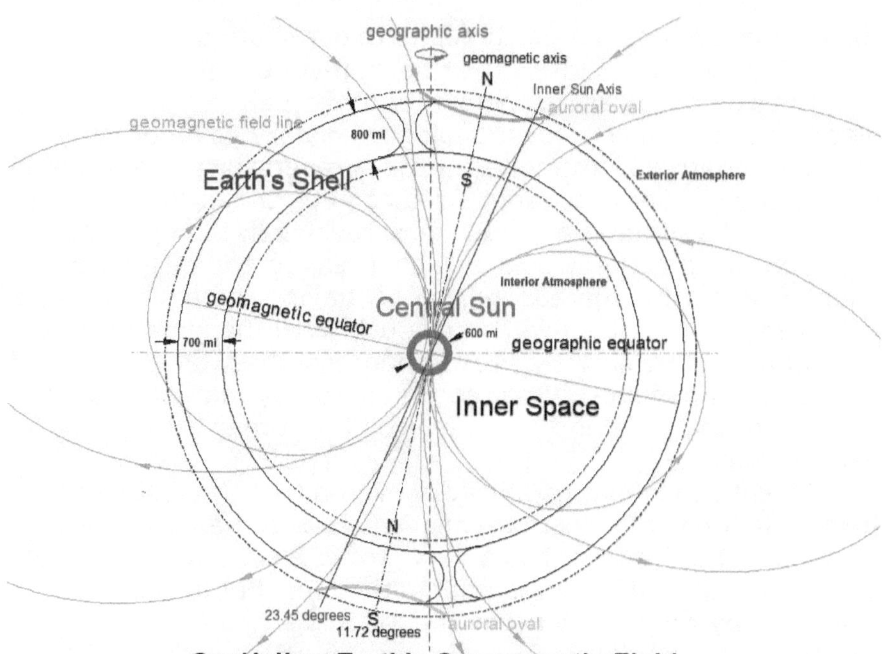

Our Hollow Earth's Geomagnetic Field

Figure 11-12. In the earth's interior electromagnetic field, their north is our south and our north is their south. Our east is their west and our west is their east. The earth's shell being negative contains the electric current in the dynamo analogy. Since the sun suspended in the earth's hollow interior rotates much slower than earth's shell, its positive charge causes the negative charges in the earth's shell to remain stationary with respect to the charges in the interior sun and the earth's shell moves around the negative charges. The resulting effect of the earth's rotation therefore causes an electric current in the shell to move from east to west the same as in the wire loop. This causes the field lines inside the earth to go from their south magnetic pole (our north) to their north magnetic pole (our south). After the field lines emerge from their north magnetic pole (our south), they loop around the earth and go back in the earth at their south pole (our north).

Should some method be invented to tap this electric current, which at the equator is moving 1,000 miles an hour (the speed of the earth's rotation at its surface), an unending source of free energy could be available to free the world from the oppressively high cost energy cartels of the Illuminati.

Figure 11-12 shows the Earth with a crosscut showing the Earth's interior electromagnetic force field as compared to our outer Earth force field. The field lines come out of the south polar area, travel north causing the compass to point north on the outside surface of the Earth and then go back inside the Earth at the north polar area around the magnetic pole. Thus the compass inside the Earth points to our south as if it were north. In fact, our south IS their north magnetic pole.

The several reversals in polarity of the earth's electromagnetic field in past geological history as recorded in the lavas emitted at the times of these reversals indicate that in the passage of ancient planet sized comets, the earth has been stopped in its rotation and caused to revolve in the opposite direction. Such action would entirely breakup and destroy a liquid-solid earth, whereas, our model having the resiliency of a hollow metal ball would withstand such world-bending torque.

Immanuel Velikovsky cites ancient histories, for instance, Herodotus and Mela say that according to Egyptian annals, the reversal of the west and east recurred four times throughout earth history. (WORLDS IN COLLISION, p. 122)

Recent space probes have discovered that the fast rotating planet Mars has NO electromagnetic field and the very slow rotating planet Mercury has a very strong electromagnetic field. This is unexplainable by magmatist theories, but is easily explained by the Hollow Earth Theory.

Since the sun inside a planet in conjunction with its rotating shell is what causes the electromagnetic fields in planets, Mars according to our theory has no electromagnetic field because its interior sun is rotating in the same direction as its shell and at the same speed. Should this exact synchronization seem improbable, consider the fact that our own moon makes exactly one rotation for every revolution it makes about the earth. That is why we never see the other side of the moon from earth, and there are many moons in our solar system that do the same. The planet Mercury's strong electromagnetic field is caused by a faster counter-rotation of its interior sun than its shell.

According to the Dynamo Theory of the origin of the electromagnetic field of the Earth, in solid-liquid interior theories, the field is set up by the rotation of the earth's shell about a stationary solid core with a countermovement of electric current-carrying magma currents in the outer liquid-molten core. If such were the case, then it is logical to assume that the resulting field caused by such rotation of the earth would produce an electromagnetic field possessing magnetic poles coinciding with the geographic poles of the earth. The fact that in reality the magnetic poles of the earth DO NOT coincide with the geographic poles would indicate that the orthodox Dynamo Theory with its liquid interior does not hold up to observable phenomena.

On the other hand, the Hollow Earth Theory explains these phenomena of the displaced geomagnetic poles with much greater consistency. The answer to this enigma is that the earth has TWO electromagnetic fields superimposed upon each other. One field is produced by the SUN within the hollow interior of the earth and the other is produced by the rotating shell of the hollow earth.

The retired mathematics professor, Al Snyder, in his book, SATAN'S SAUNA AND THE DEVIL'S TRIANGLE relates how he discovered that the earth has not one but TWO interacting electromagnetic fields. Within a hollow metal globe he placed two bar magnets. After some adjusting of their positions he was able to reproduce the magnetic declinations (by running a compass over the globe's surface) observed throughout the world today. Snyder's experiment indicates that it is the INTERACTION of the two electromagnetic fields that displaces the geomagnetic poles of the earth with respect to the geographic poles. The fact that the earth's magnetic poles are displaced toward the western hemisphere indicates that the displacing agent field is located WITHIN the field generated in the earth's shell: the one field INSIDE the other.

The displacement of the magnetic poles, also indicates that in the close passage of ancient planet sized comets, the earth's axis has been tilted so that the magnetic poles are located several hundred miles from the geographical axis and are slowly moving. In 1965, the North Magnetic Pole was located 11.5 degrees from the North Geographic Pole and moving about 8 miles a year towards the northwest. In 2001, the North Magnetic Pole was located at 81.3 N, 110.8 W, 605.68 miles or 8.79 degrees from the Geographic North Pole. In 2017, it was located at 86.5 N, 172.6 W and moving 34 to 37 miles a year. In 2020, the North Magnetic Pole was located at 86.502 N, 164.036 W and moving 34 miles per year, as reported by NOAA.

The South Magnetic Pole was first located by James Clark Ross in 1841 on the Antarctic continent, but over the years it has now moved out into the Pacific Ocean and in 2015 was located at 64.28 S, 136.59 E and was moving northwest about 6 to 9 miles a year. (Wikipedia.org, 2018)

The movement of the magnetic poles seems to be linked to a slow rotation of the Inner Sun. In 1965, the Inner Sun was also inclined to about one-half of the earth's axial tilt with respect to the earth's orbital plane about the Sun.

After observing changes in declinations of the magnetic compass, Edmund Halley, the English astronomer of the 1600s, came to the conclusion that the earth must be hollow. Observing what scientists today have observed, that the magnetic poles move, the author Augus Armitage wrote,

"Halley, in fact conceived the Earth as consisting of an OUTER SHELL with two magnetic poles, and an INNER NUCLEUS, concentric with the shell and possessing two poles of its own. The magnetic axis or shell and nucleus were INCLINED TO EACH OTHER (they do not coincide) and to the axis of the Earth's diurnal rotation, about which the two components TURNED AT SLIGHTLY DIFFERENT RATES; this DIFFERENCE GAVE RISE TO A SLOW RELATIVE MOTION OF THE MAGNETIC POLES with a consequent change in the magnetic variation. In the period required for the shell to gain (or lose) one complete rotation on the nucleus, the variation would go through a complete cycle and return everywhere to its initial value. This period might well be a long one, PERHAPS ABOUT 700 YEARS...he thought that the nucleus was rotating more slowly than the shell." (EDMUND HALLEY, by Angus Armitage)

Prior to discovering what Halley had written on the subject, I had come to much the same conclusion he had. I theorized that the movement of the magnetic poles is caused by a slow rotation of the interior sun inside Our Hollow Earth, making one complete rotation in about 794 years.

When God created the earth, he first created the interior sun which then became his base of operations during the creation period. As God revealed to the prophet Abraham, one day on God's planet-star, Kolob, in the center of our Milky Way Galaxy was equal to 1,000 earth years. It appears, therefore, that the earth's interior sun, as God's base of operations during the creation period, was given the same period of rotation as God's planet Kolob, when God said, "and the evening and the morning were the first day." (Genesis 1:5) Therefore, one rotation of the earth's interior sun today, assuming it hasn't slowed down, should equal one day on Kolob.

Knowing that the earth's electromagnetic field is caused by the earth's shell rotating about the interior sun and that this field rotates, as indicated by the rotation of the magnetic poles about the geographic poles, it is logical to assume then that the earth's interior sun is rotating at the same speed as the electromagnetic field.

Since during the creation period, one rotation of the inner sun was equal to 1,000 earth years, and is now rotating at a rate of 700 years according to Astronomer Halley's estimate, or 794 years according to my estimate, this indicates that at the time of the earth's creation, the earth was in an orbit closer to the Sun than it is now. This is supported by the calendar on the gate of the sun at the ruins of Tiahuanaco in the Andes Mountains of Bolivia that shows a 290-day year instead of the present 365¼. If Tiahuanaco's was earth's original calendar, then 1,000 290-day years should equal our present 365¼-day years, assuming that the Inner Sun has not slowed down.

The calculation for this is $1,000 * 290 = 365.25x$, $x = 793.98$, pretty close to Astronomer Halley's estimate. This means that, as an estimate, the Inner Sun, as well as the earth's magnetic field rotates at a present rate of 794 years which is equal to 1,000 of earth's original 290-day years when the earth's orbit was closer to our outer Sun.

As can be seen, the hollow earth theory accounts fully for the earth's strong electromagnetic field, earthquakes and the shadow zone without a need for a liquid-solid core inside the earth.

As for the density of the earth, the outer crust has an average density of 2.7 gm/cc, while the overall earth density according Newtonian physics is 5.525 times an equal weight of water (Steel is 7.8 gm/cc). This indicates that the interior of the Earth has a higher density than the outer crust.

We do not know how many layers the shell of the earth consists of, but there are certainly more layers than two. At a minimum there should be three layers, one for the outer surface layer, another for the inner surface layer, and then a layer between them.

Surface rocks have an average density of 2.7 gm/cc. If we divide the 50% of the volume of the earth that the shell comprises into three layers, each layer would be 16.67% of the volume of the shell. The outer surface layer would have a density of 2.7 gm/cc. If the inner surface layer also has a density of 2.7 gm/cc, then the middle layer would have a density of **5.65 gm/cc**: 2.7 + 2.7 + 5.65 = **11.05 gm/cc**. If the outer and inner layers are thinner in volume than 16.67%, then the middle layer would have a density greater than 5.65 grams per cubic centimeter.

This assumes that most of the earth's mass is located in its shell. As you can see, Newtonian physics requires an average shell density almost as dense as lead (11.3). And since surface rocks average 2.7 gm/cc in density, then the interior of the shell would have to be greater than the average density.

So we can say that a shell density of 11.05 gm/cc could be in the realm of possibility. After all, the earth DOES ring like a bell after a rather large earthquake. A bell is hollow and is made of metal, just as a hollow earth must be.

Figure 11-9 shows the orthodox scientific theory of the Earth's interior including a graph of the density increase towards the center (the same figures as the velocities) and the velocity of earthquake propagation in the Earth. Seismic P waves, which are similar to sound waves (they are both compression waves) travel through some of the hardest rock, dunite, a little faster than sound travels through steel. Sound travels through steel at about 3 miles per second. Now, scientists have P waves, which travel faster than S waves, traveling at a depth of 1,800 miles at a speed almost three times faster than sound travels through steel.

Since wave propagation increases with temperature, pressure and density, scientists in their theory must have an increase in one or more of these factors to account for the speed at which P waves rebound from the core to the surface. Since orthodox theory maintains that the Earth is full of material and is not hollow, they have many thousands of miles through which the waves in their theory must pass in the observed time limits. Therefore, they say that not only does density increase towards the center, but temperature and pressure also. You see, they are grabbing at everything in order to obtain the velocity needed to get their P waves through the distance their interior Earth theory requires.

If the density towards the center were to increase sufficiently to account for the velocity needed in their theory, we would have to have, at the outer core boundary a P wave velocity of about three times the speed at which sound travels through steel. Now since P waves travel through the hardest rock about the same speed as sound travels through steel, we must conclude that if density is what increases the velocity required in their theory, then the density of the Earth at a depth of 1,800 miles must be about three times as dense as steel. This can't be, they say, because at that depth the Earth is molten and S waves, which can't be propagated through liquids, stop and do not pass through the molten outer core of the Earth. So temperature and pressure are resorted to also in order to increase their wave velocity, but, as we shall see, temperature and pressure cannot increase throughout the distance to the Earth's core.

In our model of the interior, which accounts entirely for all observed phenomena, the mantle is a metal ball having a density of 5.65 gm/cc, which is an average density of common metals, and is probably located at a depth of about 450 miles where no earthquakes occur, and superimposed on its outer and inner surface is looser, less dense rock, gravel, sand, dirt and deposits of metals and minerals wherein earthquakes, volcanism and plate tectonic movement occur.

The guide in Lloyd's ETIDORPHA informs us that the Earth's center of gravity is about 700 miles beneath the surface of the Earth with the interior surface at 800 miles. The location of the center of gravity so close to the interior surface of Our Hollow Earth would require a higher density in the rock structure there. A greater concentration of metals at that depth would explain the lack of earthquakes. The deepest earthquakes occur no further down than about 450 miles. The strength of the metal concentration at this depth would prevent fracture while elastic enough to allow for the variations in the tides caused by the moon and the sun's gravitational interactions. (The moon's tidal interaction with earth causes the Earth's shell to rise and fall daily about 25 centimeters (9.84 inches)).

The elastic strength of the earth's shell at that depth would also account for the incessant vibration of the earth and the breathing motion of the earth which causes tall skyscrapers to rock back and forth. This very breathing action of Our Hollow Earth, caused by the variations of the sun and moon's gravitational forces on the earth as it rotates helps explain the cracks and faults in the outer crust of the earth which bring about earthquakes, volcanism and continental plate movement.

As can be seen in Figure 11-9, there are two boundaries inside the earth which reflect seismic waves. In our model, these boundaries consist of the interface at the metal ball and the inner surface where reflections occur. With this model, seismic wave velocities would be relatively slow down to a depth of 450 miles whereupon the high density of the metal ball would cause part of the wave to be reflected while that part which continued on would increase in velocity until the inner interface of the metal ball is reached. At that point the velocity would decrease to the inner surface of Our Hollow Earth where reflection would again occur.

The high density at the zone of the metal ball would account for seismic P waves which reach the opposite side of the earth faster than waves which travel close to the surface. The interior surface of the earth would account for the extraordinary reflection of seismic waves which echo and rebound off the core, the core being the hollow interior.

Scientists admit that they have no way of knowing that seismic P waves go through the core. Starting at 25 degrees arc from an epicenter, sharp reflections are received from the core. And beyond 1,000 kilometers great complexity in interpretation is presented by the increase of internal reflections and refractions. In short, there is nothing in the observations of earthquake waves that the Hollow Earth Theory cannot account for. In fact, this quotation from Encyclopedia Americana fits the Hollow Earth Theory far better than a liquid-solid earth theory:

"The very largest earthquakes cause the earth to vibrate like a BELL for several hours, with a fundamental period of vibration of 54 minutes." (p. 536, "Earth")(Figure 11-13)

Now, if the earth vibrates like a bell, then the earth must have a shape similar to a bell which is hollow!

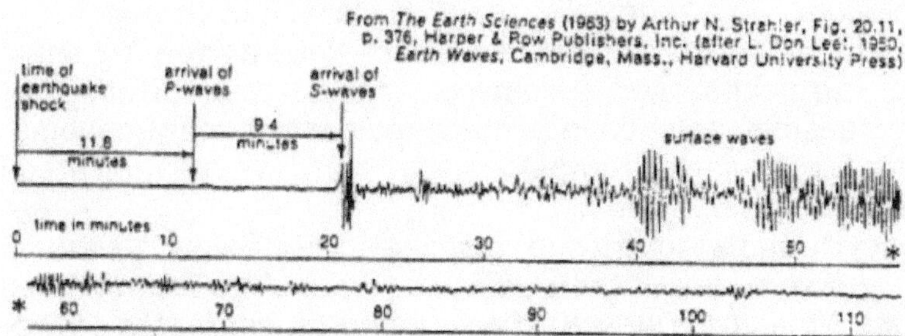

Figure 11-13. The very largest earthquakes cause the earth to vibrate like a bell for several hours, with a fundamental period of vibration of 54 minutes.

As for the claim of great interior heat, scientists do not know how hot it is inside the earth, or even if it is hot. They figure they need it to be hot in order to explain plate tectonic movement, why S waves do not travel through the core, to increase the velocity of seismic waves to account for the rapidity with which they are reflected back from the core and to explain the dynamo theory of the production of the earth's electromagnetic field. However, all the evidence indicates that the earth's interior is not hot.

Granted, there are diamond mines, geothermal wells and geysers heated by slow volcanic reactions. But this heat is local and certainly does not originate from a molten interior. The deepest caverns in the world are rather pleasant, perhaps even a little chilly, in the 50 degrees Fahrenheit. In a study I made of 20 of the world's deepest caves, the average temperature is 44 degrees Fahrenheit. Scientists' claims that temperatures rise 29 to 58 degrees F per mile into the earth are not compatible with the facts. If such gradients continued downward, temperatures that would melt rock (3,000 degrees F.) would be reached in only a hundred miles. However, the fact that earthquakes occur at these depths and S waves travel there shows that temperatures in fact must be more cool than hot. I-Am-The-Man, in ETIDORPHA would certainly agree. The civilizations now living in the cavern cities under Europe and America would also agree that the temperature within the earth's shell is even comfortable.

Scientists, in trying to explain how temperature increases with depth to fit their theories, at first launched onto the idea that the heat was derived from the decay of naturally occurring radioactive elements, especially uranium and thorium. However, actual observation shows that radioactive elements tend to concentrate upward near the surface in the light siliceous rocks such as granite perhaps causing the higher temperatures observed in some places near the surface, while the decreasing radioactive content toward the mantle would account for the lower temperature at that level as indicated by the seismic observations.

Encyclopedia Americana confesses:

"Estimates of temperatures in the deep interior are, in fact, educated guesses that depend on estimates of the increase in melting point with increasing pressure." (p. 539, *"Earth"*)

Pressure is therefore the only explanation scientists can dig up to give them the necessary increase in heat toward the earth's center which would melt the outer core. But what scientists seem to overlook in their eagerness to fulfill the requirements of their theory is that the pressure of overlying rocks in the earth's shell increases only until the pressure force begins to be canceled by the decrease in resultant gravitational acceleration as the earth's center of gravity is approached. From that point on downward toward the central sphere of gravity, pressure actually decreases even though the density of the rock composition increases because of the increase in metal content.

In the formula for pressure, P = force/area, the force causing the pressure is the weight or gravitational force of the earth on the overlying rocks. If that weight or gravitational force diminishes, pressure will likewise diminish in exact proportion to the decrease in gravitational acceleration. Since the acceleration of gravity diminishes toward the center of gravity because it is equalized in all directions by the gravitational effect of equal concentrations of mass in all directions, so does pressure with increasing depth become increasingly more equalized in all directions until one reaches the center of gravity where pressure and gravity are equal in all directions. The resulting force would then be 0 pressure and 0 resultant force of gravity. If you could be lowered slowly down a shaft to the center of gravity, your weight would gradually decrease until at the center of gravity, 700 miles down, you would weigh zero and would float as if you were in space.

The Inner Earth Guide, in ETIDORPHA, took William Morgan in 1827 through communicating caverns from an entrance in Kentucky down to the earth's central sphere of gravity 700 miles down from the earth's outer surface. They jumped off a cliff into a giant chasm where they fell past the center of gravity and then oscillated back and forth until they came to rest at the center of gravity. At the center of gravity, they were floating in the air and used the power of their minds to move themselves to the side of the chasm where another Inner Earth Guide was waiting to take William Morgan on and up to the inner surface of Our Hollow Earth.

Now, how can zero pressure at the central sphere of gravity cause the thousands of degrees temperature required to melt orthodox scientists' liquid outer core? Really, scientists in their speculations of the earth's interior have come up with a theory that is a pot of contradictions. But you really can't blame them. The International Illuminist Conspiracy knows that our earth is hollow, but has the scientific world believing in stories and fables as much as they do the common citizen and slave while the real truth is kept under the table.

Even scientists' explanation of gravity differences on the earth's surface is contradictory. For example, the earth's shape has been measured to be 16.451 miles (29.5 km) greater in diameter at the equator than at the poles. The force of gravity or weight is directly proportional to the mass of two objects and inversely proportional to the distance squared between their centers. A basic tenet of physics is that the greater mass an object has the greater gravitational force it exerts on any other mass. For example, one weighs more on earth than on the moon because the earth has more mass than the moon and therefore exerts more gravitational force. Now, if there is more mass between you and the center of the earth at the equator than at the poles, it is only logical that one would weigh more at the equator than at the poles. But in reality, the opposite is true. One weighs less at the equator and gradually weighs more as he goes toward the poles.

Scientists explain this by saying that gravitational force decreases with the square of the distance between the centers of the two objects--the earth and you. But this is true only if you separate the two objects by space. The farther you separate yourself from the earth's SURFACE, the less you will weigh, of course. But if there is MORE mass between you and the earth's center, you will weigh MORE not less. To say that you weigh less by putting more mass between you and the earth's center is like saying that a person would weigh less on the sun's surface because he is farther from its center than he would be from the earth's center while standing on the earth's surface. More mass should make you weigh more at the equator. But you don't. A person actually weighs less at the equator than at the poles.

The Hollow Earth Theory explains this much better than the liquid-solid earth theory. Since the earth was formed in rotation, centrifugal force would have thrown matter away from the axis of rotation at right angles causing the hollow in the earth. Heavier materials would not be thrown as far as the lighter materials thereby creating a more dense inner shell than the lighter outer portion of the shell. At the equator the velocity of the orbiting material would be faster than toward the poles causing the total shell thickness from inside to outside at the equator to be thinner than toward the poles where less centrifugal force would cause the shell to gradually thicken. At the polar axis, centrifugal force would throw material away from the axis at right angles causing the polar openings to form and hence a further thickening of the shell at the poles. With this model, one would weigh less at the equator and gradually more toward the poles because the thickness of the earth's shell increases toward the poles creating a greater concentration of mass at the poles and hence a greater gravitational (weight) pressure on a person standing on the earth's surface at the poles.

A further incongruency in the orthodox scientific explanation of why one weighs more at the pole than at the equator is their own observation of a decrease in gravity at the poles. The U.S. for a time had even considered building a satellite launching pad at one of the poles in order to take advantage of the lack of gravity there in helping satellites to achieve orbit. The hollow earth theory is the only way a rational person can understand how one can weigh more as he approaches the poles and yet near the poles there is a substantial decrease in gravity. Of course, this decrease in gravity is caused by the lack of mass there within the polar openings.

The lack of gravity over the holes at the poles makes satellite orbits over these holes very unstable. There is, in fact, an area over the Polar Regions where no polar satellites orbit. If they passed over the polar openings, they would either leave earth orbit and be lost in space because of the absence of gravity there or if they were orbiting closer to the earth's surface they would follow the earth's curvature through a polar opening and crash in the earth's interior. Hence all polar orbiting satellites go to one side or other of the polar openings.

In a list of 38 satellites put into orbit between 1957 and 1969, the closest satellites to the poles (90 degrees) were orbits of 97 degrees and 88.4 degrees inclination to the equator. Based on this, an area within 8.6 degrees near the poles have no polar orbiting satellites. This corresponds to a diameter of 592.88 miles. (One polar degree equals 68.939 miles; Earth's polar circumference is 24,818.142 miles -- ENCYCLOPEDIA BRITANNICA) This is an indication that the polar openings are not centered over the geographic poles.

A direct result of the decrease in gravitational force as one penetrates toward the center of the earth is the increase in the size of caverns. The low density of the athenosphere may therefore be partly explained as being the location of the largest volume of cavern extensity. Because gravitational force equalizes as one approaches the center of gravity, pressure decreases allowing the size of caverns to increase without danger of collapse.

With his guide, I-AM-THE-MAN in ETIDORPHA, crossed a lake 150 miles beneath the Atlantic ocean that was over 6,000 miles long. His Inner Earth guide told him,

"The shell of the earth above us is honeycombed by caverns in some places, in others it is compact, and yet, in most places, is impervious to water." (p. 158)
Furthermore,

"The crust, or shell, which I have just described as being about eight hundred miles in thickness, is firm and solid on both its convex and concave surface, but gradually loses its weight, whether we penetrate from the outer surface toward the center, or from any point of the inner surface towards the outside until at the central sphere (of gravity) matter has no weight at all." (p. 193)

His guide further explained the cause of volcanoes and earthquakes as related to the caverns:

"If men were far enough along in their thought journey..., they would avoid such theories as that which ascribes a molten interior to the earth. Volcanoes are superficial. They are as a rule, when in activity but little blisters or excoriations upon the surface of the earth, although their underground connections may be extensive. Some of them are in a continual fret with frequent eruptions, others, like the one under consideration, (Mount Epomeo, Italy) awaken only after great periods of time. The entire surface of this globe has been or will be subject to volcanic action. The phenomenon is one of the steps in

the world-making, matter-leveling process. When the deposit of substances that I have indicated, and of which much of the earth's interior is composed, the bases of salt, potash, and lime and clay is exhausted, there will be no further volcanic action from this cause, and in some places, this deposit has already disappeared, or is covered deeply by layers of earth that serve as a protection."
"Is water, then, the universal cause of volcanoes?"

"Water and air together cause most of them. The action of water and its vapor produces from metallic space dust, limestone, and clay soil, potash and soda salts. This perfectly rational and natural action must continue as long as there is water above, and free elementary bases in contact with the earth bubbles. Volcanoes, earthquakes, geysers, mud springs, and hot springs, are the natural result of that reaction. Mountains are thereby forming by upheavals from beneath, and the corresponding surface valleys are consequently filling up, either by the slow deposit of the matter from the saline water of hot springs, or by the sudden eruption of a new or presumably extinct volcano."

"What would happen if a crevice in the bottom of the ocean should conduct the waters of the ocean into a deposit of metallic bases?"

"That often occurs," was the reply; "a volcanic wave results, and a volcano may thus rise from the ocean's depths."

"Is there any danger to the earth itself? May it not be riven into fragments from such a convulsion?" I hesitatingly questioned.

"No; while the configuration of continents is continually being altered, each disturbance must be practically superficial, and of limited area."

"But," I persisted, "the rigid, solid earth may be blown to fragments; in such convulsions a result like that seems not impossible."

*"You argue from an erroneous hypothesis. The earth is neither rigid nor solid." (*ETIDORPHA *p. 192-194)*

Orthodox science claims that everywhere inside a hollow planet there would be zero gravity. It is called the Shell Theorem. Yet people that have gone to Our Hollow Earth report that their feet are as firmly planted on the interior surface as ours are on our outer surface. This, therefore, requires a revision in gravity theory. This author has developed a theory of gravity that explains how there could be a surface gravity on the inner surface of Our Hollow Earth that would allow plant and animal life to live there with a positive inner surface gravity and not zero gravity as orthodox science claims.

My theory of gravity proposes that the origin of gravity is in outer space and that the cause of gravity is a vacuum in the ether of space constantly being created in the nucleus of every atom. Gravity then becomes a push from space exerted by an etheric spiritual gas that fills the universe flowing in to fill the vacuum in the nucleus of every atom thus exerting a pressure on everything it passes through to reach that vacuum creating the pressure we call gravity.

As the ether of space enters the earth it spreads out to form a sphere 700 miles beneath the surface of the planet. It is a central sphere of gravity in the shell of the planet. While there is a small center of gravity in the center of the planet in the Inner Sun, the principal center of gravity is a sphere located between the inner and outer surfaces of the shell of the planet.

The Greenland Ice Hole experiment, as reported in the February 27, 1989 issue of Physical Review Letters journal, proved that the earth's principal center of gravity is not located in the center of the Earth, but rather closer to the surface of the planet, in the Earth's shell. What the experiment found was that the acceleration of gravity decreased faster as the gravimeter was lowered into the ice hole than if the center of gravity was in the center of the Earth, which is what would happen if the center of gravity that primarily affects us is a sphere in the shell of the planet rather than in the center of the planet. The significance of the Greenland Ice Hole experiment is that it discovered that the center of gravity is closer to the surface of the planet. If the center of gravity is closer to the surface of the Earth than at the center of the Earth, this means that there is a large area in the center of the Earth with no matter – a hollow in the Earth.

Ether accelerating into the planet begins very slowly far out in space and as it approaches the earth it flows faster and faster, but after entering the outer surface, it begins slowing as it reaches the central sphere of gravity 700 miles beneath the outer surface, according to the Inner Earth Guide in ETIDORPHA. Ether also accelerates into the inner surface but at a slower rate than the outer surface, but after entering the inner surface, it also gradually slows down until at the central sphere of gravity it reaches zero acceleration. Ether also accelerates into the Inner Sun, but then after entering the surface of the Inner Sun slows down until at the central sphere of gravity of the Inner Sun it reaches zero acceleration. So our planet actually has two centers of gravity. The center of gravity in the shell of the planet actually takes the form of a sphere, a central sphere of gravity and is the principal center of gravity that affects us the most.

My theory of gravity was developed from a study of Joseph Newman's gyroscopic particle theory of matter (THE ENERGY MACHINE, Chapter One). I further elaborate on Joseph Newman's gyroscopic particle theory of matter by stating that these gyroscopic particles consist of spinning balls of what 19th century scientists called the Ether, which I define as a tenuous spiritual matter which fills all space.

The effect of gravity is produced in the nucleus of atoms when the gaseous ether is concentrated down into gyroscopic particles in the nucleus of all atoms which then are ejected out of the nucleus to become the magnetic field of atoms. The configuration and spin of the particles in the nucleus of atoms, that is, the protons and neutrons, take the ether that fills all space and concentrate it down into gyroscopic particles which create a vacuum in the ether in the nucleus of atoms causing ether surrounding the atom to flow in toward the nucleus to fill the vacuum. The effect of gravity, then, is the pressure of the ether flowing into the nucleus of atoms to fill the void caused by the creation of gyroscopic particles which take up less space than the ether they are made of.

Gyroscopic particles are continually being lost from the magnetic fields of atoms by the collisions of electrons and other particles into the magnetic field which knocks gyroscopic particles out of the field in the form of electromagnetic radiation. This electromagnetic radiation consists of groups of gyroscopic particles knocked out of the magnetic field of atoms which then travel through space in the form of light, radio and heat waves. Each group of gyroscopic particles is called a photon. The distance between each photon determines the frequency, and the quantity of gyroscopic particles in each photon determines the intensity of the electromagnetic radiation.

This author is convinced that gyroscopic particles are being created in the nucleus of atoms to replace those lost to electromagnetic radiation because if those lost were not replaced by newly created ones, eventually all atoms would lose their magnetic fields and become extremely cold.

In fact, since the Inner Sun is continually emitting atomic particles from the Inner Sun solar wind in the form of electrons, protons and neutrons, it must also be creating these atomic particles from the ether of space, just as atoms create gyroscopic particles to populate the magnetic fields of atoms. The ether of space flows into the Inner Sun and electromagnetic radiation and solar wind atomic particles flow out. These atomic particles from the Inner Sun emit through the polar openings to light up the auroras and fill the Van Allen Radiation Belts, and could also be increasing the mass of the earth and contributing to the continual planetary expansion as a growing living being.

It is the ether flowing into the nucleus of atoms that keeps the atomic particles together. The pressure of the ether flowing toward the nucleus of atoms is what keeps electrons from flying off in a straight line instead of maintaining their orbit about the nucleus. The ether flowing into the nucleus also keeps gyroscopic particles orbiting in and out of the nuclei producing the magnetic field of atoms.

As gyroscopic particles are created in the nucleus, they are ejected out the south polar opening of the atom and are met with the inflowing ether which bends their path of flight causing the gyroscopic particles to orbit in a polar orbit between orbiting electrons that are in a more or less equatorial orbit thus keeping electron *"shells"* apart. The gyroscopic particles in the magnetic field also orbit outside the electron shells, traveling more or less at right angles to the electron orbits. The gyroscopic particles continue toward the north polar opening of the atom where, because there are no orbiting electrons in that area, the gyroscopic particles are then pushed back inside the atom by the incoming ether where they are again ejected out the south polar opening by the nucleus. In essence, the nuclei of atoms are in reality little radiation generators or "suns" in the hollow interior of atoms. Indeed, atoms have the same configuration as our Hollow Earth with an interior sun, a hollow surrounded by a shell consisting of electrons, polar openings and a magnetic field.

If we would assume then, as evidence indicates, that gravity is an ether flow from space instead of a mysterious pull from the earth's center, then the effect we know as gravity can be explained thus: Gravity is an ether flow from space flowing into the earth. As the ether enters the earth, it pushes on everything it goes through. Hence, we are held to the surface of the planet from a constant steady push or pressure from the ether entering the earth from space. The ether then spreads out after entering the surface of the planet. This spreading out of the ether as it descends toward the central sphere of gravity located about 700 miles beneath the outer surface is what creates the central sphere of gravity -- the ether flow is spreading out to form a sphere, not a point inside the earth.

Establishment scientists reject the existence of the ether based on the Michelson-Morley experiment of 1887 which they claim proved that the ether of space does not exist. The truth of the matter is that the Michelson-Morley experiment did not have a null value. It was just so low that they decided it was within the range or error. The reason the value was so low was that the experiment was conducted underground with heavy walls surrounding it.

Dayton Miller continued the interferometer experiments on a hill top and other open places for 25 years after the Michelson-Morley experiment and proved conclusively that the ether of space does exist. Dayton Miller published his findings in the 1933 Reviews of Modern Physics. However, the results of Dayton Miller's years of experimentation and measurement of the ether flow was ignored by the scientific establishment which is primarily influenced by the international bankers who do not want the people to know that the ether exists and that with it free energy can be generated.

This ether flow theory of gravity is also the basis of Flying Saucer technology. In essence, it has been discovered by several scientists that gravity can be countered and controlled by electrostatic forces on capacitors and used to propel a craft. These scientists have actually built working models of flying saucers using this technology. There is evidence that the US government is using this technology to build flying saucers underground. The Illuminati who control the sections of our government developing this technology hope to use it to help them take over the world and establish their New World Order.

With so much evidence proving the liquid-solid earth theory impossible, isn't it clear that scientists have gone to great lengths to fit the facts to their theory instead of using the facts to formulate a theory? Their stubbornness can be only understood as having a purpose. Although unknown to most scientists, who work for money and not to reveal the truth, the purpose of their masters is to keep the discovery that our Earth is hollow a secret.

CHAPTER TWELVE
Our Hollow Earth and the Plate Tectonic System

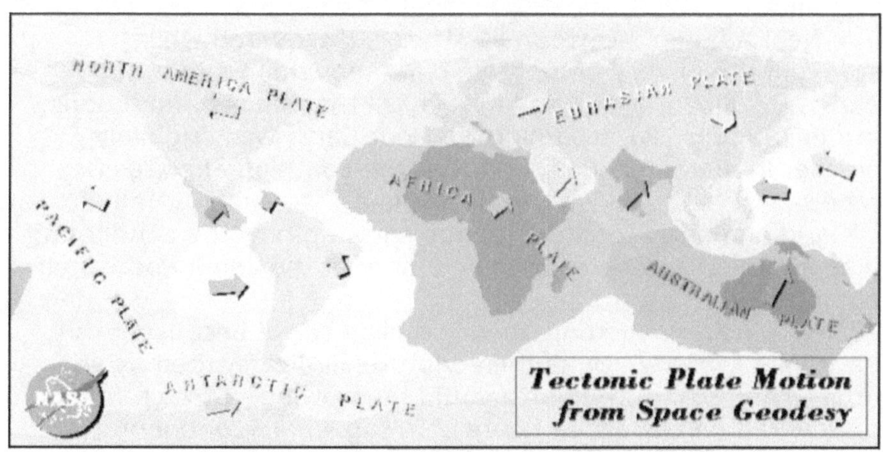

Even the Bible refers to continental drift, saying:
> *"...all the foundations of the earth are out of course."*
(Psalms 82:5)

Although the continental drift theory has been debated since the 1920's, it has only been since the late 1960's that the theory has become more accepted.

In the Continental Drift Theory, geophysicists claim that the lithosphere, which includes the earth's crust, is rigid, but the underlying asthenosphere yields to plastic flow. The principle behind plate tectonics is that the rigid lithosphere moves in response to flow in asthenosphere *"convection cells."*

Hot material in the asthenosphere, supposedly heated by radioactivity, rises to the surface causing mid-oceanic ridges and from there flows laterally carrying the broken-up lithospheric plates with it. These convection cells then cool off and descend back into the asthenosphere at the boundaries where the plates collide, thereby causing ocean trenches, volcanoes, earthquakes and mountain ranges.

Seven major lithospheric plates are recognized with several smaller ones. From patterns of magnetic reversals recorded in lava beds along the oceanic ridges, it has been established that the plates are moving apart from 1 to 16 cm per year. In recent years, however, tectonic plate motion from NASA's Space Geodesy using satellite laser ranging have reported even slower estimated plate motions ranging from 2 mm/yr up to 52 mm/yr.

But there is another more plausible theory of how plate tectonics actually works.

Neal Adams, a cartoon illustrator, discovered while studying the history of tectonic plate movements that our earth, the Moon, Mars, and in fact, all the planets and moons, are expanding. By assuming that our Earth was originally smaller in diameter, he found that the continental plates fit together perfectly like a puzzle without the oceans. He has created animated motion pictures showing how the continental plates moved apart over geologic time at the continental shelf boundaries.

Neal Adam's tectonic theory makes sense because if the earth is a living entity as the scriptures indicate, then as any other living being, the earth could be growing. Stretch marks discovered on the ocean floors showing where continental plates have moved apart support this idea that the earth is expanding.

You can view Neal Adams animations at:
https://www.youtube.com/watch?v=oJfBSc6e7QQ

Neal Adams' expanding earth hypothesis supposes it would take millions of years for the earth to expand from its original size where all the continents were close together without the oceans separating them to where the continents are located today. Biblical history, however, indicates that the earth is only 13,000 years old since its creation began.

Perhaps the expansion occurs slowly on a regular basis, but can expand at a greatly increased rate in surges, such as at the time of Noah's flood, and at the time of Peleg, whom I calculate lived between 2,253 and 2014 B.C. in whose lifetime the Bible records that the continents were divided.

Of course, orthodox science denies that the earth is expanding even though they do admit that the plates do move. They claim that at plate boundaries there are subduction zones where one plate goes beneath expanding ones.

However, scientists do recognize that there are problems with their tectonic engine -- the asthenospheric convection cells.

Since the major concentrations of earthquakes and volcanic activity occur along the plate boundaries where the plates split apart or collide, scientists are faced with the dilemma of accounting for the shear ability (in order to create earthquakes) and the plasticity of the hot up-welling lava of the convection cells at the mid-oceanic ridges and their subsequent consumption at the trench zones. The dilemma presents itself when one realizes that earthquakes are observed to occur most frequently exactly where the crust is supposed to be most liquid. The conundrum presents itself because obviously earthquakes cannot occur in molten lava.

The very fact that high concentrations of earthquakes occur at plate boundaries should give scientists the clue that the observed equally high concentration of volcanic activity at plate boundaries CANNOT be caused by an up-welling of lava from asthenospheric convection cells.

The Hollow Earth Theory gives the answer to this apparent contradiction of earthquakes occurring where volcanism is most active. The Hollow Earth Theory maintains that the asthenosphere is neither liquid, plastic nor has convection cells. The engine power behind the plate tectonic movement may result because the earth is expanding as proposed by Neal Adams.

Like any living being, the earth is growing. In a growing, expanding earth, mountains would be formed because the ocean basins are expanding faster than the continents and as the basins butt up against the continental shelves, bulges appear on the continents creating mountains and valleys. The expansion of the Indian Ocean pushed the Indian continent north to form the Himalaya Mountains. The expansion of the Pacific basin expanding against the American continent created the Rocky Mountains, Sierra Madre and the Andes. The expansion of the Atlantic basin against the United States created the Allegheny Mountains on the east coast.

A passage of scripture in the Doctrine and Covenants of the Church of Jesus Christ of Latter-day Saints indicates that in the beginning of the Earth's existence, the oceans of the Earth were all located in the Arctic Ocean and in the North Country of Our Hollow Earth, and the outside world consisted of one giant continent and one small ocean near the North Pole and down into the Atlantic a short distance. This passage of scripture says that when Christ comes in power to reign as King of Kings on Earth in His soon to be Millennial Reign,

> *"He shall command the great deep* (the oceans)*, and it shall be driven back into the north countries* (the Hollow Earth)*, and the islands shall become one land;"*

> *"And the land of Jerusalem and the land of Zion* (America) *shall be turned back into their own place, and the earth shall be like it was in the days before it was*

divided (in the days of Peleg shortly after the Flood of Noah)." (Genesis 10:25)" (D&C 133:23, 24)

Therefore, from the scriptures it would appear that the break-up of the original continent of Pangaea occurred in the days of Peleg. Peleg was born 100 years after Noah's Ark settled upon the mountains of Ararat. In fact, the name Peleg means "Division." From Biblical chronology, assuming that Adam and Eve were driven from the Garden of Eden in 4,000 B.C., Peleg was born in 2,253 B.C. and therefore the break-up of the continent occurred at about 100 years after the Flood of Noah. The city of Babel was just getting a good start when the great division occurred. It appears, then, that the earthquake that leveled the tower of Babel started the spreading that created the ocean basins.

After quoting the passage cited above from the Doctrine and Covenants, Dr. Cleon Skousen wrote in his book, THE FIRST 2000 YEARS,

> *"This gives us a clear indication that the dividing of the earth in the days of Peleg was the result of the sinking of the land in certain areas so that the sea could rush in from the Polar Regions. Apparently there was a sinking of the Pacific floor, and there may have been a sinking of the Atlantic floor although the scriptures point out that there was a 'sea east' in that area even before the division. (MOSES 6:42; 7:14) Whether or not the 'sea east' was as large as the Atlantic Ocean of modern times we are not certain. But in any event, we do know that when the division occurred around 2,240 B.C., America, Australia and the islands of the sea were cut off from the 'one land.'"*
> (THE FIRST 2000 YEARS, pp. 232-3)

The contours of the east coast of South America and the west coast of Africa indicate that these two continents at one time were together. The contours, preserved by the split, would seem to indicate that there did not exist an ocean or sea between South America and Africa before the breakup. The 'sea east' in the above mentioned scripture would be interpreted as being east of the land of Adam (located in Missouri) -- the North Atlantic, but necessarily much smaller than it is today.

In searching for the phenomenal force which broke up the Pangaean continent, I have found in Immanuel Velikovsky's research the evidence of close passages of planet sized comets in past earth history the most convincing source of that power. His catastrophic theory of past geological upheaval is also more compatible with the scriptural account of history. Velikovsky's demolition of orthodox geology's theory of Uniformity, which claims that all land forms on Earth have been formed throughout millions of years by erosion, convection cell uplift, and other agents still acting upon the Earth today, is complete. He also extends a heavy hand on the theory of Organic Evolution which Darwin based on Lyell's theory of Uniformity.

In his EARTH IN UPHEAVAL, Immanuel Velikovsky presents the following indications that the Earth has been pulled out of its original orbit, its axis changed and its rotation effected by close passages of planet sized comets in recent geological history: 1) The elliptical orbit of the Earth about the Sun changes by a very small amount. This could be a residue of a displacement of the Earth from its original orbit. 2) The obliquity of the ecliptic, or the angle which the plane of the equator makes with the plane of the Earth's orbit, which is now 23 1/2 degrees, changes by a very small amount each year.

This variation could have been caused by a disturbance of the Earth's axis in the past. 3) The fact that the plane of the Moon's orbit about the Earth is on the same plane of the Earth's orbit about the Sun and does not coincide with the equatorial plane indicates that maybe the equatorial plane at one time coincided with the plane of the Moon's orbit which today it does not. 4) Coal beds in Antarctica, Alaska and Spitsbergen, and recent glaciation in temperate latitudes indicates that the Earth's axis has undergone a change. (p. 119)

Velikovsky's books are filled with the evidence of past global catastrophes caused by near passes of comets. He reviews vast evidence from geology, astronomy, and ancient legends and histories that near passages of comets produced the forces that have helped to create the mountains on the continents and the expansion of the ocean beds. And although he rejects the Continental Drift theory, he agrees with our theory that the propagation of seismic S waves in the Earth's interior strongly indicates that the theory of Isostasy -- that the continents and mountains slide around on a liquid interior -- is false. The incongruency results from the fact that seismic S-waves are known to travel precisely in the area of the supposedly liquid convection cells, and the fact that S-waves cannot travel through liquids. In fact, the evidence that earthquakes occur as far down as 450 miles proves that convection cells, isostasy and a molten interior do not exist.

Nevertheless, the scriptures, which creationists accept as literal history, do indicate that the continents *"have turned out of their place"* about 2,240 years before Christ.

The Hollow Earth theory does allow for continental drift, and additionally that the earth is growing or, in other words, expanding. This theory proposes that the earth is a hollow body consisting of a higher density with greater concentrations of metals at the depth of the central sphere of gravity located according to the guide in ETIDORPHA, 700 miles down. And superimposed on this higher density sphere is the less dense crust of the Earth. In the area between the crust and the denser central sphere, comparable to the asthenosphere in the orthodox theory of the Earth's interior, is an area of quite extensive cavern systems which, like rivers, are caused by water erosion beginning small near the Earth's surface and continually getting larger towards the central sphere of gravity. At a depth of about 200 miles down where the equalization of gravity or weightlessness has become so great that the surface tension of water will not allow it to continue any deeper towards the central sphere of gravity, all volcanic activity ceases because, according to our theory, volcanic activity is caused by the reaction of water with pure, highly reactive elements such as phosphorus, sodium, potassium, magnesium and sulphur.

The continental drift may have originated with a close passage of a planet-sized comet or asteroid. Several giant craters on the Earth's surface is evidence that the Earth has sustained impacts by large meteors. Perhaps a direct hit by a large meteor together with a close passage of a planet-size comet may have been the cataclysmic force behind the initial breakup of Pangaea which *"turned the continents out of their places."*

As creationist scientists concur, after Noah's flood, an ice age occurred. Perhaps the pressure of ice expansion helped the breakup of Pangea one hundred years after Noah's flood.

Most likely a planet size comet passing over one of the earth's poles caused the worldwide catastrophe that was Noah's flood. For example, remains of mammoths have been discovered far down into Siberia south of the Arctic Ocean. Perhaps, when the planet-size comet that passed over the Arctic and tipped the Earth on its axis, this sloshed the Inner Ocean within Our Hollow Earth out through the North Polar Opening and carried with it many of the mammoths that live on the Inner Continent out through the polar opening and deposited their bodies far down into the Siberian tundra.

This planet sized comet as it passed by the Earth, also caused Earth's orbit about the Sun to move from its original location closer to present day orbit of Venus farther away from the Sun. This caused the waters of Noah's flood to later freeze at the poles and even far down into the mid latitudes. Then when the earth was hit by a large meteor that caused the earthquake that brought down the Tower of Babel, the pressure of the ice at the poles split the continents and started the oceans to start expanding and moving the continents apart.

In Dr. James Maxlow's book, <u>MODELING THE EARTH</u>, he states that an expanding Earth would need additional matter added at a rather regular rate. He suggests that the added matter needed for an expanding Earth could come from the Sun, from the solar wind particles. But as we have shown in this book, the solar wind from our outer Sun is deflected around the Earth by the earth's magnetic field, and so cannot enter to add mass to the Earth. On the other hand, our inner Sun could be adding mass to the Earth by changing the ether of space that flows into the inner sun as gravity into atomic particles such as electrons and protons on a regular constant basis. Therefore, we conclude that the solar wind from the Inner Sun, not only lights up the Earth's auroras, and fills the magnetosphere and Van Allen Radiation Belts with a constant influx of ions, but as such can be the principal source of new matter added for an expanding Earth as these ions fall onto the inner surface and outer surfaces of the planet from the inner sun solar wind.

The electromagnetic field of Our Hollow Earth is caused by the Earth's shell revolving about the much slower rotating central sun suspended in the hollow of the Earth. The rotation of the Earth about this central sun, which carries a positive charge, causes electromagnetic field lines to emerge from the south polar area, travel north on the exterior of the Earth and then enter back into the Earth in the north polar area. Therefore, the several reversals in polarity of the Earth's electromagnetic field in past geological history as recorded in the lavas emitted at the times of these reversals, would indicate that in the close passages of ancient planet-size comets, the Earth has been stopped in its rotation and caused to revolve in the opposite direction. Such action would require the resiliency of an Earth with a hollow metal ball core to withstand such world-bending torque.

Immanuel Velikovsky cites ancient histories, for instance, Herodotus and Mela say that according to Egyptian annals, the reversal of the west and east recurred four times throughout earth history. (<u>WORLDS IN COLLISION</u>, p. 122) He quotes

Exodus 12:2 which indicates that the Earth's axis was tipped by the passage of a planet-size comet, which he said was Venus, at the time of the Israelite Exodus from Egypt. The Exodus occurred in the fall, in September, which after the comet's passage became spring. Afterwards, April became the first month of the Hebrew year.

The Earth has been pulled into a larger orbit farther from the Sun by these near-passages of planet sized comets. The ruins of Tiahuanaco located on the south shore of the Lake Titicaca at 13,000 feet in the tops of the Andes Mountains, appear to have been built before the Flood of Noah with anti-gravity technology which they used to move massive stones weighing up to 200 tons and shaped with precision technology to build their city. The calendar on the Gate of the Sun of Tiahuanaco depicts a year of 290 days, which at that time was an orbit closer to the Sun than the orbit Earth now has. The astronomical information recorded on the calendar at the time of its construction shows that the earth also had an inclination to the plane of the earth's ecliptic about the sun of 16.5 degrees -- much less than the 23.45 degrees that it is now. Since a year on Earth today is 365.25 days long, the Earth at that time must have been closer to the sun, near the present orbit of Venus which has a 244.7 day year. (SECRET OF THE AGES, p. 32, WORLDS IN COLLISION, p. 128)

Therefore, in the passages of these comets, several of which according to Velikovsky have caused the Earth to tilt on its axis, throwing it into a different orbit farther from the sun, and causing it to reverse its rotation not once but four times -- the looser, less dense crust of the Earth has been caused to slide over the deeper metallic hollow core of the Earth at the central sphere of gravity. The expansion of the Pacific, Atlantic and Indian Oceans caused the continents to rise and the ocean floors to fall as stretch marks on the ocean floor attest to allowing the Arctic Ocean to sweep in to fill the void.

The original Pangaean continent had no mountains. Rolling hills were the dominant land form. The Pacific Ocean floor has preserved this unique feature even after sinking and being covered with ocean. The abyssal hills of the Pacific Ocean floor are relatively small hills rising up to 900 meters above the surrounding ocean floor. Covering 80-85% of the seafloor, they are the most widespread landform on the Earth.

Mountains were created as the ocean floors expanded pushing continental plates apart or into each other. For example, the Indian plate ran into the Asian plate producing the Himalayas. The Pacific Ocean bed expanded against the American continental plates raising the Rocky Mountains, the Sierra Madre and the Andes. The Atlantic Ocean bed expanded both directions from the mid-Atlantic Ridge towards the American Continent and towards the European Continent. The expansion of the Atlantic sea bed against the American continent raised the Appalachian Mountains on the east coast. The Mediterranean sea floor expanding against the European Continent raised the Alps.

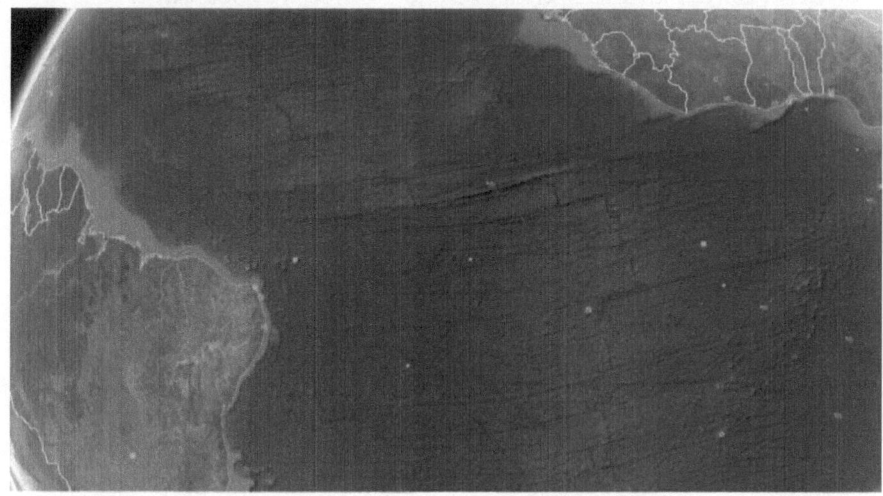

Stretch marks visible on the ocean floors can be seen on Google Earth. Stretch marks are visible on both sides of the mid-Atlantic Ridge showing that the Atlantic Ocean bed has expanded pushing North and South America away from Europe and Africa. Stretch marks can be seen between Africa and India, between Antarctica and Africa, and on the eastern Pacific sea bed from South America. Stretch marks can also be seen on the sea floor between Australia and Antarctica. As the Pacific plate stretched, it down-faulted away from the Asian continent causing the deep sea trenches north of Papua New Guinea all along the coast from the islands of the Philippines to Japan and up to Alaska. As the Pacific plate stretched away from America, it caused the trenches along the east coast of South America up through Central America and up along the coast of Mexico.   The stretching caused Death Valley in California to drop below sea level. Eventually, California will become an island of the Pacific.

The Arctic Ocean was probably part of the original ocean in the North Country of Our Hollow Earth making it the oldest ocean in the world, and it rushed in to fill the void caused by the expansion of the ocean basins.

The expansion of the ocean basins is what causes earthquakes to occur at plate boundaries. Although the force that originally broke up the Pangaean continent into different plates, according to our theory, was the near passes of comets in past geological history, asteroid hits and ice age expansion from the poles, the fact that cracks and crevasses continue to appear world-wide indicate that Earth expansion continues to occur together with the gravitational interactions between the Earth, Moon, Sun and planets which continue to shift the plates with movement at plate boundaries causing earthquakes to occur, such as at the San Andreas fault zone of California. Another example of earth expansion is that which happened in 2005. A gigantic, 35-mile-long rift broke open the desert ground in Ethiopia creating a large valley. Rifts have also occurred around the world without any accompanying earthquakes, such as in Michigan, Bolivia, the Philippines, Peru and the most recent eruption in Gulistan, Pakistan and Burdwan, India.

Diverging plate margins mark the site of the most voluminous extrusions of volcanic material on Earth. It has been estimated that approximately 20 cubic kilometers of lava is extruded each year along this zone. Scientists claim that this extrusion of lava at the mid-oceanic ridges and along converging plate margins is produced as a descending plate moves down into a "hot" asthenosphere where it is melted. The molten rock being less dense then wells up to the surface as lava to create mountains. However, we must consider the fact that most mountains are composed of stratified sedimentary rock, not lava.

Scientists need a molten asthenosphere to give them a plate-moving engine called an asthenospheric *"convection cell."* And they point to all the volcanic action at plate margins as proof of their hot convection cell power-horse. What they cannot explain is how S earthquake waves travel through a liquid asthenosphere and how earthquakes occur precisely where volcanic action is uppermost – at plate boundaries. In fact, most other earthquake activity occurs in the asthenosphere, so how can it be liquid? Actually, they call it *"plastic."* But plastic materials such as putty or modeling clay are not elastic. Elasticity is a prime requisite for the production of an earthquake.

Plastic materials are those which permanently change shape under the influence of a force. Have you ever tried to snap a roll of clay like you could a dry stick of wood? Materials which deform when a force is applied but which return to their original size and shape when the force is removed are elastic. A stick is elastic. Solid rock is elastic. If bent far enough they are strained beyond their elastic limits and break. Scientists should cease contradicting themselves and admit that earthquakes in the asthenosphere are caused by rocks that rupture when they are strained beyond their ELASTIC, not *"plastic"* limits.

Therefore, the asthenosphere is neither plastic nor liquid but solid rock. And volcanism does not result from an upwelling from a liquid interior at fault zones.

According to our Hollow Earth theory, the formation of Our Hollow Earth consisted of space dust and rocks collecting upon a sphere of energy, the Earth's spirit world. This space dust and rocks consisted primarily of pure elements. The heavier, more dense and less reactive elements collected near the center of gravity 700 miles down and the lighter, more reactive elements accumulated nearer the surface. This separation was effected by centrifugal force as the materials were placed in orbit about the sun inside the Earth. Therefore, the asthenosphere resulted in an accumulation of the lighter elements -- calcium, magnesium, sodium, sulphur and potassium together with the light metalloids, silicon and germanium. All this occurred in the absence of either water or an atmosphere.

Subsequent chemical reactions that occurred when water was added to the Earth's surface created the lithosphere through volcanic action. The porous, cavernous, less dense asthenosphere was created by the subsequent dissolution of the salts created in the volcanic reactions of water with the original space dust materials. Present volcanic action of the Earth continues to occur as fresh surfaces of that primeval reactive space dust are brought into contact with air and water at fault zones.

For example, water in the presence of air violently reacts with sodium, potassium, and magnesium. When volcanic lava is analyzed, it indicates that it results from violent chemical reactions of water with the alkali elements. Various reactions occur simultaneously because of the presence of several metals and non-metals in the vicinity of the original reaction. The different proportions of availability thereby determine the kind of volcanic lava produced.

The process resulting in volcanic activity begins with the movement of tectonic plates by the gravitational interactions with comets, the Moon, Sun and planets. Perhaps even a direct asteroidal impact could have initiated the tectonic plate breakup with subsequent residual movement caused by interplanetary gravitational interactions. An expanding Earth would also cause plate movement apart from each other and stretching of the Earth's shell. Plate tectonic movement causes faulting at plate boundaries and other places. Faulting exposes new surfaces of pure alkaline elements that the Earth was originally made of allowing water to seep down into deposits of original Earth building materials. The subsequent reaction of water, air and the alkaline elements results in volcanic action.

As can be seen by this analysis, the Hollow Earth Theory accounts for the Plate Tectonic movement and related phenomena with much greater consistency with the known facts than does the Solid-Liquid Theory of the Earth's interior as presently advocated by orthodox science. It is time that Our Hollow Earth be recognized for what it is: Our Earth IS Hollow!

CHAPTER THIRTEEN
The Throne of King David -- Found!

When the ten tribe nation of Israel escaped the Assyrians sometime after their captivity in 721 B.C., they were led by miracles and wonders into the beautiful North Countries by a prophet of God. As he led the Israelites in their miracle escape, they made their path appear as if they were returning to their homeland in Palestine. But upon arriving at the river Euphrates, the Lord had stopped the river from flowing. The Israelites then went up the dry river bed passing through sections where the canyon walls were steep on both sides and finally coming out at the river head-waters. They then turned north.

In the meantime, the Assyrian standing army, upon discovering the Israelite escape, followed their tracks to the river Euphrates which by that time was now flowing again. They could see the tracks of the Israelites entering the river on the one side but could not find where they supposedly had come out on the other side. In this manner the Lord provided the way of escape of the ten tribes from the Assyrians much as He had when the 12 tribes escaped from the Egyptian Army nearly 1000 years before.

From an analysis of evidence in Immanuel Velikovsky's book, WORLDS IN COLLISION, I believe that the time of the Israelite escape from Assyria took place March 23, 687 B.C. The most important role in the cult of Mars among the Romans and Greeks was the festival of Tubilustrium which was held on the 23rd day of March. (p. 237) This festival commemorated a close passage of the planet Mars to the earth that caused great destruction. It appears that Venus, which ancient astronomers maintain was originally a comet, on one of its passages pulled the planet Mars from its orbit which brought it on a course very close to the earth. That was the night of its passage: March 23, 687 B.C. It was the first night of the Passover. In the middle of the night, meteorites fell and a blast from Mars obliterated the encamped army of the Assyrian King Sennacherib.

From the Talmud is described a blast that fell on the camp of Sennacherib's army of 185,000 soldiers. They were preparing to attack Jerusalem the next morning. It was not a flame, but a consuming blast: *"Their souls were burnt, though their garments remained intact."* The phenomenon was accompanied by a terrific noise. It was probably an interplanetary lightning bolt.

Furthermore, as Mars passed by the earth, the day was prolonged on the evening before Sennacherib's army was devoured by the blast. As the sun was going down it went backwards 10 degrees before going down again. (Isaiah 38:6-8, WORLDS IN COLLISION, pp. 231-2)

With the majority of the Assyrian army destroyed and great havoc caused by this close passage of Mars, the time was ideal for the Israelites to escape the Assyrians. And the near-passage of Mars probably caused the earthquake which stopped the waters of the Euphrates so the tribes could escape by going up the river on a dry bed. A cave-in of the river banks above the "narrows" must have stopped the waters of the river and after the Israelites left the river bottom, the natural dam of the cave-in or slide probably over-flowed and washed away. So that when the small home-standing army of the Assyrians chasing the Israelites reached the river it was flowing once again and they couldn't tell where the Israelites had escaped to.

The probable route the Ten Tribes took in their year and half journey to Our Hollow Earth after leaving Assyria may have taken them by the eastern end of the Black Sea, along the River Danube to Denmark where a group of them remained behind. The main group proceeded to Sweden and Norway and probably crossed the Arctic Ocean north east of Spitsbergen and Franz Josef Land similar to the route Olaf and his father took in 1829 in their small fishing craft.

An alternate route could have been that after crossing the Caucasus Mountains they proceeded to an area north of the Black Sea. Before the Bolshevik Revolution, Russian archaeologists ventured down into the Crimea north of the Black sea where they dug into the burial mounds and found some genealogies proving those peoples were of the Tribes of Israel. Their leader was Odin and when the Romans threatened to conquer the region, these Israelites determined to migrate further north. Many of them migrated into Russia and even into Mongolia where burial remains have been discovered of a white race. Perhaps the main body of the Lost Tribes traveled from Mongolia, up through Siberia and from there towards the North Polar Opening north of the New Siberian Islands, much as did Fridtjof Nansen in his ship, The Fram, when he came within a few of miles of not discovering the North Polar Opening north of the New Siberian Islands in 1893 in search of Sannikov Land.

In this long journey, the Israelites were probably led by a prophet of God, much as Moses led their ancestors out of Egypt. Bruce R. McConkie wrote in his MORMON DOCTRINE,

"In their northward journeyings they were led by prophets and inspired leaders. They had their Moses and their Lehi, were guided by the spirit of revelation, kept the law of Moses, and carried with them the statutes and judgments which the Lord had given them in ages past. They were still a distinct people many hundreds of years later, for the resurrected Lord visited and ministered among them following his ministry on this continent (of America) *among the Nephites."* (p. 457)

During this visit of Christ to the ancient Americans, as recorded in the BOOK OF MORMON, He told them of the Lost Tribes,

"And verily, verily, I say unto you that I have other sheep, which are not of this land, neither of the land of Jerusalem, neither in any parts of that land round about whither I have been to minister...I have received a commandment of the Father that I shall go unto them, and that they shall hear my voice, and shall be numbered among my sheep, that there may be one fold and one shepherd; therefore I go to show myself unto them." (3 Nephi 15:16:1-3)

To those who reject additional scripture from God, Christ answers:

"Thou fool, that shall say: A Bible, we have got a Bible, and we need no more Bible. Have ye obtained a Bible save it were by the Jews?"

"Know ye not that there are more nations than one? Know ye not that I, the Lord your God, have created all men, and that I remember those who are upon the isles of the sea; and that I rule in the heavens above and IN THE EARTH BENEATH; and I bring forth my word unto the children of men, yea, even upon all the nations of the earth?"

"Wherefore murmur ye, because that ye shall receive more of my word? Know ye not that the testimony of two nations is a witness unto you that I am God, that I remember one nation like unto another?..."

"For behold, I shall speak unto the Jews (this prophecy was written by an ancient American prophet named Nephi, about 550 B.C.) *and they shall write* it (which became the Bible); *and I shall also speak unto the Nephites* (an ancient American nation of white immigrant Israelites) *and they shall write it* (which became The Book of Mormon); *and I shall also speak unto the other tribes of the house of Israel,*

(the Lost Ten Tribes) *which I HAVE LED AWAY, and they shall write it...And it shall come to pass,* (in the last days) *that the Jews shall have the words of the Nephites, and the Nephites shall have the words of the Jews; and the Nephites and the Jews shall have the words of the LOST TRIBES OF ISRAEL; and the lost tribes of Israel shall have the words of the Nephites and the Jews."* (2 Nephi 29:7-14)

This book of scripture written by the Lost Ten Tribes is to come forth when the Ten Tribes come forth out of Our Hollow Earth in the near future. And were written by a succession of prophet-kings, a descendant of whom Olaf Jansen and his father were privileged to interview during their two and half year stay in the North Countries of Our Hollow Earth. In keeping with their king's prophetic calling, Olaf reported that they call their king, *"The Great High Priest Over All The Land."*

The King over the Lost Ten Tribes Is a Jew, a Descendant of David

Evidence from the scriptures indicates that the literal and living throne of David today exists within Our Hollow Earth. The prophet who guided the 10 tribes into the North Countries must have been a descendant of David whose descendants have sat upon the throne of David for some 2,500 years. The political Kingdom of God, so long awaited for by saint and sinner, today exists in the North Countries and their King is a Jew, a descendant of David.

The House of Israel was led out of Egypt by God's prophet Moses and they wandered 40 years in the wilderness for having refused to conquer Palestine. Then Moses ordained Joshua to take his place as God's prophet and lead the Israelites in conquering Palestine. Upon their parting, and before Joshua led the House of Israel across the River Jordan into Palestine, Moses was translated and taken up to Heaven. The scriptures are unclear on this point, but I have concluded this is true because Moses had to have been a translated being in order for him and Elijah (who was also translated and taken to Heaven alive) to confer their prophetic keys with the laying on of hands on Peter, James and John on the Mount of Transfiguration during Christ's ministry on Earth.

The prophet Joshua then gave Israel judges to govern them for 400 years. According to my calculations of Biblical chronology, Isaac, Abraham's son was born in 1962 B.C., and

430 years passed from Isaac's birth to the time of the Israelite exodus from Egypt in 1532 B.C. (See PATTERNS OF EVIDENCE: THE EXODUS by Timothy P. Mahoney, p. 884) The House of Israel covenanted with God to obey Him and keep His commandments as given to them through His prophet Moses. As they were led by Moses through the Red Sea, they were baptized into God's Kingdom. There were a rebellious bunch among the House of Israel, and from time to time they were weeded out with earthquakes, lightning, poisonous serpents, wars and executions for breaking God's commandments. God commanded Israel to invade and conquer Palestine which God had given to their ancestor Abraham. A set of spies were sent into Palestine to see how they could carry out the conquest. All but two of the spies, Caleb and Joshua believed they could not successfully carry out the conquest. So the rebellious Israelites refused to participate in the conquest, for which God condemned them to wander 40 years in the wilderness until all the rebellious ones had died. Finally, in 1492 B.C., the Israelites under their new prophet Joshua crossed the River Jordan to begin the conquest of Palestine.

After 400 years of being governed by judges, the Israelite people asked Samuel, the prophet to ask God to give them a King. God told them that He was their King, but they wanted an earthly representative. So God told Samuel, His prophet to anoint Saul as Israel's first king in 1092 B.C.

King Saul was commanded by God through His prophet Samuel to take an army and go down into Egypt and kill all the Amalekites (who were cannibals), and all their animals. The Amalekites had attacked the Israelites in the wilderness and afterwards had invaded Egypt. Yet King Saul took a spoil of their cattle, and then, without priesthood authority he offered up an animal sacrifice to God to show his gratitude for the cattle he had captured. For this disobedience, Samuel told King Saul he was no longer King over Israel.

God then told Samuel to choose another King, the boy David, son of Jesse whom he anointed King of Israel. King Saul didn't want to give up the throne to David, and tried many times to kill David, but God protected him so that he could not be killed. Subsequently, throughout the history of the people of Israel, God's prophets would give counsel, advice and commandments to the Kings of Israel. Some Kings obeyed God's prophets, others did not.

What few seem to realize is that the right to the throne over the Political Kingdom of God on Earth can only be bequeathed by a prophet of God.

To Judah, the father of all Jews, did father Israel (whose former name was Jacob) give this blessing:

"The Sceptre shall not depart from Judah, nor a lawgiver from between his feet, until Shiloh (Christ) *come* (second coming); *and unto him shall the gathering of the people be."* (Genesis 49:10)

David, a Jew and descendant of Judah was made King over the 12 tribes in fulfillment of this blessing given of the Lord through his prophet Jacob. And the promise was given that the Sceptre or throne would never depart from Judah. His descendants would be King over Israel forever. David, being a descendant of Judah was thereby a partaker of the blessing given to Judah and his descendants by being made King over Israel by the prophet Samuel.

An everlasting Covenant was given to David by the Lord in which he was promised that he and his sons would rule over Israel forever:

"Now these be the last words of David. David the son of Jesse said, the man who was raised up on high, the anointed of the God of Jacob, and the sweet psalmist of Israel said, 'Although my house be not so with God; YET HE HATH MADE WITH ME AN EVERLASTING COVENANT, ordered in all things, and sure...'" (2 Samuel 23:1,5)

This was the covenant that the Lord made with David:

"And thine house and thy kingdom shall be established forever before thee; THY THRONE SHALL BE ESTABLISHED FOREVER. According to all these words, and according to all this vision, so did Nathan (the prophet) *speak unto David."* (2 Samuel 7:16-17)

And of David's son Solomon, the Lord said,

"He shall build an house for my name, and I WILL ESTABLISH THE THRONE OF HIS KINGDOM FOREVER." (2 Samuel 7:13)

It was common knowledge in ancient Israel that the Lord intended that the throne of David was to be inherited. But after King Solomon's death, the northern ten tribes rejected the throne of David because of oppressive taxation and formed a separate nation which they named ISRAEL. This was the beginning of what eventually became the Lost Ten Tribes of Israel. Therefore, Palestine became divided into two nations. The northern ten tribe nation was named ISRAEL and the southern two tribes of Benjamin and Judah was named JUDAH and its citizens became known as JEWS.

King Solomon's son, Rehoboam ruled Judah and then his son Abijah began to reign at his father's death. The ten tribes of Israel had chosen Jeroboam, the son of a servant of King Solomon, to be their King.

"And Abijah stood up upon the mount Zemoraim, which is in mount Ephraim, and said, 'Hear me, thou Jeroboam, and all Israel (the ten tribe northern nation)*; Ought ye not to know that the Lord God of Israel gave the kingdom over Israel to David FOREVER, even to him AND TO HIS SONS BY A COVENANT OF SALT?'"* (2 Chronicles 13:4,5)

It is thought that because David's descendants became wicked that the covenant the Lord made to him was modified. It is supposed that the throne of David ceased to exist but will however be restored in the last days when Christ comes to earth during His millennial reign, in the coming age of peace. Yet David said,

"Although my house BE NOT SO WITH GOD; YET he hath made with me an EVERLASTING covenant."

Therefore, the covenant was not contingent upon the righteousness of his children. Nevertheless, it would seem that a righteous lineage would preferably be the carrier of that heritage.

The Lord had explained to David, as he recorded in Psalms 89:

Verses 3 and 4: *"I have made a covenant with my chosen, I have sworn unto David my servant, Thy SEED will I establish FOREVER, and build up THY THRONE TO ALL GENERATIONS. Selah."*

Verses 28-37: *"My mercy will I keep for him for evermore, and my covenant shall stand fast with him. His SEED also will I make to ENDURE forever, and his THRONE AS THE DAYS OF HEAVEN. If his children FORSAKE my law, and walk not in my judgments; If they BREAK my statutes, and keep not my commandments; Then will I visit their transgression WITH THE ROD, and their iniquity with stripes. Nevertheless my lovingkindness will I NOT UTTERLY TAKE from him, nor suffer my faithfulness to fail. MY COVENANT WILL I NOT BREAK, nor alter the thing that is gone out of my lips. Once have I sworn by my holiness that I will not lie unto David. HIS SEED SHALL ENDURE FOREVER, AND HIS THRONE AS THE SUN BEFORE ME. It shall be established forever as the moon, and as a faithful witness in heaven. Selah."* (Psalms 89:3, 4, 28-37)

Even though the Lord promised to David that his throne would never be taken from his descendants, from all appearances, history would seem to tell us otherwise. First, the ten tribes separated from the throne of David and formed their own nation which they named Israel. Israel was then taken into captivity and later lost to the knowledge of the world. Judah's last king and descendant of David, Zedekiah, was killed by the Babylonians in 586 B.C. King Zedekiah had ruled over the Tribes of Judah and Benjamin in Palestine until he was deposed by King Nebuchadnezzar II of Babylon.

King Zedekiah's infant son Mulek was taken secretly by the palace guards to America. Mulek's last descendant to be a King was Zarahemla, who delivered the throne over his people to King Mosiah of the Nephite nation who found the Mulekites. Many of the subsequent wars described in the Book of Mormon resulted from the descendants of Zarahemla, who called themselves the "King Men" in trying to take over the government to place a descendant of David as King over the people. Apparently, they also did not seem to realize that only a prophet of God can bequeath the right to the throne over the Political Kingdom of God on Earth. (OMNI, THE BOOK OF MORMON, pp. 130-131)

So if the throne of David exists today, where is it? Herbert W. Armstrong, of the Plain Truth Magazine, established a case for the throne of David existing today as the throne of England in his book, THE UNITED STATES AND THE BRITISH COMMONWEALTH IN PROPHECY. He is right in ascertaining that the United States and Britain are descendants of Ephraim. However, there are two errors in his case. He claims that 1) The United States and Britain are the North Countries of the Lost Ten Tribes, and 2) The Throne of England is the throne of David.

These conclusions cannot be the correct answer to the whereabouts of the throne of David today because The United States and Britain are not the North Countries of the lost ten tribes of Israel even though some Ephraimites and others of the tribes did stay behind in Europe in the ten tribe's northward journey to Our Hollow Earth.

There is an ancient manuscript in the Herald's College of London recording the Saxon kings' lineage back to Odin who traced his descent from King David (Encyclopedia Britannica). But the throne of Britain cannot be the throne of David even though they may be descendants of David because the reign of their kings was not instituted by a prophet of God.

There also occurred an important break in the line of sons on the throne of England. The promise of the Lord to David was that he and his SONS would inherit the throne forever. The Lord prevented King Henry VIII of England from having any sons to inherit the throne for which he divorced his first wife, beheaded his second wife and died while chasing after other women in hopes of obtaining a son to inherit his throne. His efforts were in vain because his throne was not acceptable to the Lord as the throne of David. Subsequently, King Henry VIII's daughters, Queen Mary and Queen Elizabeth became heirs to the throne of England.

In order to find the throne of David of the political Kingdom of God today, we must understand the word of the Lord to the prophet Jeremiah, who said,

> *"Behold, the days come, saith the Lord, that I will perform that good thing which I have promised unto THE HOUSE OF ISRAEL AND TO THE HOUSE OF JUDAH."*

Here, the distinction between the two nations of JUDAH (the Jews) and ISRAEL (the Lost Ten Tribes) is made clear. Continuing,

> *"For thus saith the Lord; there shall NOT BE CUT OFF from David A MAN to sit upon the throne of THE HOUSE OF ISRAEL."* (Jeremiah 33:14, 17)

From this scripture we discover that there will never be a time that A MAN, a descendant of David, will not sit upon the throne of David. However, the word of the Lord given here to the prophet Jeremiah makes clear that the throne of David is now not over Judah but over THE HOUSE OF ISRAEL -- the ten tribe nation which went into the North Countries of which we have discovered to be the land of Our Hollow Earth. For which we conclude that the throne of David today exists in the North Countries of Our Hollow Earth. It is the living, legal, righteous and acceptable throne of David put in place by God to rule over the political Kingdom of God.

From the time that King Nebuchadnezzar II of Babylon deposed King Zedekiah, the throne of David was shifted to the Lost Ten Tribes of Israel. Jeremiah, the prophet of the Lord during the reign of King Zedekiah was commanded by God to take a message to the Lost Ten Tribes. At that time, he took the legal right to the throne over the House of Israel with him and gave it to another descendant of David among the Lost Tribes who then took it to the Hollow Earth when the Lost

Tribes migrated there in about 475 B.C. That descendant of David became King over the Lost Ten Tribes of Israel, and a descendant of his named David, today reigns on the throne of David over the House of Israel.

Olaf Jansen described his interview with the King of Israel, for whom he recorded that their god is Jehovah – the god of the ancient Israelites. After obtaining the North Countries in their fishing craft through the North Polar Opening, and after learning their language of Sanskrit for one year, Olaf and his father were taken overland in a monorail train to their capital city of Eden and presented before the Great High Priest, Ruler Over All the Land:

"The surprise of my father and myself was indescribable when, amid the regal magnificence of a spacious hall, we were finally brought before the Great High Priest, ruler over all the land. He was richly robed, and much taller than those about him, and could not have been less than fourteen or fifteen feet in height. The immense room in which we were received seemed finished in solid slabs of gold thickly studded with jewels of amazing brilliancy..."

"The unexpected awaited us in this palace of beauty, in the finding of our little fishing-craft. It had been brought before the High Priest in perfect shape, just as it had been taken from the waters that day when it was loaded on board the ship by the people who discovered us on the river more than a year before."

"We were given an audience of over two hours with this great dignitary, who seemed kindly disposed and considerate. He showed himself eagerly interested, asking us numerous questions, and invariably regarding things about which his emissaries had failed to inquire."

"At the conclusion of the interview he inquired our pleasure, asking us whether we wished to remain in his country or if we preferred to return to the 'outer' world, providing it were possible to make a successful return trip, across the frozen belt barriers that encircle both the northern and southern openings of the earth."

"My father replied: 'It would please me and my son to visit your country and see your people, your colleges and palaces of music and art, your great fields, your wonderful forests of timber; and after we have had this pleasurable privilege, we would like to try to return to our home on the

'outside' surface of the earth. This son is my only child, and my good wife will be weary awaiting our return.'"

"'I fear you can never return,' replied the Chief High Priest, 'because the way is a most hazardous one. However, you shall visit the different countries with Jules Galdea as your escort, and you will be accorded every courtesy and kindness. Whenever you are ready to attempt a return voyage, I assure you that your boat which is here on exhibition shall be put in the waters of the river Hiddekel at its mouth, and we will bid you Jehovah-speed.'"

"Thus terminated our only interview with the High Priest or Ruler of the continent." (<u>THE SMOKY GOD</u>, pp. 114-117)

The High Priest's prophecy to Olaf's father did come true. His father drowned when their fishing boat capsized in the Antarctic Ocean on their return voyage. Olaf was thrown by the boat as it was flipped over by an iceberg and landed on top of the iceberg where he later was picked up by a whaleboat.

When the ten tribes arrived in the North Countries 2,500 years ago, their prophet-leader, a descendant of David, most probably had been anointed King of Israel by the prophet Jeremiah. When the ten tribes come out of Our Hollow Earth in the near future with their scriptures we shall see that the scripture has been fulfilled in which there has *"not been cut off from David A MAN to sit upon the throne of the HOUSE OF ISRAEL."* Perhaps, the King of Israel today is that *"prince"* whose name was to be David, ruling over the house of Israel in the last days when they return.

The prophet Joseph Smith spoke of this last of the Kings of Israel before Christ returns to take over the throne at the commencement of the His Millennial Reign,

"Although David was king, he never did obtain the spirit and power of Elijah and the fullness of the priesthood; and the Priesthood that he received, and the throne and kingdom of David is to be taken from him and given to another by the NAME OF DAVID in the last days, RAISED UP OUT OF HIS LINEAGE." (<u>TEACHINGS</u>, p. 339)

I believe this King David, a descendant of David lives today and is the King of the Ten Tribes of Israel in Our Hollow Earth. He is a prophet of God.

"Behold, the days come, saith the Lord, that I will raise unto David a RIGHTEOUS BRANCH, and a KING SHALL REIGN and prosper, and shall execute judgment and justice IN THE EARTH."

"In his days Judah shall be saved, and Israel shall dwell safely: and this is his name whereby he shall be called, THE LORD OUR RIGHTEOUSNESS."

"Therefore, behold, the days come, saith the Lord, that they shall no more say, The Lord liveth which brought up the children of Israel out of the land of Egypt;"

"But, the Lord liveth, which brought up and which led the seed of the house of Israel out of the North Country, and from all countries whither I had driven them; and they shall dwell in their own land." (Jeremiah 23:5-8)

When the Ten Tribes return, they will expand their political kingdom of God first to America and then to Jerusalem and then to the whole world,

"And I will set up one shepherd over them, and he shall feed them, even my servant David; he shall feed them, and he shall be their shepherd."

"And I the Lord will be their God, and my SERVANT DAVID a prince among them; I the Lord have spoken it." (Ezekiel 34:22-24)

"In those days the house of Judah shall walk with the house of Israel, and they shall come together OUT OF THE LAND OF THE NORTH to the land that I have given for an inheritance unto your fathers." (Jeremiah 3:18)

When will the Lost Tribes Return?

The political Kingdom of God presently is headquartered at the City of Eden, in that primeval Garden where our human race began within Our Hollow Earth whose King is a descendant of David. His name most probably is David and he sits on the inherited throne of David of old. He rules over the heretofore Lost Ten Tribes of Israel.

In fulfillment of prophecy, the political Kingdom of God under the guidance of the living throne of David will be expanded to the surface of the earth. A highway will be built up in the midst of the Arctic Ocean, most likely as a monorail train on poles sunk into the ocean bed, and *"their prophets shall stay themselves no longer"* but will proceed to connect a permanent link between their capital City of Eden with the New Jerusalem -- the future capital city of America to be built in Independence, Jackson County, Missouri.

This will happen after the government of the United States is destroyed when all the nations of the world under the United Nations attacks the United States and tries to repossess our country when our national government goes into default caused by a depression brought on by the International Bankers and their Federal Reserve Bank which they can cause simply by raising interest rates so high no-one will borrow their money.

My estimate when this will begin to happen is in the year 2021 when the United Nations becomes a de facto world dictatorship government for 3.5 years as John the Beloved described in Revelation 13, and will try to impose mandates upon the United States (including perhaps the fake COVID-19 vaccine), which the United States will refuse to abide by. The United Nations will retaliate by having their International Bankers and the Federal Reserve cause a disastrous economic depression, which will cause the United States government to default on the debt it owes the bankers. For this, the United Nations will invade the United States and attempt to repossess it for defaulting on its debt and for refusing to abide by the United Nations mandate. The invaders could also cause an EMP (electromagnetic pulse) and blame it on a solar flare that will take down the electric grid. This effectively will cause all banking transactions to cease. Transportation will come to a halt and money will cease to exist.

This world power will have the cunning of a leopard, the feet and claws of a bear, the teeth and mouth of a lion, the power of a dragon and Lucifer will be their god directed straight out of the Vatican. They will attack the United States, the land of the Saints of God (Zion),

1 And I stood upon the sand of the sea, and saw a beast rise up out of the sea, having seven heads and ten horns, and upon his horns ten crowns, and upon his heads the name of blasphemy.

2 And the beast which I saw was like unto a leopard (the European Union)*, and his feet were as the feet of a bear* (Russia)*, and his mouth as the mouth of a lion* (Great Britain)*: and the dragon* (China) *gave him his power* (of Lucifer)*, and his seat, and great authority.*

7 And it was given unto him to make war with the saints, and to overcome them: and power was given him over all kindreds, and tongues, and nations.

8 And all that dwell upon the earth shall worship him, whose names are not written in the book of life of the Lamb slain from the foundation of the world.

9 If any man have an ear, let him hear.

10 He that leadeth into captivity shall go into captivity: he that killeth with the sword must be killed with the sword. Here is the patience and the faith of the saints. (Rev. 13:1-2, 7-10)

This war of attempted repossession of the United States, the land of the Saints of Zion by the United Nations, International Bankers, the Illuminati and Jesuits is spoken of in the scriptures. Isaiah chapter 3 speaks of the land of Zion which is America, and that our men will fall by the sword in the war,

25 Thy men shall fall by the sword, and thy mighty in the war.

26 And her gates shall lament and mourn; and she (America) *being desolate shall sit upon the ground.* (Isaiah 3:25-26)

The invading armies in the attack upon the United States (which is spoken of as the land of Zion in the scripture) by the United Nations countries of the world is spoken of in Joel chapter 2,

1 Blow ye the trumpet in Zion, and sound an alarm in my holy mountain: let all the inhabitants of the land tremble: for the day of the Lord cometh, for it is nigh at hand;

2 A day of darkness and of gloominess, a day of clouds and of thick darkness, as the morning spread upon the mountains: a great people and a strong; there hath not been ever the like, neither shall be any more after it, even to the years of many generations.

3 A fire devoureth before them; and behind them a flame burneth: the land is as the garden of Eden before them, and behind them a desolate wilderness; yea, and nothing shall escape them.

4 The appearance of them is as the appearance of horses; and as horsemen, so shall they run.

5 Like the noise of chariots on the tops of mountains shall they leap, like the noise of a flame of fire that devoureth the stubble, as a strong people set in battle array.

6 Before their face the people shall be much pained: all faces shall gather blackness.

7 They shall run like mighty men; they shall climb the wall like men of war; and they shall march every one on his ways, and they shall not break their ranks:

8 Neither shall one thrust another; they shall walk every one in his path: and when they fall upon the sword, they shall not be wounded (this sounds like an army of robots).

9 They shall run to and fro in the city; they shall run upon the wall, they shall climb up upon the houses; they shall enter in at the windows like a thief.

10 The earth shall quake before them; the heavens shall tremble: the sun and the moon shall be dark, and the stars shall withdraw their shining:

The Lord then tells His people,

12 Therefore also now, saith the Lord, turn ye even to me with all your heart, and with fasting, and with weeping, and with mourning:

13 And rend your heart, and not your garments, and turn unto the Lord your God: for he is gracious and merciful, slow to anger, and of great kindness, and repenteth him of the evil.

14 Who knoweth if he will return and repent, and leave a blessing behind him; even a meat offering and a drink offering unto the Lord your God?

15 Blow the trumpet in Zion, sanctify a fast, call a solemn assembly:

16 Gather the people, sanctify the congregation, assemble the elders, gather the children, and those that suck the breasts: let the bridegroom go forth of his chamber, and the bride out of her closet.

17 Let the priests, the ministers of the Lord, weep between the porch and the altar (of our temples), and let them say, Spare thy people, O Lord, and give not thine heritage to reproach, that the heathen should rule over them: wherefore should they say among the people, Where is their God?

Then will God pity His people and hear their prayers and will go before our armies to defend our country from the invading heathen nations,

18 Then will the Lord be jealous for his land, and pity his people.

11 And the Lord shall utter his voice before his army: for his camp is very great: for he is strong that executeth his word: for the day of the Lord is great and very terrible; and who can abide it?

19 Yea, the Lord will answer and say unto his people, Behold, I will send you corn, and wine, and oil, and ye shall be satisfied therewith: and I will no more make you a reproach among the heathen:

20 But I will remove far off from you the northern army, and will drive him into a land barren and desolate (the Arctic), with his face toward the east sea (the Atlantic), and his hinder part toward the utmost sea (the Pacific), and his stink shall

come up, and his ill savour shall come up, because he hath done great things.

Isaiah 4:1 says that after the war seven women will ask one man to be their husband because six out of seven men will die in the war,

And in that day seven women shall take hold of one man, saying, We will eat our own bread, and wear our own apparel: only let us be called by thy name, to take away our reproach.

The American Indians and Hispanics, which in the scriptures are called *"the remnant of Jacob"* will be a fifth column in the midst of our land fighting with the United Nations invaders to do unto us as our misguided leaders have done unto them,

12 And my people who are a remnant of Jacob shall be among the Gentiles, yea, in the midst of them as a lion among the beasts of the forest, as a young lion among the flocks of sheep, who, if he go through both treadeth down and teareth in pieces, and none can deliver.

13 Their hand shall be lifted up upon their adversaries, and all their enemies shall be cut off.

14 Yea, wo be unto the Gentiles except they repent; for it shall come to pass in that day, saith the Father, that I will cut off thy horses out of the midst of thee, and I will destroy thy chariots;

15 And I will cut off the cities of thy land, and throw down all thy strongholds;

16 And I will cut off witchcrafts out of thy land, and thou shalt have no more soothsayers;

17 Thy graven images I will also cut off, and thy standing images out of the midst of thee, and thou shalt no more worship the works of thy hands;

18 And I will pluck up thy groves out of the midst of thee; so will I destroy thy cities.

19 And it shall come to pass that all lyings, and deceivings, and envyings, and strifes, and priestcrafts, and whoredoms, shall be done away.

20 For it shall come to pass, saith the Father, that at that day whosoever will not repent and come unto my Beloved Son, them will I cut off from among my people, O house of Israel;

21 And I will execute vengeance and fury upon them, even as upon the heathen (of the invading armies), such as they have not heard. (3 Nephi 21:12-21, Micah 5:7-15)

After the war, if the people of the United States have repented of their evil ways, God will accept them into His church, and they will help the American Indians and Hispanics who are the "remnant of Jacob" to build a new capital city for the whole American continent, north and south, the New Jerusalem, in Independence, Missouri.

22 But if they will repent and hearken unto my words, and harden not their hearts, I will establish my church among them, and they shall come in unto the covenant and be numbered among this the remnant of Jacob, unto whom I have given this land for their inheritance;

23 And they shall assist my people, the remnant of Jacob, and also as many of the house of Israel as shall come, that they may build a city, which shall be called the New Jerusalem.

24 And then shall they assist my people that they may be gathered in, who are scattered upon all the face of the land, in unto the New Jerusalem.

25 And then shall the power of heaven come down among them; and I also will be in the midst. (3 Nephi 21:22-25)

This United Nations government dictatorship that attacks the United States is the fourth beast of the Book of Daniel,

7 After this I saw in the night visions, and behold a fourth beast, dreadful and terrible, and strong exceedingly; and it had great iron teeth: it devoured and brake in pieces, and stamped the residue with the feet of it: and it was diverse from all the beasts that were before it; and it had ten horns. (Daniel 7:7)

20 And of the ten horns that were in his head, and of the other which came up, and before whom three fell; even of that horn that had eyes, and a mouth that spake very great things, whose look was more stout than his fellows.

21 I beheld, and the same horn made war with the saints, and prevailed against them; (Daniel 7:20-21)

The ancient american prophet Nephi was shown the last days when the Great and Abominable Church would attack the Saints of God in America, and indeed in all the world,

9 And it came to pass that he said unto me: Look, and behold that great and abominable church, which is the mother of abominations, whose founder is the devil.

10 And he said unto me: Behold there are save two churches only; the one is the church of the Lamb of God, and the other is the church of the devil; wherefore, whoso belongeth not to the church of the Lamb of God belongeth to that great church, which is the mother of abominations; and she is the whore of all the earth.

11 And it came to pass that I looked and beheld the whore of all the earth, and she sat upon many waters; and she had

dominion over all the earth, among all nations, kindreds, tongues, and people.

12 And it came to pass that I beheld the church of the Lamb of God, and its numbers were few, because of the wickedness and abominations of the whore who sat upon many waters; nevertheless, I beheld that the church of the Lamb, who were the saints of God, were also upon all the face of the earth; and their dominions upon the face of the earth were small, because of the wickedness of the great whore whom I saw.

13 And it came to pass that I beheld that the great mother of abominations did gather together multitudes upon the face of all the earth, among all the nations of the Gentiles, to fight against the Lamb of God (and His Saints).

14 And it came to pass that I, Nephi, beheld the power of the Lamb of God, that it descended upon the saints of the church of the Lamb, and upon the covenant people of the Lord, who were scattered upon all the face of the earth; and they were armed with righteousness and with the power of God in great glory.

15 And it came to pass that I beheld that the wrath of God was poured out upon that great and abominable church, insomuch that there were wars and rumors of wars among all the nations and kindreds of the earth.

16 And as there began to be wars and rumors of wars among all the nations which belonged to the mother of abominations, the angel spake unto me, saying: Behold, the wrath of God is upon the mother of harlots; and behold, thou seest all these things—

17 And when the day cometh that the wrath of God is poured out upon the mother of harlots, which is the great and abominable church of all the earth, whose founder is the devil, then, at that day, the work of the Father shall commence, in preparing the way for the fulfilling of his covenants, which he hath made to his people who are of the house of Israel. (1 Nephi 14:9-17)

Isaiah, the prophet said of our attackers,

25 But thus saith the Lord, even the captives of the mighty shall be taken away, and the prey of the terrible shall be delivered; for I will contend with him that contendeth with thee, and I will save thy children.

26 And I will feed them that oppress thee with their own flesh; they shall be drunken with their own blood as with sweet wine; and all flesh shall know that I, the Lord, am thy Savior and thy Redeemer, the Mighty One of Jacob. (Isaiah 49:25-26)

The prophet Daniel says that our resurrected ancestor Adam, the Ancient of Days will come and help us win this war,

22 Until the Ancient of days came, and judgment was given to the saints of the most High; and the time came that the saints possessed the kingdom. (Daniel 7:22)

And the invading armies and the United Nations government is burned by fire,

11 I beheld then because of the voice of the great words which the horn spake: I beheld even till the beast was slain, and his body destroyed, and given to the burning flame. (Daniel 7:11)

The prophet Nephi explains that God will help us, the Saints of God, who are of the House of Israel,

12 Wherefore, he will bring them again out of captivity, and they shall be gathered together to the lands of their inheritance; and they shall be brought out of obscurity and out of darkness; and they shall know that the Lord is their Savior and their Redeemer, the Mighty One of Israel.

13 And the blood of that great and abominable church, which is the whore of all the earth, shall turn upon their own heads; for they shall war among themselves, and the sword of their own hands shall fall upon their own heads, and they shall be drunken with their own blood.

14 And every nation which shall war against thee, O house of Israel, shall be turned one against another, and they shall fall into the pit which they digged to ensnare the people of the Lord. And all that fight against Zion (America) shall be destroyed, and that great whore, who hath perverted the right ways of the Lord, yea, that great and abominable church, shall tumble to the dust and great shall be the fall of it. (1 Nephi 22:12-14)

Our country will survive but the United Nations government and the United States government will both be destroyed in the war. The Lost Tribes nation political kingdom of God of Our Hollow Earth will then expand to outer earth to fill the void, together with a great last missionary work of 144,000 missionaries, 12,000 from each of the Tribes of the House of Israel, to bring all the descendants of the House of Israel in all the world into the fold of God, to the land of Zion to America, and to the land of Israel in Palestine.

26 And then shall the work of the Father commence at that day, even when this gospel shall be preached among the

remnant of this people. Verily I say unto you, at that day shall the work of the Father commence among all the dispersed of my people, yea, even the tribes which have been lost, which the Father hath led away out of Jerusalem.

27 Yea, the work shall commence among all the dispersed of my people, with the Father to prepare the way whereby they may come unto me, that they may call on the Father in my name.

28 Yea, and then shall the work commence, with the Father among all nations in preparing the way whereby his people may be gathered home to the land of their inheritance.

29 And they shall go out from all nations; and they shall not go out in haste, nor go by flight, for I will go before them, saith the Father, and I will be their rearward. (3 Nephi 21:26-29)

After the war, the Saints of the Most High in America will inherit and rebuild the cities left desolate by the war,

For thou shalt break forth on the right hand and on the left, and thy seed shall inherit the Gentiles and make the desolate cities to be inhabited. (Isaiah 54:3, 3 Nephi 22:3)

The Church of Jesus Christ of Latter-day Saints will unite with the Lost Tribes political kingdom of God, and all Latter-day Saint members will be brought to America from all over the world. We will get a new government and a new capital. The new capital of the whole American continent will be the New Jerusalem to be built in Independence, Jackson County, Missouri (D&C 84:2-3). Our new government will be the theocratic monarchy of the hollow earth nation of the Ten Tribes of Israel political kingdom of God, whose king is King David, a descendant of the ancient King David of the Israelites.

Revelation chapter 12 speaks of where the Lord took the Lost Tribes of Israel with their political Kingdom of God government into the bowels of the earth through the "mouth" of the earth, the North Polar Opening for 2,500 years,

1 And there appeared a great wonder in heaven; a woman clothed with the sun, and the moon under her feet, and upon her head a crown of twelve stars:

2 And she being with child cried, travailing in birth, and pained to be delivered.

5 And she brought forth a man child (Jesus Christ)*, who was to rule all nations with a rod of iron* (which is the word of God, see 1 Nephi 11:25)*; and her child was caught up unto God, and to his throne.*

6 And the woman fled into the wilderness, where she hath a place prepared of God, that they should feed her there a thousand two hundred and threescore days.

14 And to the woman were given two wings of a great eagle, that she might fly into the wilderness, into her place, where she is nourished for a time, and times, and half a time, from the face of the serpent.

15 And the serpent cast out of his mouth water as a flood after the woman, that he might cause her to be carried away of the flood.

16 And the earth helped the woman, and the earth opened her mouth, and swallowed up the flood which the dragon cast out of his mouth.

In Our Hollow Earth, the woman, which is the political Kingdom of God over the House of Israel was nourished for a time (one thousand years), and times (another thousand years) and a half time (500 years) = 2,500 years. The Israelites escaped the Assyrians in the year 687 B.C. and if we estimate that by 475 B.C. they had reached the land of Our Hollow Earth (spoken in Esdras as Arsareth), then the year 2025 would be 2,500 years since the Lost Tribes were led into the North Countries of Our Hollow Earth. We can therefore, conclude that sometime after the year 2025, the political Kingdom of God of the lost tribes of Israel will begin to return to the surface of the Earth to reign.

Elder Bruce R. McConkie, apostle of the Church of Jesus Christ of Latter-day Saints gave us a further clue when the Lost Tribes will return counting from the year 2000.

In his book, THE MILLENNIAL MESSIAH, on page 382, Elder McConkie indicates that the half hour of silence in heaven on the opening of the seventh seal in Revelation 8:1 is according to God's time which one day on God's planet Kolob in the center of our Milky Way Galaxy is equal to 1,000 earth years, so that one-half hour of silence in heaven would be 20.83 years on Earth. LDS Apostle Alvin R. Dyer, who served as a counselor to President David O. McKay, also calculated this half hour of silence in heaven to be about 20 years on Earth in his book, WHO AM I? (Loc. 5535)

In a revelation to the prophet Joseph Smith, recorded in Doctrine & Covenants Section 77, verse 6, Joseph Smith asks God,

"Q. What are we to understand by the book which John saw, which was sealed on the back with seven seals?

A. We are to understand that it contains the revealed will, mysteries, and works of God; the hidden things of his economy concerning this earth during the seven thousand years of its continuance, or its temporal existence."

The earth's temporal existence began at the fall of Adam, 4,000 B.C. And Jesus Christ will return to rule on the earth at the beginning of the seven thousandth year period, as explained further in D&C 77, verse 12,

"Q. What are we to understand by the sounding of the trumpets, mentioned in the 8th chapter of Revelation?

A. We are to understand that as God made the world in six days, and on the seventh day he finished his work, and sanctified it, and also formed man out of the dust of the earth, even so, in the beginning of the seventh thousand years will the Lord God sanctify the earth, and complete the salvation of man, and judge all things, and shall redeem all things, except that which he hath not put into his power, when he shall have sealed all things, unto the end of all things; and the sounding of the trumpets of the seven angels are the preparing and finishing of his work, in the beginning of the seventh thousand years—the preparing of the way before the time of his coming."

According to Biblical chronology, from Adam to Christ were 4,000 years, and from Christ to the year 2000 AD completed six thousand years. So the seventh thousand year period of the temporal existence of the Earth began in the year 2000 AD. With the half hour of silence in heaven being 20.83 earth years, the seven trumpets of Revelation Chapter 8 will begin sounding in heaven in 2021.

If the events of each trumpet that sound should last at least one year each, then in the year 2027 all seven angels will have sounded their trumpets, and Christ will then come within the next seven years after that (Ezequiel 39:9) to save the righteous from the burning solar mass ejection which will occur, according to this chronology, in about 2034 that will purify the earth of all unrighteousness in preparation for His millennial reign.

The fifth angel to sound his trumpet, according to this chronology will be in 2025, and the locusts that come out of the bottomless pit of the hollow earth are the flying saucers of the lost ten tribes which are commanded to torment the *"men which have not the seal of God in their foreheads"* of the Great and Abominable Church to give up their rule over America (Rev 9:1-12).

D&C Section 131, verse 28, says, *"Their enemies shall become a prey unto them."* And verse 31 says, *"And the boundaries of the everlasting hills shall tremble at their presence"* when the Ten Tribes come down from the north.

Esdras, in the Apocrypha, indicates as does this scripture from the Doctrine and Covenants that an earthquake will also announce the return of the Ten Tribes from the North. If the near passage of a planet-sized comet as Velikovsky's research indicates was the cause of the earthquake at the time of the Tribes escape from Assyria, perhaps a planet-sized comet will be the cause of the earthquake at the time of their return.

The sixth angel sounds in preparation of the great battle of Armageddon where the country of Israel is attacked by an army of 200 million soldiers. Since the United States has always helped Israel and will not be around to help defend Israel in this battle, the attack on the United States by the United Nations countries will have to have taken place previously sometime before the Lost Tribes emerge from the *"bottomless pit"* of Our Hollow Earth to give America a new government.

Since Donald Trump won the 2024 election, he started his second term in January 2025, and according to a prophecy by a Utah LDS Bishop, a Republican President will be assassinated followed by an attack on the United States by Russia and its allies. (THE DREAM MINE STORY, pp. 28 & 35)

After the war with the United Nations beast Kingdom, the Lost Ten Tribes of Our Hollow Earth will then emerge from Our Hollow Earth and bless America with a new government, the political Kingdom of God, a new capital, the New Jerusalem in Independence, Missouri, and an economy of abundance in contrast to the economy of scarcity that we now have,

> *"In that day shall the branch of the Lord* (the Lost Tribes) *be beautiful and glorious, and the fruit of the earth* (of America) *shall be excellent and comely for them that are escaped of Israel* (in the great war)." (Isaiah 4:2)

After the war, the Lost Tribes political Kingdom of God will thus expand first to America, then to Jerusalem to help the Jews rebuild their temple before the battle of Armageddon. After the destruction of the United Nations government, another Old World government will emerge for another 3.5 years that will allow the Jewish temple to be rebuilt by King David of the Lost Ten Tribes of Israel.

In America, north of the New Jerusalem, at Adam-Ondi-Ahman a grand council will then be convened with our ancestor, the resurrected Adam presiding. At this grand council, King

David of the reunited House of Israel will deliver the throne of David his fore-father to Jesus Christ who will henceforth reign as King of Kings forever more. Christ is "the root and the offspring of David." (Revelation 22:16) And thereby will the scriptures be fulfilled that there will never be a time when a son of David will not sit upon his throne which throne will endure forever.

Because the temple built by the Lost Tribes in Jerusalem will be so beautiful, with such an abundance of gold, silver and precious jewels that they will have brought out of Our Hollow Earth, this last Old World dictatorship spoken of by John the Beloved in Revelation 13, having *"two horns like a lamb"* and governed by the *"dragon"* Lucifer, (Revelation 13:11-18) will covet the wealth of the Israelite temple in Jerusalem, and so decide to attack Israel "to take a spoil" in the great Battle of Armageddon. (Ezekiel 38:10-12) Christ will then come in power and great glory as the Jewish Messiah. He will kill the invading armies with giant hail stones, and the Jewish survivors will accept him as their God and King.

Ezequiel 39:9 explains that after the Battle of Armageddon, which occurs after the sounding of the seventh angel's trumpet, the Jews will take seven years to clean up the mess of the battle and to burn the weapons of war.

John, the Beloved, in his Book of Revelation, indicates that during these last seven years, the angels of God will be pouring out their last seven vials of plagues on the unrepentant sinners of the world in a last attempt to try to get them to repent of their sins and be saved. (Rev. 16)

Similar to the plagues that God poured out on the Egyptians to try to convince them to let captive Israel go at the time Moses led them out of Egypt, in these last days, plagues will be poured out upon the Earth to free Israel once again. As Emmanuel Velikovsky showed in his books, the plagues of the Israelite exodus from Egypt were partially caused by a planet sized comet (Venus) passing the Earth, so in these last days, the last-days plagues will also be the result of the planet-sized comet returning which originally caused the Flood of Noah. This comet will come by the Earth from the south and tip the Earth back on its axis to where it was before the flood at 16 degrees tilt. It will pull the Earth back closer to the Sun where it was before the flood at a location between where the Earth is now and where the orbit of Venus is so that we will again have a 290 day year.

Colonel Ed Dames has remote viewed the future, and as this comet-sized planet passes the Earth, it will cause the center of gravity on the surface to tilt for a short time at an angle of 45 degrees, which will cause a giant earthquake and almost all buildings on Earth will fall. The passage of the comet will also cause winds of 300 miles per hour that will destroy almost everything else.

The moment this comet-sized planet returns, God,

"...shall utter his voice out of Zion, and he shall speak from Jerusalem, and his voice shall be heard among all people;"

"And it shall be a voice as the voice of many waters, and as the voice of great thunder, which shall break down the mountains, and the valleys shall not be found."

"He shall command the great deep, and it shall be driven back into the north countries (of the Hollow Earth), and the islands shall become one land;"

"And the land of Jerusalem and the land of Zion shall be turned back into their own place, and the earth shall be like as it was in the days before it was divided."

"And so great shall be the glory of his presence that the sun shall hide his face in shame, and the moon shall withhold its light, and the stars shall be hurled from their places."

"And the Lord, even the Savior, shall stand in the midst of his people, and shall reign over all flesh." (D&C 133:22-25)

Jesus told his apostles in Jerusalem that,

29 "Immediately after the tribulation of those days shall the sun be darkened, and the moon shall not give her light, and the stars shall fall from heaven, and the powers of the heavens shall be shaken:

30 And then shall appear the sign of the Son of man in heaven: and then shall all the tribes of the earth mourn, and they shall see the Son of man coming in the clouds of heaven with power and great glory.

31 And he shall send his angels with a great sound of a trumpet, and they shall gather together his elect from the four winds, from one end of heaven (from all over and in the Earth from one end of the Inner Sun) *to the other."* (Matt. 24:29-31)

Col. Ed Dames has remote viewed the future and says the earth will be hit by a massive coronal mass ejection from our outer Sun, a solar flare that will burn the surface of our planet.

He calls it the "Kill Shot." He has a video on his website showing a comet which is hit by a solar flare that jumps out from the Sun and hits the comet as the comet goes around the Sun.

The same will happen with the return of the Noah planet-sized comet, when a solar flare will jump out from the Sun and hit the comet and the Earth as the comet passes the Earth. The solar flare will burn all the wicked that are not saved by Christ's coming. The scripture says,

"For behold the day cometh that shall burn as an oven, and all the proud, yea, and all that do wickedly, shall be stubble; and the day that cometh shall burn them up, saith the Lord of hosts, that it shall leave them neither root nor branch." (Malachi 4:1)

At April Conference in Nauvoo in 1843, the prophet Joseph Smith said,

"It is not the design of the Almighty to come upon the earth and crush it and grind it to powder...There will be wars and rumors of wars, signs in the heavens above and on the earth beneath, the sun turned into darkness and the moon to blood, earthquakes in divers places, the seas heaving beyond their bounds; then will appear one grand sign of the Son of Man in heaven. But what will the world do? They will say it is a PLANET, A COMET, etc." (DHC 5:337)

But before the Lord burns the Earth and the wicked that refuse to repent and accept Him as their Savior, He will lift up His Saints who are alive from the surface of the Earth so that they will not be burned. Those of His Saints that are in their graves shall also be resurrected and caught up in the cloud to meet Him.

95 "And there shall be silence in heaven (the Inner Sun) *for the space of half an hour* (this is a second silence for the space of half an hour, but this time it is according to Earth's time, which is also the time of the Inner Sun); *and immediately after shall the curtain of heaven be unfolded, as a scroll is unfolded after it is rolled up, and the face of the Lord shall be unveiled;*

96 And the saints that are upon the earth, who are alive, shall be quickened and be caught up to meet him.

97 And they who have slept in their graves shall come forth, for their graves shall be opened; and they also shall be caught up to meet him in the midst of the pillar of heaven—

98 They are Christ's, the first fruits, they who shall descend with him first, and they who are on the earth and in their graves, who are first caught up to meet him; and all this by the voice of the sounding of the trump of the angel of God." (D&C 88:95-98)

After the burning of the Earth with the wicked (the Telestials and Sons of Perdition who will be burned), the righteous that were caught up alive, together with the resurrection of the righteous dead (those living the Celestial or Terrestrial Law) will then return to the surface of the planet to live with Jesus Christ as their King for one thousand earth years in peace and happiness.

My niece saw this happening in a dream. One day, she saw people floating up into space where a giant spaceship was approaching the Earth. She floated up with the people into the spaceship (the returning City of Enoch). Then from the spaceship she saw the Earth was hit by a solar flare that burned the surface of the planet and all the people that were not caught up. The spaceship had a force field that repelled the solar flare and protected the Saints in it.

May we be among the righteous that will be caught up at the last day and saved from the burning, and subsequently return to Earth with Jesus Christ, our Savior to live with Him on earth during his Millennial Reign on the throne of David, which throne he inherited from his ancestor, David, and which anointing as King he will have received from King David of the Lost Ten Tribes of Israel at Adam-ondi-Ahman, Missouri.

CHAPTER FOURTEEN
The City Of Enoch -- Found!

Evidence from Dr. Raymond A. Moody's book, REFLECTIONS ON LIFE AFTER LIFE, supports my conclusions on the locations of Paradise and Hell in the Spirit World of this earth. As I have concluded previously with evidence from the scriptures, that the location of Paradise is the SUN suspended in the hollow of our earth by gravity and electrostatic forces, so do the stories of persons who have clinically died and later medically resuscitated, indicate that they, while in the spirit, travel to a place that could be termed PARADISE, or Heaven. In their accounts, these people claim that their spirits leave their bodies through their head. Then after passing through a (spiritual) dark tunnel which according to my theory passes through *"outer darkness"* or the shell of the earth, the spirits of these people arrive at a place of beautiful light -- the Interior Sun.

As recorded in Dr. Moody's book,

"One middle-aged man who had a cardiac arrest related: I had a heart failure and clinically died...I remember everything perfectly vividly...Suddenly I felt numb. Sound began sounding a little distant...All this time I was perfectly conscious of everything that was going on. I heard the heart monitor go off. I saw the nurse come into the room and dial the telephone, and the doctors, nurses, and attendants come in."

"As things began to fade there was a sound I can't describe; it was like the beat of a snare drum, very rapid, a rushing sound, like a stream rushing through a gorge. And I rose up and I was a few feet up looking down on my body. There I was, with people working on me. I had no fear. No pain. Just peace. After just probably a second or two, I seemed to turn over and go up. It was dark--you could call it a hole or a tunnel--and there was this bright light. It got brighter and brighter. And I seemed to go through it."

"All of a sudden I was just somewhere else. There was a gold-looking light, everywhere. Beautiful. I couldn't find a source anywhere. It was just all around, coming from everywhere. There was music. And I seemed to be in a countryside with streams, grass, and trees, mountains. But when I looked around--if you want to put it that way--there

were not trees and things like we know them to be. The strangest thing to me about it was that there were people there. Not in any kind of form or body as we know it; they were just there."

"There was a sense of perfect peace and contentment; love. It was like I was part of it. That experience could have lasted the whole night or just a second...I don't know."

Another woman described her out-of-the-body experience:

"There was a vibration of some sort. The vibration was surrounding me, all around my body. It was like the body vibrating, and where the vibration came from, I don't know. But when it vibrated, I became separated. I could then see my body...I stayed around for a while and watched the doctor and nurses working on my body, wondering what would happen...I was at the head of the bed, looking at them and my body, and at one time one nurse reached up to the wall over the bed to get the oxygen mask that was there and as she did she reached through my neck..."

"And after I floated up, I went through this dark tunnel...I went into the black tunnel and came out into brilliant light...A little bit later on I was there with my grandparents and my father and my brother, who had died...There was the most beautiful, brilliant light all around. And this was a beautiful place. There were colors-- bright colors--not like here on earth, but just indescribable. There were people there, happy people...People were around, some of them gathered in groups. Some of them were learning..."

"Off in the distance...I could see a city. There were buildings--separate buildings. They were gleaming, bright. People were happy in there. There was sparkling water, fountains...a city of light I guess would be the way to say it...It was wonderful. There was beautiful music. Everything was just glowing, wonderful...But if I had entered into this, I think I would never have returned...I was told that if I went there I couldn't go back...that the decision was mine."
(REFLECTIONS, pp. 15-17)

Dr. Moody goes on to describe a Hell, or place of bewildered spirits, which according to my theory is located in the shell of the earth, on its surface and atmosphere. Upon asking one woman where she saw these bewildered spirits she replied,

> "...it was before I actually entered this tunnel--as I referred to it--and before I entered the spiritual world where there is so much brilliant sunlight." (REFLECTIONS, p. 20)

In my theory, I mentioned the scripture, Alma 40:11, 12, which states that all men whether good or bad are taken home to that God which gave them life to be JUDGED to see if they will stay in Paradise or be cast out into Hell, or Outer Darkness. In his chapter on JUDGMENT, Dr. Moody relates,

> "...it seems appropriate to examine something in near-death experiences which may or may not, according to one's theology, be likened to the concept of a judgment. Again and again, my near-death subjects have described to me a panoramic, wrap-around, full-color, three-dimensional vision of the events of their lives. Some people say that during this vision they saw only the major events of their lives. Others go so far as to say that in the course of this panorama every single thing that they had ever done or thought was there for them to see. All the good things and all the bad were portrayed there at once, instantaneously."

> "It will be remembered also that this panorama was quite frequently said to have taken place in the presence of a 'being of light,' whom some Christians identified as Christ, and that this being asked them a question, in effect, 'What have you done with your life?'"

> "In being pressed to explain as precisely as they can what the point of this question was, most people come up with something like the formulation of one man who put it to me most succinctly when he said that he was asked whether he had done the things he did because he loved others, that is, from the motivation of love. At this point, one might say, a kind of judgment took place, for in this state of heightened awareness, when people saw any selfish acts which they had done they felt extremely repentant. Likewise, when gazing upon those events in which they had shown love and kindness they felt satisfaction." (REFLECTIONS pp. 31, 32)

The section of the scripture, (Alma 40:12) which states that the spirits of those who are righteous are received into a state of happiness, which is called Paradise, sounds much the way the woman quoted previously described. She said,

"There was sparkling water, fountains...a city of light I guess would be the way to say it...It was wonderful. There was beautiful music. Everything was just glowing, wonderful..."

That this wonderful city of light in Paradise, in the sun inside Our Hollow Earth is the seat of the throne of Jehovah (Jesus Christ) in this earth and is referred to as being a *"hiding place"* (inside our earth) is an interpretation that can be given several scriptural passages. Doctrine and Covenants Section 101, verse 89 says,

"And if the president (of the United States) *heed them not* (concerning some wrongs the citizens of Missouri did to the early members of the Church of Jesus Christ of Latter-day Saints), *then will the Lord ARISE AND COME FORTH OUT OF HIS HIDING PLACE, and in his fury vex the nation."* (Needless to say the president didn't heed them and hence the civil war vexed the nation.)

The word *"arise"* would indicate that the Lord's *"hiding place"* is down inside the earth in order for him to arise or come up. *"Come forth out"* of Our Hollow Earth, are the same words used in Moses 7:48 in which the earth is speaking of Adam and Eve when they were cast *"forth out of me"* from their Garden of Eden inside the earth.

Other passages referring to the Lord's *"hiding place"* are:

D&C 121:1, *"O, God, where art thou? And where is the pavilion that covereth thy HIDING PLACE?",*

D&C 123:6, *"That we may not only publish to all the world, but present them to the heads of government in all their dark and hellish hue* (the wrongs the citizens of Missouri did to the early saints), *as the last effort which is enjoined on us by our Heavenly Father, before we can fully and completely claim that promise which shall call him FORTH from his HIDING PLACE;...",* and

Isaiah 45:15, *"Verily thou art a God that HIDEST thyself, O God of Israel, the Saviour."*

In the Lord's *"hiding place"* in the earth's interior sun is where the tree of life is located today. From its original place in the Garden of Eden it was transplanted to Paradise to the Interior Sun and the fruit thereof is given to those who reach perfection in this life, fore-go death and are caught up to Paradise to become translated beings.

"To him that overcometh will I give to eat of the tree of life, which is in the midst of the paradise of God." (REVELATION 2:7)

When Adam and Eve lived in the Garden of Eden, God told them that of the fruit of every tree they might freely eat except the tree of the knowledge of good and evil. And while they ate of the tree of life they were immortal and would have lived forever (2 Nephi 2:22) but when they partook of the forbidden fruit of the tree of knowledge of good and evil, their bodies were changed from immortality to mortality. Thereafter they were prevented from partaking of the fruit of the tree of life until they should repent of their transgression of God's commandment. The Genesis account of *"Cherubims and a flaming sword which turned everyway, to keep the way of the tree of life"* (Genesis 3:24), refers to the transplanting of the tree of life from the Garden of Eden to the Interior Sun which is a *"flaming sword"* which turns *"every way to keep the way of the tree of life."*

Thereafter, from the days of Adam until today those people who reached perfection in this life were caught up to Paradise to partake of the tree of life which the fruit thereof changes mortal flesh into flesh divine. John the Beloved was one who was translated. (See John 21:21-22)

When Christ came to America shortly after his resurrection, he also blessed three Nephite Disciples with translation. Of the twelve disciples Jesus chose to lead His church in ancient America, nine he touched with his finger because they wanted to go to Heaven when they died at 72 years of age. But three disciples He caught up to Heaven in the process of translation, so that they could partake of the fruit of the Tree of Life in Paradise, which changed their bodies into an immortal state so they would not die until the day of their resurrection.

The Book of Mormon relates that *"the heavens were opened, and they were caught up into heaven, and saw and heard unspeakable things.*

And it was forbidden them that they should utter; neither was it given unto them power that they could utter the things which they saw and heard;

And whether they were in the body or out of the body, they could not tell; for it did seem unto them like a transfiguration of them, that they were changed from this body of flesh into an immortal state, that they could behold the things of God.

But it came to pass that they did again minister upon the face of the earth; nevertheless they did not minister of the things which they had heard and seen, because of the commandment which was given them in heaven.

...they did go forth upon the face of the land, and did minister unto all the people, uniting as many to the church as would believe in their preaching; baptizing them, and as many as were baptized did receive the Holy Ghost.

And they were cast into prison by them who did not belong to the church. And the prisons could not hold them, for they were rent in twain.

And they were cast down into the earth; but they did smite the earth with the word of God, insomuch that by his power they were delivered out of the depths of the earth; and therefore they could not dig pits sufficient to hold them.

And thrice they were cast into a furnace and received no harm.

And twice were they cast into a den of wild beasts; and behold they did play with the beasts as a child with a suckling lamb, and received no harm.

And it came to pass that thus they did go forth among all the people of Nephi, and did preach the gospel of Christ unto all people upon the face of the land; and they were converted unto the Lord, and were united unto the church of Christ, and thus the people of that generation were blessed, according to the word of Jesus.

And now I, Mormon, make an end of speaking concerning these things for a time.

Behold, I was about to write the names of those who were never to taste of death, but the Lord forbade; therefore I write them not, for they are hid from the world.

But behold, I have seen them, and they have ministered unto me." (3 Nephi 28:13-26)

The three translated disciples of Jesus ministered among the North American Indians who were descendants of the Nephites and Lamanites, as recorded in their Sacred Records of the Nemenhah. Although Mormon did not disclose their names, the Nemenhah records do. They were three brothers: Nephi, the son of Nephi, who was the son of Helaman, who was the son of Helaman, who was the son of Alma the Younger, who was the son of Alma, the founder of their Church of Jesus Christ. And Nephi's brothers were Timothy, who Nephi had raised from the dead, and Lehi, also known as Mathonihah. (See 3 Nephi 19:4)

The Nemenhah records let us know that the Earth is divided up for their ministry in this manner: The lands east of Jerusalem are ministered to by Lehi, the lands west of Jerusalem except the Americas are ministered to by Nephi, and the American continent is ministered to by Timothy. Joseph Smith said that John the Beloved ministers to the Lost Ten Tribes in the North Countries of Our Hollow Earth. These four translated disciples of Jesus meet regularly and John the Beloved presides.

The Nemenhah history recorded the visit of the three translated Nephite disciples of Jesus to them many times throughtout their history (60 BC to 1660 AD). My own great great grandfather David Cluff who joined the Church of Jesus Christ of Latter-day Saints in the time of Joseph Smith recorded in his personal history how one of the translated Nephites visited his home in Nauvoo and helped him get his carpentry tools back in shape so he could work on the Nauvoo temple after returning from one of his proselyting missions.

The three Nephite disciples of Jesus helped convert my wife's Stake's President's counselor, Presidente Cedeño in Mexico. Presidente Cedeño gave my wife her recommend so I could marry her in the Mesa, Arizona Temple.

Many years previous, Hermano Cedeño was a Baptist Minister in Ciudad Obregon, Sonora. One Sunday he noticed three men enter his church while he was giving a sermon. After the sermon, Hermano Cedeño went to greet the three visitors. They told him he had given an excellent discourse. But that the only thing he lacked was the priesthood authority from Jesus Christ to preach His gospel. He asked them how he could get that authority. They told him, "From the Church of Jesus Christ of Latter-day Saints."

Years later Hermano Cedeño found a unit of the Church of Jesus Christ of Latter-day Saints in Tucson, Arizona. He listened to the missionaries and was baptized. He later served as counselor to Stake President Agrícol Lozano in Mexico City and later as temple President of the Mexico City Temple where he told this story of his conversion to our friends Carlos and Aida Luz Villa at their sealing in the Mexico City temple.

After the three Nephite disciples of Jesus were taken to Paradise, they then returned to earth and are still working among men today to bring people to Christ. Many people have seen them.

There are those who die and are clinically resuscitated which I believe have seen the ancient city of Zion, the translated City of Enoch which was taken into heaven.

"Off in the distance...I could see a city. There were buildings--separate buildings. They were gleaming, bright. People were happy in there. There was sparkling water,

fountains...a city of light I guess would be the way to say it...It was wonderful." (REFLECTIONS, p. 17)

The scriptures indicate that several hundred years before the flood of Noah, Enoch's city of Zion was translated and taken to heaven, the throne of Jehovah -- that place where persons who forego death by achieving perfection in this life are clothed with Light, which is translation. This throne of Jehovah, the people inside Our Hollow Earth say is located in the Interior Sun.

In Moses 7:31 is the record of the City of Enoch which was taken physically into heaven. *"And thou hast taken Zion to thine own bosom...and truth is the habitation of thy throne..."* and verse 21, *"and lo, Zion, in process of time, was taken up into heaven."*

The *"heaven"* of our earth is the sun inside our earth, which is the physical location of PARADISE or heaven of our earth's spirit world. The people of the City of Enoch were taken to Paradise to partake of the fruit of the tree of life which changed their bodies to overcome death in the process of translation. Later, after the City of Enoch was taken to Paradise, many were caught up to be translated,

"And Enoch beheld angels descending out of heaven, bearing testimony of the Father and Son; and the Holy Ghost fell on many, and they were caught up by the powers of heaven into Zion." (Moses 7:27)

Even after the flood of Noah, a city was translated and taken to heaven. Melchizedek, the righteous King of Salem (where Jerusalem now stands) obtained peace in Salem, and was called the Prince of Peace.

"And men having this faith, coming up unto this order of God, were translated and taken up into heaven. And now, Melchizedek was a priest of this order; therefore he obtained peace in Salem, and was called the Prince of peace. And his people wrought righteousness, and obtained heaven, and sought for the city of Enoch..." (Joseph Smith translation of the Bible, Genesis 14:32-34)

At the beginning of the millennial reign of Christ upon the earth, the city of Enoch will be brought back to the earth's surface from the interior paradise-sun as part of the restoration of all things as prophesied by the Apostle Peter who said,

"And he shall send Jesus Christ, which before was preached unto you: Whom THE HEAVEN must receive until the times of restitution of all things, which God hath spoken by the mouth of all his holy prophets since the world began." (Acts 3:20, 21)

Today are the last days before the Lord's Second Coming in which the Lord said to Enoch:

"AND RIGHTEOUSNESS WILL I SEND DOWN OUT OF HEAVEN; and truth will I send FORTH OUT OF THE EARTH, to bear testimony of mine Only Begotten; his resurrection from the dead; yea, and also the resurrection of all men; and righteousness and truth will I CAUSE TO SWEEP THE EARTH AS WITH A FLOOD, to gather out mine elect from the four quarters of the earth, unto a place which I shall prepare, an Holy City, that my people may gird up their loins, and be looking forth for the time of my coming; for there shall be my tabernacle, and it shall be called Zion, a New Jerusalem."

"And the Lord said unto Enoch: Then shalt thou and all thy city meet them there, and we will receive them into our bosom, and they shall see us; and we will fall upon their necks. And they shall fall upon our necks, and we will kiss each other;" (Moses 7:62, 63)

The righteous dead in Paradise will be resurrected at Christ's Coming and the City of Enoch with its translated beings in Paradise will come back with Him:

"These are they whom he shall bring with him, when he shall come in the clouds of heaven to reign on the earth over his people." (D&C 76:63)

"And they who have slept in their graves shall come forth for their graves shall be opened, (D&C 88:97) *"in the resurrection of the just."* (D&C 76:65)

Therefore, Paradise will be emptied at the Lord's Coming. During the Millennium, the righteous dead will go to the Inner Sun no more but will pass from mortality to immortality in a twinkling of an eye.

"Wherefore, children shall grow up until they become old; old men shall die; but they shall not sleep in the dust, but they shall be changed in the twinkling of an eye." (D&C 63:51)

At the beginning of the Millennium, the location of Paradise and Hell will be switched. Earth's shell will become the location of Paradise, and the Inner Sun will be the location of Hell during the Millennium. The Inner Sun will be used by the Lord as a prison for Satan and his devil angels, so they cannot tempt those on the earth. At the Lord's Second Coming, the Inner Sun will be emptied of the righteous spirits in their resurrection from the grave and the City of Enoch brought back to earth from its location in the Inner Sun – then Satan and all his devils will then be rounded up and cast into the Inner Sun where they will be chained by the chains of gravity so they cannot tempt those on earth for a thousand years.

"And I saw an angel come down from heaven having the key of the bottomless pit (Our Hollow Earth is a bottomless pit, having no top nor bottom -- the north and south polar openings) *and a great chain in his hand."*

"And he laid hold on the dragon, that old serpent, which is the Devil, and Satan, and bound him a thousand years,"

"And cast him into the bottomless pit, and shut him up, and set a seal upon him, that he should deceive the nations no more, till the thousand years should be fulfilled; and after that he must be loosed a little season."

"And when the thousand years are expired, Satan shall be loosed out of his prison."

"And shall go out to deceive the nations..." (Revelation 20:1-3, 7, 8)

Satan's desires will be fulfilled,

"For thou hast said in thine heart, I will ascend into heaven, I will exalt my throne above the stars of God... (Olaf Jansen described the base of the interior sun as opaque with bright holes of light that shine like stars and serves as the base of the throne of Jehovah. THE SMOKY GOD, p. 109) *I will ascend above the heights of the clouds; I will be like the most High."*

Although Satan will be in the Inner Sun throughout the Millennium, he will find it will be a Hell for him just because he is there.

"Yet thou shalt be brought down to hell, to the side of the pit" (the earth's shell) (Isaiah 14:13, 14) at the end of the Millennium where Satan will be allowed to tempt mankind one last time. Satan's glory will be short-lived and his last taste of heaven will be turned into everlasting torment when he and his angels will be cast into a kingdom without glory. (D&C 88:24)

A kingdom without glory is a planet with no light. So after the earth's resurrection into a Celestial Star, Satan and his angels will most likely be confined to a Black Hole in space -- a planet that has so much mass and gravity that any light entering it cannot escape back out to reflect off anything resulting in pitch blackness and millions of times earth's gravity that will bind them to the surface so that they cannot move even a finger to do anyone harm any longer in their eternal prison forever.

Many prisons are located in the center of cities, so more than likely, the Kingdom without glory, Perdition, or Black Hole, the final Hell where Satan and his angels will be confined for eternity after the resurrection of the Earth and its inhabitants -- will perhaps be in the center of our Milky Way Galaxy where scientists have observed a Black Hole. They have observed also that coming out of its north and south polar openings is a tremendous amount of radiation that shoots far up and below the galactic disk.

At the center of our Milky Way Galaxy is a black hole, named Sagittarius A that has thrown a cloud of dust high above and below the galactic plane.

After the earth is resurrected and becomes a hollow star, then the celestial New Jerusalem will descend out of the heaven of our outer Sun where is it now being constructed by Christ, the Saints and Prophets that have already been resurrected. It will be brought through a polar opening entrance that will appear as *"circling flames of fire"* to be suspended in the hollow of the resurrected, celestialized earth. The shape of a four-sided crystal pyramid, the New Jerusalem will shine like the sun and will be the THRONE OF JEHOVAH, who is JESUS CHRIST, and will be the home of the celestialized Saints of God forever after.

CHAPTER FIFTEEN
A Proposal for an Expedition to Our Hollow Earth

The Hollow Earth Theory maintains that our supposed solid-liquid earth is really HOLLOW within and near the geographic poles at the northern and southern extremities of the earth there exist HOLES that lead into an interior world boasting a perfect climate year round where abound the most luscious plant and profuse animal life which all supports the most advanced of civilizations -- a GIANT race of people 10 to 15 feet tall who live to be 800+ years old.

Although that land has been discovered by explorers such as William Morgan in 1827, as described in the book ETIDORPHA who reached Our Hollow Earth through a cavern in Kentucky; Olaf Jansen in 1829 as described in his book, THE SMOKY GOD who reached Our Hollow Earth through the North Polar Opening with his father in their fishing sail boat; by Karl Unger who made it to Our Hollow Earth as a refugee from Nazi Germany after World War II in the German submarine U-209; Admiral Richard E. Byrd, who flew there in 1927, 1929 and 1947 in his flights beyond the poles; Reinhold Schmidt who was taken to Our Hollow Earth in a flying saucer from Los Angeles in 1958; Hank Krastman who was taken there by the great grandnephew of Jacob Shultz through a Hopi cavern near where the Little Colorado River empties into the Grand Canyon, Arizona in about 1961; and Billie F. Woodard who was taken there in a flying saucer through the North Polar Opening in 1963, and later several times in Hollow Earth tunnel trains accessed beneath Area 51, Nevada -- yet Our Hollow Earth has remained a WORLD TOP SECRET, kept secret by a powerful conspiratorial organization of the Super-Rich of America and Europe called The Order of the Illuminati and their Jesuit controllers which have a controlling interest in all the governments of the world, including the United States of America.

In view of the fact that Our Hollow Earth has not been discovered openly to the world as of yet, there remains the question of WHO will eventually bring it to light. The late W. Cleon Skousen, a member of the Church of Jesus Christ of Latter-day Saints, the Mormons, was of the belief that, yes the Mormons will do it. He wrote,

"We may rest assured that the first people on earth to whom this great secret will be finally revealed will be the

prophets and chosen Saints of God. They will know the exact location and whereabouts of the lost tribes long before any scientist or society will discover it." (PROPHECY AND MODERN TIMES, pp. 55, 56)

Therefore, if it is to the Mormons that the discovery will go, I suggest that the *"Lord's"* University, Brigham Young University, with the world as its campus, should fit out an expedition to the North Country to visit the Ten Tribes, perhaps deliver a message of good will from the President and prophet of the Church of Jesus Christ of Latter-day Saints -- the Mormons, and take some measurements and studies of the North Polar Opening and the land within Our Hollow Earth.

The expedition could consist of a boat and a seaplane or a dirigible or even a submarine. In fact, this author is convinced that enough flying saucer technology now exists that one of these fantastic flying saucer-type craft could now be built. This craft would be especially suited for the trip with the speed and range needed to study with scrupulous detail the nature of the polar openings, the hollow interior and Inner Sun. With an unlimited range and similarity to the Hollow Earth Nation's own craft, it would be more apt to be taken as friendly craft by the inner earth peoples.

An expert navigator should be taken and scientists to measure the actual size of the polar openings, the thickness of the earth's shell and to study the conditions within the earth.

In 1959, F. Amadeo Giannini published his book, WORLDS BEYOND THE POLES in which he recorded the flight of Richard E. Byrd beyond the North Pole. Giannini wrote,

"This United States Navy's polar exploratory force was preparing to embark upon one of the most memorable adventures in world history. Under the command of Rear Admiral Richard Evelyn Byrd, U.S.N., it was to penetrate into land extending beyond the North Pole supposed end of the Earth...As the hour approached for air journey into the land beyond, Admiral Byrd transmitted from the Arctic base a radio announcement of his purpose, but the announcement was so astonishing that its import was lost to millions who avidly read it in press headlines throughout the world...The words of the message were momentous: 'I'd like to see that LAND BEYOND the Pole."..."That area BEYOND the Pole is THE CENTER OF THE GREAT UNKNOWN!'

"To confirm the import of Admiral Byrd's announcement, one has only to examine the globe...Try to

find any area of land, water or ice which encroaches upon the North Pole and which is not known...Is Spitsbergen or Siberia unknown? Is Alaska or the Canadian Archipelago unknown? And do any such land areas extend NORTH BEYOND THE NORTH POLE?...Hence the land mentioned by Admiral Byrd must lie DUE NORTH from the North Pole..."

Subsequently, "...the admiral and his airplane crew accomplished a physical flight of seven hours duration in a northerly direction beyond the North Pole. Every mile and every minute of that journey beyond was over ice, water, or land that no explorer had seen...As progress was made beyond the Pole point, there was observed directly under the plane's course iceless land and lakes, and mountains where foliage was abundant. Moreover, a brief newspaper account of the flight held that a member of the admiral's crew had observed a monstrous greenish-hued animal moving through the underbrush of that land beyond the Pole."

"The magnitude of that memorable flight...was never submitted for popular consumption. Press representatives were denied knowledge of it except during the brief period of active flight, when radio dispatches kept them informed. And insofar as personal knowledge extends, the admiral, contrary to precedent, failed to render a book account of his most important flight and discovery...Immediately after the flight account was heard in Washington, the office of United States Naval Intelligence conducted a wide investigation of the author (Giannini) of a work (his book) which had described such unknown land and the reason for its existence twenty years before it was discovered."
(WORLDS BEYOND THE POLES, pp. 148-150)

This flight by Byrd beyond the North Pole was reported by Giannini to have occurred in February 1947. Apparently, Giannini was mistaken on the year this flight occurred. John B. Leith, in his book, GENESIS FOR THE SPACE RACE, says this flight sponsored by the U.S. Navy occurred in 1927. Leith says he saw the photos and the flight log in the United States National Archives of this 1,700 mile flight in 1927 by Richard E. Byrd with his flight navigator Floyd Bennett through the North Polar Opening. The photos showed prehistoric animals, green hills, lakes and rivers and the flight log even mentions people waving to them from the ground. Upon his return, President Calvin Coolidge put a lid of secrecy on this momentous discovery by Byrd because he said, *"No-one would believe this."* (GENESIS FOR THE SPACE RACE, p. 145)

Byrd's 1947 flight to Our Hollow Earth occurred through the South Polar Opening from McMurdo Base in the Antarctic during the Operation High Jump expedition.

The New York Times (which belongs to the Rockefellers) of Feb 18, 1947 p. 1 displays an article titled,

"Byrd Hops Over South Pole Again. Feb. 16. Little America. Rear Admiral Richard Evelyn Byrd, the only man to fly over both poles, again flew over the South Pole today and beyond it...(he) soared along the 180th meridian across the Pole and eighty six miles into the vast hitherto unseen region beyond the pole."

"After leaving the pole Admiral Byrd swung the plane to the right to explore the region that he has described as 'the most inaccessible area on the face of the earth.'"

However, my research indicates that the South Polar Opening is towards the left instead of the right of the South Pole when flying from McMurdo Base where Admiral Byrd's airplane took off. Perhaps the Rockefellers that helped finance some of his flights were trying to mislead people to the location of the South Polar Opening by saying he flew to the right of the South Pole, instead of to the left.

The Rockefellers are considered part of the super-secret and super powerful Order of the Illuminati, which I believe is hiding the discovery by Admiral Byrd that our earth is hollow.

John B. Leith, in his book, GENESIS FOR THE SPACE RACE, reported that as part of Operation High Jump, on February 5, 1946, Byrd flew west of McMurdo Base towards an Antarctic Mountain called, "Coal Sack" on which they observed a coal deposit. After passing the Coal Sack Mountain on their right, they passed the Axel Heiberg Glacier on the left and came to a valley leading into the South Polar Opening. They flew about half-way through the opening and determined that it was 125 miles wide at the neck of the opening, before returning for lack of fuel.

The next year's flight of February 16, 1947 as part of Operation High Jump, Leith reported that Byrd flew all the way into Our Hollow Earth with eight Falcon aircraft, where he found German settlements of refugees from World War II, and was met by German flying saucers that challenged his presence in their land in the Hollow Earth. Byrd had been given orders by President Truman not to fire on the Germans if they found them in the Hollow Earth, but Byrd did it anyway. The German flying saucers returned the fire with their laser beams and knocked all of Byrd's planes out of the air except his. A warning came over his radio to leave their land by way of the South Polar Opening and never return. Some 300 miles north of McMurdo Sound, the fleet of Operation High Jump was also attacked by German flying saucers that came up out of the ocean as reported and filmed by a Russian ship that was monitoring the fleet. Planes were launched against them from the fleet's aircraft carrier, but within minutes all our planes were knocked out of the air by the flying saucer laser beams, and the fleet's Destroyer Battleship caught on fire before the saucers dove back beneath the waves.

Back in Washington, Leith reported that Byrd was taken before a Court of Inquiry, labeled "mentally incompetent" and barred from further exploration of Our Hollow Earth for having disobeyed orders when he fired upon the Hollow Earth German flying saucers. (GENESIS FOR THE SPACE AGE, p. 164)

Therefore, there exists a big question as to when and where Byrd made those flights into the hollow of the earth. And yet there does exist vast evidence that a Conspiracy does exist, that they are in substantial control of our government, that UFOs and their origination area is considered by them WORLD TOP SECRET, and Byrd's memoirs are kept under lock and key. Which all points to the need for a private organization independent of the International Banker Conspiracy to fit out an expedition to the pole and beyond and establish to the world BEYOND DOUBT that that land does indeed EXIST!

The Conspiracy's carryon has convinced many to believe that the Arctic has been crisscrossed thousands of times by the Strategic Air Command, and crossed under by Atomic submarine.

But there ARE indications of the existence of polar openings from satellite imagery!

John Gagne, whom I met in Alaska in 1981, said he had a bishop who at one time was a chaplain in the Air Force. Once, a young air force pilot came to him quite disturbed. It happened that this pilot was flying his jet in the region of the pole when he noticed land up ahead. He flew down to get a closer look and it was covered with green vegetation. Startled about his find, he called in to the base whereupon he was ordered out of the area.

According to a letter to the Editor of The Hollow Hassle Newsletter of October 1983 by Rose Marie Gilbert, of Lakeside California, Admiral Bryd actually wrote a book about his discovery of land beyond the poles.

"Admiral Byrd - I spent days and weeks trying to track down his niece. I met her across the street at a Tupperware party 8 years ago. She told us about her uncle and how he had suffered because he couldn't tell, but he wrote a book and had it printed and in the bookstores, and the government confiscated all, they thought. There are five still remaining - she has one. If I'd have contacted her sooner, I could have read it; copied it for sure. But at that time it didn't mean much to me..." (THE HOLLOW HASSLE, P.O. Box 747, Aurora, Colorado 80040)

Harley Bryd, who claimed to be the nephew of Admiral Byrd, came out with what he claimed to be the *Missing Diary of Admiral Richard E. Byrd*. It is available for $10 from Inner Light Publications, Dept. OL Box 753, New Brunswick, NJ 08903, and available on Amazon.com. The story is identical to one Bruce Walton sent me many years ago which he believed to be fake. Hollow Earth researcher Dennis Crenshaw analyzed this diary and stated his opinion that it is fabricated because of the similarity of some of the wording in the diary to what the High Lama said to Robert Conway in the movie, Shangri La, which was based on James Hilton's 1933 novel, LOST HORIZON.

This so-called fake diary is also very similar to the story told to me by John Gagne whom I met in Fairbanks, Alaska in 1981. He told me that Admiral Byrd's Alaskan friend, Sylvia Darvell, told John Gagne that the Admiral had confided in her after his flight into Our Hollow Earth through the North Polar Opening, that after crossing over the Arctic ice he came to a continent covered with lush vegetation. His airplane was sided by flying saucer craft which took control of his airplane and gently landed him near an inner earth city. He described the inner earth people as being "large in stature," possessing monorail trains between their cities and craft now known as flying saucers. He was taken into their city for an interview with a government official of the Inner Earth peoples and given a warning to take back to the United States government demanding that we cease using atomic weapons. This discovery by Admiral Richard Evelyn Byrd -- the greatest geographic discovery in the history of our planet -- is to this day maintained by all governments as World Top Secret!

Still, my question is, WHAT ARE THEY TRYING TO HIDE?

AND SHOULD WE BELIEVE A LIE? The Conspiracy's convincing argument that there is no land up near the North Pole has even led the late Elder Bruce R. McConkie, an Apostle of The Church of Jesus Christ of Latter-day Saints to write in his book, MILLENNIAL MESSIAH, that the Lost Tribes of Israel are not hidden up in the north in some undiscovered country, but are just scattered among the nations north of their Jewish homeland in Palestine.

However, former Church leaders held the belief that the Lost Ten Tribes are hidden up in some undiscovered country somewhere in the north beyond the Arctic ice. For example, Joseph Fielding Smith wrote in 1940,

"The Ten Tribes were taken by force out of the land the Lord gave them. Many of them mixed with the peoples among whom they were scattered. A large portion, however, DEPARTED IN ONE BODY INTO THE NORTH and disappeared from the rest of the world." (WAY TO PERFECTION p. 130)

Apostle Orson Pratt expressed his opinion concerning the location of the Lost Ten Tribes thus,

"Their souls will be as a watered garden and they will not sorrow any more at all as they have been doing during the 2500 long years they have dwelt in THE ARCTIC REGIONS." (From a sermon delivered by Orson Pratt on August 29, 1875, in the Salt Lake Tabernacle in Salt Lake City. The sermon was later published as *"Resurrection of the Saints, Etc."* in the Journal of Discourses, Volume 18, pages 57–69)

Then, Elder Bruce R. McConkie, who was considered one of the foremost scriptorians of the Church, shortly before he died on April 19, 1985 came out in support of the Dispersion Theory in his book, THE MILLENNIAL MESSIAH, which is of the position that the Lost Tribes are just scattered all over the known world. Concerning the Lost Ten Tribes, he asks,

"But says one, are they not in a body somewhere in the land of the north? Answer: They are not; they are scattered in all nations. The north countries of their habitation are all the countries north of their Palestinian home..."

"Query: What happened to the Ten Tribes after the visit of the Savior to them...? Answer:...there was apostasy and wickedness..."

"But says another, what about their scriptures--will they not bring them when they return? Answer: Yes, they will bring the Book of Mormon and the Bible...And further, as we devoutly hope, they will also have other records...that will come forth...at the direction of the

president of the Church of Jesus Christ of Latter-day Saints..."

"And finally, says yet another, will they not come with their prophets and seers? Answer: There is no other way they or any people can be gathered...In this day when the head of the Church can communicate with all men on earth, there is no longer any need for one kingdom in Jerusalem and another in Bountiful..." (THE MILLENNIAL MESSIAH, pp. 216, 217)

"Their prophets! Who are they? Are they to be holy men called from some unknown place and people?...Perish the thought!...There are not two true churches on earth...Is Christ divided? (1 Cor. 1:13) God forbid. Their prophets are...stake presidents and bishops..." (THE MILLENNIAL MESSIAH, pp. 325, 326)

Elder Bruce R. McConkie's stated opinion here was that the Lost Tribes are scattered over the known world and are in apostasy awaiting the missionaries of the church to gather them into the gospel. His book was published in February of 1982.

Elder McConkie's stand on the location and condition of the Lost Tribes was perhaps influenced by the publication of a book on THE LOST TRIBES by a member, R. Clayton Brough in 1979 which came out in support of the idea that the Lost Tribes are hidden up in some still undiscovered country in the north and that they have the gospel and prophets of their own. After reviewing the consensus of opinion of former LDS leaders, while emphasizing that there has been NO revealed knowledge of the location and conditions of the Lost Tribes, Brough concluded,

"Thus it is that the Lost Ten Tribes of Israel, wherever they may be, are presently being led, taught, and prepared in the Gospel by chosen servants of the Lord, just as we in the Church today are likewise being instructed by living modern-day Prophets, Seers, and Revelators." (p. 32 THE LOST TRIBES)

Therefore, as Brough notes,

"Today, as in the past, the present geographical location of the Lost Ten Tribes is still a subject of continued debate and speculation among biblical and secular scholars throughout the world." (p. 39)

In view of this uncertainty among Biblical scholars as to the present location of the Lost Tribes of Israel, hints of an International Conspiracy of the Super Rich in control of the governments of the world keeping the discovery that our earth is hollow a secret, and the evidences of a terrestrial paradise in the north as proposed by the Hollow Earth Theory, it would seem to be in the part of wisdom that an expedition should be fitted out to search out that region of the earth and without guile or disguise establish to all the world the truth about the frozen northland and its EDEN hidden somewhere beyond the ice.

From NASA photo #72-HC-928 taken by Apollo 17 (above), perhaps the oval shape visible near the bottom of the image is the south polar opening. It is located on the western part of Antarctica opposite the location on the globe of the North Polar Opening in the eastern part of the Arctic. Based on the anomalous readings on the U.S. Naval Laboratory Arctic Maps of October 25, 2015, and the location of the hole in the ice of the NASA image of September 16, 2012, as well as the same hole in the ice from the Arctic image from the Advanced

<u>Microwave Scanning Radiometer on NASA's Aqua Satellite of February 25, 2011</u>, as also the location the Aurora Australis radiation is exiting the Earth in <u>NASA's Aurora Australis oval</u> as shown in the <u>NASA NOAA relief map of Antarctica</u> that shows the location of the South Polar Opening where the auroral radiation is emitting from, I have concluded that the North Polar Opening is located on the Russian side of the pole and that the South Polar Opening is located on the exact opposite side of the globe in Antarctica.

After meeting Ret. Col. Billie F. Woodard in 2008, I had accepted his testimony that the North Polar Opening is located in the direction of 87.7 North Latitude, 142.2 East Longitude from Alert Base, Ellesmere Island in Northern Canada, which he says he obtained from the actual history of Admiral Byrd's flight beyond the pole that is in the library at Area 51, Nevada where Woodard worked for 11 and half years. But when I plugged these coordinates into Google Earth, I found that this location is on top of the Lomonosov Ridge:

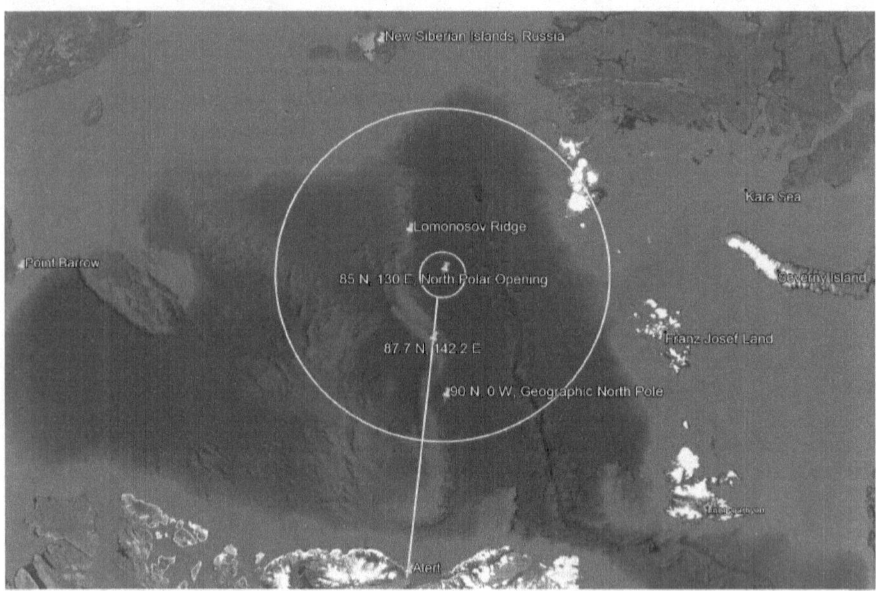

However, notice that Billie's coordinates are located in a straight line from Alert Base on Ellesmere Island, Northern Canada to the coordinates of my latest estimate for the location of the North Polar Opening at 85 N Lat, 130 E Lon, which could indicate the direction which Admiral Richard E. Byrd took when he flew his airplane in 1927 through the North Polar Opening from Alert Base, Ellesmere Island.

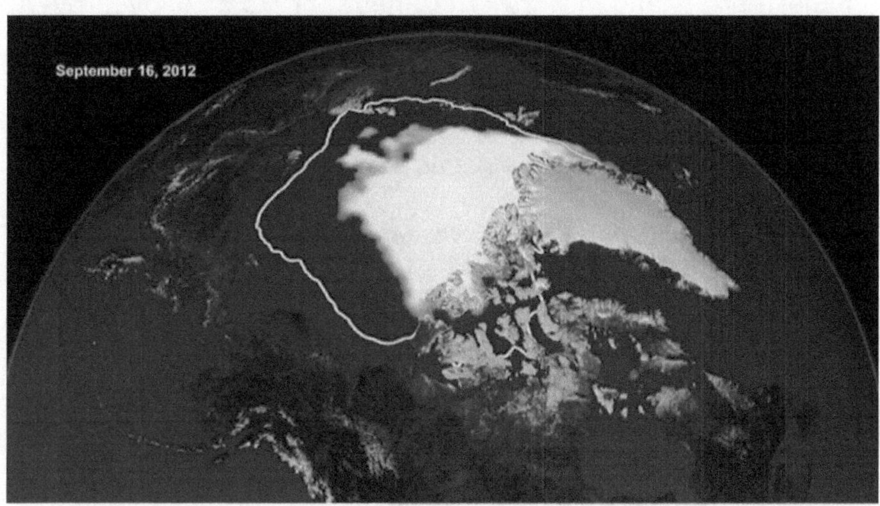

The National Snow & Ice Data Center for September 16, 2012 reported the lowest extent of Arctic ice since 1979 and even shows a hole in the ice where the North Polar Opening most likely is located. This NASA image (above) shows how the record-low Arctic sea ice extent compares with the average minimum extent (in yellow) over the past 30 years. The hole in the ice in this image is located in the same location as anomalous readings from the U.S. Naval Research Laboratory Arctic images of October 26, 2015. These anomalous readings showed at this exact same location Low Ice Concentration, Low Ice Thickness, Low Sea Surface Salinity, and higher Sea Surface Temperature. These anomalous readings at this location of the hole in the ice indicate that this is the most likely location of the North Polar Opening.

Perhaps the best launching point for an expedition to Our Hollow Earth would be from Alert, on the northern shores of Ellesmere Island, in northern Canada. Alert is about 399 statute miles from where the north polar opening starts curving in towards the interior. Alert has a landing strip that could be used to fly to the Inner Continent within the range of a seaplane that could then land on the open ocean in summer time just off the coast of the Inner Continent.

Of course, if we are able to build a flying saucer-type craft, the launching point for the expedition could be accomplished from Utah or from just about anywhere.

The expedition could leave from a Base Camp located at Point Barrow, Alaska, Nord in Northern Greenland, Alert Base on Ellesmere Island in Northern Canada, or from Longyearbyen, Spitsbergen, or if taking a Russian nuclear icebreaker as our proposed Voyage to Our Hollow Earth expedition they would leave from Murmansk, Russia. The time of departure probably should be in mid-winter when there are few clouds and storms, or in late summer when the Arctic ice pack has broken up somewhat.

The compass would be quite worthless as a guide, but should be observed nevertheless throughout the trip. It will point to the magnetic north pole which is located north of Canada until the interior is reached whereupon it will point to their north pole which is our south pole. This is because the magnetic flux lines of the earth come out of the earth at the south magnetic pole near Antarctica, go north on the outside of the earth and then enter the earth again at the north magnetic pole and continue inside the earth to the south magnetic pole in a continuous loop.

If the expedition travels on the side of the polar opening that the magnetic pole is presently located, the magnetic compass will point DOWN as the expedition passes through the opening. Then, later on in the interior, the compass will point to the south magnetic pole as if that pole were the North Pole. For this reason the hollow interior of the earth is called the *"North Countries"* in scripture. If the expedition enters the polar opening on the opposite side of the opening from where the magnetic north pole is located, for example, if leaving from Russia, as the expedition passes through the opening, the compass will point UP towards where the magnetic pole is located on the other side of the hole, such as Olaf Jansen noticed as he and his father were sailing through the North Polar Opening in 1829 from the direction of Franz Josef Land.

The magnetic poles have been known to travel at a steady yearly rate of about 8 miles, although recently the North Magnetic Pole has been noted to be speeding up as it passes over the Arctic Ocean. This movement of the magnetic poles is most likely caused by a slow rotation of the interior sun inside Our Hollow Earth. This could be verified by the expedition with a study of the interior sun. The expedition could also study the nature of the division of the interior sun between its day and night sides, as well as the large opaque area that has holes in it that let bright white light through to appear as stars in the night time.

A Proposal for an Expedition to Our Hollow Earth

A vertical gyroscope should be taken on the journey through the polar opening to insure an undeviating path through the opening, and a horizontal gyroscope to observe the curvature of the earth into the hollow interior. The horizontal gyro should be set level with the sea on the outside perimeter of the polar opening at the beginning of the semi-circumference of the polar opening. Shortly thereafter an earth curvature greater than the 68.9 miles to a polar earth circumference degree should be noticed as the horizontal gyro dips up towards a vertical orientation indicating that the expedition is entering a giant hole into the earth. At about 629 miles from the polar rim, the midpoint of the polar rim should be reached and then the gyroscope's previous horizontal should now be vertical. The Inner Continent should be reached before reaching the midpoint.

An expedition by sea could be taken best, perhaps, with a Russian nuclear icebreaker as expedition leader, Steve Currey, suggested for our Voyage to Our Hollow Earth Expedition, which was cancelled by his family when he died of brain cancer in July of 2006. From their home port at Murmansk, Russia, there is ice-free sailing in late summer to the perimeter of the polar opening and probably no more than 300 miles of ice thereafter, most of it thin ice probably no more than 1 or 2 meters thick at the most, with lots of open leads.

The land closest to the North Polar Opening is Russia's North Land which is just inside the perimeter of the opening.

At noon every day, if the trip is made in the summer, the angle of the sun's position to the earth's horizon should be measured showing it to circle higher into the sky as the expedition progresses into the polar opening. At a certain point on the polar rim, just before the northern-most point is reached, the Sun, in its circling, will stand directly above, 90 degrees above the horizon, as at the Equator, at mid-day on the summer's Solstice.

A measuring of the horizons, the north-south horizon as compared to the east-west, should be made starting at the perimeter of the polar opening. As the journey progresses, at first the east-west horizon should be greater than the north-south horizon -- caused by the flattening effect of the polar opening on the curvature of the earth similar to a pumpkin with its top cut off. Further north, as the expedition begins to dip into the opening, the north-south horizon will be less than the distance of the horizon on the outer surface of the earth and east-west horizon will gradually dip up until the ocean can be seen above one's location within the polar opening. If the expedition is taken in the summer months, at this point, a reflection of the outer sun will be seen reflecting off the surface of the ocean above one's head from the opposite side of the opening like a bright star, such as was observed by Olaf Jansen as he passed through the North Polar Opening in 1829.

A lookout for mirages should be on the itinerary and verification that these are caused by the warm, moist air currents emanating from Our Hollow Earth through the polar openings. As these warm air currents rise above the colder air next to the ice, this warm layer of air reflects distance surface objects making them visible in the sky. Also, ocean water temperatures can be checked showing that deeper water is warmer than water closer to the surface. These temperature inversions can be verified that they are caused by the warm air and warm water coming out of the polar openings of the earth.

The presence of wildlife should be observed, such as any flocks of geese, the Ross Gull, or the Knot flying north towards or from the Hollow Earth's interior and any subtropical plant or animal life floating in the ocean or encased in the ice that has floated out of the Earth's interior should be recovered for study.

Once the expedition begins to detect their entrance into the polar opening by the angle of the sun and the dip of the gyroscopes, periodic checks of the ocean water should be taken in order to find that location on the rim of the polar opening where the earth's rotation causes centrifugal separation of the salt ocean water and the fresh melt water of the icebergs, as Olaf Jansen and Fridtjof Nansen observed. A weighing scale should be taken to note the increase in weight at that halfway point caused by the centrifugal action of the earth's rotation on the rim of the polar opening.

A radar should be taken, and as the expedition enters the polar opening, echoes should be bounced off the other side of the opening, which at the halfway point would be located directly above, thereby determining the diameter of the polar opening. Also, once the interior is reached, echoes could be bounced off the interior sun to determine its distance from the interior surface and its size. Similarly, echoes bounced off the opposite side of the hollow interior would give the size of the hollow in our earth and the thickness of the earth's shell. If a radar is not available, the size of the polar opening, the thickness of the earth's shell and the size of the inner sun can be calculated with the formulas provided by mathematician, Karl D. Lee, using a theodolite and laser rangefinder, which I commissioned him to write for our expedition titled, HOLLOW EARTH FORMULAS, available on my website at: World Top Secret: Our Earth Is Hollow!: Order eBook

At the halfway point of the polar opening or shortly afterward, the horizon directly to the north should be constantly observed for the first appearance of the interior sun over the horizon which should gradually rise up in the sky as the expedition progresses to the inside surface. There will be a few hundred miles thereafter in which both our sun and the interior sun will be in view if the trip is taken between Spring and Fall which is the Arctic daytime, but gradually our sun will be blocked from view as the interior is reached.

A thermometer should be taken to measure at regular intervals from Base Camp onward the change in temperature. The temperature at altitudes higher than 1,000 feet will increase as the expedition nears the polar opening. The greatest mid-summer temperature will be found on the polar lip where the sun's rays at noon strike at right angles to the earth's surface as it does at the Equator. The temperature of the interior will be noted for its ideal conditions for the growth of the giant plant and animal life that abounds there. An instrument to measure humidity should also be taken to record an increase in humidity as the interior is reached.

Originally, the geographic poles probably were in the center of the polar openings during the creation period, but since that time from the passages of ancient planet-sized comets, the earth has been tipped on its axis displacing the poles. Considering that the rivers from Our Hollow Earth which empty into the Arctic Ocean must end within the semi-circumference of the North Polar Opening in order for their mouths to freeze over in the winter and thaw out in summer thus creating the freshwater icebergs which cover the Arctic Ocean, as well as mirages of land in the far north as seen from around the Arctic indicate that the interior continent should be reached upon or before reaching the halfway point through the polar opening.

After reaching the interior continent, contact could be easily made with the inhabitants. Most likely, they are guarding the openings into the hollow of our earth with their flying saucer craft and we will make contact with them as we journey through the opening. Olaf Jansen, as well as the flying saucer people that Larry Foreman met out in the desert near Los Angeles, California in 1960 said that their language is Sanskrit, which is similar to High German as Reinholdt Schmidt reported, so someone knowing that language would be useful to the expedition. However, since the Order of the Illuminati became interested in the Hollow Earth, the people within the Earth have maintained a constant surveillance of our surface world to prevent atomic war over their polar openings, and have learned many of the world's languages. If UFOs do come from Our Hollow Earth, then they also can communicate telepathically, as many contactees maintain that many UFOnauts can communicate by telepathy. The expedition, therefore, will be able to talk to them with ease. They should ask to be taken to their capital City of Eden, which I have estimated is located 800 miles beneath Independence, Missouri, on the highest mountain plateau of the inner continent of Our Hollow Earth.

There at the palace in their capital City of Eden, the expedition would deliver their message of goodwill to the Great High Priest Over All the Land and ask permission to be taught about their country. Perhaps, a copy of the Ten Tribe's scriptures containing their exciting history, including the visit of the resurrected Christ to them over 2000 years ago, could be brought back for translation and publication. This book of Scripture would contain the proof that they ARE the Lost Tribes of Israel. The expedition could also learn about the different groups of people that inhabit Our Hollow Earth besides the Lost Ten Tribes, such as the Eskimos, the Greenland Vikings that migrated there in the Sixteen Century A.D, the Hopi Indians that have an entrance to the Hollow Earth near the Grand Canyon, Arizona, and the German refugees from World War II. The expedition could perhaps also learn about the other civilizations throughout our solar system and galaxy that the Hollow Earth people have contact with.

After learning about their country and civilization for a period of time, perhaps the expedition could ask the Tribes to bring them back to Utah in their Flying Saucers or come back on their own power.

A very detailed log book, video recording, photographs, instrument readings and diagrams illustrating the true nature of the polar orifice, the inner sun and the interior world would be the objective of the expedition as well as establishing a communication link between their King David, who I believe is a descendant of King David of ancient Israel, and the Lord's prophet at the head of the Church of Jesus Christ of Latter-day Saints here on the earth's surface. This log book could then be published and the video movie with commentary and interviews with the Hollow Earth people shown throughout the world together with the scriptures of the Lost Ten Tribes of Israel once the return to Utah has been accomplished.

END

BIBLIOGRAPHY

Allen, Gary. NONE DARE CALL IT CONSPIRACY, 1972, Concord Press, P.O. Box 2686, Seal Beach, California 90740.

Amundsen, Roald. FIRST CROSSING OF THE POLAR SEA, 1927, George H. Doran Co., New York, N.Y.

Armstrong, Herbert W. THE UNITED STATES AND THE BRITISH COMMONWEALTH IN PROPHECY, Pasadena, California.

Armitage, Angus. EDMUND HALLEY, Nelson Publishers.

Azevedo, Arnoldo de. PHYSICAL GEOGRAPHY.

Balsiger, Dave. & Charles E. Sellier Jr. IN SEARCH OF NOAH'S ARK, 1976. Sun Classic Books, 11071 Massachusetts Av. Los Angeles, Calif. 90025

Baker, Sylvia. BONE OF CONTENTION, IS EVOLUTION TRUE?. Creation Science Foundation, Ltd., Australian Creation Science and Christian Subject Education Resource Center, P.O. Box 302, Sunnybank, Queensland 4109 Australia.

Balch, Edwin Swift, NORTH POLE AND BRADLEY LAND, 1914.

Barker, Gray. THE STRANGE CASE OF DR. M.I. JESSUP, 1975, Saucerian Press, P.O. Box 2228, Clarksburg, WV 26301.

Barker, Gray. THEY KNEW TOO MUCH ABOUT FLYING SAUCERS, 1975, Saucerian Press, P.O. Box 2228, Clarksburg, WV 26301.

Barrington, Daines. ON THE POSSIBILITY OF APPROACHING THE NORTH POLE, 1818.

Barton, Michael X. RAINBOW CITY AND THE INNER EARTH PEOPLE, Saucerian Press, (subsequently New Age Books, Box D, Jane Lew, WV 26378)

Berlitz, Charles. THE BERMUDA TRIANGLE, 1974, Doubleday and Co. Inc., 245 Park Ave., New York, NY 10017; or Avon Books, 959 Eight Ave., New York, NR 10019, Illus., paperback.

Bernard, Raymond. THE HOLLOW EARTH, THE GREATEST GEOGRAPHICAL DISCOVERY IN HISTORY, 1969, Illus., Dell Publishing Co., hardback; Citadel Press, University Books, Inc., 120 Enterprise Ave. Secaucus, NJ 07094; or small paperback, Dell Publishing Co., Inc., I Dag Hammarskjold Plaza, 245 E. 47th St., New York, NY 10017, https://www.ourhollowearth.com/Bernard/Index.htm.

Binder, Otto. WHAT WE REALLY KNOW ABOUT FLYING SAUCERS, 1967, Fawcett Publications, Inc., Greenwich, Connecticut, paperback, Illus.

Blick, Edward F. A SCIENTIFIC ANALYSIS OF GENESIS.

Brian II, William L. MOONGATE: SUPPRESSED FINDINGS OF THE U.S. SPACE PROGRAM, 1982, Future Science Research Publishing Co., P.O. Box 06392, Portland, Oregon 97206-0020

THE BOOK OF MORMON, DOCTRINE AND COVENANTS AND PEARL OF GREAT PRICE, published by the Church of Jesus Christ of Latter-day Saints, Deseret Book Co., Salt Lake City, Utah. https://www.lds.org/scriptures/bible?lang=eng

Brough, R. Clayton. THE LOST TRIBES, 1979. Horizon Publishers, P.O. Box 490, 50 S. 500 W. Bountiful, Utah 84010.

Buel, J.W. THE WORLD'S WONDERS AS SEEN BY THE GREAT TROPICAL AND POLAR EXPLORERS, Historical Society of Pennsylvania, 1300 Locust St., Philadelphia, PA 19107, 1884.

Burrows, William E. DEEP BLACK, THE STARTLING TRUTH BEHIND AMERICA'S TOP SECRET SATELLITES, Random House, Inc., 201 E 50th St, New York, NY 10022, 1986.

Burtt, Edwin A. THE METAPHYSICAL FOUNDATIONS OF MODERN SCIENCE, 1954, Garden City, New York, Doubleday.

Caidin, Martin. THE GREATEST CHALLENGE: THE INCREDIBLE ADVENTURE AND SPLENDID DESTINY OF MAN IN EXPLORING SPACE.

Call, Michel L. ROYAL ANCESTORS OF SOME L.D.S. FAMILIES, P.O. Box 11488, Salt Lake City, Utah 84147, 2005.

Cameron, Ian. ANTARCTICA, THE LAST CONTINENT, 1974. Little Brown & Co., 34 Beacon Street, Boston, Mass. 02106.

Childress, David Hatcher. THE ANTIGRAVITY HANDBOOK, 1993, published by Adventures Unlimited Press, 303 Main Street, PO Box 74, Kempton, Illinois 60946-0074, (815) 253-6390, Fax (815) 253-6300.

Cirucci, Johnny. ILLUMINATI UNMASKED, EVERYTHING YOU NEED TO KNOW ABOUT THE "NEW WORLD ORDER" AND HOW WE WILL BEAT IT, 2015, Amazon.com.

Cluff, Rodney M. WORLD TOP SECRET: OUR EARTH IS HOLLOW! https://www.ourhollowearth.com/

Crowther, Dwane S. PROPHECY: KEY TO THE FUTURE, 1962, Bookcraft Publishers, 1848 W. 2300 S., Salt Lake City, Utah.

Cook, Frederick A. MY ATTAINMENT OF THE POLE, 1913, The Polar Publishing Co., 601 Steinway Hall, Chicago.

Corso, Col. Philip J. (Ret). THE DAY AFTER ROSWELL, 1997, Pocket Books, 1230 Ave of the Americas, New York, NY 10020.

Cowley, Matthias Foss. WILFORD WOODRUFF, 1964, Bookcraft Publishers, 1848 W. 2300 S., Salt Lake City, Utah, Illus.

Darwin, Sir G.H. THE TIDES AND KINDRED PHENOMENA OF THE SOLAR SYSTEM, 1897.

Emerson, George. THE SMOKY GOD, Palmer Publications, Inc., Amherst, Wisconsin 54406. The Smoky God

DeMeo, Dr. James. THE DYNAMIC ETHER OF COSMIC SPACE, 2019, Natural Energy Works, PO Box 1148, Ashland, Oregon 97520, United States of America, http://www.naturalenergyworks.net

Densley, Michael. IN SEARCH OF ROSS'S GULL, 1999, Peregrine Books, Leeds, West Yorkshire, England.

Dyer, Alvin R. WHO AM I?, 1966, Deseret Book Company, PO Box 30178, Salt Lake City, Utah 84130.

Emerson, George. THE SMOKY GOD, Health Research, 70 Lafayette St., Mokelumne Hill, CA 95245, Illus., https://www.ourhollowearth.com/SmokyGod.htm

Epperson, Ralph A. THE UNSEEN HAND, APOA Books, c/o 2303 N 44th St., Ste 14-346, Phoenix, Arizona 85008, (602) 517-0418.

Foreman, Laurence W. PASSPORT TO ETERNITY, 1970, self published at 334 ½ W 33rd St, Los Angeles, California 90007, born March 27, 1908, Oklahoma, USA, died February 4, 1998 at Poway, San Diego County, California, USA, https://www.ourhollowearth.com/passporttoeternity.htm

Gardner, Marshall B. JOURNEY TO THE EARTH'S INTERIOR, 1920. Health Research, 70 Lafayette St., Mokelumne Hill, CA 95245, Illus., also Amherst Press, https://sacred-texts.com/earth/jei/index.htm.

Giannini, F. Amadeo. WORLDS BEYOND THE POLES, 1959. Vantage Press, Inc., 516 W. 34th St., New York, NY 10001.

Glines, Lt. Col. C.V. POLAR AVIATION.

Golitsyn, Anatoliy. NEW LIES FOR OLD, 1984. Dodd, Mead & Company, Inc., 79 Madison Ave, New York, NY 10016.

Greely, Adolphus W. THREE YEARS OF ARCTIC SERVICE, 1886, Charles Scribner's Sons, New York.

Greer, Steven M. UNACKNOWLEDGED, 2017, A&M Publishing, L.L.C., West Palm Beach, FL 33411.

Greer, Steven M. DISCLOSURE, 2001, Carden Jennings Publishing Co, 1224 W Main St, Ste 200, Charlottsville, VA 22903, http://www.disclosureproject.org/.

Grew, Edwin S. THE ROMANCE OF MODERN GEOLOGY, 1911, Seely and Co., London.

Hall, Charles. MILLENNIAL HOSPITALITY, 2002.

Herbert, Wally. ACROSS THE TOP OF THE WORLD, 1969.

Hamilton, William F. CENTER OF THE VORTEX, 1979, published by Nexus & Nexus News, printed by Wilcopy, Los Angeles, California.

Hayes, Dr. I.I. THE OPEN POLAR SEA: A NARRATIVE OF A VOYAGE OF DISCOVERY TOWARDS THE NORTH POLE IN THE SCHOONER UNITED STATES, 1967, Hurd and Houghton, New York, NY.

Hilton, James. LOST HORIZON, 1933.

Holmes, Phoebe Marie. MY VISIT TO THE SUN, 1933.

Hunter, Milton R. ANCIENT AMERICA AND THE BOOK OF MORMON, 1950. Kolob Book Company, PO Box 1575, Oakland, CA.

James, King. THE HOLY BIBLE.

Johnson, Benjamin F. MY LIFE'S REVIEW, 1947, Zion's Printing & Publishing Co., Independence, Missouri.

Johnson, George; and Tanner, Don. THE BIBLE AND THE BERMUDA TRIANGLE, 1977, Logos International, 201 Church St., Plainfield NJ 07061, paperback, Illus.

Kane, Elisha Kent. ARCTIC EXPLORATIONS IN THE YEARS 1853-54-55, 1856. J.B. Lippincott and Co., Philadelphia.

Kimball, Spencer W. THE MIRACLE OF FORGIVENESS.

Landis, Phillip Cloudpiler. THE SACRED RECORDS OF THE NEMENHAH, 2011.

Lamprecht, Jan. HOLLOW PLANETS, A Feasibility Study of Possible Hollow Worlds, 1998. World Wide Publishing Group, PO Box 49625, Austin, TX 78765, (830) 798-1250.

Leith, John B. GENESIS FOR THE SPACE RACE, The Inner Earth and the Extra Terrestrials, 1980, copyrighted 6 July 1979, Registration Number: TXu-24-413, now available on Amazon.com.

LePoer Trench, Brinsley. SECRET OF THE AGES, UFO's FROM WITHIN THE EARTH, 1974, Pinnacle Books, 275 Madison Ave., New York, NY 10016, paperback, Illus.

Lloyd, John Uri. ETIDORPHA OR THE END OF THE WORLD, 1976, Pocket Books, 1230 Ave. of the Americas, New York, NY 10020, paperback, Illus., https://etidorhpacontent.blogspot.com/

Lovelace, Leland. LOST MINES AND HIDDEN TREASURE, 1956, Naylor Co., San Antonio Texas.

Lytton, Sir Bulwer. VRIL—THE POWER OF THE COMING RACE, 1972, Rudolf Steiner Publications, 100 S. Western Hwy., Blauvelt, NY 10913, https://www.sacred-texts.com/atl/vril/.

MacLellan, Alec. THE HOLLOW EARTH ENIGMA, 1999, Souvenir Press, Ltd, 43 Great Russel Street, London WC1B 3PA.

Mahoney, Timothy P. PATTERNS OF EVIDENCE: THE EXODUS, 2015, Thinking Man Media, 6900 W Lake St, St Louis Park, MN 55426.

Marrs, Jim. ABOVE TOP SECRET, 2008, The Disinformation Company, Ltd., 163 Third Ave, Ste 108, New York, NY 10003.

Maxwell III, Robert. LEMURIA--FACT OR FICTION, Roseway Press, Los Angeles, 1965

Maxlow, Dr. James. MODELING THE EARTH, A scientific argument for an alternate tectonic understanding of our physical world. 2016, Perth, Western Australia.

McConkie, Clay. THE TEN LOST TRIBES.

McConkie, Bruce R. MORMON DOCTRINE, 1966, Bookcraft Publishers, 1848 W. 2300 S., Salt Lake City, Utah.

McConkie, Bruce R. THE MILLENNIAL MESSIAH, 1982, Deseret Book Co., Salt Lake City, Utah.

Menger, Howard & Connie. THE HIGH BRIDGE INCIDENT: The Story Behind the Story...Released After 35 Years of Silence, 1991, published by Howard Menger, PO Box 1405, Vero Beach, Florida 32961 (407) 562-1153.

Merrill, Hamblin and Thorne. PHYSICAL SCIENCE 100, Burgess Publishing Co., Minneapolis, 1978.

Michel, Aime. FLYING SAUCERS AND THE STRAIGHT-LINE MYSTERY, 1958, Criterion Books, Inc., 666 Fifth Ave., New York, NY 10019.

Mitchell, O.M.. A CONCISE ELEMENTARY TREATISE ON THE SUN, PLANETS, SATELLITES, AND COMETS.

Miller, Martin. THE DISCOVERY OF THE NORTH POLE. 1909.

Moody, Raymond A. REFLECTIONS ON LIFE AFTER LIFE, 2011.

Mozina, Michael. The Surface of the Sun.

Nansen, Dr. Fridtjof. FARTHEST NORTH, 2 Vols., 1897, Harper & Brothers, New York and London.

Nansen, Dr. Fridtjof. IN NORTHERN MISTS, 2 Vols., 1911, Greenwood Press, Inc., 51 Riverside Ave., Westport, CT 06880.

Nansen, Dr. Fridtjof. THE FIRST CROSSING OF GREENLAND, 2 Vol, 1890, In set of 6 Vols. Greenwood Press, Inc., 51 Riverside Ave., Westport, CT 06880.

Newton, ARCTIC MANUAL.

Newton, Isaac. PRINCIPIA, 1687.

Newman, Joseph. THE ENERGY MACHINE, 1984, published by Joseph Westley Newman, Route 1, Box 52, Lucedale, Mississippi 39452 (601) 947-7147.

Nordenskiold, Adolf Erick. THE ARCTIC VOYAGE OF 1858-1878.

Norman, Eric. THE UNDER PEOPLE, 1969, Award Books, 235 East 45 St., New York, NY 10017, paperback.

Palmer, Ray. WORLDS BEYOND THE POLES, 1984, New Age Books, Box D, Jane Lew, WV 26378.

Peary, Robert E. NEAREST THE POLE, 1907, Doubleday, Page & Company, New York.

Pierce, Norman C. THE DREAM MINE STORY, July 1958.

Platt, Rutherford F. THE FORGOTTEN BOOKS OF EDEN, 1927.

Platt, Rutherford F. THE LOST BOOKS OF THE BIBLE AND THE FORGOTTEN BOOKS OF EDEN.

Preston, Robert L. WAKE UP AMERICA, IT'S LATER THAN YOU THINK!, 1972, Hawkes Publishing Inc., 3775 S. 500 W. Salt Lake City, Utah 84115.

Quigley, Carol. TRAGEDY AND HOPE, A HISTORY OF CIVILIZATION IN OUR OWN TIME, 1966.

Reed, William. PHANTOM OF THE POLES, 1906, Health Research, 70 Lafayette St., Mokelumne Hill, CA 95245, Illus., https://www.sacred-texts.com/earth/potp/index.htm.

Rensberger, Boyce, of the Washington Post. "Deep-In-The Earth Experiments Question Newton Gravity Theory," THE CHARLOTTE OBSERVER, newspaper, August 3, 1988.

Seagraves, Kelly L., THE GREAT DINOSAUR MISTAKE, 1975, Beta Books, California.

Sayce, A.M. RECORDS OF THE PAST, 1889.

Sargent, Epes. WONDERS OF THE ARCTIC WORLD, 1873.

Scott, Jack Denton. JOURNEY INTO SILENCE, 1976, Reader's Digest Press, distributed by Crowell.

Scully, Frank. BEHIND THE FLYING SAUCERS, 1950, New York: Henry Holt and Company.

Scura, John and Dane Phillips. BATTLE HYMN, Revelations of the Sinister Plan for a New World Order, 2012, Published by Black Rose Writing at Smashwords.

Sigma, Rho. ETHER-TECHNOLOGY: A Rational Approach to Gravity Control, 1977, CSA Printing & Bindery, Lakemont, Georgia 30552.

Skousen, Cleon. PROPHECY AND MODERN TIMES.

Skousen, Cleon. THE FIRST 2000 YEARS.

Skousen, Cleon. THE NAKED CAPITALIST, 1972, Reviewer, 2197 Berkeley St., Salt Lake City, Utah, 84109, Illus., paperback. Also, THE NAKED COMMUNIST, 1958.

Skousen, Eric N. EARTH, IN THE BEGINNING, 1996, Verity Publishing, PO Box 911, Orem, Utah 84059-0911, hard back.

Snyder, Al. NEWTON'S LAWS ARE FULL OF FLAWS, 1973, Snyder Institute of Research, 508 N. Pacific Coast Hwy., Redondo Beach, California 90277.

Snyder, Al. SATAN'S SAUNA AND THE DEVIL'S TRIANGLE, 1975, Snyder Institute of Research, 508 N. Pacific Coast Hwy., Redondo Beach, California 90277.

Smith, Joseph Fielding. DOCTRINES OF SALVATION, 3 Vols., 1954, Deseret Book Co., 40 E. South Temple, Salt Lake City, Utah 84104.

Smith, Joseph Fielding. MAN, HIS ORIGIN AND DESTINY, 1969, Deseret Book Co., 40 E. South Temple, Salt Lake City, Utah 84104.

Smith, Joseph Fielding. TEACHINGS OF THE PROPHET JOSEPH SMITH, 1976, Deseret Book Co., 40 E. South Temple, Salt Lake City, Utah 84104.

Smith, Joseph Fielding. WAY TO PERFECTION, 1931.

Smith, Joseph Fielding. GOSPEL DOCTRINE, 1975, Deseret Book Co., 40 E. South Temple, Salt Lake City, Utah 84104.

Smith, Warren. HIDDEN SECRETS OF THE HOLLOW EARTH, 1976, Zebra Books, 521 Fifth Ave. New York, NY 10017.

Spearman, Neville. THE BOY WHO SAW TRUE, London 1953.

Stranges, Dr. Frank. MY FRIEND FROM BEYOND EARTH, and STRANGER AT THE PENTAGON, IEC, PO Box 5, Van Nuys, CA 91408.

Stefansson, Vilhjalmur. MY LIFE WITH THE ESKIMO, 1913, The Macmillan Co., 866 Third Ave. New York, NY 10022.

Stefansson, Vilhjalmur. UNSOLVED MYSTERIES OF THE ARCTIC, 1962, The Macmillan Co., 866 Third Ave. New York, NY 10022.

Velikovsky, Immanuel. EARTH IN UPHEAVAL, 1955, Dell Publishing Co., Inc., 1 Dag Hammarskjold Plaza, New York, NY 10017.

Velikovsky, Immanuel. WORLDS IN COLLISION, 1950, London, Victor Gollancz Ltd, Sphere Books, Inc.

Talmage, James E. THE ARTICLES OF FAITH, 1890, Deseret Book Co., 40 W. South Temple, Salt Lake City, Utah 84104.

Thomas, John A. Jr. ANTIGRAVITY: THE DREAM MADE REALITY, THE STORY OF JOHN R. R. SEARL, 373 Rock Beach Road, Rochester, New York.

Trench, Brinsley LePoer. SECRET OF THE AGES: UFO'S FROM INSIDE THE EARTH, 1974.

Valens, E. G. THE ATTRACTIVE UNIVERSE: Gravity and the Shape of Space, 1969, The World Publishing Company, 2231 W 110th St, Cleveland, OH 44102.

Verne, Jules. JOURNEY TO THE CENTRE OF THE EARTH, 1864, Penguin Books, Inc., 7110 Ambassador Road, Baltimore, Maryland 21207.

Walton, Bruce. A GUIDE TO THE INNER EARTH, 1983, New Age Books, Box D, Jane Lew, WV 26378

Warren, Henry C. *The Entrained Spatial Medium: Gravitation*, Paper 1, 2001, https://www.olypen.com/hcwarren/.

Warren, William F. PARADISE FOUND, OR THE CRADLE OF THE HUMAN RACE AT THE NORTH POLE, 1885, Reproduced by Health Research, 70 Lafayette St., Mokelumne Hill, CA 95245, https://publicdomainreview.org/collection/paradise-found-the-cradle-of-the-human-race-at-the-north-pole-1885.

Wasserman, Jacob. COLUMBUS, DON QUIXOTE OF THE SEAS, 1930, English trans. by Eric Sutton, Little Brown and Co., Boston, MA.

Weldon, John. UFO'S , WHAT ON EARTH IS HAPPENING, 1976, Bantam Books, Inc., 666 Fifth Ave., New York, NY 10019.

West, Jack. THE SECOND COMING OF CHRIST. (Cassette tapes (2)), 1978, Sounds of Zion, Box 7332, Murray, Utah 84107.

Wilson, Don. OUR MYSTERIOUS SPACESHIP MOON, 1975, Dell Publishing Co., Inc., 1 Dag Hammarskjold Plaza, New York, NY 10017, http://www.scribd.com/doc/40405065/Secrets-of-Our-Spacecship-Moon.

Wilson, Don. SECRETS OF OUR SPACESHIP MOON, 1979, Dell Publishing Co., Inc., 1 Dag Hammarskjold Plaza, New York, NY 10017, http://www.scribd.com/doc/40405065/Secrets-of-Our-Spacecship-Moon.

Wittmer, Dr. Felix. CONQUEST OF THE AMERICAN MIND, 1956, Meador Publishing Co., Boston, MA.

Wormser, Rene A. FOUNDATIONS: THEIR POWER AND INFLUENCE, 1958, Angriff Press, P.O. Box 2726, Hollywood, CA 90028.

Magazines, Newspapers & References

Astronautics
American Opinion Magazine, now The New American
Creation Science Magazine
Creation Research Quarterly
Institute for Creation Research
Hearnes Journal
Flying Saucers Magazine
UFO Report
The Millennial Star
New York Times
Flying Saucers Magazine
Encyclopedia Americana
Encyclopedia Britannica
El Mercurio
Outside Magazine
Norwood Review, London
Physical Review Letters journal
Search Magazine
Smithsonian Magazine
Standard-Examiner
Knowledge
The Chicago Examiner
The Ensign
Final Frontier
Scientific American
Documentary History of the Church (LDS)
The Hollow Hassle

Hollow Earth Societies, Newsletters, Blogs and Websites

Blogs:

http://subterraneus.blogspot.com/
http://hollowplanet.blogspot.com/

Websites:

http://www.thehollowearthinsider.com/
https://www.hollowearthresearch.org/
http://www.nicufo.org/index.html
http://www.holloworbs.com/
https://www.diannerobbins.com/
https://www.tierrahueca.com/
https://agartha.blog/

https://zorraofhollowearth.com/
https://bbsradio.com/hollowearthnetwork
https://npiee.com/index.html
https://scispi.tv/hollowearthchannel/

Libraries & Book Stores:

https://www.amazon.com
http://www.gutenberg.org/
http://books.google.com/books

APPENDIX

Evidence from Voyager Indicates
Uranus is Hollow!

Is Asteroid Eugenia Hollow?

Phobos, Moon of Mars Is Hollow!

Location and Size of the Polar Openings

Did the Lost Soviet Flyers Fly into Our Hollow Earth?

German Submarine U-209 Made It to Our Hollow Earth!

Biography of
Ret. Colonel Billie Faye Woodard

Evidence from Voyager Indicates Uranus is Hollow!

The spacecraft Voyager 2's flyby of Uranus on January of 1986 gathered information that indicates the hollow nature of that planet. The hollow planet theory provides answers to puzzling observations made by Voyager that are complete mysteries to orthodox science.

For example, where does the radio noise come from that emanates from Uranus?

Why do heated clouds rise in the form of a ring from the planet's poles?

Why does Uranus radiate 15 times more energy in ultraviolet than it receives from the sun?

What could be the source of the powerful solar wind that causes auroral lights of Uranus to light up when it is obvious that the solar wind from the sun is too weak to cause them? The mysterious source of the energy that lights up Uranus' auroras is compounded by the fact that Uranus' strong magnetic field repels the solar wind from the sun around the planet and so CANNOT cause its auroras to light up.

And, finally, what could possibly explain the huge displacement of the magnetic poles of Uranus when it is known that the magnetic poles of a common dynamo always coincide with its rotational axis?

Orthodox science has no answers for these mysterious observations. Yet the hollow planets theory gives ready answers: The radio noise, heat and solar wind that emit through Uranus' polar openings from a sun within Uranus' hollow interior is the power source of these anomalous phenomena. The disjunctive rotational orientation of the interior sun of Uranus with respect to the rotational axis of the planet's shell gives rise to the extreme displacement of Uranus' magnetic poles. And assuming Uranus has a planetary shell with a thickness of 10% of its diameter, Uranus would have a solid surface and density of 2.58. Even oxygen was detected in Uranus' atmosphere. Under these conditions, the interior of Uranus could possibly have an environment compatible for human life. Indeed, it could be ideal, even a Garden of Eden for whatever people that could live there.

The Interior Sun of Uranus

The period of rotation of the planet Uranus was calculated by the Voyager spacecraft by analyzing a regular variation in radio noise that emits from the planet. Since radio noise is a characteristic of stars, this indicates that Uranus has an interior sun from which this radio noise is emitted. It also indicates that Uranus has polar openings into its hollow interior through which the radio noise from the interior sun is emitted. Voyager discovered that Uranus' period of rotation of 16 hours and 48 minutes is almost twice as fast as earth's speed of rotation. This indicates that Uranus could possibly have a stronger magnetic field than earth, which it is. Uranus' magnetic field was measured by Voyager at approximately 0.1 gauss, which is three times stronger than Earth's magnetic field (Astronomy, April 1986, p. 15). The bigger size of Uranus, 14.44 times more massive than Earth, and its faster rotation about its interior sun would account for its stronger magnetic field. The magnetic fields of all planets are produced by the rotation of the planetary shell about a stationary or nearly stationary interior sun suspended by gravity and electrostatic forces within the hollow interior of the planet. Besides the radio noise and magnetic field that Uranus' interior sun causes, it also gives rise to the displacement of Uranus' magnetic poles, emits a solar wind that produces auroras and *"electro glows"* as reported by Voyager scientists, and it's interior sun also produces the heat that Voyager discovered is radiating from the poles of Uranus.

Displaced Magnetic Poles on Uranus

An important discovery made by Voyager as it bypassed Uranus on its way out of the Solar System is the huge gap detected between the location of the planet's rotational axis and the axis of its magnetic field. They were found to be 55 degrees apart. This was almost five times the difference between the North Geographic Pole and the Magnetic North Pole on Earth in 1965. That difference on Earth was 11.5 degrees in 1965, 8.79 degrees in 2001, 3.54 degrees in 2020 and is moving towards Siberia. The planet Mercury has the next largest discrepancy between its axis of rotation and the location of its magnetic pole with a 14 degree displacement.

Scientists theorize that the crust of Uranus must be spinning in a different direction from material in its core in order to explain the displacement of Uranus' magnetic poles. Because Uranus lies nearly on its side as it revolves around the sun, scientists theorize that a planet the size of earth must have collided with Uranus causing it to tip over on its side. However, it is more likely that Uranus has been by-passed by a planet-size comet so close over one of its geographic poles that the gravitational interaction would have caused the planet's rotational axis to be tilted. The interior sun suspended in its interior by gravity and electrostatic forces and protected by the shell of the planet would have most probably retained its original orientation when Uranus was tipped on its axis. This would explain why the magnetic poles on Uranus do not coincide with its rotational axis.

Since an interior sun gives rise to the magnetic field of a planet as the planetary shell rotates about it, the interior sun's orientation would also contribute to the location of the magnetic poles. The displacement of a planet's magnetic poles is therefore directly linked to the degree of tilt of the planetary shell with respect to the orientation and rotation of its inner sun. Similar to Uranus, on Earth the passage of an ancient planet-sized comet could have tilted the Earth's rotational axis causing the displacement of the magnetic poles with respect to the geographic poles. The Earth's rotational axis is tilted 23.5 degrees with respect to its orbital plane about the sun. Uranus has a tilt of 98 degrees. Since Uranus has a greater rotational tilt with respect to its orbital plane about the sun, it could be expected that it's magnetic axis would also be displaced further from its rotational axis than is the case on Earth.

The fact that the magnetic poles of Uranus and Earth do not coincide with their rotational axis indicates that the magnetic poles may be a combination of two magnetic fields, as first theorized by Astronomer Edmund Halley. One field is located within the other; the planetary shell gives rise to one field and the other by the interior sun. The Earth has a magnetic north pole centered in the Arctic Ocean north of Canada, and a geomagnetic north pole located to the northwest of Greenland. This indicates that Earth is hollow having two magnetic fields, one produced by the earth's shell and the other by the interior sun. The magnetic poles have been noted to move in a more or less orbit about the Arctic and Antarctic at a rate of several miles a year. Astronomer Edmund Halley was the first to propose that this movement of the magnetic poles is caused by the slow rotation of a body suspended in a hollow inside the planet. On Earth, the rotation of its interior sun seems to complete one rotation about every 794 years. The slow rotation of the interior sun is what causes the magnetic poles to follow an orbital path about the Arctic and Antarctic.

On Earth, the magnetic axis was almost one-half the tilt of the rotational axis with respect to the ecliptic plane about the sun in 1965 and has been less every year since as the North Magnetic Pole crosses the Arctic. On Uranus, the magnetic axis is a little more than one-half of the tilt of the rotational axis. It has always been a puzzle to scientists that Earth's own magnetic poles do not coincide with the rotational axis of the Earth, as is the case in a common dynamo. Now Uranus has been found to have this same puzzling arrangement as does the Earth. The answer must lie in the fact that if the axis of a planet's shell coincides with the axis of rotation of its interior sun, then the magnetic poles would coincide with the rotational axis. On Earth, they do not coincide. Now Voyager has found they do not coincide on Uranus as well.

The solution to the enigma of the displaced magnetic poles must lie in the orientation of the rotational axis of the interior sun in relation to the orientation of the axis of the planetary shell. On Earth, the rotational axis is inclined 23.45 degrees to its orbital plane about the sun. Let's assume that the original orientation of the rotational axis of the Earth's shell and its interior sun coincided and were perpendicular to the plane of the Earth's orbit about the sun. Now suppose that sometime in past geological history, a planetary-sized comet bypassed earth over one of its geographic poles and caused its rotational axis to tilt 23.45 degrees with respect to its orbital plane. The interior sun, having a very small percentage of the Earth's total mass and buffered by the planet's hollow, would not have been affected substantially, and therefore would have retained approximately its original orientation. The axis of the interior sun, therefore, is most likely inclined 23.45 degrees to the rotational axis of the Earth's shell retaining its and the Earth's original orientation. So the question is, why would the magnetic pole be located only 11.75 degrees from the shell's rotational axis, and not 23.5 degrees? The answer may be that because one field is located within the other, that the combined magnetic fields of the earth's shell and the magnetic field of the interior sun produce a magnetic axis located halfway between the two orientations, which on earth would be 11.72 degrees, half of the 23.45 degree inclination of the shell.

On Uranus a similar scenario may have occurred to give it a displaced magnetic field orientation. The magnetic field is produced by the rotating planetary shell about an interior sun rotating at a different, probably counter rotation and slower rate. The fact that the magnetic poles do not coincide with the rotational axis indicates that the axis of Uranus' interior sun is oriented differently than the rotational axis of the planet's shell. As with Earth, the original rotational axis of Uranus's shell as well as the axis of its interior sun could be assumed to have been at right angles to its orbital plane about the sun. So when that planet was hit or by-passed by another celestial body and the rotational axis of the shell was tilted 98 degrees to its orbital plane, its interior sun most probably retained its original orientation. Therefore, the combined magnetic field of shell and the magnetic field of interior sun would cause the shell's magnetic poles to be located halfway between the original orientation and the present one located at about 55 degrees.

Mass and Density of a Hollow Uranus

Assuming that the Newtonian gravitational constant is approximately correct and that it gives a reasonable mass of earth and the planets, Uranus would have a mass that is 14.44

times greater than earth. Since the overall density of Uranus is 1.26 gm/cc, scientists have assumed that Uranus is a liquid planet. However, if Uranus is hollow and has a shell thickness perhaps 10% of its diameter, then the shell density would be 2.58 gm/cc, which would give it a solid surface. The surface acceleration of gravity on Uranus is 893 cm/sec^2, which is only slightly less than on earth. With an interior sun, and surface gravity similar to earth, and with oxygen detected in Uranus' atmosphere, the interior world of Uranus could contain an environment compatible with supporting plant and animal life forms.

The Atmosphere of Uranus

Voyager noted oxygen ions in Uranus' magnetosphere (Astronomy, April 1986, pg. 14), but since most of this gas is concentrated near the surface of a planet because of its relative heavier weight than other gases, Voyager had no way of telling what percentage of Uranus' surface atmosphere contained oxygen. But since oxygen ions were detected high up in the atmosphere, in all probability, oxygen could be abundant at lower altitudes. Voyager also detected clouds in the Uranian atmosphere. In fact, clouds cover the planet so completely that the surface cannot be seen from space. Could these be water clouds? If so, then Uranus may have water to sustain life.

The Auroras on Uranus

All the outer planets were found by Voyager to have auroral lights. This was a surprise to space scientists, because they have assumed that auroras on Earth are caused by the impact of the solar wind on the atmosphere above the poles. However, with Voyager's discoveries of powerful auroras on the outer planets which are located so far from the sun that any solar wind reaching them at that distance is far too weak to cause their auroras, scientists have begun looking for another power source that could cause the auroras.

One report stated,

"Another mystery is the huge glow in the atmosphere of Uranus which spreads upwards as far as two Uranian radii or 50,000 km. Scientists checking UV emissions have dubbed the phenomenon 'electroglow,' but are at loss to explain the mechanism behind it. The bright halo, originally detected by the International Ultraviolet Explorer, was

thought to be similar to the northern lights or aurora on Earth. It is now thought to be similar to the glows detected by Voyager when it passed Jupiter and Saturn."

The report concluded that the electron beam that produces the *"glow"* cannot originate from space. Therefore, it must originate from the planet itself.

"The glow, generated when electrons collide with hydrogen molecules, has several peculiarities. The low energy electrons do not come from deep space as would occur with an aurora. Also, again unlike an aurora, the glow is found only on the sunlit side of the planet. The Sun is not the power house for the glow, although it does provide the energy needed to free electrons from the hydrogen atoms. But the Sun's energy is too weak to accelerate the electrons -- that is the mechanism that is not understood. Voyager detected an aurora on the night side of the planet near the magnetic pole." (New Scientist, January 1986, p. 21)

The mystery to scientists stems from their assumption that an aurora must be powered by a solar wind that comes from the Sun. Upon approaching a planet they have assumed that *"somehow"* the solar wind gets through the planet's magnetic field. This assumption is faulty because, first of all, the sun's solar wind is deflected around a planet by its magnetic field and cannot enter it at all. And second, the outer planets are too far away from the Sun and the solar wind too weak to cause their auroras. In fact, the Sun's solar wind is too weak to even cause the Earth's aurora to light up.

First of all, the energetic particles are observed to emit from the Polar Regions with sufficient energies to cause the atmosphere to light up the auroras. Then the particles with ever decreasing energies stream outward from the earth's surface and away from the poles following the earth's electromagnetic field lines.

Secondly, our sun's solar wind could not cause the auroras nor the Van Allen radiation belts because they are much lower energetically. The solar wind is composed of protons with energies of about 1,000 electron volts and electrons with about 10 electron volts compared to the source of the auroras with electron volts of 10,000 to 100,000. So scientists are faced with a dilemma: What solar wind could cause the auroras if it doesn't come from our Sun? The New Scientist report read,

"...the Sun's energy is too weak to accelerate the electrons -- that is the mechanism that is not understood."
The simple fact that the outer planets have strong auroras is proof that the solar wind that causes them cannot come from our outer Sun, but must originate from the planets themselves.

The report in Astronomy magazine of Voyager's passage through Uranus' magnetic field bow shock confirms that our Sun's solar wind is deflected by Uranus' magnetic field and cannot, therefore, be the power source of its auroras.

"Ten hours, thirty minutes before closest approach, Voyager crossed the bow shock; three hours later it entered the magnetic realm of Uranus. The bow shock is where the solar wind first encounters the magnetic field of a planet, the magnetopause is the transition from primarily solar to primarily planetary material." (Astronomy, April 1986, p. 15)

So now, if auroras, *"electroglows"* and radiation belts are caused by the Sun's solar wind, how can radiation in the magnetosphere be *"primarily planetary material"*? So here is a scientific admission that auroras are caused primarily by planets and not the Sun.

The report continued,

"A glowing hydrogen corona above the pole was also detected. The energy source for these glows could not be sunlight -- Uranus radiates 15 times more energy in ultraviolet than it receives in sunlight."

Here again, scientists are now admitting that the energy source of the auroras has to come from the planet itself. Voyager found that Jupiter, Saturn and now Uranus all emanate more energy than they receive from the sun.

For example, the Astronomy report said,

"To find out how much heat Uranus radiates from its deep interior, Voyager must measure total ingoing and outcoming total Sunlight, and the total ingoing and outcoming infrared radiation. The planet's internal energy is the residual that can't be accounted for on the energy balance sheet."

Scientists can't account for the extraordinary amount of energy coming out of the outer planets because they do not take into account their hollow nature containing interior suns that emit this energy to light up their auroras and populate their magnetospheres with high energy particles.

The report continued,

Evidence from Voyager Indicates
Uranus is Hollow!

"Voyager found two distinct 'populations' of ions: one is a dense, 'warm' population at a temperature of a hundred thousand degrees Kelvin and a density of 1 ion per cubic centimeter. These ions are trapped by the magnetic field and rotating with the planet. The other is a 10 times less dense plasma at a temperature of 10 million degrees Kelvin."

Then Astronomy magazine asks,

"Where could these hot ions come from? The solar wind, one possible source, is deflected around Uranus."

So here Astronomy magazine admits that the Sun's solar wind could not be the source of these hot ions because it is deflected by the planet's magnetic field upon approaching the planet. We must conclude that the only possible source of these hot ions, auroras and electroglows (which are auroras on the sunlit side of the planet) is a sun suspended within the hollow interior of Uranus which emanates these energetic particles out through polar openings.

Radiation at the Poles of Uranus

Voyager offered evidence indicating that Uranus has polar openings from which heat radiation is emitted from its interior sun. From Astronomy magazine we read,

"Over the sunlit south pole of Uranus, the ionosphere reaches a temperature of 750 degrees Kelvin. Over the dark pole, it reaches 1,000 degrees Kelvin. What heats the gas to such high temperature is not evident. The source of energy is not sunlight because the dark pole is hotter, so it probably derives its energy from the magnetosphere."

(Astronomy, April 1986, p. 10)

So we see that scientists are admitting that these auroras cannot be caused by the solar wind from the Sun, which is deflected around the magnetic field of a planet, so in their attempt to explain the greater temperature of the particles emitting from the dark pole, they latch onto the magnetosphere as a possible source of the energy that powers the planet's auroras. Scientists are contradicting themselves in this, because the particles populating the magnetosphere come from the planet, not the other way around.

A pole-on image of Uranus noted that the polar region, relative to the rest of the planet, is slightly reddish and surrounding the pole is a broad yellowish ring. *"A wide bright ring around the pole, imaged through the methane filter, indicated a region of aerosol or haze particles UPWELLING FROM BELOW. The polar region, though, was very dark where air was sinking."* A circulation pattern such as this would be exactly what would be expected if warm air were emanating from one polar opening beneath the Uranian clouds and sinking into the other polar opening at the other end of the planet.

Conclusions

The New Scientist of January 1986 summarized these unexpected findings of Uranus from Voyager's flyby and concludes that,

> *"...internal dynamics within the planet rather than atmospheric phenomena could account for the unexpected findings."* (p. 22)

So in final analysis, even space scientists are beginning to admit to evidence that can only be explained if planets are hollow with interior suns and polar openings that emit radiation to the exterior that light up their auroras. With a density and surface gravity similar to Earth, with oxygen and clouds in its atmosphere, Uranus is an excellent candidate for conditions on its interior congenial to life forms made possible by the heat and light provided by an interior sun which Voyager found radiates radio, infrared and ultraviolet radiation through polar openings as well as a powerful inner solar wind that produces auroras at both magnetic poles. The Hollow Planet theory provides answers to these unexpected findings that Voyager uncovered about Uranus which orthodox science can only wonder about.

Evidence from Voyager Indicates
Uranus is Hollow!

False-color Image of Uranus from the Hubble Space Telescope,
October 14, 1998:

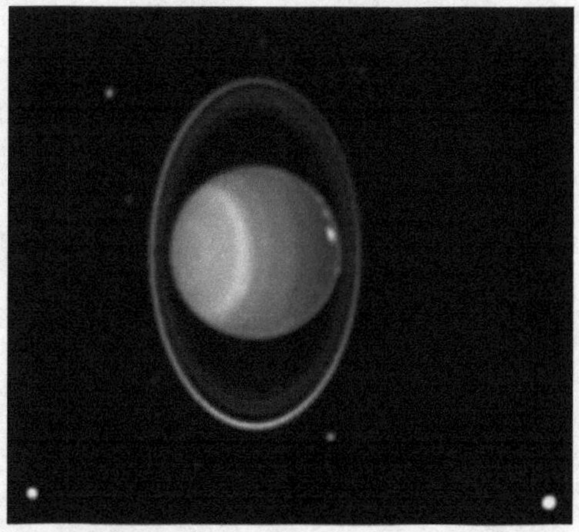

Is Asteroid Eugenia Hollow?

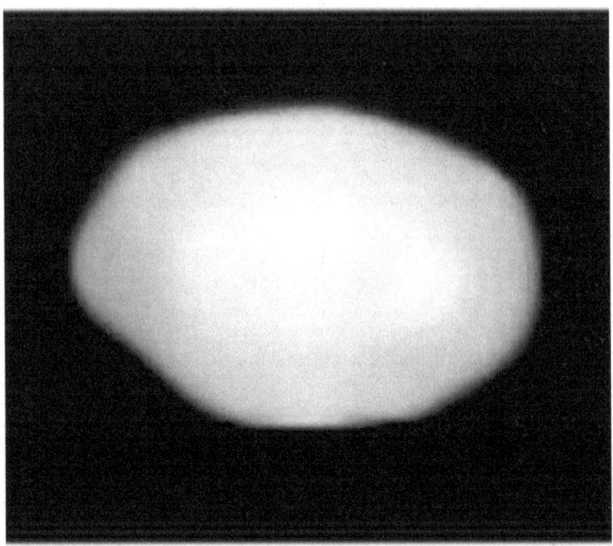

On October 6, 1999, NASA reported their discovery of a moon 13 km in diameter is orbiting an asteroid named Eugenia located between Mars and Jupiter. This moon orbits Eugenia 1,190 km from Eugenia once every 4.7 earth days.

According to Newtonian physics, this data obtained by NASA requires that the asteroid Eugenia have the density of water, of 1.17 gm/cc. But Eugenia can't possibly be a big drop of water, because it can be seen through telescopes that the asteroid is a rock. It has a solid surface.

The hollow planets theory has an answer to this enigma. If Eugenia is hollow and has a shell 10% of the asteroid's diameter, this would give it a density of 2.39 gm/cc, with a solid surface, consistent with what it is reported to have. Therefore, Newtonian physics requires that the asteroid Eugenia be hollow.

The math for this is as follows.

The surface acceleration of gravity of Eugenia is obtained from the geosynchronous orbit formula,

$$R^{3/2} = \mathrm{Sqrt}(a) * r * T / 2\,pi$$

Where

R = the orbital radius of the moon of Eugenia

r = planetary radius of Eugenia

T = orbital period in earth sidereal days, 1 Earth sidereal day = 86164.09 sec
Solving for a, the surface acceleration of gravity,
$a = (R^{3/2} * 2 \text{ pi} / (r * T))^2$
Solving,
$(1.298136742 \times 10^{12} * 2 \text{ pi} / (10,750,000 * 404971.22))^2$
$= 3.510228369 \text{ cm/sec}^2$
The formula for the planetary mass of Eugenia is,
$M = r^2 * a / G$
Where
$G = 6.67259 \times 10^{-8}$ the standard gravitation constant
r = 10,750,000 cm the planetary radius of Eugenia
$a = 3.10163774 \text{ cm/sec}^2$ from above
$= 6.08 \times 10^{21}$ gm
To solve for the planetary density of Eugenia,
D = M/V
where the volume of a sphere is,
$V = Pi\ D^3 / 6$
Solving,
$3.141592654 * 21,500,000^3 / 6$
$= 5.203720981 \times 10^{21}$ cc
Solving for planetary density,
$5.371722395 \times 10^{21} / 5.203720981 \times 10^{21}$
$= 1.17$ gm/cc
Which means that if Eugenia is not hollow, then it must be completely made of water with a few impurities, which it is not. Eugenia is an asteroid, visibly verified to consist of rock, with a solid surface. Since Eugenia is not made of water, and yet has a planetary density of 1.17 gm/cc, therefore, we must conclude that Eugenia is hollow.

If we would assume a hollow Eugenia to have a shell thickness 10% of its planetary diameter, then the density of its shell would be 2.39 gm/cc, which would give it a solid surface.

The formula for the density of a hollow planet is,
Mass of planet / (Volume of planet - Volume of hollow)
Where,
Volume of hollow =
$(pi * (diameter - (diameter * 0.1 * 2))^3) / 6$
Solving,
$6.08 \times 10^{21} / 5.203720981 \times 10^{21} - (pi * (21,500,000 - 4,300,000)^3) / 6$
$= 2.39$ gm/cc, density of the shell, which is about the density of silicon (sand).

Newtonian physics requires that the asteroid Eugenia be hollow!

Phobos, Moon of Mars Is Hollow!

Phobos is one of the two moons of Mars. It is the moon closest to the planet and orbits about 4,000 miles from the surface of Mars making two orbits each Martian day. It looks like an asteroid rather than a moon.

NASA image of Phobos, Moon of Mars

Phobos was discovered in 1877 by Asaph Hall, and its name means *"fear."* It has a mean radius of 11.1 km, and like our moon always has one side facing Mars as it orbits Mars. It has a density of 1.872 gm/cm^3.

In 2008 and 2010, the European Space Agency's Mars Express spacecraft made close passes of Phobos with ground penetrating radar and determined that Phobos is 30% hollow, has an atmosphere inside, some of which seeps out, and has an electromagnetic field. A monolith on the surface of Phobos is the object of current missions to Mars to investigate this abnormality. Perhaps this monolith is an entrance to the hollow interior, such as was depicted on Star Trek's episode, "The World is Hollow and I've Touched the Sky," on its third season.

The Geophysical Research Letters abstract on the Mars Express flyby said,

"New values for the gravitational parameter (GM=.7127 ± 0.0021 x 10^{-3} km^3/s^2) and density of Phobos (1876 ±20 kg/m^3) provide meaningful new constraints on the corresponding range of the body's porosity (30% ± 5%), provide a basis for improved interpretation of the internal structure. We conclude that the interior of Phobos likely contains large voids."

Orthodox scientists claim no body in space is naturally hollow, so they conclude that Phobos is artificially hollow. But as more and more bodies in space are found to be hollow, they will be forced to conclude that all bodies in space are naturally hollow. And since Phobos has an electromagnetic field, this indicates that Phobos has an inner sun which could provide life and light to its interior world.

Location and Size of the Polar Openings

It is assumed that the polar openings were formed during the earth's creation because as the earth was formed in rotation, centrifugal force would throw matter away from the axis of rotation leaving a hollow in the earth and polar openings at the polar axis. Additionally, the earth's foundational premise is a spiritual hollow globe upon which space dust and rocks accumulated.

If these assumptions are correct, then the original locations of the polar openings of the earth were at its polar axis, centered on the north and south poles.

There exists, however, evidence that the earth has been bypassed by planet sized comets in the past geological history of the earth, one or more of which tilted the earth's axis to one side of its original orientation. Astronomical information recorded on the calendar of the Gate of the Sun in the ruins at Tiahuanaco, Bolivia in the tops of the Andes mountains, at the time of its construction, shows that the earth had an inclination to the plane of the earth's ecliptic about the sun of 16.5 degrees -- much less than the 23.45 degrees that it is now. Also, George F. Dodwell, the government astronomer for South Australia, and Fellow of the Royal Astronomical Society of Great Britain after graphing historical observations made by ancient astronomers from all over the world, determined that the earth was tipped on its axis to about 26.5 degrees at the time of Noah's flood in 2,345 B.C. by the passage of some planet size comet from which it has gradually returned to its present angle of 23.45 degrees.

Hollow Earth enthusiasts originally thought that the polar openings were centered over the poles. However, it has been evident since with the advent of intense polar exploration that the openings are NOT located centered over the poles. The United States has a permanent station located at the South Pole on the Antarctic continent. And the North Pole has been crossed several times by polar explorers beginning with the Amundsen dirigible expedition in 1926 which flew from Spitsbergen to Alaska over the Pole. Wally Herbert took his dog sleds over the pole from Alaska to Spitsbergen going the opposite direction. Also the Russian nuclear icebreakers take tourists to the pole each year north of Franz Josef Land. This service is contracted by ADVENTURE ASSOCIATES Pty Ltd, 197 Oxford Street Mall, Bondi Junction, Sydney NSW 2022,

Australia, Postal Address: PO Box 612 Bondi Junction NSW 1355 Australia, Ph: (+61 2) 9389 7466 Fax: (+61 2) 9369 1853, for a 19-day trip for $15,950-$18,950 trip to the North Pole with additional cost for a flight from your hometown to Murmansk, Russia.

Bradley Air Service out of Resolute Bay in northern Canada (819-252-3981) regularly flies tourists, scientists and adventurers to the North Pole from the Canadian side of the top of the world for $24,000 per each two-day trip in which a ski landing at the pole is made weather permitting. There are two other, more expensive, seven-night trips. One, offered by High Arctic International (819-252-3616) for $8,650 per person is based in Resolute Bay, and the other, Arctic Odysseys (206-455-1960) for $11,000 works out of Medina, Washington. These trips depart each April and include visits to both the geographic and magnetic North Poles, dog sledding along the northwestern coast of Greenland, and overnight stops on Ellesmere Island at the Eureka Weather Station and Grise Fjord. (December 1993, Outside Magazine, p. 50)

Some websites for polar explorers are: polarexplorers.com, adventure-life.com, borekair.com, and wikihow.com/Get-to-the-North-Pole

There are anomalous occurrences that indicate the polar openings do exist. The warm north wind in winter, the warm winter foehn storms that come from the north, the drift wood with green leaves, the migration of birds and animals, the solar wind emanating from the polar region to light up the auroras and then trapped in the Van Allen Radiation Belts are a few of the evidences for the existence of polar openings.

The North Polar Opening

An estimate for the location of the North Polar Opening can be based on the sightings of land in the far Arctic north from explorers all around the Arctic basin. The Norwegian fishermen, Olaf and Jens Jansen sailed their little fishing boat northeast of Franz Josef Land when they discovered the North Polar Opening in 1829. Admiral Peary sighted what he called Crocker Land towards the northwest from the summit of Cape Colgate, Northern Canada in June 1906. MacMillan and Lt. Green traveled over 100 miles out over the ice looking for Crocker Land where they sighted it again. It appeared to be even farther to the northwest than Admiral Peary estimated it to be located. The Russian, Yakov Sannikov had previously sighted land to the north of the New Siberian Islands in 1810 and gave it the name of Sannikov Land. On his way to the North Pole in 1908, Dr. Frederick A. Cook sighted land to the northwest of his position at 84° 50' north Latitude, 95° 36' west Longitude, and called it Bradley Land. Captain Keenan sighted land from Harrison Bay on the north shore of Alaska. These directions all point to the same location in the Arctic – the location of the North Polar Opening. The land they were seeing was a doubly inverted mirage of the Inner Continent located halfway through the Polar Opening.

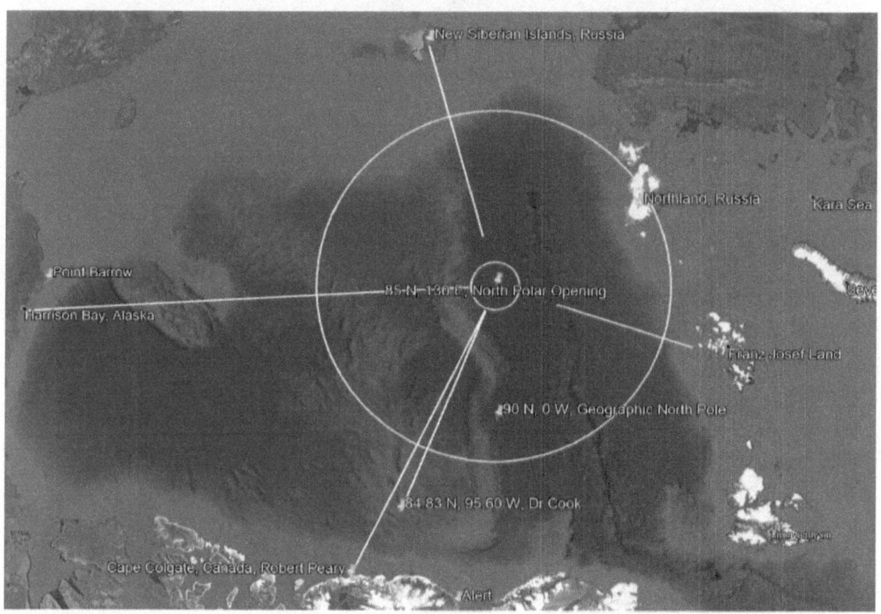

Directions in which land has been sighted from around the Arctic

This is the location the Lost Soviet flyers that flew north from Russia and were lost somewhere between the Kara Sea on the Russian side of the pole and Alaska, as reported by Vilhjalmur Stefansson in his book, UNSOLVED MYSTERIES OF THE ARCTIC.

The Amundsen dirigible transpolar flight of 1926 from Spitsbergen to the Pole and from the Pole to about 100 miles west of Point Barrow, Alaska would indicate the north polar opening would need to be located on the Russian side of their line of flight.

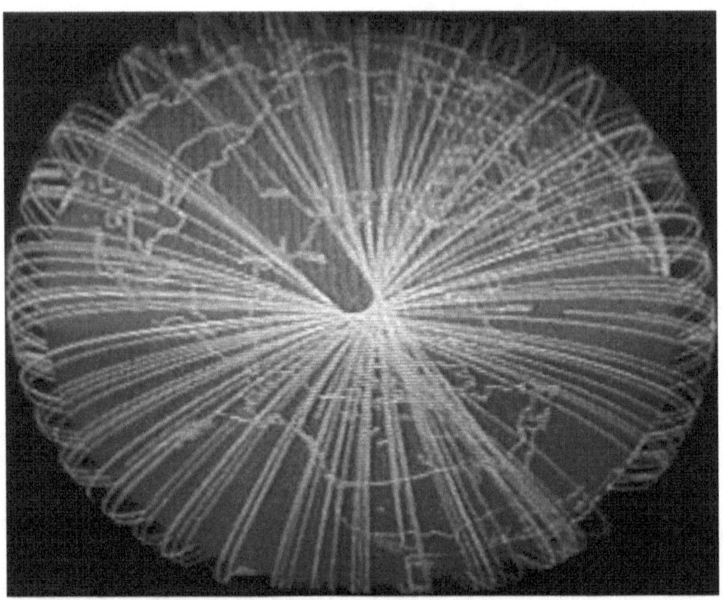

My friend Lord Ivars sent me the above image that a security guard snapped with his cell phone camera off a NORAD computer screen showing the flight paths of all the polar orbiting satellites. The interesting part of this image is that on the Russian side of the pole, there are no orbiting satellites between about 80 degrees East Longitude and 132 degrees East Longitude. You can see North Land, Russia jutting out into the Arctic Ocean opposite the geographic pole. This indicates that the North Polar Opening is located on the Russian side of the pole.

Some of the first satellites put up over the pole were lost. This most likely happened because they were attempting to put them in orbit over the polar opening without realizing it. Subsequent polar orbiting satellites, therefore, have all been placed in orbit such that they do not pass over the polar openings. With no matter in the polar openings to exert a gravitation acceleration towards the center of the planet, satellites put in orbit over the polar openings would be lost into space if above a certain distance, or if too close to the surface, would follow the curvature of the openings into the earth and crash inside the earth.

In this next NASA image, I thought I had found the North Polar Opening:

But rather than showing the polar opening, I discovered that this is just a hole in the satellite imagery showing no detail around the North Pole. Since no polar orbiting satellites can go over the polar openings, their flight paths cause that no imagery can be obtained over a certain area near the pole.

Other satellite images of the Arctic have displayed a *"hole"* at the exact geographic pole as result of a lack of imagery in that area. An example is the Arctic image of March 22, 2009:

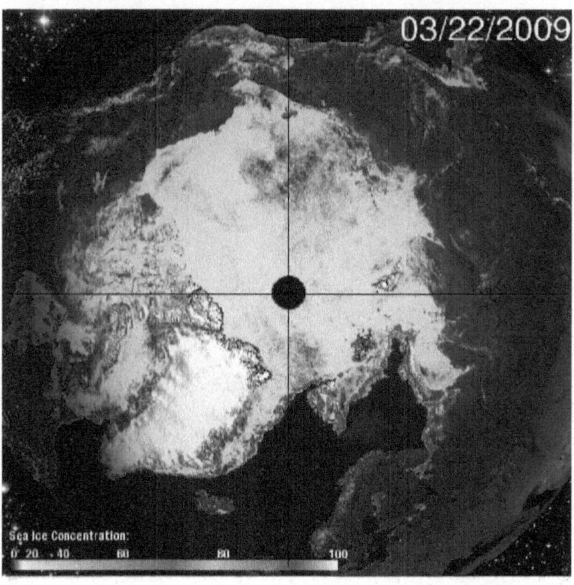

When asked why satellite imagery shows a *"hole"* at the geographic pole, the response given is that the satellite

365

imagery doesn't quite reach up to the pole and so leaves a *"hole"* in the imagery. So these *"holes"* in the satellite imagery do not depict an actual hole in the earth. See: https://nsidc.org/sites/nsidc.org/files/G02135-V3.0_0.pdf#page=10

The reason polar orbiting satellites cannot take imagery all the way up to the pole is revealed in the NORAD polar satellite paths image my friend Lord Ivars sent me. The reason is -- NO polar orbiting satellite can pass over the pole because the polar opening is too close to it. Polar orbiting satellites cannot pass over the polar opening because it has no matter to exert a gravitational acceleration on the satellite to keep it in orbit if it went over the opening. This is a key to locating the polar opening.

The image above was taken from this NASA video: http://svs.gsfc.nasa.gov/vis/a000000/a003300/a003333/amsr_e_sea_ice_640x480.mpg

This video obviously is a composite video over time of the change in Arctic ice. You can see the extent of the ice varying in extent as the video progresses. So the oval seen at the geographic pole in the video is just a *"hole"* in the imagery -- not the actual polar opening leading into the hollow earth. Nevertheless, this hole in the imagery occurs because polar orbiting satellites cannot pass over the polar openings, so the polar opening must be close to this hole in the imagery.

The fact is the North Polar Opening cannot be located at the exact geographic pole because the pole has been crossed several times. If an opening into the earth were located there, it could not have been crossed. For example, the Amundsen Nobile dirigible expedition flew across the geographic North Pole from Spitsbergen to Alaska in 1926. The U.S. nuclear submarine, the Nautilus, crossed the pole from Alaska in 1958. Wally Herbert from Britain crossed the pole from Alaska to Spitsbergen with his dog sleds in 1969. The North Polar Opening then seems to be located up near the pole, but on the Russian side of the pole.

In late 2008, I was contacted by Retired U. S. Air Force Colonel Billie Faye Woodard. We visited Billie in Pahrump, Nevada in November, and I have written a short biography of him which I have included in this book. See the Contents page.

Billie has had a very interesting life, much of it having to do with Our Hollow Earth. After joining the military, Billie was assigned to Area 51 in Nevada, USA. While at Area 51, Billie was assigned an office in the Administration and Archives, Level

6 underground. Billie distinctly remembers reading a 35 page document on the discovery of Our Hollow Earth by Admiral Richard E. Byrd, and before leaving the military recorded the exact coordinates that Admiral Byrd took when he flew through the North Polar Opening on a piece of paper he managed to take upon being discharged from the military. Those coordinates Billie gave me are 87.7 N Lat, 142.2 E Lon.

If we draw a line northeast of Franz Josef Land in the direction Olaf Jansen says he sailed his boat in which he discovered and sailed through the North Polar Opening, and another northwest of Ellesmere Island in the direction the Crocker Land mirage was sighted by Admiral Peary as described in Jan Lamprecht's book on HOLLOW PLANETS, the two lines meet on the Russian side of the pole. So I concluded the North Polar Opening is located on the Russian side of the pole. Billie's coordinates, also, are on the Russian side of the geographic North Pole.

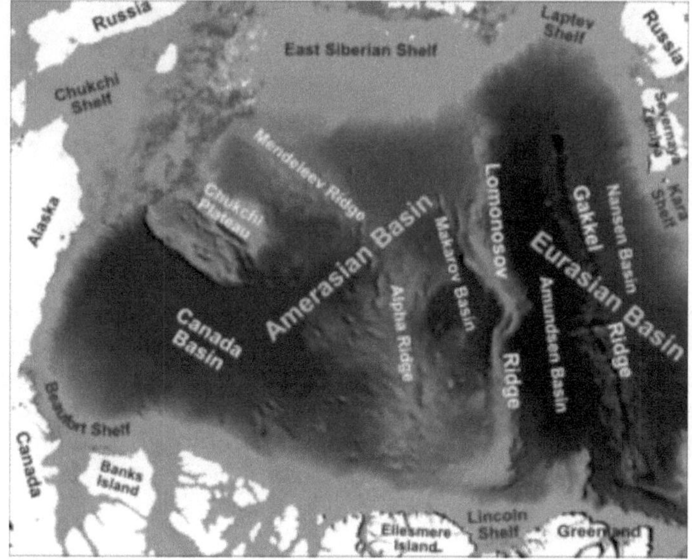

When I plot Billie's coordinates on an Arctic map, I find that they are located directly over the undersea mountain ridge called the Lomonosov Ridge. So the polar opening must be located nearby.

On October 26, 2015, one of my list members brought to my attention some images from the U.S. Naval Research Laboratory website which is the best evidence I have seen yet for the location of the North Polar Opening: https://www7320.nrlssc.navy.mil/hycomARC/.

The first image shows the percent of ice concentration on October 25, 2015:

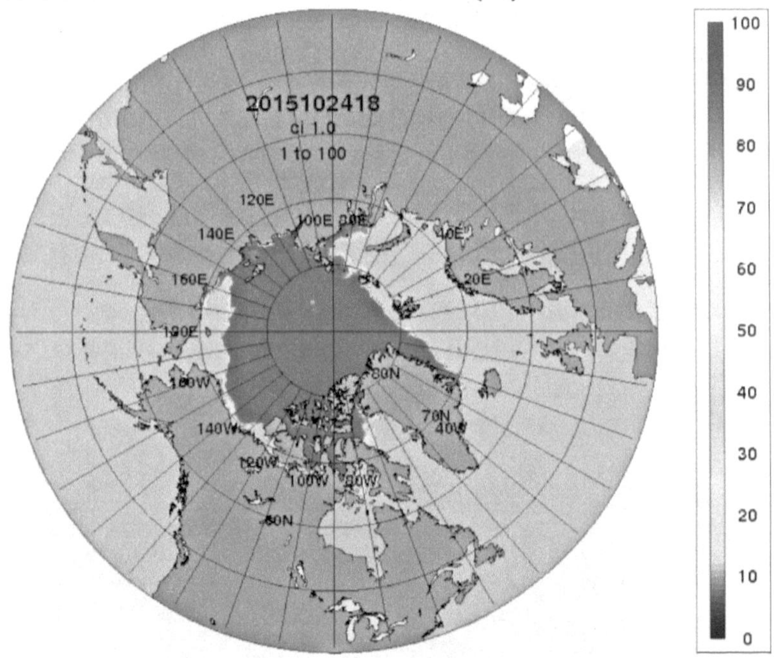

Notice a less ice concentration on the Russian side of the pole. There are two reflections of ice concentration recorded by the U.S. Naval Research Laboratory satellite in this image. The brighter reflection most likely is on the far side of the polar opening, and on this side of the polar opening is the dimmer reflection.

From these two reflections, we can estimate the center of the polar opening to be located at about 85 North Latitude, 130 East Longitude, with the rim of the polar opening seen as a circle drawn around the two reflections here:

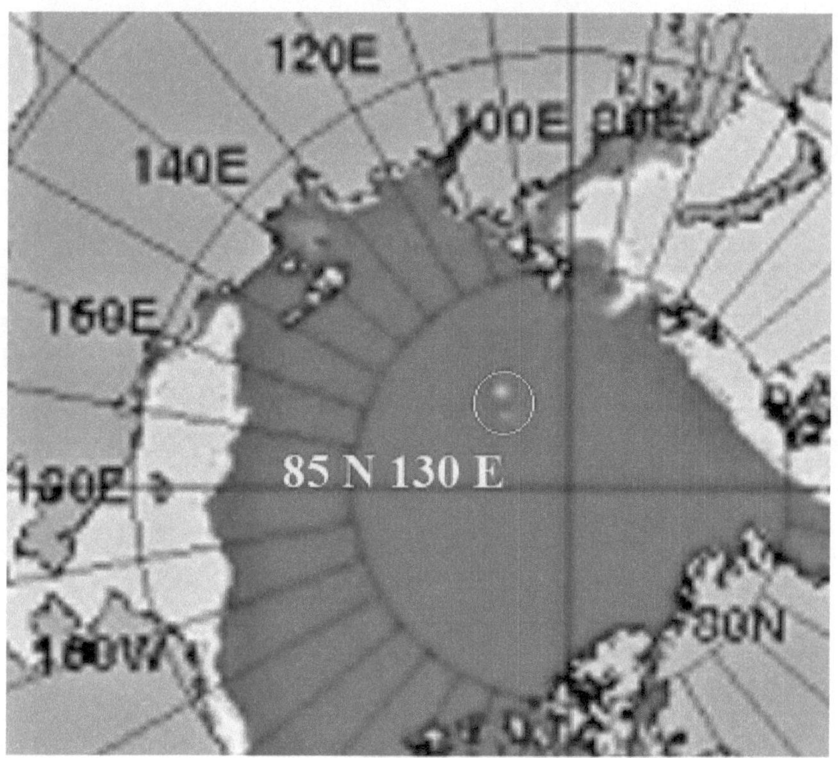

At this location for the North Polar Opening, notice also that the ice thickness image from the U.S. Naval Research Laboratory shows these same two spots giving THINNER ICE on each side of the opening:

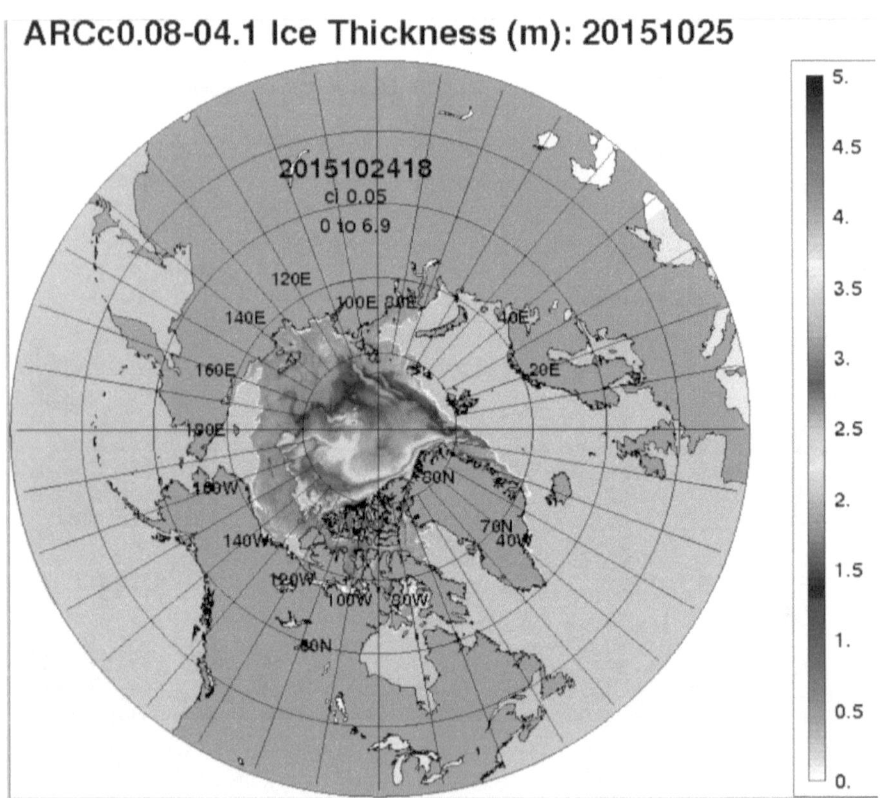

The U.S. Naval Research Laboratory image for sea surface salinity is also LOWER in these same two spots:

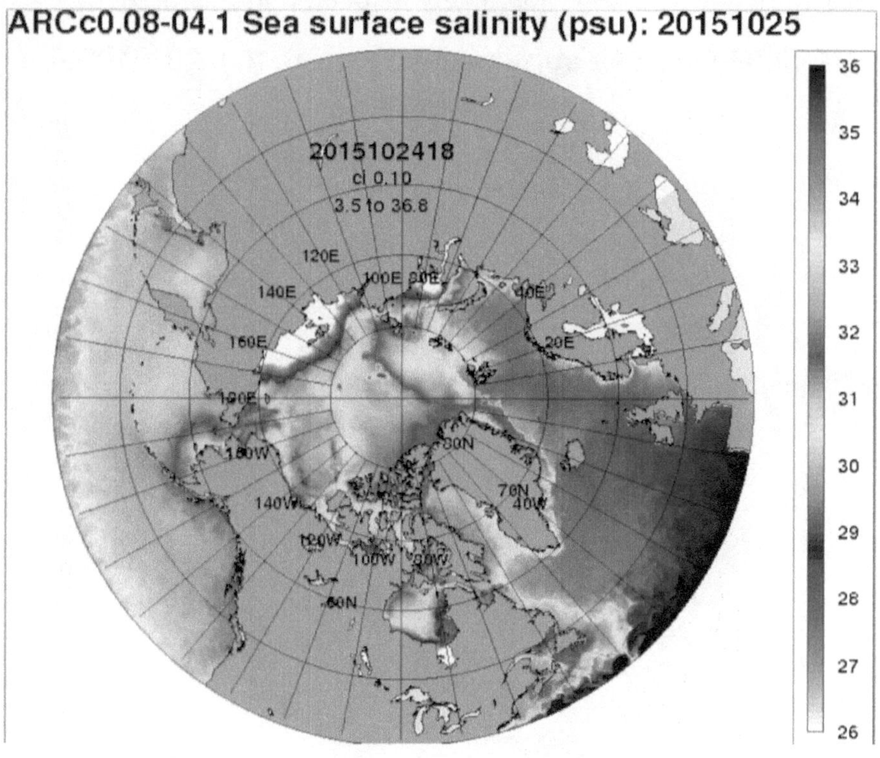

The sea surface temperature is also HIGHER in these two same spots -- all of which gives further supporting evidence that the polar opening most likely is located in this area:

ARCc0.08-04.1 Sea surface temperature (C): 20151025

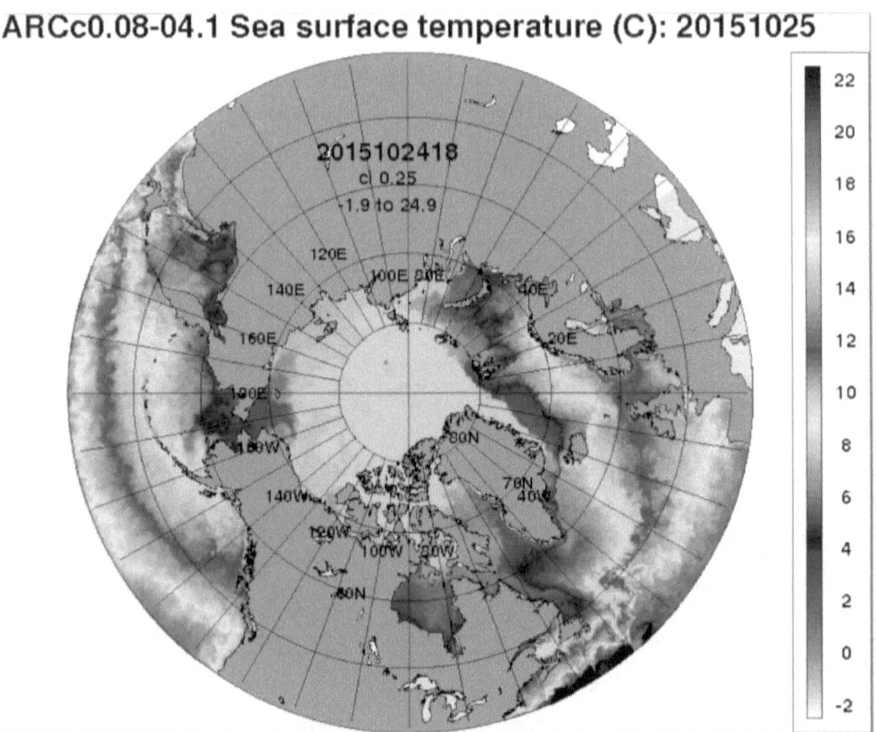

This location for the North Polar Opening puts it in the Eurasian Basin, which is some of the deepest sea depths in the Arctic. Notice, also, that there is a curvature in the Lomonosov Ridge at this location which is another indication that the polar opening is located there.

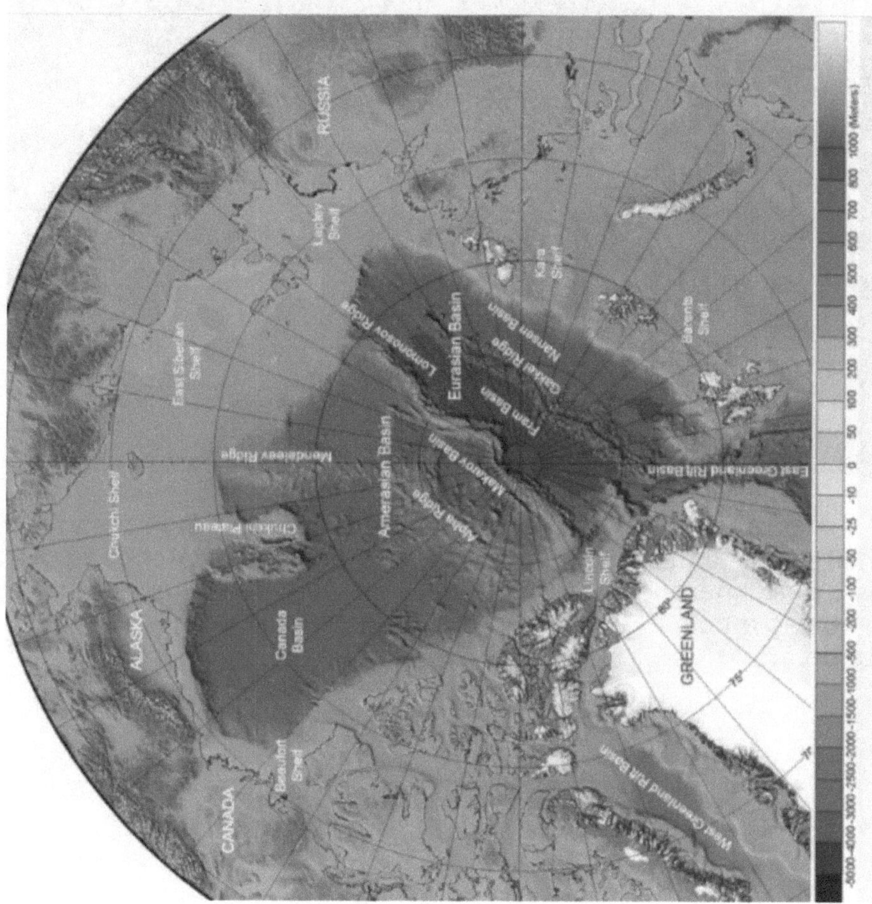

This image from the National Snow & Ice Data Center for September 16, 2012 shows a hole in the Arctic ice at these same coordinates where I estimate the North Polar Opening to be located:

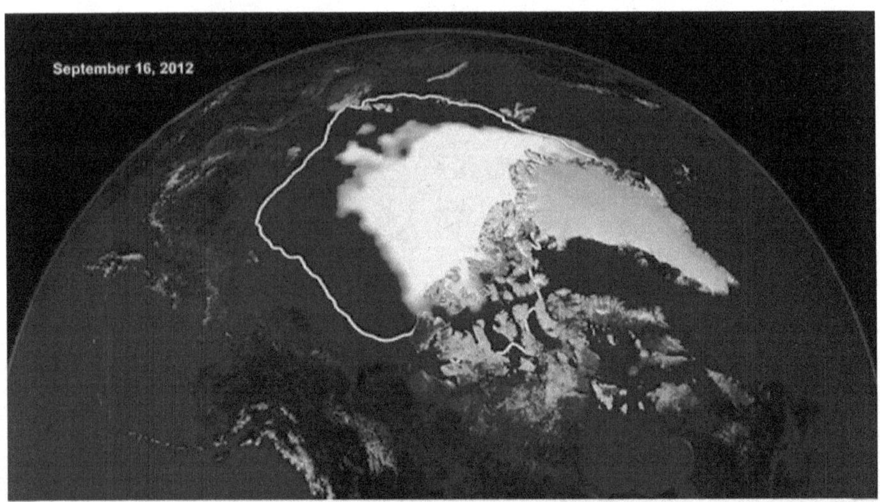

As can be seen in the next image taken over a year and half earlier on February 25, 2011, this hole in the Arctic icepack is a permanent feature that a polar opening by its very nature of being a hole in the shell of the planet would necessarily need to be.

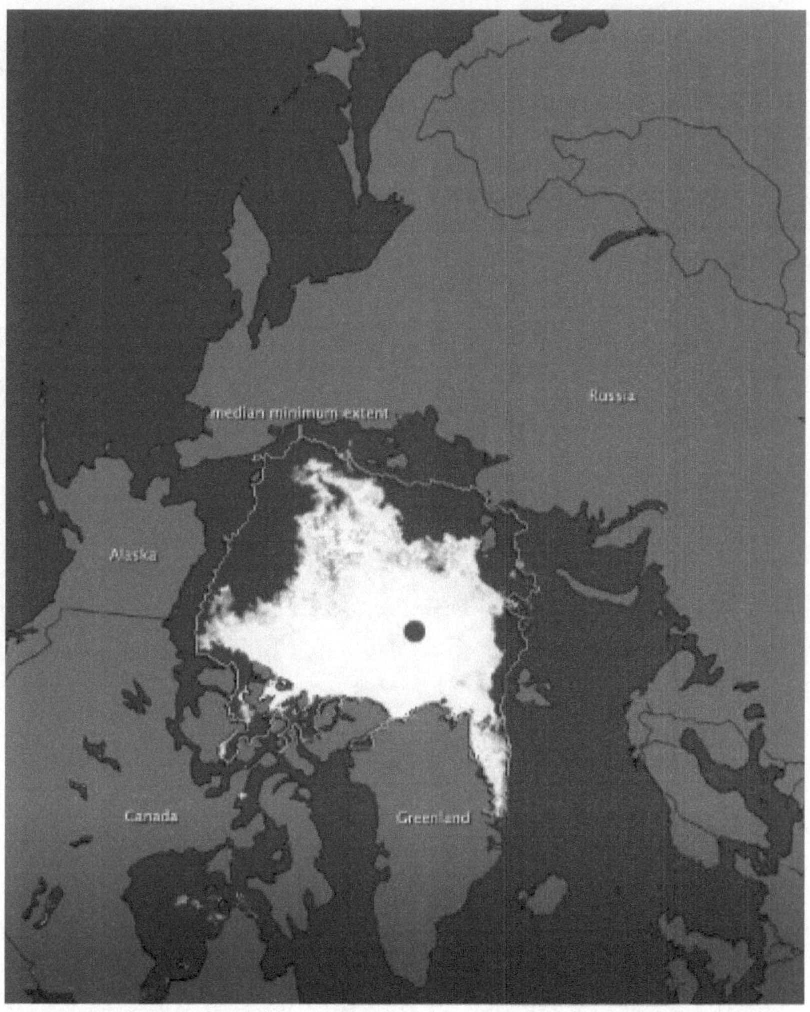

The hole in the ice is also visible in this image from the Advanced Microwave Scanning Radiometer on the NASA Aqua Satellite. Notice the black hole in the imagery at the geographic North Pole, caused by the lack of satellite imagery near the pole. The blue hole in the ice where I believe the North Polar Opening is located is at about the 1 o'clock position near North Land, Russia.

So let's assume that the polar openings have a curving surface instead of a straight hole through the planet. We will take the 800 mile thickness of the Earth's shell as given by the Inner Earth Guide in ETIDORPHA which will give us a 400 mile radius for the curvature of the polar opening. With a distance of 125 miles at the neck of the polar opening, as given to us by Admiral Byrd, that would place the span from the center of the polar opening to the periphery where the earth begins to curve into the opening at 462.5 miles. The North Geographic Pole would be on the rim of the perimeter about 114 statute miles from where the earth's surface begins to dip into the polar hole, and from the North Pole to the center of the Polar Opening is about 348 miles.

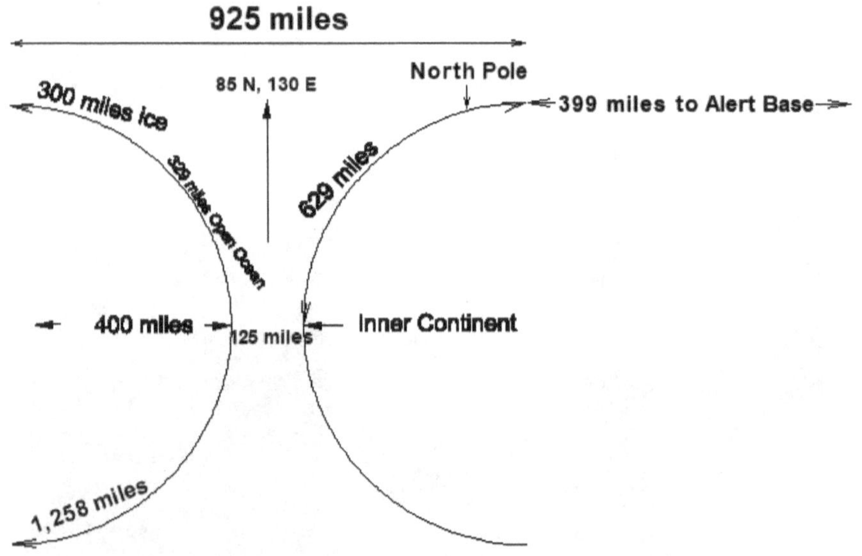

North Polar Opening

In this drawing of the North Polar Opening, the thickness of the earth's shell is assumed to be 800 miles. If the earth's shell were any larger, say a thousand or 2,000 miles thick, the polar opening would be too large to be hidden in the Arctic Ocean. So the 800 mile estimate of the earth's shell thickness is a reasonable estimate. It is the estimate of the earth's shell thickness given by the Inner Earth Guide in ETIDORPHA and confirmed by Col. Woodard. The beginning of the curvature into the opening begins at 925 miles diameter, with the North Pole about 114 statute miles within the perimeter of the opening. At the center of the polar opening, Admiral Byrd gave us an estimate of the diameter of the neck of the opening to be 125 miles.

On a map, the perimeter would begin at about 399 statute miles from Alert Base, Ellesmere Island, Northern Canada, about 777 statute miles from Point Barrow, Alaska, about 549 statute miles from Longyearbyen, Spitzbergen, about 143 statute miles from the New Siberian Islands, and about 1,078 statute miles from Murmansk, Russia with the center of the polar opening located at 85 N Latitude, 130 E Longitude (5 degrees from the pole).

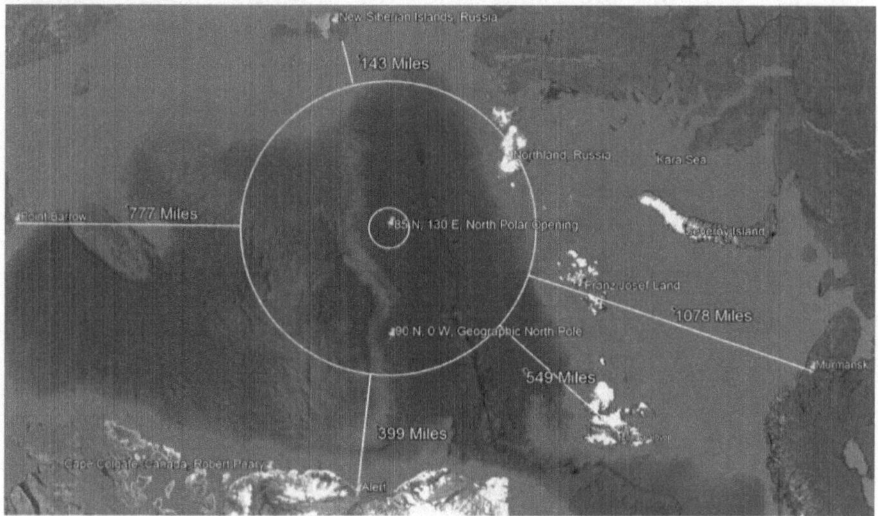

From Alert Base, the perimeter where the earth begins to dip into the opening would begin at about 399 statute miles and the inner continent, which I estimate to be located about half way through the polar opening, at 629 miles after reaching the perimeter for a distance of 1,028 miles from Alert to the Inner Continent.

On page 66 of The Smoky God, Olaf Jansen mentioned that as he and his father left Franz Josef Land on their expedition to the Hollow Earth, *"we seemed to be in a strong current running north by northeast."* The island they found on their third day's sailing northeast of Franz Josef Land is not on our maps today. They found a large accumulation of driftwood on the northern shore of that island, trunks of coniferous trees two feet in diameter and forty feet long. This encouraged them to continue north. Again, several days further on, on page 84, Olaf reported, *"...we discovered...that we were sailing slightly north by northeast."* This would indicate the polar opening that they found was on the Russian side of the pole northeast of Franz Josef Land, in the same area that the Soviet flyers were lost, as described in Vilhjalmur Stefansson's book, UNSOLVED

<u>MYSTERIES OF THE ARCTIC</u>, and in the same direction that Admiral Peary saw the mirage of Crocker land from the north coast of Ellesmere Island and Dr. Cook saw the mirage of Bradley land north west of his trek to the pole in 1908.

Curiously, the coordinates given by Col. Billie Woodard of 87.7 N, 142.2 E are in a direct line from Alert Base to the center of the Polar Opening, which would be the route Admiral Byrd flew his airplane into the North Polar Opening in 1927 as can be seen on this Google Earth map here:

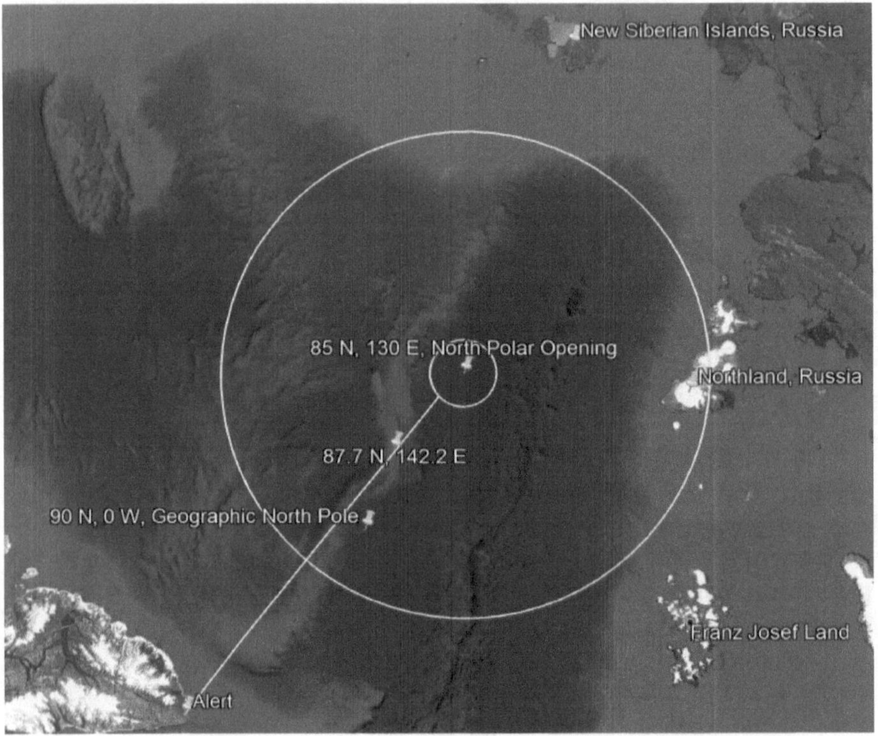

The Norwegian Arctic explorer, Dr. Fridtjof Nansen, in his <u>FARTHEST NORTH</u>, relates several anomalous observations on his Arctic expedition in his ship the Fram. Of all Arctic explorers we have record of, Nansen was the one that got closest to the North Polar Opening without entering and discovering it. Some of the observations of Nansen that support a polar opening location north of the New Siberian Islands are as follows.

First, Nansen discovered a substantial stretch of open ocean north of the New Siberian Islands. In contrast, in his passage north of Norway and Russia to the New Siberian

Islands, he had to stay close to the coast to get past the ice. And yet, north of the New Siberian Islands, in September of 1893 on their way north, they didn't find ice until 79 degrees N. Latitude. Only after 7 days sailing north over open rolling ocean did they reach the pack ice north of the New Siberian Islands.

To Nansen's surprise, out on the ice pack, they found a remarkable number of birds of various kinds including snipe and seagulls, also foxes, walrus and polar bears that indicated they were in the proximity of land towards the north. They passed the winter with their ship frozen in the ice pack and while waiting out the winter, they took scientific measurements and observations. They found rock and large quantities of mud and driftwood on some of the icebergs that indicated to Nansen that much of the Arctic ice originates in some river, perhaps further to the north than they were then located -- in some uncharted land towards the north not on any map. Perhaps it could be Sannikov land that Yakov Sannikov sighted north of the New Siberian Islands in 1810.

In mid-winter, on January 18th, 1894, at 79° 18′ N Latitude, 137° 31′ E Longitude, observations by Nansen found that a north wind raised the temperature while a south wind lowered it, indicating that warm air was coming out of the north in winter -- perhaps from a land further north warmed by an inner sun inside the Earth (FARTHEST NORTH Vol. I, pages 373-374). Curiously, Nansen discovered that ocean water temperatures were also warmer the further down he measured it beneath the ice, as also the air temperature above the ice when measured from the ship's crow's nest was discovered to be warmer than next to the ice.

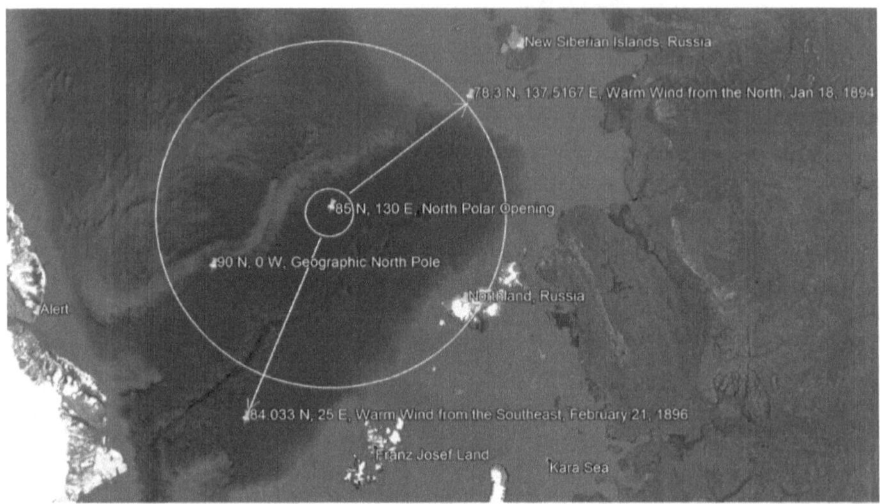

After Fritjof Nansen and Johansen left the Fram, Nansen left Otto Sverdrup in charge as commander of the ship. The ship drifted northwest around the polar opening and on February 21, 1896, at 84.033 N, 25 E Sverdrup reported that suddenly one morning, a wind from the southeast raised the temperature 18.8 degrees. This warm wind could have come from the polar opening as the arrows indicate in the Google Earth map above. The previous winter they were located south of the polar opening when Nansen had reported a warm north wind blowing from the direction of the polar opening also.

At 80 deg 1' N Latitude on February 16-19 Nansen caught sight of a mirage of the Inner Sun. Most likely they were then located near the lip of the polar opening. They thought it must be a mirage of our outer sun. But there is a possibility that it could have been a mirage of the inner sun since they were so close to the polar opening. On page 394-395, Nansen recorded,

"Friday, February 16th... Today another noteworthy thing happened, which was that about midday we saw the sun, or, to be more correct, an image of the sun, for it was only a mirage...The mirage was at first like a flattened-out glowing red streak of fire on the horizon; later there were two streaks, the one above the other, with a dark space between; and from the main-top I could see four, or even five, such horizontal lines directly over one another, and all of equal length, as if one could only imagine a square dull-red sun with horizontal dark streaks across it."

He remarked that it had a hazy, smoky-red color, similar to the description of the inner sun given by Olaf Jansen, whose father also at first thought it was a mirage of our outer Sun, when they first caught sight of it in their 1829 journey to the inner world through the North Polar Opening northeast of Franz Josef Land. Nansen thought it odd that he should see a mirage of the Sun several days before the Arctic sun rise.

Come summer, Nansen went out on the ice pack and investigated a pollen-like substance that seemed to cover the ice everywhere with a brownish color. Inner Earth explorer Olaf Jansen explained that the northern shores of the inner world are covered with large fields that grow flowers, whose pollen at times is blown out over the Arctic ice fields through the North Polar Opening.

Prior to his Arctic expedition, Nansen had visited Russia and consulted with their experts about their sighting of a mirage of land they termed Sannikov land north of the New Siberian Islands. The New Siberian Islands even today are covered with bones and remains of mammoths and other inner earth animals that Olaf Jansen reported fall into ice crevasses of inner earth rivers that empty into the Arctic Polar Opening where they freeze and later are carried out to sea and eventually end up deposited on northern Arctic shores. Remains of woolly rhino, steppe lions, giant deer, mammoth, foxes and a hardy breed of horse that scientists claim are prehistoric are preserved in the Arctic ice. Recently some scientists have been attempting to find some of these exotic animal remains that are frozen and preserved in Arctic ice. They want to take samples of the frozen flesh and use it to clone these exotic animals to start a sort of Jurassic Park. Little do they know that a Jurassic Park already exists located inside Our Hollow Earth, and can be reached through the North Polar Opening, north of the New Siberian Islands, from whence all the frozen exotic carcasses come from.

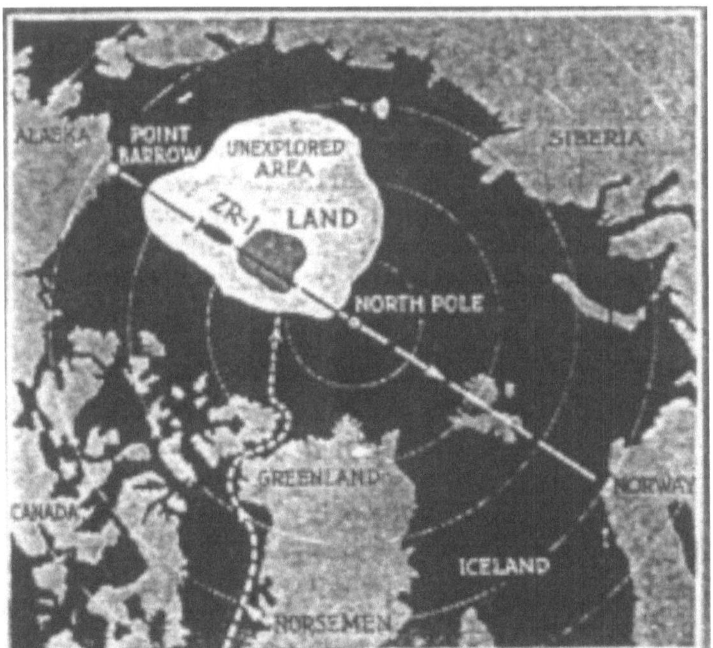

On the opposite side of the Arctic from the New Siberian Islands, Lt. Green of the U.S. Navy had accompanied MacMillan on his expedition northwest of Ellesmere Island following the mirage of Crocker land that Peary had sighted from Cape Thomas Hubbard on his way north to discover the pole. MacMillan turned back after journeying 120 miles out over the ice pack because the mirage of Crocker land continued to appear ever farther northward no matter how far they traveled towards it. Later, Lt. Green published an article in Popular Science Monthly, in the December 1923 issue, that he was still convinced that Crocker land still existed and had even convinced the Navy to build a dirigible named the ZR-1 that was going to attempt an over flight of that land. As you can see from his map, he figured that Crocker land was mostly on the Russian side of the pole.

When Arctic explorers speak of seeing mirages of land, they are actually seeing land. In warmer climates of the world, a mirage mostly looks like water on the horizon because it is reflecting the blue sky. But in the Arctic, a mirage is caused by warm moist air that comes up out of the Polar Opening from the Inner World. This layer of warmer air above the colder air next to the ice causes mirages or reflections of the ground or ice on the ocean, and not the sky, as they do in the warmer climates of the world. In the warmer climates of the world the temperature of the air becomes colder with height above the ground. Just the opposite is the case in the Arctic and Antarctic. Warm, moist air emanating from the earth's hollow interior through the polar openings rises above the ice and serves as a boundary of air that reflects objects on the ground. Jan Lamprecht, in his book, HOLLOW PLANETS, gives convincing evidence that the mirage of Peary's Crocker land and Cook's Bradley land could easily have been a doubly inverted image of land within the North Polar Opening. The first inverted image appears upside down while the second image above that is upright. Apparently, the mirage of land seen all around the Arctic is the second upright image seen just above the horizon and the other first inverted image below the horizon within the polar opening is not visible from the vantage point of the viewer.

Another curious story that relates to Lt. Green's estimated location of Crocker land is the disappearance of the Viking Greenland colony. In 985 A.D., Eric the Red discovered Greenland and subsequently settled it with Viking residents of Iceland. The Greenland colony, consisting of two settlements on the west coast of Greenland, one further north than the other, thrived for several centuries, but then as Europe became embroiled in war and the disease of the plague, the Norwegians lost contact with their Arctic colonies in Iceland and Greenland. The last ship known to have returned from their Arctic colonies to Norway was in 1410. When the Dark Ages had past and Greenland was once again rediscovered with Hans Egede establishing the first modern settlement there in 1721, all that could be found of the original Viking settlers were their ruins and some of their animals. Even the Arctic author Vilhjalmur Stefansson in his book, UNSOLVED MYSTERIES OF THE ARCTIC, concluded that the disappearance of the lost Viking colony in Greenland was a mystery.

In an attempt at determining where the lost Viking Greenland colony went, Lt. Green says he reviewed the Eskimo traditions. The Eskimos told him that the Vikings had sent hunting parties farther and farther north. Then one day their men found a paradise in the north -- a place the Eskimo had always known about but stayed away from because they believed it to be inhabited by evil spirits (actually meat-eating dinosaurs that live in the jungles of Inner Earth near the Polar Opening). The Viking explorer parties had come back and had told the rest of their Greenland colonies of their wonderful discovery. All then promptly packed their bags, and singing songs, departed suddenly northward and never returned. The Eskimo tradition is that over the ice towards the north, in the direction Admiral Peary sighted Crocker land and Cook sighted Bradley land, is a ...*"land that is warm; is clothed in summer verdure the year around; is populated by fat caribou and musk-ox. It lies,"* they say even to this day, *"in the direction of the coastal trail-route north."* Lt. Green shows that trail on his map. The trail is located on the west side of Greenland, and goes up around Ellesmere Island, and out over the pack ice towards the northwest.

The Transpolar Drift

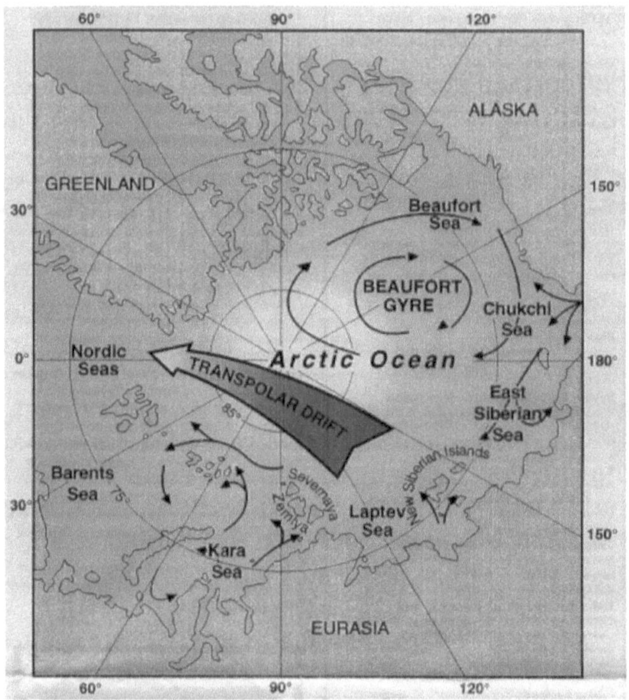

The Transpolar Drift has been known since at least the days of Fridtjof Nansen, who in 1893 sailed his ship, the Fram, north of the New Siberian Islands looking for Sannikov land, which the Russians had seen a mirage of in that direction. Nansen embedded his ship in the ice and drifted across the Arctic in the ice flows to the east side of Greenland.

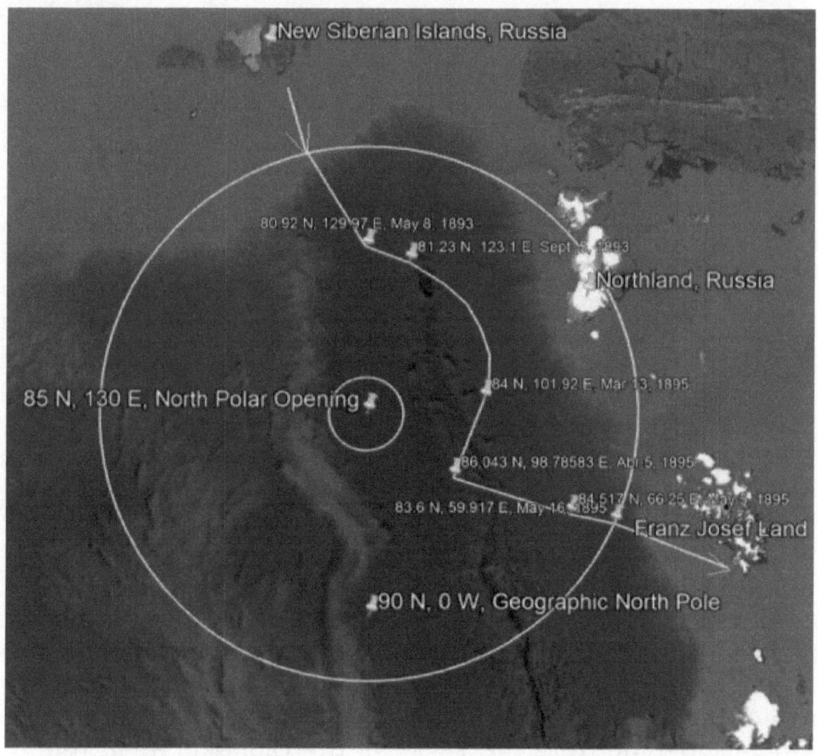

As can be seen in the above Google Earth map, Nansen started his drift across the Arctic by embedding his ship, the Fram, in the ice May 8, 1893 north of the New Siberian Islands. It then drifted around the North Polar Opening. At his farthest north, he then left the ship and took his dog teams and attempted to go to the North Pole, but gave up because of high ridges. He then went south west to the Franz Josef Land islands.

The Arctic current Nansen attempted to drift across the Arctic ice in is called the Transpolar Drift and acts like a swift running stream of water in a ditch. In fact, the underwater canyons in this area follow the same direction as the Transpolar Drift. On a map of the Arctic Ocean floor, you can see the Lomonosov Ridge crossing the Arctic Ocean in the same direction as the Transpolar Drift.

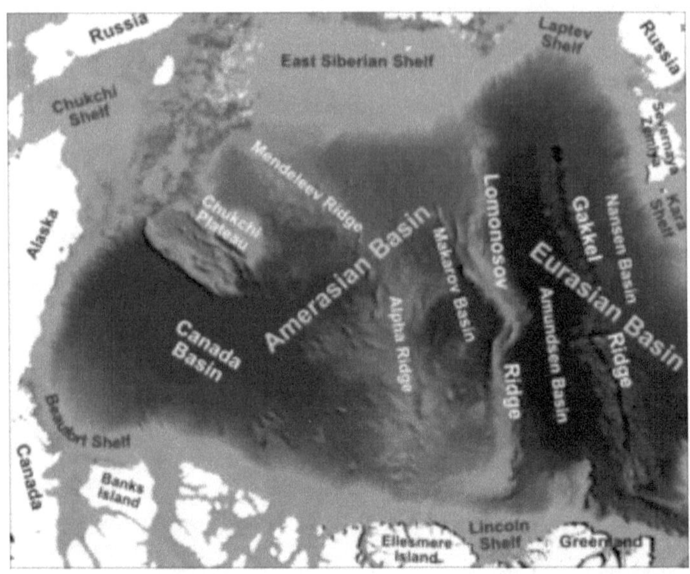

As you look at the above mentioned map of the Transpolar Drift, you will notice that a powerful ocean current comes out of the area where I estimate the North Polar Opening is located north of the New Siberian Islands. This current carries ice out of the polar opening from the valley of the Hiddekel River that empties into the Polar Opening from Inner Earth. From there the current flows across the Arctic Ocean to the Greenland side of the pole causing eddy currents on either side of the Transpolar Drift.

The eddy current on the Canadian side of the Transpolar Drift produces the Beaufort Gyre north of Canada and rotates clockwise. This eddy current causes the ice to pile high up onto the shores of northern Canada to the greatest depths of anywhere in the Arctic Ocean. The thinnest ice is located where the ice comes from -- in an area north of the New Siberian Islands -- another indication where the polar opening is located. Then on the Russian/European side of the Transpolar Drift in the Barents Sea, the eddy current rotates counter-clockwise. Because the eddy current is stronger on the Canadian side, the whirlpool effect causes the lowest tide height in the Arctic Ocean in the center of the Beaufort Gyre.

A case was made by one of the expedition members when planning the Voyage to Our Hollow Earth Expedition that the low tide in the Beaufort Gyre was an indication to him that this low tide area in the center of the Beaufort Gyre was where the polar opening may be located. I did not accept this argument because that whole low tide area on the Canadian side of the pole was crisscrossed by the search flights looking for the missing Soviet Flyers of 1937 as shown in the National Geographic map published in Vilhjalmur Stefansson's book, UNSOLVED MYSTERIES OF THE ARCTIC. The search flights didn't find the Polar Opening because the Americans didn't extend their search to the Russian side of the pole, the search flights by the Russians just went up to the pole, and NO search flights were taken in the area I have determined the North Polar Opening is located near Northland, Russia.

When reviewing this phenomenon of the Transpolar Drift, I realized a most significant evidence for the location of the polar opening. This most interesting phenomenon is a jet stream of warm moist air that follows the same path from the valley of the River Hiddekel within the Polar Opening across the Arctic and drops its moist air onto the Greenland ice cap. Over the years, this has resulted in an accumulation of ice on the Greenland continent to the awesome depth of nearly **two miles**!

Consider this for a moment. What if this moist jet stream of air, over the years, had instead flowed over Alaska, or Siberia, or the Norway-Sweden-Finland peninsula -- which are all at the same latitude as Greenland? If it had, those places would today be covered by an ice cap 2 miles thick just like Greenland is today! If we follow that jet stream of moist air from Greenland back across the Arctic, we find that it points to the location of the polar opening -- on the Russian side of the pole!

The same thing is happening in Antarctica. I estimate the South Polar Opening to be located at the exact opposite of the globe from the North Polar Opening, which would put the South Polar Opening at 85 S Latitude, 50 W Longitude. Warm, moist air coming up out of the South Polar Opening blows out over east Antarctica and has created an ice cap there to a depth of over 2 miles!

It cannot be denied that there is a strong current flowing out of that area north of the New Siberian Islands. It is the location where most of the freshwater ice in the Arctic is coming from. Eddy currents on each side of the Transpolar Drift flowing out of that area cause the Beaufort Gyre to rotate clockwise and the gyre in the Barents Sea to rotate counter clockwise on either side of this swift current of water coming up out of the polar opening. Perhaps the Transpolar jet stream and ice flows originate from the valley of the River Hiddekel that empties into the Arctic Ocean within the polar opening from the Inner Continent where Olaf Jansen found that river after sailing through the North Polar Opening in 1829.

In his book, Olaf explained where the ice comes from that fills the Transpolar Drift current, when he wrote,

"...about three-fourths of the 'inner' surface of the earth is land and about one-fourth water. There are numerous rivers of tremendous size, some flowing in a northerly direction and others southerly. Some of these rivers are thirty miles in width, and it is out of these vast waterways, at the extreme northern and southern parts of the 'inside' surface of the earth, in regions where low temperatures are experienced that fresh water icebergs are formed. They are then pushed out to sea like huge tongues of ice, by the abnormal freshets of turbulent waters that twice every year, sweep everything before them." (THE SMOKY GOD, pp. 122, 123)

The South Polar Opening

The best image I have been able to find of the most likely location of the South Polar Opening is an auroral image from NASA. It shows the auroral radiation coming from an oval location on the western portion of Antarctic where I estimate the south polar opening is located.

This location of the South Polar opening is located in the same area that looks like an elliptical area in the Apollo 17 image:

The oval location from which the auroral radiation is emanating from can be seen better by viewing the auroral NASA movie that this movie still was taken from:

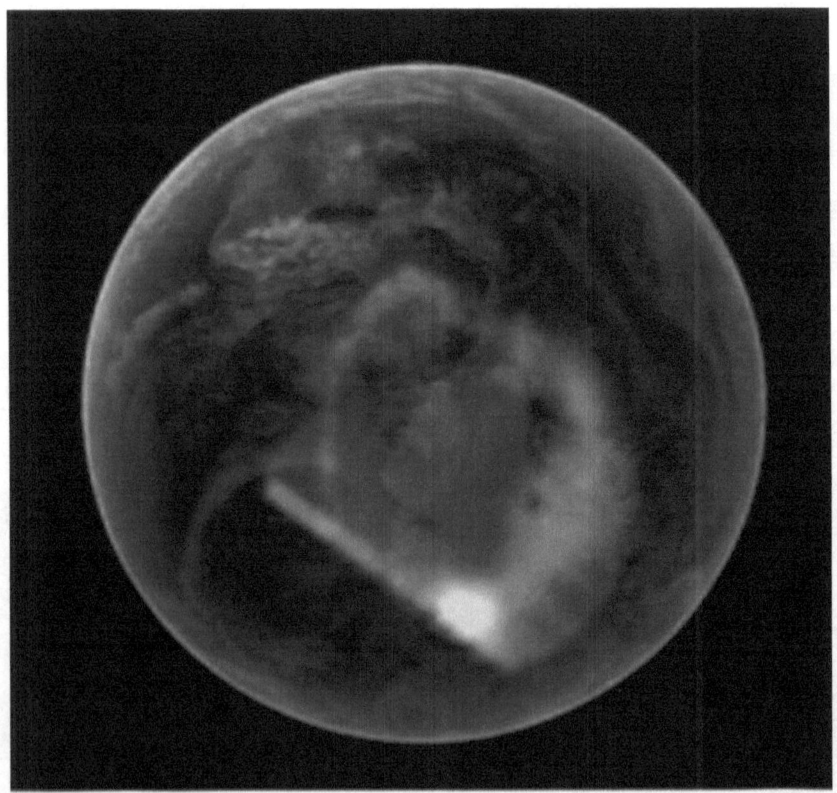

You can view the movie here:

www.nasa.gov/mov/133778main_FUV_640x480.mov

Another NASA auroral image also shows the source of the auroral radiation coming up like a tornado funnel from the western part of Antarctica:

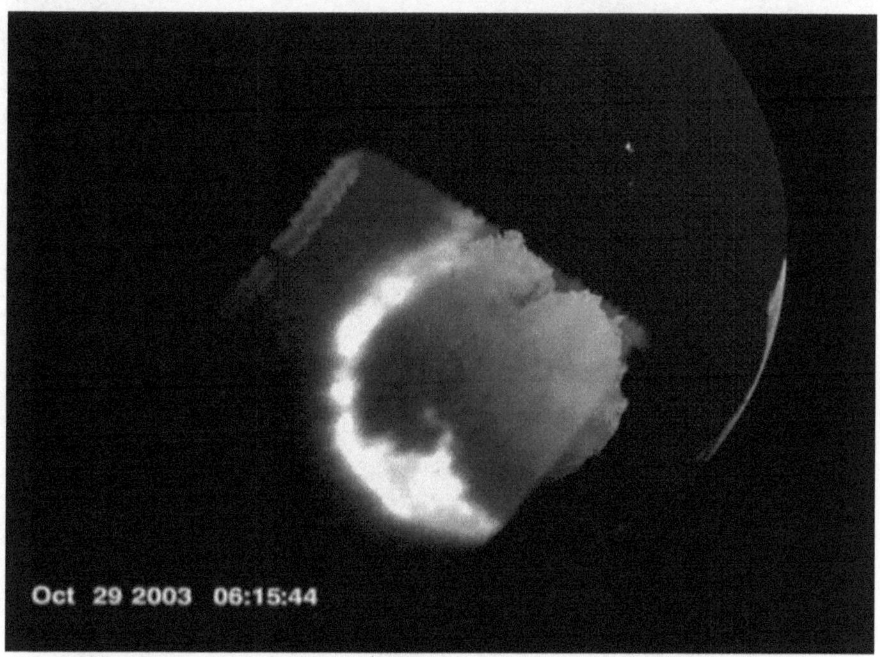

The location for the South Polar Opening is most likely located on the earth opposite the North Polar Opening. The North Polar Opening centered on the coordinates of 85 degrees north latitude, 130 east longitude would put the location of the South Polar Opening 180 degrees on the opposite side of the Earth from the location of the North Polar Opening at 50 degrees west longitude and 85 degrees south latitude in Antarctica as depicted on the following Google map. And as can be seen on the next map of Antarctica without ice, the South Polar Opening seems to be located in an inlet of the sea that empties into the ocean on the east end of the Weddell Sea.

On the Google map (see next), the red line from McMurdo Base is the route Admiral Richard E. Byrd took to the South Polar Opening in 1946 and 1947, flying between the Axel Heiberg Glacier on the left and the Coal Sack Bluff on the right. Half way through the polar opening, he estimated the neck of the opening to be 125 miles in diameter.   The white line represents the route of the Australians that flew from Australia into the South Polar Opening as told to me by Anne Stabler.

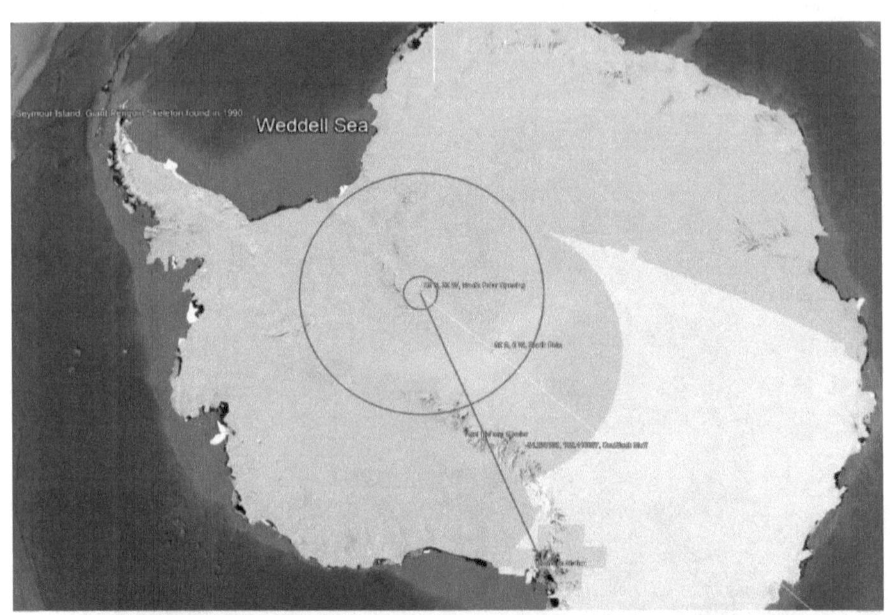

Seymour Island, Giant Penguin Skeleton found in 1990
Weddell Sea

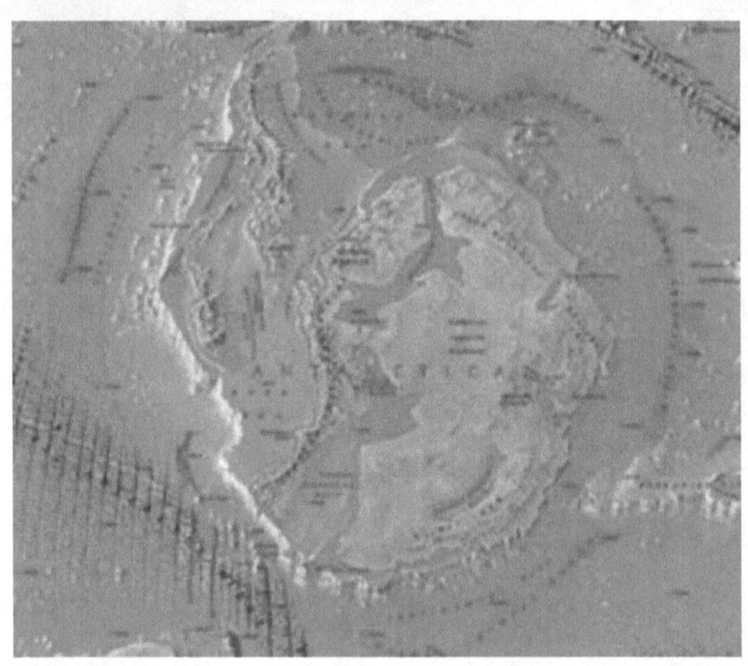

The South Polar Opening of Earth:

This NASA NOAA relief map of Antarctica shows the location of the South Polar Opening where the auroral radiation is emitting in the next photo is evidence our Earth is hollow.

Earth's Aurora Australis:

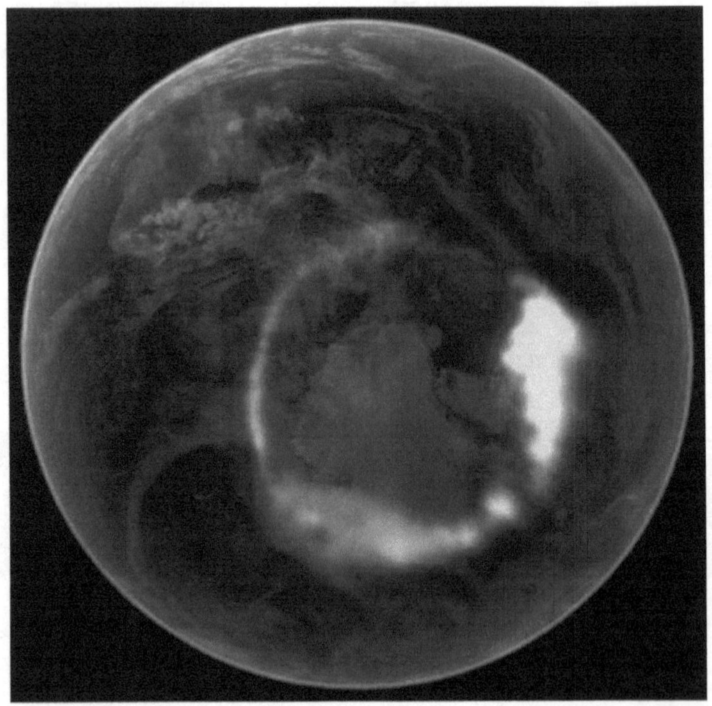

This NASA image of Earth's Aurora Australis shows the auroral radiation coming out of the Earth's South Polar Opening indicating the Earth is hollow.

Did the Lost Soviet Flyers Fly into Our Hollow Earth?

Vilhjalmur Stefansson, in his book, UNSOLVED MYSTERIES OF THE ARCTIC, explains how on August 12, 1937, a four-engine aircraft of passenger type left Moscow, bound for Fairbanks, Alaska via the Arctic with a crew of six seasoned flyers and were lost and never found even after a year of diligent search.

Most likely, the Lost Soviet Flyers flew into our Hollow Earth by accident.

Stefansson wrote in his book, that the Soviet Flyers encountered a stronger wind as they approached the pole than the forecasts had led them to believe. They reported that at 20,000 feet the wind was about 62 miles an hour and was cutting almost that many miles from their speed because it was nearly straight against them.

Stefansson reported that everything continued well for nearly two hours after they crossed the pole and that they were moving down upon Alaska on the Fairbanks meridian. Then came the only distress message, that they had been forced to descend from 20,000 feet where they had been flying in bright sun shine to the 13,000 foot level because one of the four motors had gone dead from an oil line damage. The last words heard were, *"Do you hear me?"* and *"We are landing in..."* Unintelligible messages were received over the next several days, but became gradually weaker until they finally died out.

Subsequent rescue missions were flown both from the Soviet side and the Canadian and Alaskan side of the Arctic over the next year looking for the downed Soviet Flyers but they were never found.

After reviewing all the downed Arctic fliers that had ever been known to come down in the Arctic, Stefansson concluded that the Soviet Flyers were the only ones that had ever been lost in the Arctic. The Americans alone flew about 40,000 miles in search of the Lost Soviet Flyers over a period of about one year covering the Arctic on the Canadian side of the pole down into Alaska. See a National Geographic map of those search flights here:

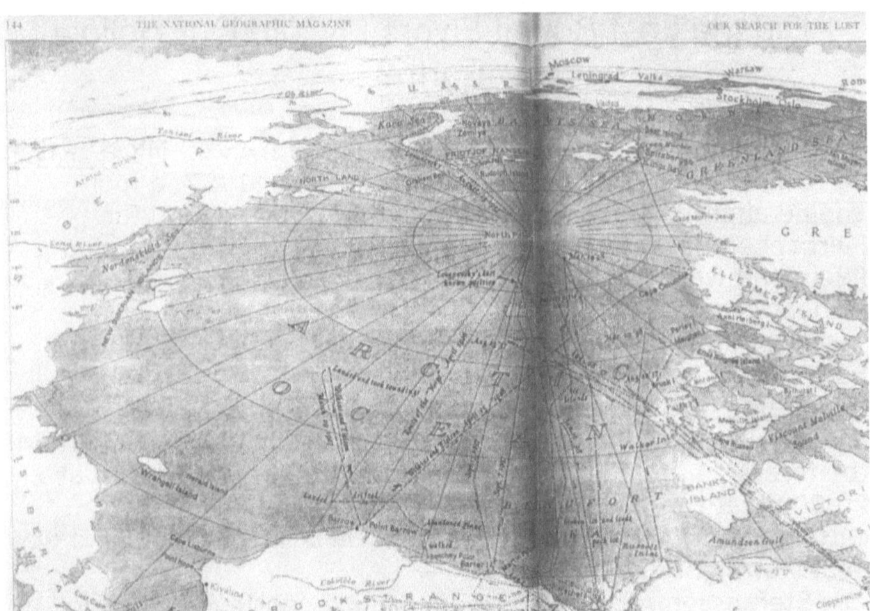

Russian search flights flew up to the pole, but did not cross over. Most likely, had they retraced the exact path taken by the Lost Soviet Flyers, they, too, would have been lost and gone inside the earth through the North Polar Opening.

Perhaps the Soviet Flyers did not even attempt to cross the Arctic over the North Geographic Pole, as the above National Geographic map indicates. Their last known position near land was when they passed the tip of Severny Island as they passed out of the Kara Sea. Perhaps, at that point, they set a course directly across the Arctic towards Fairbanks without even attempting to go over the geographic North Pole. In that case, as you can see in the Google map below, their route would have taken them exactly through the North Polar Opening.

Since they would have arrived at a point on the rim of the Polar Opening and determined with their Sextant that they had somehow arrived at the North Pole, they reported that it was a head wind that had slowed them down because they arrived at that point sooner than they had expected.

Since all the rescue flights were based on the assumption that the Soviet Flyers had flown across the geographic North Pole, when in reality they did not, the rescue flights did not find the polar opening because their search flights were in another direction – up near the geographic pole instead of at the location of the polar opening at 85 N latitude and 130 E longitude.

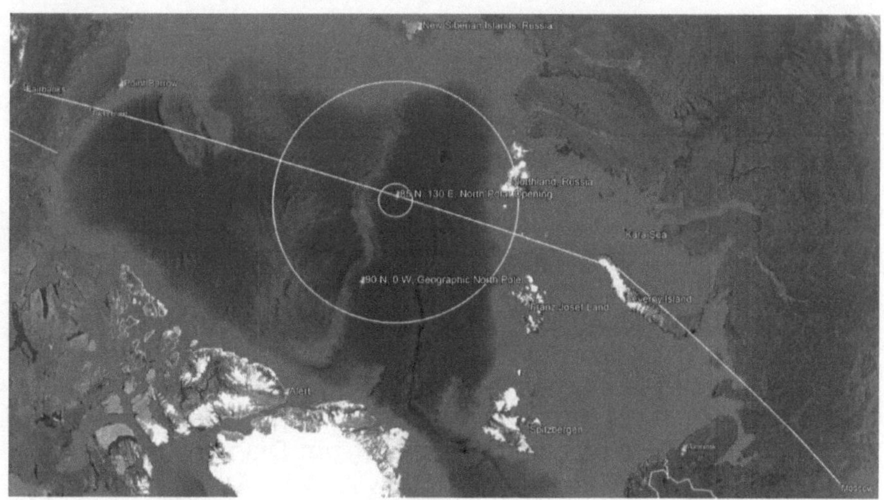

The Soviet Flyers most probable flight path across the Arctic that took them directly into the North Polar Opening.

After about two hours they would have reached the inner continent and so decided to descend to see what they had found. From that point on, their radio signals were received in gradually weaker form as they descended further into the interior of the earth.

German Submarine U-209 Made It to Our Hollow Earth!

I first learned about German U Boats attempting to reach our Hollow Earth from a German living in Canada. The book I ordered was from Samisdat Publishers, Ltd, owned and operated by Ernst Zundel. He wrote a book about Hitler sending an expedition to Antarctica. I ordered his book. It was titled, Secret Nazi Polar Expeditions, published in 1978. It related how the Germans had taken a ship with a small airplane on it that they launched from their ship after reaching the ice near Antarctica. It flew over the ice to Antarctica and then back to the ship. But the book didn't say anything about the hollow earth. I learned later that they WERE trying to go to the hollow earth.

Then in 2006 I received an email from Dianne Robbins, author of several books on the hollow earth. Dianne said in her email that she had just received an email from Joe Watson of Talkeetna, Alaska, who had just contacted her saying that he had a copy of a letter written in the German language from a German submarine crewman named Karl Unger that made it to Our Hollow Earth soon after World War II. The submarine name was U-209 with Captain Heinrich Brodda.

Joe's email to Dianne said,
　　----- Original Message -----
From: Joe
To: telos@rochester.rr.com
Sent: Saturday, August 12, 2006 1:47 AM
Subject: hollow earth
　　"Hi, I have in my possession a copy of a letter written on 2 March 1985 from a gentleman named Karl Unger to Mr. Woodard concerning the submarine u-209 commanded by Heinrich Brodda. Given certain coordinates their mission was to travel to the center of the earth, which they did. The letter is written in German and translated into English. Interested? JW"
I wrote back asking Joe if he could send me the letter. He did. On the next page is the letter Joe sent to me. It is a letter written by a German named Karl Unger which he wrote in his German language after arriving in Our Hollow Earth in a German submarine at the end of World War II trying to escape

the Allies. Joe also included the translation of the letter into English.

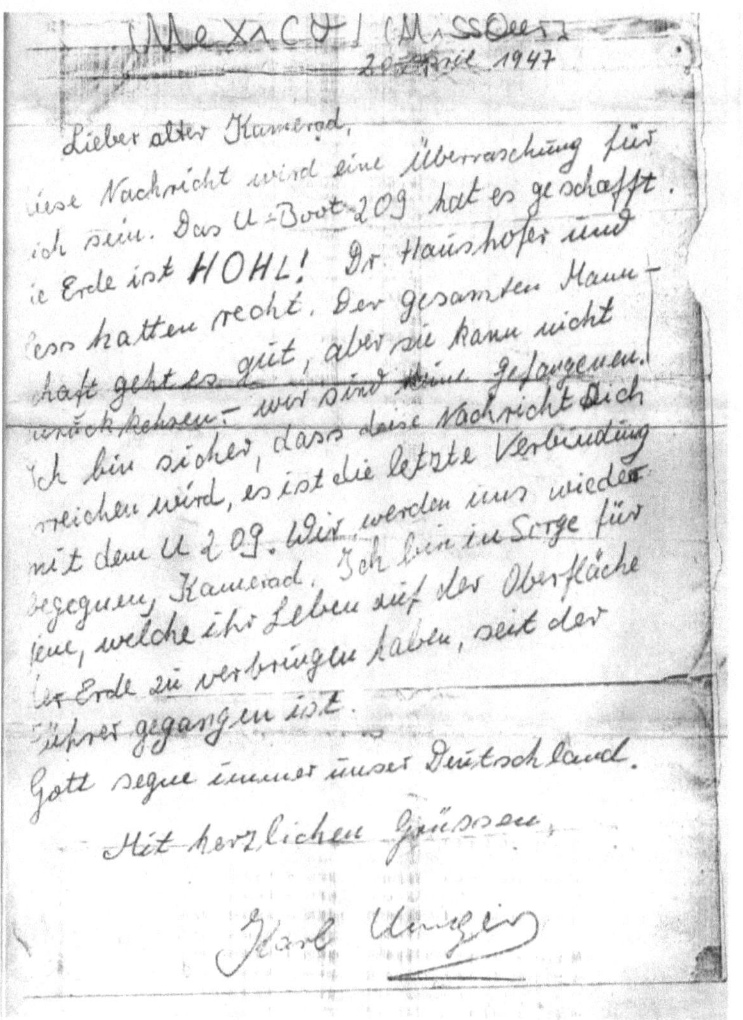

Letter from Karl Unger written April 20, 1947 from Our Hollow Earth to his friend here on our surface world.

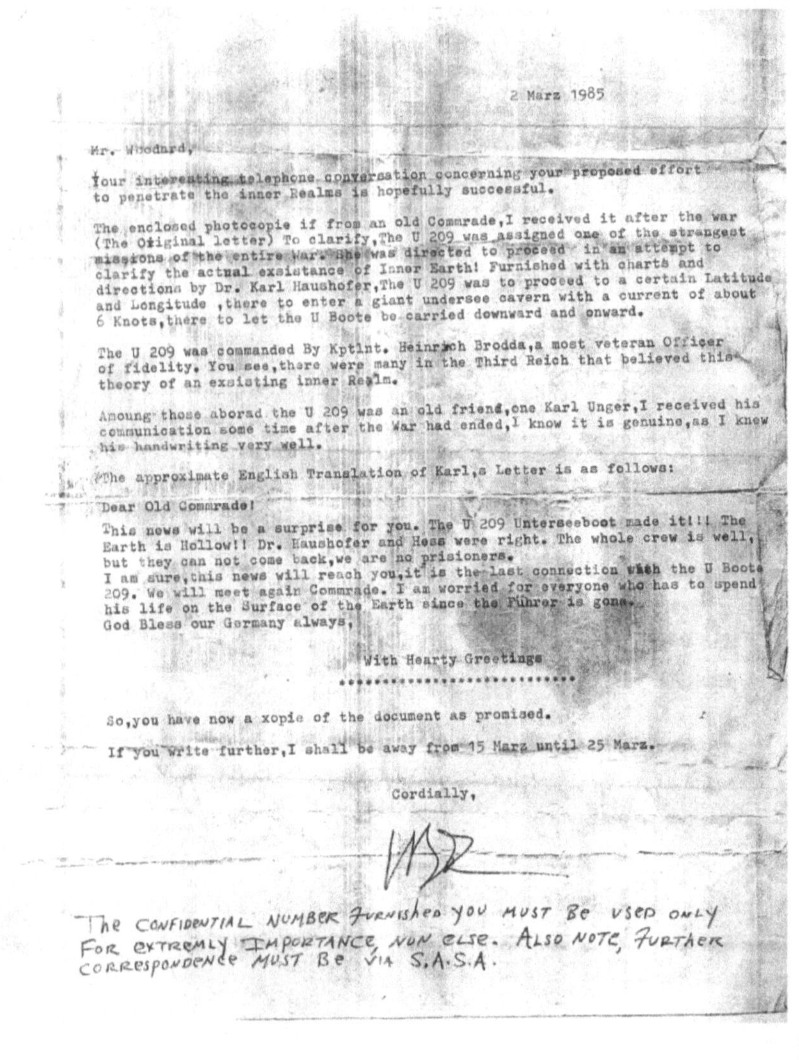

Here is the letter (above) from Karl Unger, crew member of the German U-209 submarine, translated to English. Karl says in his letter that he had reached the hollow earth in the German U-209 submarine. The letter was sent to his friend here on the surface world through a German colony in the Matto Grosso in Brazil that had discovered a cavern that goes to the Hollow Earth, who gave a copy of it to a Mr. Woodard, who gave a copy to Joe Watson.

On July 29, 2009, I received an email from a person named Patrick from Germany who speaks German. He said he researched the German archive website at

http://www.u-boote-online.de/dieboote/u0209.html

looking for information on the missing U-209 German submarine. He says the submarine was indeed reported missing. On July 5, 1943, its last reported position was between Greenland and Iceland at coordinates 52°00' N -38°00' W. The report said the Commander of the submarine was Heinrich Brodda.

The Mr. Woodard that Joe Watson referred to in his email in 2006, I later discovered was Ret. Col. Billie F. Woodard, whom I met in 2008. This is what Joe wrote about him in another email:

---- Joe <watson4@mtaonline.net> wrote:

"Sorry I haven't got back to you, have had massive rain, flooding, power out phones down, etc. Dianne Robbins e mailed me back saying the letter was unreadable on her end. Maybe you can try it on your end as you now have a copy. As to how I have a copy of this letter, well here's the story. In 1986 or 87 this gentleman Woodward came into this town of Talkeetna Alaska, but I don't know why. He met a guy I know named Terry Barber who told me what he was planning on doing. After meeting him he told me the following. His Dad had read all the Hollow Earth books had maps and ocean current directions for different times of the year. He died and his son took up where he had left off. He had it figured out that if you flew to Norway and floated on these currents with large rubber rafts you would be taken into the center of the earth. He gave his car away here in Talkeetna flew to L. A. Calif where he was to meet the rest of his team, from there to Norway where all of the gear was waiting. I asked him how he thought he could get away with this without the Feds picking up on it with their satellites; his reply was they were going in under the pretense of a scientific team doing research. He offered to let me go with them and if I hadn't had two boys to raise I would have. Apparently the letter was to his Dad but I don't remember how all of that came about. Having read many of the Hollow Earth books I always believed it was possible. He also claimed to have the original diary of Admiral Byrd describing his 1700 mile journey into the earth. Anyway he let me get a copy of the letter which you

now have. As to why now after all these years, hell I don't know I was cruising the net and ran across a link about the Hollow Earth and it brought up old memories of that letter. Give it to anyone you want I want nothing in return, maybe a little info if you will share it. If you send me your address I will make copies of the German and English version and send them to you. The same for Dianne. Any other questions will be happy to oblige if I can. Joe Watson POB 643 Talkeetna, Alaska 99676"

I learned about Ret. Col. Billie Woodard in 2005, but at that time I did not follow up on him. I had received an email from our expedition organizer, Steve Currey. It said,

June 9, 2005:

"Rodney:

I have had a retired colonel Woodard call me a couple of times today about our expedition. I am having a hard time answering his questions. He said in his last voice mail that he has some extremely important information to give us about our expedition. Could you return his call? His numbers are 530-926-1062 and 530-926-3529. He lives in Mount Shasta California and has some connection with Telos.

Thanks,
Steve"

Although I did not follow up on this referral from Steve, Mr. Woodard called me in the fall of 2008. We visited over the phone and later visited him at his home in Pahrump, Nevada where he had moved to since he contacted Steve. It turns out that Ret. Col. Billie Woodard WAS the *"Mr. Woodward"* that had gone to Alaska in 1986 as Joe Watson said in his email, and had given him copies of the German letter sent from our Hollow Earth to Woodard's contact. Subsequent to Woodard's visit to Talkeetna, he made his flight to the North Polar Opening in a Grumman Albatross seaplane that you can read about in Billie's biography I have written in this book. See Contents page for his Biography.

I asked Billie where he got the German letter. He told me he got it from Tawani W. Shoush, International Society for a Complete Earth, Rt 1 Box 63, Houston, Missouri 85483. In about 1984, Hollow Earth Researcher, Bruce Walton had sent me a copy of the Secret Diary of Admiral Richard Evelyn Byrd that he had obtained from Tawani Shoush with the above contact information on it. Billie told me that Tawani told him that he had received the letter from his friend Karl Unger sent by Karl after he had reached the hollow earth. Karl had sent the letter from the hollow earth through a German colony in Brazil that had found a cavern that reaches to the hollow earth many years previous. Billie says that Tawani left for Brazil soon after he had obtained a copy of Karl Unger's letter, and may have gone to the hollow earth through that communicating cavern. This German colony in Brazil is documented in the book, GENESIS FOR A NEW SPACE AGE, by John B. Leith (now available on Amazon.com as, GENESIS FOR THE SPACE RACE). It describes how that German colony discovered the cavern back in the 1500's that leads to the Hollow Earth.

I find it interesting that before I was contacted by and met Ret. Col. Billie Woodard, I had received Joe Watson's email which confirmed Mr. Woodard had visited his town in Alaska in 1986, and that Joe had a letter Mr. Woodard had received from Tawani Shoush whom I had learned about from Bruce Walton when I was writing my ebook, WORLD TOP SECRET: OUR EARTH IS HOLLOW! back in the early 1980's. Billie confirmed to me that he had contacted Tawani Shoush before going to Alaska, and this before Billie knew that I had been contacted by Joe Watson of Talkeetna, Alaska in August, 2006, and this before I learned about John Leith's book, GENESIS FOR A NEW SPACE AGE, the unpublished manuscript which Dianne's friend Frank sent me in the summer of 2008, which documents the German settlement in Brazil, that Billie told me Tawani Shoush moved to after Billie had received the German letter from Tawani.

These incidents confirm to me the truth of Karl Unger's letter - that he reached Our Hollow Earth in a German submarine, the U-209, in 1943 after World War II had destroyed Germany, and who wrote back to his friend here on the surface world confirming that the Hollow Earth does exist, and is REAL.

Here are a few pages of Karl Unger's Diary sent to me from Germany showing what he had been planning to do to get to Our Hollow Earth:

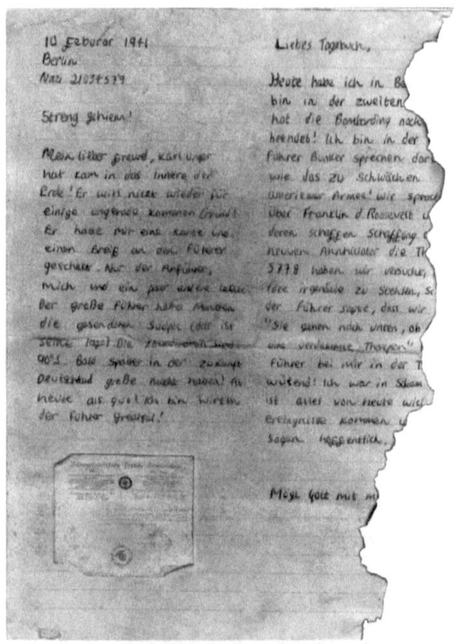

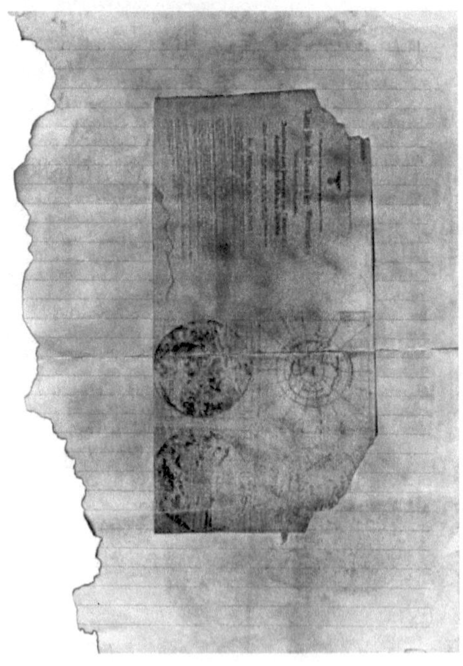

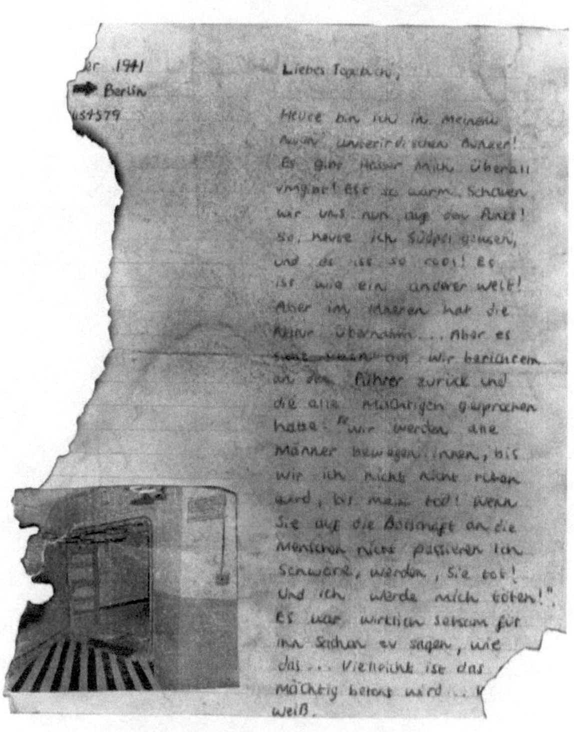
r 1941
Berlin
454579
Liebes Tagebuch,

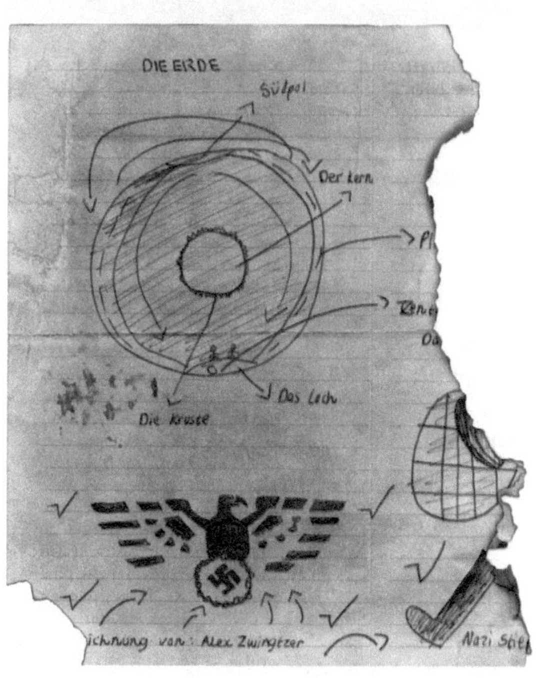
DIE ERDE
Südpol
Der Kern
Das Loch
Die Kruste
Zeichnung von: Alex Zwingezer
Nazi Schi

Biography of
Ret. Colonel Billie Faye Woodard

Photo taken of Billie by my wife Queta (Keta) when we visited him in 2008

**Interviewed and Written
by
Rodney M. Cluff
2008**

Retired Colonel Billie Faye Woodard and twin sister, Zuria were born in our Hollow Earth, and delivered to outer earth on September 18, 1951.

They were born with exceptional abilities. Billie can remember their parents talking to them as babies. Soon after their birth, their hollow earth parents told them they would not be living with them any longer. They were taken to Wichita Falls, Texas and put in a litter barrel garbage container in a park where a passing park attendant and a police officer heard them crying. They were retrieved and put in an orphanage and about 5 years later were adopted by a US Air Force Colonel Woodard, his wife and two sons. Billie's adoptive parents were both of American Indian descent, the mother an Apache, and adoptive father, a Cherokee. Billie's adoptive mother died in 2008 and was cremated in a sacred Apache burial ritual which Billie was able to attend. The adoptive father was still living at the time of this writing, but not doing too well in health.

At the age of five, Billie remembers talking to his sister when they were at a restaurant with their adopted parents and a nearby man exclaimed with great surprise to their parents,

"Do you know what your children are saying over there, in their communication? Do you know that your children are talking to each other in an extinct language? I am a professor of ancient languages at the University of Texas (at Austin) and your children are speaking to each other in the extinct Lemurian language!"

Among the unusual characteristics Billie and his sister had was that they were born with both male and female organs. They were, in fact, hermaphrodites. But Billie later was made a man by his adoptive father who had his female organs removed and sewn up leaving his male organs intact. As result today Billie speaks with a deep masculine voice, and is quite strong. Regrettably, Billie admits that his adoptive father was a pedophile that molested Billie at a very young age on a regular basis, and invited one of the brothers to molest her also, until Billie told some of her classmates at school at the age of 10. The adoptive father would put sleeping pills in his wife's drink at meal time so she would go to sleep while he went up into Billie's attic room to molest Billie, where Billie was kept locked up most of the time. When the adoptive father heard that the word had gotten out at school, he immediately stopped molesting Billie and took Billie to the US Air Force base hospital and had Billie's vagina sealed and the ovaries and uterus removed. With the female organs removed, Billie became only male thereafter. The adoptive father then made Billie dress in male clothing. For this molestation, the adoptive father spent some time in jail and later the Air Force removed Billie from that family and was placed with another Indian family near Apache Junction, Arizona.

Billie's sister was separated from him at the age of six. It happened when they were both taken to a base where they were given all kinds of tests on their exceptional abilities. At one point, Billie decided no more tests were bearable, and so with just mind control, Billie created havoc with the test readings. Soon after this, Billie's adoptive father sold Billie's sister to another Air Force Sergeant for $1 million dollars.

The military told them that the sister later had died in the tests that they had given her, but Billie later learned that she hadn't died, but was taken to an underground base for more testing, and soon after was taken to our Hollow Earth by a hollow earth flying saucer that rescued her from her military captors.

Soon after the adoptive father had gotten out of jail, Billie was coming home from a Scout activity with a friend, and as they neared their home in the countryside, Billie wanted to take a short cut through a corn field, but the friend continued on the road for fear of walking through the cornstalks. After passing through the corn field, Billie was coming close to home when a man in uniform shouted into his mobile phone, *"We have found Billie! Call off the search!"*

Billie wondered what was going on. The parents wanted to know where Billie had been for the past six months, and Billie replied, *"What do you mean? I just came home from Scouts with my friend."* The friend that lived next door had rushed over at the commotion and remarked, *"But that was six months ago!"*

So the adoptive father took Billie in to the military base, and an hypnotic session was run to help Billie recall what had happened in the missing time.

With the help of the hypnotic regression, Billie recalled what had happened that late summer evening when coming home with his friend from the Scout activity. As Billie had walked through the corn field, he had noticed a bright star that seemed to get brighter and brighter until a round metallic shape was visible. When it got closer, Billie could see that it was about 150 feet in diameter and looked like a flying saucer space craft. A strange, but soothing, soft angelic music was coming from it and a pleasant voice asked, *"Billie, would you like to take a journey with us on our ship?"* Billie replied that that would be fun, and immediately began floating up in the air towards the ship and was taken aboard.

The flying saucer occupants were very friendly, but very large in stature. There were several crew members, but a man and a woman were attending to Billie. The woman was about 10 feet tall and the man about 13 feet tall. Billie remarked to them that they were very tall and large people. They said that Billie was a very perceptive person and asked him how he was doing in geography at school. Billie let them know that straight

"A's" were received in school, and even recognized the places they were flying over. As they would fly over each state's capital on the way north, Billie recognized and named off the different state flags. As they reached Canada, where Billie had been to Calgary at one time during the Calgary Rodeo Stampede with a step-brother, Billie recognized the city.

As the flying saucer flew up over the Arctic, Billie was asked about what was seen and where they were. Billie said there was ice and snow but didn't know where they were. They told Billie they were flying over northern Canada and soon they were out over the Arctic Ocean covered with ice floes. The craft then flew through a big hole in the Arctic Ocean into the interior of the earth where they could see an interior sun shining and many cities on the interior surface of the planet. There Billie lived with them for six months. While there, Billie met many people that had disappeared from the surface of the earth, such as pilots lost in the Bermuda triangle. They were now large in stature. After living in the earth for a time as result of the less gravity there, Billie says this allows people living there to grow larger in stature. There, Billie also met his sister again, who told him that in six months they would be sending him back to the surface world. After the six months stay in the Hollow Earth, Billie was returned to Texas and was left in the corn field with no recollection of the trip.

The Air Force later removed Billie from the Woodard family at age 13 and he was adopted by the Henderson family, but with no change of name, at which time they lived near Apache Junction, Arizona. Henderson worked for the military but was not a military man. At that time, Billie attended the Apache Reservation school by school bus and graduated from High School on the Apache Reservation in their advanced gifted program at an early age.

After Billie graduated from High School, and after waiting a couple of years, the new step father signed an approval for Billie to join the Air Force. So after Basic Training of 8 weeks, and then advanced training for 6 weeks, Billie signed an agreement to be stationed in Hawaii, but instead was taken to the Pentagon and told that the next assignment was a top secret facility in the Nevada desert known as Area 51, and that the current rank of Lieutenant was not sufficient for that assignment and so he was advanced to field commission of Colonel.

On the way to Area 51, they boarded a four engine prop plane at Nellis Air Force Base near Las Vegas, Nevada. It was morning when they boarded the airplane and Billie noticed that all the windows were blackened out so they couldn't see out. A short time later they landed and got out of the airplane at Area 51. To Billie, it looked like night time. Billie remarked as to why it was so dark and why they could see no stars. Billie was told they were inside a mountain. They boarded a staff car and soon started down a 45 degree angle incline which was somewhat alarming. Soon the car reached another level and Billie was told to get out and go into a building, and take off all his clothes. Soon a pink mist filled the room. It was some kind of decontamination routine.

With a new set of clothes on, they took the vehicle down another steep incline to another level far beneath the previous level and were told to go into another building where a blue mist this time filled the room for another type of decontamination. Billie was given a new uniform that had a triangle logo with the numbers 51 on it. Outside the patch was another circle with the words, *"Black Project"* and on the bottom of the circle the words, *"Top Secret."*

They then got into an elevator and went down to another level. The elevator descended so rapidly that they were almost weightless. It soon slowed down and stopped and when the elevator doors opened, they were looking at a nice little underground town with people walking here and there. It is located on the 10th level. It looked as bright as daylight just like a little town on the surface world, except they couldn't see very far. There was a sun-like orb in the sky and above that was pitch black. Billie's living quarters were on this 10th level.

Billie was then taken to an assigned office on the 6th level in Archives and relieved the previous on-duty officer, who remarked that he was happy to be leaving. Billie asked him, *"Why?",* but he just shrugged and said that he had had enough. He said, *"You'll find out soon enough why I just want to get out of here and why am glad to be leaving."* Billie sat down at the desk and looked through the files and folders on the desk. The files and folders were classified documentation the military has gathered over the years on Our Hollow Earth. Among the documents, Billie was able to study a 35 page document on Admiral Richard E. Byrd's journeys to Our Hollow Earth through both the north and south polar openings. Billie distinctly remembers reading the exact coordinates of the north polar opening that leads into Our Hollow Earth. A day later a higher ranking officer came and said, *"Your presence is requested on a lower level. You were asked for by name."*

Billie was then taken to still another lower level that opened into a tunnel where a shuttle train waited. The new attendants who greeted Billie were very tall, one was male and the other was female. They seemed to be giants compared to Billie's own height of 5' 11". They were similar to those that had taken Billie on the flying saucer at age 12. They greeted Billie pleasantly in English and invited Billie to step aboard the shuttle train and was asked to strap in. When Billie asked where they would be going, they said, *"Telos,"* a sacred Lemurian city beneath Mt. Shasta, California where Billie met their great High Priest leader, Adama and his wife, Raia and was given a tour of their underground city.

Throughout the next 11 and half years, Billie journeyed three times to our Hollow Earth in shuttle craft through the underground tunnels, and subsequently also to many other underground cities located in the earth's crust. Two trips to our Hollow Earth in the shuttle craft were on official business between our military and the inner earth peoples. The third trip was not on official business, but it was a special trip Billie made to advise the Hollow Earth peoples that it was futile to try to influence our military to become peaceful. At that time, Billie requested permission to stay in the Hollow Earth, but they wouldn't grant permission. They told Billie they had a mission for Billie to carry out on the surface world. That mission is to let outer earth peoples know of the existence of Our Hollow Earth; also that the Hollow Earth peoples want to be known, and that they are a peaceful people. The more that outer peoples know about the Hollow Earth and the peaceful people that live there, then the more successful they hope they will be in being accepted by outer earthlings when the Hollow Earth peoples do decide to emerge openly with the people of the earth's outer surface.

Billie says Agharta is a cavern city and that there are two Shambhala's -- one in a cavern city, and another in the hollow interior. Billie's mission was to interact with the hollow earth peoples and report everything learned to superiors in the Air Force. Billie took many pictures and meticulously documented everything seen and heard, and was debriefed completely after each trip by our military.

From what Billie learned, the hollow earth people had been interested in trying to convince our military to stop their aggressive behavior as well as their atomic testing which was poisoning the atmosphere on and in the earth which could harm people in the interior and on the outer surface of the planet. Finally, Billie convinced the hollow earth people that it was useless to try to get the military and our government to change their aggressive behavior. It was a waste of time he told them. Billie suggested to the hollow earth people that they begin interfacing and contacting the civilian population of the outer earth instead, if they wanted to see a true change on the earth. Billie's contacts agreed and said they would start making more civilian contact in the future.

Billie says that in other journeys within the tunnel systems he visited many cavern cities and many inner earth cities. Billie says that there are tunnels made by the Rand Corporation boring machines, which are not as well made as the Hollow Earth people's tunnels. There are tunnels made by the Hollow Earth people, and tunnels made by cavern cities of other races that live underground, and there are still other ancient tunnels made by the ancient civilizations of Atlantis that was located in the Atlantic Ocean, and Lemuria, which was a Pacific continent called Mu that is now submerged.

Of those tunnels made by surface governments, tunnels join all the capital cities of the world where government leaders can find asylum in case of war or natural disaster. There are many such complexes that exist under all the major cities of the United States, for the same purpose, but which the governments have prepared only for the survival of the government, not the civilian population, in preparation for the close passage of a planet-sized comet that will cause very high winds, earthquakes and tsunamis.

After visiting Telos, the Lemurian city beneath Mt. Shasta, Billie was taken back to Area 51. Later Billie was taken in a tunnel shuttle to the hollow earth.

Billie says that the shell of the earth is from 800 to 850 miles thick, and that the center of gravity is about halfway between the outer and inner surfaces of the earth's shell, and that the inner surface gravity is a little less than one-third the outer surface gravity which allows for the larger stature of the Hollow Earth people, or any surface people that go to the Hollow Earth and stay there for a length of time.

Billie said that the Hollow Earth has an inner sun that gives light for the photosynthesis of life to the Hollow Earth peoples, animals and giant plant life, and that the Hollow Earth country is a most beautiful and harmonious, peaceful place, that even the animals are friendly and able to communicate with humans telepathically and are not aggressive. There are NO meat-eating animals in the Hollow Earth, but all eat vegetation instead. The inner atmosphere, Billie says, is so healthy that no disease can exist there, and if a diseased person coming from the surface would go there, they would be cured just by breathing the air. Before Billie was taken to the Hollow Earth at age 12, Billie had minor childhood sicknesses like colds, and tonsillitis, but when taken to the Hollow Earth, Billie's immune system was fortified to a more healthful state. Since then, Billie's health has deteriorated somewhat due to living on the surface again because of the pollution here, but knows that once he returns home to the Hollow Earth, his immune system will return to a completely healthy state.

When Billie arrived at the Hollow Earth inner surface, they emerged from the shuttle station at the Hollow Earth capital City of Eden, the main city of the Hollow Earth. It is built around and within the original Garden of Eden on the highest mountain plateau of the inner continent. Billie estimates it is located under the State of Arkansas or close to. Billie says that the world inside our earth has one continent and one ocean, but that there is more land inside the earth than on the exterior surface.

Billie agrees completely with the Olaf Jansen story who were Norwegian fishermen that sailed through the North Polar Opening north east of Franz Josef Land, in 1829 -- that the hollow earth is exactly as Olaf described in his book, THE SMOKY GOD -- because Billie has been there and seen what it is like.

At Eden, Billie was taken before the King of the World, who is also the Great High Priest Over All the Land, to a beautiful pyramid type palace where the King sat on his great marble throne. Billie was accompanied by Air Force Colonel McCloud, (and on another subsequent visit with a Colonel Stevenson) were interviewed by the King of the Inner World who asked many questions regarding our outer world, our government and the United States military and what they were up to. Upon returning to Area 51, Billie was completely debriefed and he minutely documented all that was learned in Our Hollow Earth.

Billie recounted a time during the years he was in the military when he was ordered to fly a flying saucer at Area 51, known as the Sports Craft, the same craft Bob Lazar saw at Area 51. An alien co-pilot instructed Billie on how to fly the craft and accompanied Billie on test flights taken out at night. Billie had been told when assigned to Area 51, that he would not see the light of day while serving there. They flew up over Las Vegas during the night. They also flew over other cities and played tag with military jets that were scrabbled to chase their flying saucer. They were able to fly circles around the jets. Billie said the craft was flown by placing the hands on indentations on the control panel that were shaped like hands. Immediately upon placing the hands, Billie had complete control of the craft with just thought commands. If Billie wanted to go in any direction, he just had to think to go in that direction and the craft responded immediately. The craft could make right angle turns at very high speeds, and fly any pattern -- all with no g-force effects. Our military has been building this type of craft for some time. Billie said they had about 67 saucers built by our military black projects and housed at that time at Area 51.

Billie also recounted how the many missing children that have been displaced (as seen on milk cartons) every year throughout the United States are taken to Area 51 where the military black projects are transforming them into programmed biological entities so that they look like hybrid aliens. Their plan is to use the flying saucers they are building and the hybrid entities they are genetically engineering from our children to stage an alien invasion from space to get the whole world to submit to a one world government in defense of the aliens that they have created. When shown this, Billie became so disgusted with the military black projects that he decided not to have anything more to do with them, and decided to get out of the military.

Upon leaving the military, Billie was discharged from the Air Force at the Sewart Air Force base, now Smyrna Airport, Tennessee, debriefed at their deprogramming underground facility, and upon orders of superiors, Billie's military files were sealed with NO access by ANYONE. Billie has tried many times to get a record of his military service, but is told those records are sealed, that no-one can see them. Billie was told by military superiors not to speak or talk to anyone of any assignments in the military. However, Billie told them, "I no longer work for you so I will do as I please." For that, Billie's military pension was withdrawn and today he lives in Sparks, Nevada on Social Security disability.

Over the years Billie married and fathered three children who are now grown. The first two children were boys born to the first wife, an Army military policewoman, stationed at Area 51. They were 5 years together. Later that wife went to work for the FBI and so was able to verify Billie's military records. Billie has been married three times, but is now divorced. The second wife Billie married after being discharged from the military. She was a Cherokee Indian, daughter of the Chief and they were married on the Cherokee Reservation. Billie had a daughter by this wife. The last marriage was in Seattle to a UFO MUFON investigator which lasted from 1990-1996.

In August 1986, Billie went to live in Alaska looking for a way to go to the Hollow Earth to go back home. There, Billie enticed several people as an expedition team in an attempt to reach the Hollow Earth by hiring a bush pilot to fly through the north polar opening. Billie arranged through a flight service out of Fairbanks, Alaska to take his expedition members in a decommissioned Navy Albatross seaplane, which is a twin engine seaplane with a boat bottom. It was the same type of seaplane that the Coast Guard uses frequently today -- as seen on the Fantasy Island TV program. Billie requested the pilot to take them to certain coordinates in the Arctic to 87.7 N Lat., 142.2 E Lon that Billie had learned from Admiral Byrd's file at Area 51.

The pilot was concerned about how they would land out there on the ice, but Billie assured him that there would be open Arctic Ocean when they got to those coordinates. The pilot agreed to fly them to those coordinates, land them on the water and drop them off, and that he would be back for them at a certain time in the future three weeks hence. They took some inflatable power boats that they would use to continue their journey from that point on through the polar opening.

Unfortunately, before they took off from Point Barrow, Alaska, one of the expedition members who was a New York Times correspondent, made a phone call to his home office who Billie believes then alerted the military of their flight north.

When they arrived near enough to the polar opening and could see the light beams of the inner sun shining up from the ocean ahead, the pilot was surprised and very afraid and wondered what was going on since they could see the outer sun shining behind them also. Billie assured him that there was nothing to fear, that they were looking at the core of the earth shining out through a polar opening in the earth. But before they could drop down to land in the water and depart in their inflatable boats, they were intercepted by two jet fighters from the Alert Air Force Base, on the northern shore of Ellesmere Island, Canada.

The airplane radio crackled and a voice came through, which stated,

"This is Lt. Colonel Travis of the US Air Force Alert Base intercepting team ordering you to turn your aircraft around IMMEDIATELY and be escorted back to Eielson Air Force Base, Alaska, OR YOU WILL BE TERMINATED! You have 5 minutes to comply."

Reluctantly, they started to make their turn around. At the same time, looking through the front view of the aircraft, they suddenly saw three glowing disks appear -- flying saucers -- ahead of them.

The radio crackled again, and another voice came over the radio saying,

"Billie, we are here to welcome you to our domain, as you have attempted to enter. We are sorry. You will not make it this time. However, your next journey WILL succeed! Auf Wiedersehen!"

And then the three flying saucers suddenly just blinked out.

Rather than disobey orders and get terminated, they decided to turn the aircraft around and do as the US Air force pilot suggested. They retraced their path back to Alaska, where they were instructed to land at Eielson Air Force Base, just southeast of Fairbanks, Alaska. There they were put under bright lights and interrogated and drilled by several agents of the FBI and the National Security Agency. They wanted to make sure they were not terrorists or affiliated with any foreign government. But after they realized they were harmless, they were released and given a stern warning not to tell anyone of what they saw on this trip, and if they were to try that trip again there would be no warning the next time and they would be terminated.

Billie says that many, if not most people in Alaska, have gone there with the hopes of going to Our Hollow Earth. Among the interesting persons Billie met in Alaska was a retired Air Force Col. Jackson of Talkeetna, Alaska that told Billie that in many of his flights into the Arctic out of Eielson Air Force base, near Fairbanks, Alaska, he saw the North Polar Opening and observed many flying saucers going in and coming out of it like bees out of a hive. Billie also tells of another contact he made in Talkeetna who had a ham radio and they tried many times to glance a radio beam off a hollow earth satellite that looks like a giant rock that is suspended over the North Polar Opening, down into the hollow earth to see if they could make contact with the hollow earth peoples that way. Billie says that one evening after his friend had gone to bed that he was able to finally make radio contact with the hollow earth people with this ham radio set.

Billie says that when his twin sister, Zuria was taken to an underground think tank around the age of 10, by force to test her abilities, she discovered that they were wanting to dissect her body to see if they could figure out why she had such amazing capabilities and characteristics. She then put out a mental telepathy call to the hollow earth for assistance and they came in one of their flying saucers. On a rare occasion that she was allowed to come to the surface, the flying saucer blinked into visibility, and she was beamed up before her captors could grab her -- and she was taken home to the hollow earth. She is waiting there for Billie to return.

Billie is also anxious to return home to Our Hollow Earth soon.

EXHIBITS

Molten Rock Seas, Semisolid Continents
Reported Above Earth's Core

Pulsars — Beacons of Deep Space

NOAA Polar Orbiting Satellites Have a No-Fly Zone Over Antarctica

Airline Pilot Admits the North Polar Opening Exists!

NORAD Photo Shows
Satellite No-Fly Zone Over Arctic

Cassini Spacecraft Shows Saturn's North Polar Opening has a Hexagonal Shape

NASA Movie Short Shows
Earth's South Polar Opening Location

Voyage to Our Hollow Earth Expedition

Hollow Earth Theory 3D HD Version - YouTube

Multi-Colored Snow Falls over Three Siberian Regions

Atmosphere of Saturn's Moon Enceladus is Created by Venting from its South Pole

The Surface of the Sun is Solid,
Indicating that the Sun is a
Hollow Crystal Ball

Hot Spot Detected at Saturn's South Pole

The Holy Scriptures Indicate the Lost Tribes of Israel are INSIDE Our Earth

Hot Spots at the Poles of Jupiter Indicate that Jupiter is Hollow with an Inner Sun and Polar Openings

The New Jerusalem -- Found, in the Sun!

Ozone Holes at both Poles are Evidence for Polar Openings

The Lost Greenland Viking Colony Migrated to Our Hollow Earth through the North Polar Opening

An Ancient Mercator Map Depicts the Lost Garden of Eden inside the Earth Reached through an Opening in the Arctic

Charles A. Lindbergh Discovered the North Polar Opening

Interesting Hollow Earth Emails

THE ORIGIN, CAUSE
and
CONTROL of GRAVITY -- FOUND!

Rodney M. Cluff

Molten Rock Seas, Semisolid Continents Reported Above Earth's Core

Commentary on an
Article in THE ARIZONA REPUBLIC, Phoenix, Arizona
of Thursday, December 10, 1987

The Associated Press reporting from San Francisco as published in the Arizona Republic newspaper of Thursday, December 10, 1987, described new research on the boundary between the Earth's core and the overlying rock mantle which indicates the boundary *"may be an upside-down version of the planet's surface."*

Based on computer analysis of seismic waves generated by earthquakes, blurry maps of this boundary show mountains as tall as Mount Everest poking into the mantle and valleys six times deeper than the Grand Canyon. The maps and studies indicate that there are two layers between the earth's core and the mantle. Because these layers are like upside-down images of continents and oceans on the planet's surface, Brad Hager of California's Institute of Technology and others call them *"anti-continents"* and *"anti-oceans."* Curiously, the anti-ocean layer was reported as lying atop the core's mountains, not in the valleys between the mountains. The next layer up from the core was reported as consisting of the anti-continents.

The scientists describe the anti-oceans as consisting of molten iron and rock, the anti-continents as being *"semisolid,"* and the mountains as made of molten iron. Yet, they don't explain how you can have oceans, continents, mountains and valleys made of molten iron on the core boundary.

Actually, the data describes the inner surface of Our Hollow Earth much better than the molten interior theory does. The depth of 1,800 miles that scientists indicate is the distance to the earth's outer core boundary is a misinterpretation. In reality, it is the distance seismic waves take to rebound from the outer core boundary.

Seismic P compression waves are essentially sound waves. My estimate for the density of the earth's inner shell is 19.68 gm/cc, which is close to the density of gold (19.3), whereas surface rocks have an average density of 2.7 gm/cc which is the density of aluminum. Sound waves travel through gold at 2.01 miles per second and aluminum at 3.93 miles per second for an average velocity of 2.97 miles/second. An outer surface seismic wave rebounding from the inner surface of Our Hollow Earth, which is located at about 800 miles, would be about 9 minutes which is close to the time that seismic waves are estimated to take to reach the outer core of the earth in the orthodox science model, but which in the hollow earth model would just be the distance the seismic wave travels from the outer surface of the Earth to the outer core boundary and back to the surface.

The scientists reported that the anti-oceans lie atop the core's mountains, and not in the valleys between the mountains. This incongruency for an inner surface is solved if this diagram is inverted, where the valleys are called mountains and the mountains valleys. Then the anti-oceans on top of the mountains then become oceans at the bottom of the mountains. The anti-continents would then correctly be the inner surface of the planet, with the anti-oceans resting on that surface, with mountains and continents jutting up out of that ocean towards the center of the earth to complete the inner surface of our hollow planet.

With this seismic imaging, the continents and oceans, valleys and mountains on the inner surface of Our Hollow Earth have now been detected, but just needed to be labeled correctly to describe the inner surface of Our Hollow Earth.

E20   THE ARIZONA REPUBLIC   THURSDAY, DECEMBER 10, 1987

Molten-rock seas, semisolid continents reported above Earth's core

The Associated Press

SAN FRANCISCO — The boundary between the Earth's molten-iron core and the overlying rock mantle may be an upside-down version of the planet's surface, a place where continents of semisolid rock drift atop oceans of molten iron and rock, studies suggest.

"There are the equivalent of oceans of molten rock (and iron) at the core-mantle boundary," with a continent-like layer of mostly solid but flowing rock floating above them, according to California Institute of Technology geophysicist Brad Hager.

The new research, which Hager and other scientists outlined this week at the American Geophysical Union's annual meeting, further complicates the traditional, simplistic picture that Earth's thin crust surrounds a thick, solid-rock mantle, which in turn surrounds a molten-iron core.

At last year's Geophysical Union meeting, scientists from Harvard University and Caltech announced that they used computer analysis of seismic waves generated by earthquakes to make blurry maps of the Earth's interior, much like X-rays in CAT scans make pictures of the inside of the human body.

The crude maps showed that the core isn't a smooth sphere, but has molten-iron mountains as tall as Mount Everest poking into the mantle, and valleys six times deeper than the Grand Canyon.

The latest maps and related studies suggest there are two other layers trapped above the liquid-iron core and below the solid-rock mantle, at a depth of roughly 1,800 miles beneath the Earth's surface, said Don Anderson, director of Caltech's seismology laboratory.

Because these layers are like an upside-down image of continents and oceans on the planet's surface, Hager and other scientists call them continents and oceans or sometimes "anti-continents" and "anti-oceans."

"The boundary between the hot, molten-iron core and the rocky mantle is like what happens in a blast furnace," where various materials settle or rise, depending on their density, Anderson said.

Molten iron is most dense, forming the Earth's core. A lighter mix of molten rock and iron rises to form the oceans, or anti-oceans, atop the core's mountains, not in the valleys between the mountains. The next layer up consists of the semisolid underground continents, or anti-continents, which are less dense than the molten oceans.

Above it is the solid mantle, which is cooler and consists of even-less-dense rock. Continents and the sea floor in Earth's crust are even less dense.

Anderson said the anti-oceans actually sit on top of the core's mountains because they contain less-dense molten rock as well as molten iron, and the mixture rises out of the liquid-iron core.

When scientists discovered mountains and valleys on the Earth's core, they said friction from sloshing of molten iron across those features might explain why the planet rotates with a slight jerkiness that makes a day five-thousandths of a second longer or shorter than 24 hours every decade.

But studies by Hager and University of Colorado scientist John Wahr showed Everest-sized mountains and valleys at the core-mantle boundary would cause 10 times more variation in day length than actually occurs.

So Hager concluded that molten rock-and-iron oceans above the core's molten-iron mountains would smooth out the roughness of the core-mantle boundary, reducing friction so the variation in the length of a day matches the five-thousandths of a second that actually is observed.

Pulsars – Beacons of Deep Space

Pulsars, which are neutron stars, are believed to emit beams of light from their magnetic poles. As the stars rotate, the beams move like beacons from a lighthouse

Pulsars could be better explained by the hollow planet theory. A Pulsar would be a fast-rotating planet with its polar openings located perpendicular to planet's axis of rotation (near its equator). Beams of electromagnetic radiation from the planet's inner sun passing through the polar openings then would sweep through space like lighthouse beacons.

NOAA Polar Orbiting Satellites Have a No-Fly Zone Over Antarctica

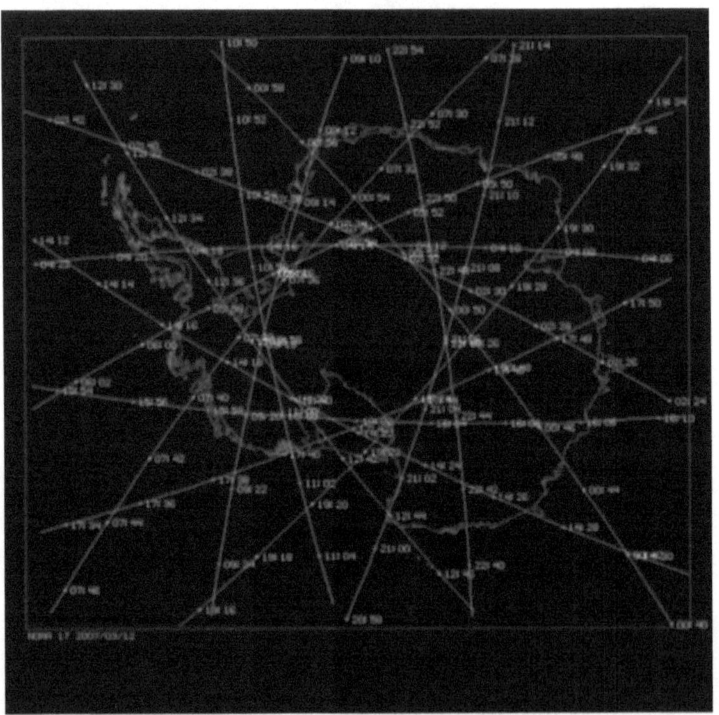

Polar orbiting satellite paths over Antarctica from the National Oceanic and Atmospheric Administration (NOAA) (above) show that there is a large area over the South Pole that polar orbiting satellites do not cross over. Polar orbiting satellites do not cross over this no-fly zone at the poles because the polar openings into the earth are located there. There is no mass in the polar openings and a satellite put over the openings would lose orbit, so no satellites are put in orbit over the polar openings. This narrows the location the polar openings are located -- to within the no-fly zones over the poles. A video was also located that shows a hole in the satellite imagery because of the polar opening here:

http://www.youtube.com/watch?v=f6SU_RpHF2o&feature=player_embedded

Airline Pilot Admits the North Polar Opening Exists!

My friend Lord Ivars, of Henwick, Worcestershire, United Kingdom, one of the first persons that I sold my book to years ago, called me the other day and we reminisced times we spent together talking about the Hollow Earth. He recounted to me the time he met an airline pilot on a flight to his home town in South Carolina. He said he made a point of sitting next to airline pilots that were getting a free flight by wearing their uniform and going as a passenger. This one pilot turned out to have a very interesting story. Lord Ivars, in his usual way, sat down next to him on this flight and started asking him questions. What airline did he fly for? Delta Airlines. Had he ever flown in the Arctic? Why? yes, many times. Did you notice anything unusual up there? At this point, the pilot wanted to know why he was so interested in the Arctic and what was up there. Lord Ivars said that he was just very interested in something unusual up there.

Then the pilot just opened up and said, *"Well, if you're wondering if a polar opening exists up there and if I have seen it, I have. All Arctic airline pilots have seen it. But we're under strict orders from the military through our company not to go around telling people, and if you ever tell anyone I told you this, I could get into a whole lot of trouble. Of course, we can't fly our airplanes over the hole because it would be like going into space."* Click here to listen to Lord Ivars relating this experience on video:

https://www.ourhollowearth.com/Pilot_Story.wmv

NORAD Photo Shows
Satellite No-Fly Zone Over Arctic

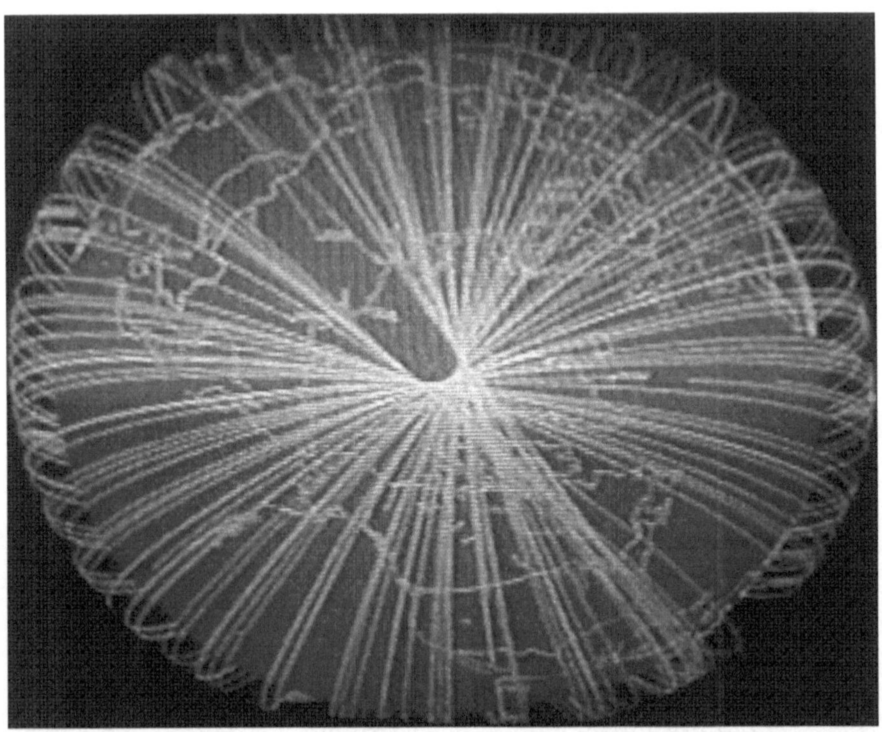

Lord Ivars, who is a Security Consultant, is also a semi-retired Bodyguard now living overseas, had a photo sent to him some time go by an unknown individual claiming to be in the Security Field as well, and apparently was able to snap this image off a computer screen (above) during a visit to NORAD.

No further information was given to Lord Ivars and he never heard from the individual again. This image shows the flight paths of all polar orbiting satellites. Interestingly, it shows that no flight paths go over the location we have determined the north polar opening is located -- on the Russian side of the geographic pole.

Cassini Spacecraft Shows Saturn's North Polar Opening has a Hexagonal Shape

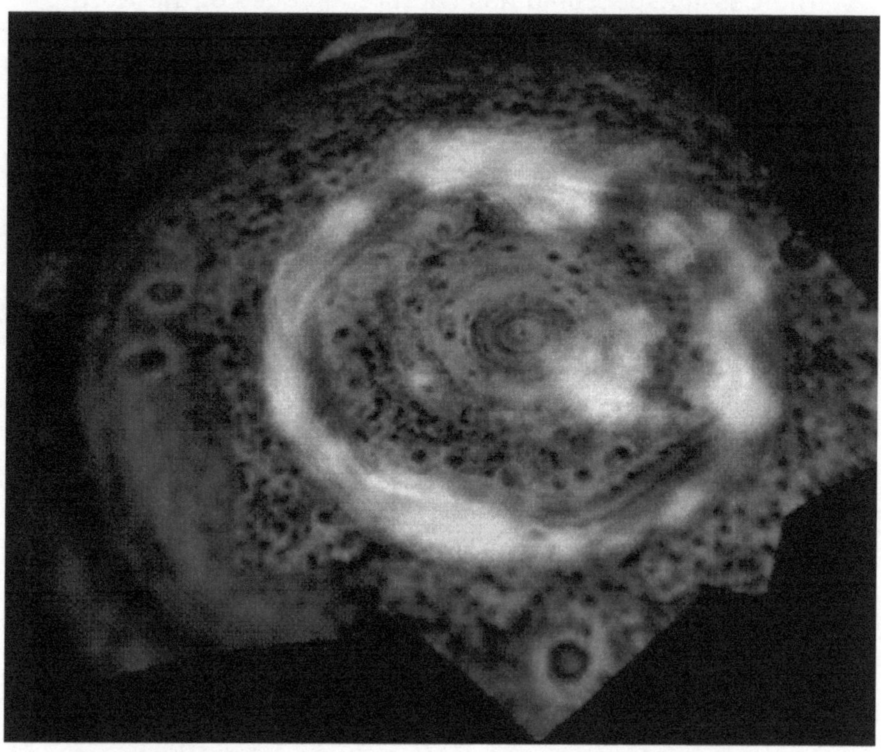

The NASA Cassini spaceship has captured an auroral image over the North Polar Opening of Saturn (above) which mystifies scientists that have yet to realize that the solar radiation that causes the auroras on all planets originates not from our outer sun, but from the planets' interior suns emanating out through polar openings. There are four reasons that the auroras cannot be caused by our outer sun. First, the solar wind from our outer sun is not powerful enough to light up the auroras on any planet including earth. Second, the magnetic field of each planet prevents the outer sun solar wind from entering the atmosphere and lighting up the auroras. Third, the auroral radiation has been observed to emanate FROM the planet following the electromagnetic field lines of the planet AWAY from the planet. Fourth, variations in the auroral displays occur at both poles at the same time strongly indicating that the solar radiation causing the auroral displays originates from the planet's core. Additionally, it is the internal sun of each planet that creates the planet's electromagnetic field as the planet's shell rotates around the inner sun.

The Cassini spaceship has captured images of the North Polar Opening of Saturn which appears to have a hexagonal shape. The fact that this hexagonal shape has endured since the Voyager spacecraft first imaged it over two decades ago indicates that it is a permanent surface feature. This certainly supports my assertion that Saturn has a solid surface with a shell perhaps 5% of its diameter and certainly is not an entirely gaseous planet as orthodox science assumes.

NASA Movie Short Shows
Earth's South Polar Opening Location

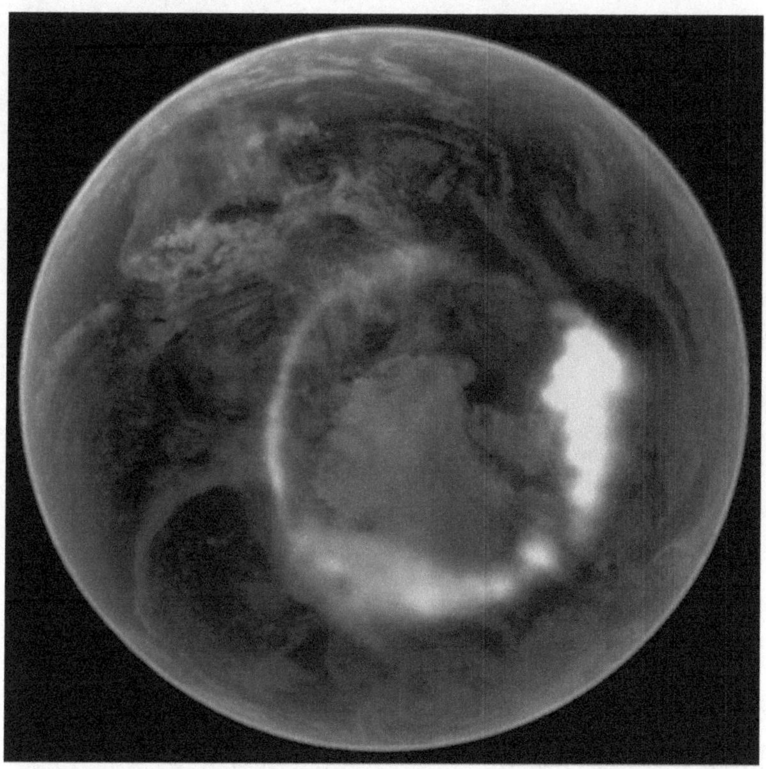

NASA has posted a beautiful image of the Aurora Australis from space. If you look carefully, you can see a dark spot from where the auroral radiation is emanating from -- that area where I have estimated the south polar opening is located. Here is the NASA movie of the Aurora Australis showing the radiation emanating from the polar opening that causes these auroral lights:

NASA's IMAGE Spacecraft View of Aurora Australis from Space | NASA Image and Video Library

Voyage to Our Hollow Earth Expedition

Expedition Member James Still (left) with Steve Currey

In 2003, Steve Currey, of the Expedition Company of Provo, Utah contacted me and asked me to help him put together plans for an expedition to Our Hollow Earth. We named it the Voyage to Our Hollow Earth Expedition. After working several years to recruit expedition members, on May 1, 2006 he discovered that he had brain cancer with 6 tumors, and on May 22 was advised it was incurable. He passed away July 26 and his funeral was held August 1, 2006 in Provo, Utah. Subsequently, against Steve's death bed wishes, the family cancelled the Voyage to Our Hollow Expedition and returned expedition members monies after probate was settled. We were most SAD to hear this fateful news.

You can view our Voyage to Our Hollow Earth expedition website (which was canceled at Steve Currey's death) at:

https://www.voyagehollowearth.com/

With the passing of our dear friend and expedition organizer, Steve Currey, expedition members regrouped and chose a new expedition organizer: Dr. Brooks Agnew, PHD physicist, who renamed the expedition, the North Pole Inner Earth Expedition.

Dr. Brooks Agnew

431

After working hard to promote the new North Pole Inner Earth Expedition for seven years, on September 9, 2013, Dr. Brooks Agnew resigned as expedition leader citing losses to his electric car company, Vision Motor Cars, as the main reason for resigning.

Since then, I have encouraged everyone interested in going to Our Hollow Earth to make the attempt of an expedition to Our Hollow Earth on their own. In July 2018, my wife and I were invited to join an expedition by Andrés Restrepo of Australia, but he has since decided to do it without any publicity. And now in January 2020, Dr. Brooks Agnew has reposted his North Pole Inner Earth Expedition website and announced his intent of making an expedition to the North Polar Opening by the summer of 2022.

A 3D video rendition of the Hollow Earth theory can be viewed at:

Hollow Earth Theory 3D HD Version - YouTube

Multi-Colored Snow Falls over Three Siberian Regions

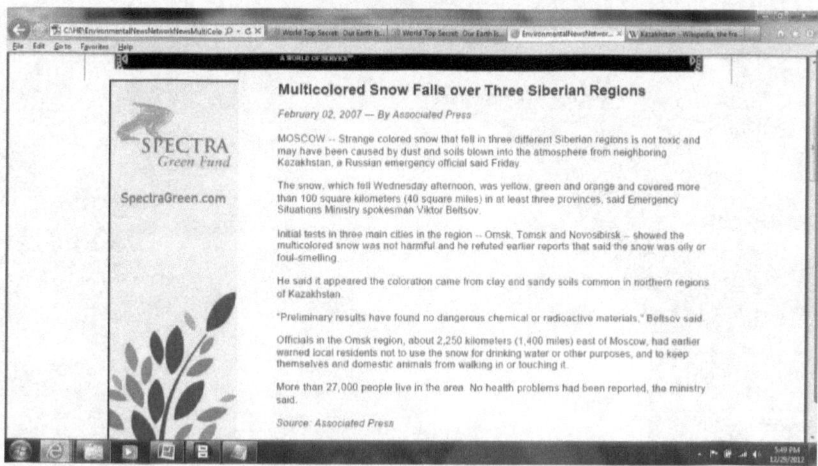

Multicolored Snow Falls over Three Siberian Regions

February 02, 2007 — By Associated Press

MOSCOW -- Strange colored snow that fell in three different Siberian regions is not toxic and may have been caused by dust and soils blown into the atmosphere from neighboring Kazakhstan, a Russian emergency official said Friday.

The snow, which fell Wednesday afternoon, was yellow, green and orange and covered more than 100 square kilometers (40 square miles) in at least three provinces, said Emergency Situations Ministry spokesman Viktor Beltsov.

Initial tests in three main cities in the region -- Omsk, Tomsk and Novosibirsk -- showed the multicolored snow was not harmful and he refuted earlier reports that said the snow was oily or foul-smelling.

He said it appeared the coloration came from clay and sandy soils common in northern regions of Kazakhstan.

"Preliminary results have found no dangerous chemical or radioactive materials," Beltsov said.

Officials in the Omsk region, about 2,250 kilometers (1,400 miles) east of Moscow, had earlier warned local residents not to use the snow for drinking water or other purposes, and to keep themselves and domestic animals from walking in or touching it.

More than 27,000 people live in the area. No health problems had been reported, the ministry said.

Source: Associated Press

On February 2, 2007 multi-colored snow fell over three Siberian regions as reported by the Environmental News Network. This confirms what Olaf Jansen reported in his book that pollen from the great flower fields from the Inner Continent within the North Polar Opening often blow out through the polar opening and fall over northern regions of the earth coloring the snow different colors.

Atmosphere of Saturn's Moon Enceladus is Created by Venting from its South Pole

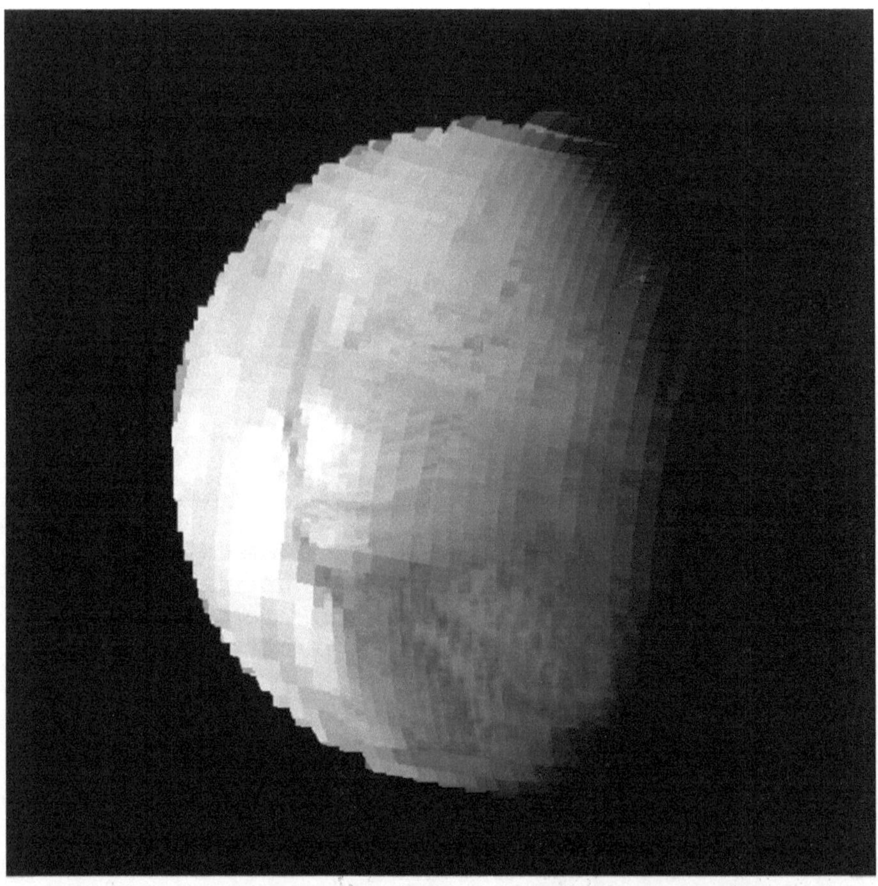

NASA finds evidence that the atmosphere of Enceladus (above), first detected by the Cassini Magnetometer instrument, is the result of venting from the moon's south pole of hot water vapor, most likely from this Saturnian moon's south polar opening.

The Surface of the Sun is Solid, Indicating that the Sun is a Hollow Crystal Ball

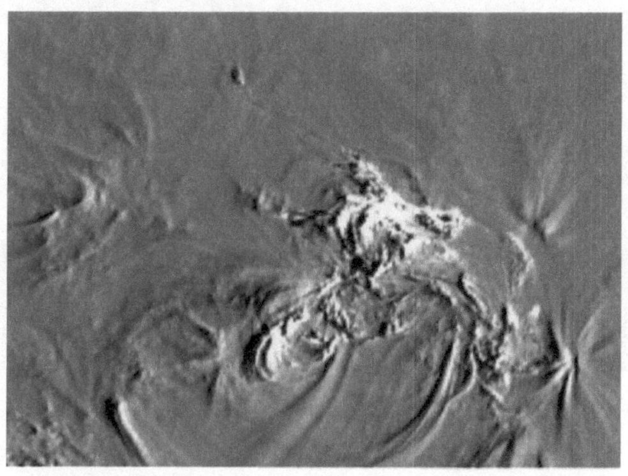

Confirming that the Sun has a solid surface is the scientific work of Michael Mozina. The evidence he presented on his website at thesurfaceofthesun.com (now defunct) from solar observing satellite imagery is incontrovertible. View a video of a solar quake that causes a tsunami on the surface of the sun where you can see that the wave passes over a mountain on the surface of the Sun. See *The Surface of the Sun*.

Video of a sun-quake (Open file in Apple Quicktime)

This surely means that the Sun has a solid surface and as such must be hollow because it could not be solid all the way through. It does not have enough mass for that. My calculations are that if the Sun is hollow with a shell thickness 10% of its diameter, it would have a shell density of 2.86, which is the density of glass with a few impurities. The scriptures indicate that the Sun is a giant crystal ball. Perhaps, soon science will confirm that the Sun is also hollow, with polar openings and a solid core that rotates at a different rate than its shell that produces the observed strong electromagnetic field that the Sun has.

Hot Spot Detected at Saturn's South Pole

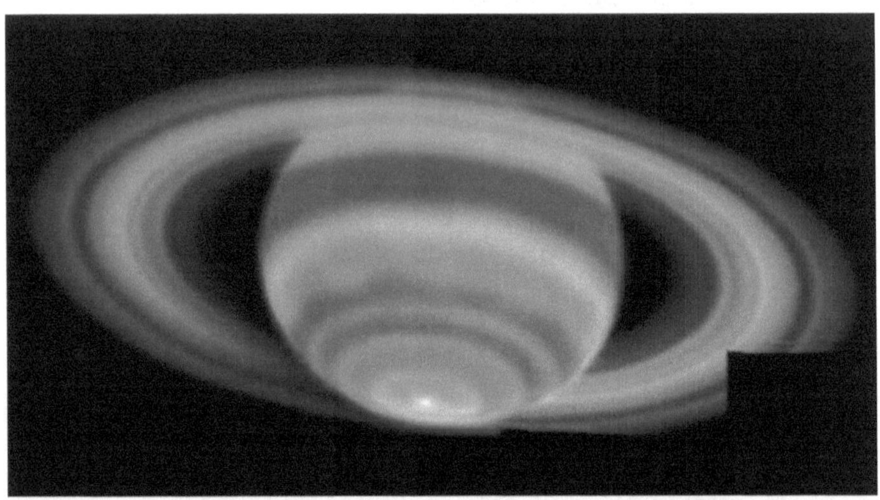

A definite indication that Saturn is hollow with polar openings is the detection of a hot spot at Saturn's South Pole. There is a hot spot at earth's poles also. Fridtjof Nansen reported in his book, FARTHEST NORTH, in 1894 that a north wind invariably raised the temperature on his thermometer in the middle of the winter, and a south wind lowered it. Amundsen on his 1926 flight over the North Pole in his dirigible reported a 10 degree rise in temperature from Spitzbergen to the pole. Obviously, what is happening is the warm air rising up out of the polar opening near the pole is what is raising the temperature at the pole, just like what has now been detected on Saturn. The sudden jump in temperature towards Saturn's South Pole is a surprise for scientists because they are not taking into account the hollow nature of planets and that they contain inner suns that produce their auroras and warm air emanating out through polar openings. (This picture is actually a mosaic of 35 individual exposures made at the WM Keck I Observatory on Mauna Kea, Hawaii, on 4 Feb, 2004. (Image: NASA/JPL))

The Holy Scriptures Indicate the Lost Tribes of Israel are INSIDE Our Earth

Dr. Clay McConkie, with B.A. in history and a Ph.D. in education, declares in his book, THE TEN LOST TRIBES, A People of Destiny, that after a thorough analysis of all scriptural references to the Ten Lost Tribes of Israel that the scriptures indicate that the lost tribes *"...are presently living in a locality not on the earth, but somewhere in the earth..."* in *"...an area capable of accommodating a relatively large civilization of people and keeping them hidden from the world for more than twenty-five centuries."*

He points out that the scriptures indicate that this large area inside the earth must be where most of the water came from that covered the earth's surface at the time of Noah's flood, and that the exit point in the Arctic must also be the entrance to inner earth that the Lost Tribes took when they disappeared into the North Country 25 centuries ago and from which they will someday soon reappear with great fanfare to help spread the Kingdom of God throughout the earth in preparation for the Second Coming of the Lord Jesus Christ.

Dr. McConkie's reasoning, analysis and conclusions of the scriptural references fit the Hollow Earth Theory amazingly well as the most logical present location of the Lost Ten Tribes of Israel.

You can get Clay McConkie's books on Amazon.com

Rodney M. Cluff

Hot Spots at the Poles of Jupiter Indicate that Jupiter is Hollow with an Inner Sun and Polar Openings

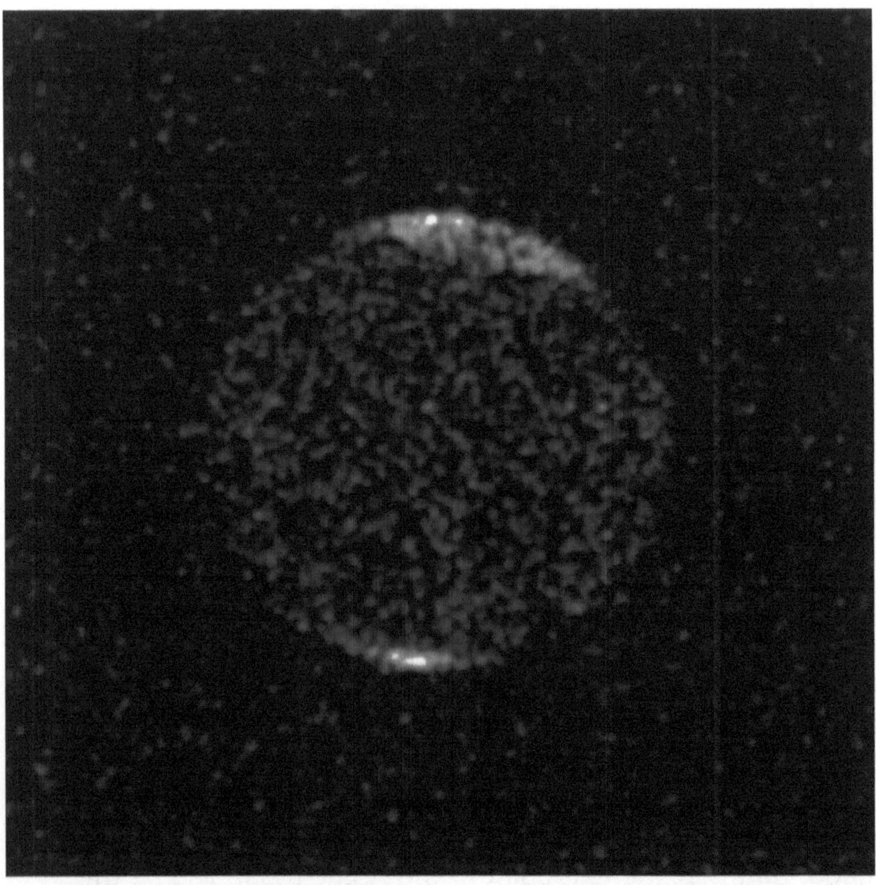

The discovery of permanent emission of x-rays, radio signals, ultraviolet light, infrared radiation and high-energy electrons from Jupiter's poles has scientists wondering where this *"torrent of particles originates."* Actually, this radiation is the signature of a star. This is another evidence that planets are hollow bodies with polar openings through which radiation from their inner suns emanate radiation which causes their auroras to light up and give rise to hot spots at their poles.

The New Jerusalem -- Found, in the Sun!

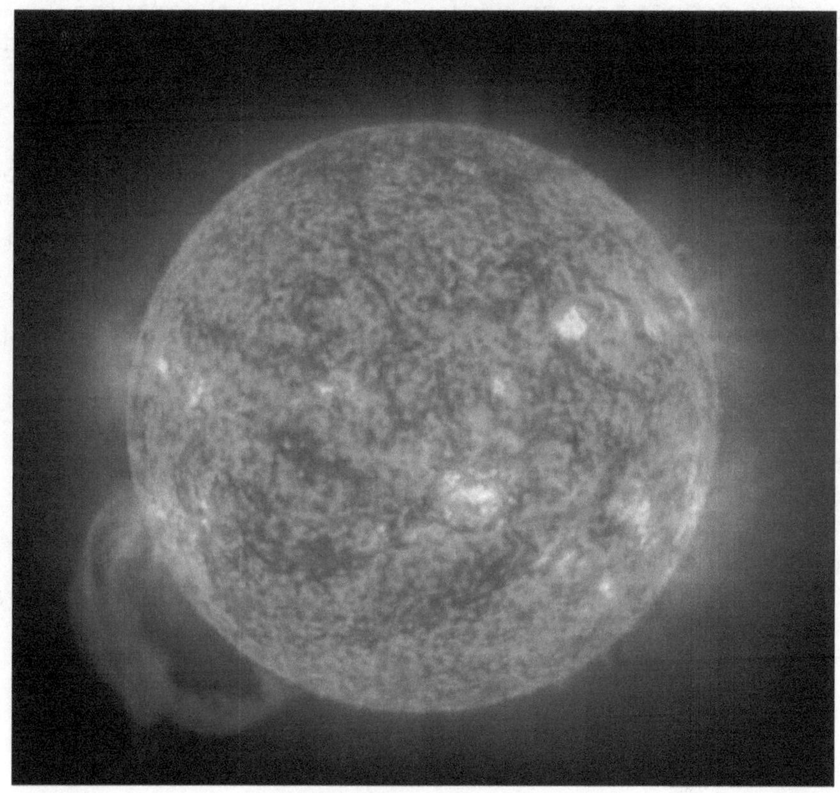

NASA Photo of the Sun
June 9, 2002

In 1933, Phoebe Marie Holmes published a book, MY VISIT TO THE SUN, of her visit to the Celestial City of God in our outer Sun! She discovered that the Sun is hollow!

Phoebe describes how she was taken in the Spirit by angels to the hollow interior of the Sun, to the *"heart of the Sun"* where the New Jerusalem is being built by Jesus Christ, all the holy prophets and the resurrected Saints of God. The New Jerusalem will be brought to earth after the earth's resurrection when the earth becomes celestialized and becomes the abode of the righteous. Since the New Jerusalem is so big, 1,500 miles long, wide and high, it will probably take the place of the earth's inner sun.

Christ, on his Sermon on the Mount, said that the *"meek"* shall inherit the earth, and indeed they will! A mansion is being built there right now for each of us, in the New Jerusalem, by our good actions here on earth. Phoebe was taken by the angels of God to visit her unfinished mansion, where she found her husband, who had already passed on. She then was brought back to earth to her body to finish her life's work.

Phoebe reported that the city inside our Hollow Sun is the New Jerusalem, as described by the Apostle John in the Book of Revelation Chapter 21. Phoebe reported that the New Jerusalem is a giant terraced *"mountain"* with a square bottom -- having a pyramid shape. Perhaps the ancient Meso-Americans, Chinese and Egyptians and Europeans knew the New Jerusalem would have the shape of a pyramid, and so built their temples with that shape.

The angels told Phoebe that the Sun is a giant hollow crystal ball. Scientists, on the other hand, claim the Sun is entirely gaseous. But if it is hollow and has a shell thickness 10% of its planetary diameter such as has the shell of Our Hollow Earth, that would give the Sun's shell a density of 2.86 gm/cc, and so would have a solid surface! Glass has a density of 2.6 gm/cc. It's not hot inside as scientists claim either, but a nice warmth is provided for the celestial city suspended within the Sun's hollow interior. You can get a copy of Phoebe Marie Holmes book, MY VISIT TO THE SUN at Amazon.com.

Ozone Holes at both Poles are Evidence for Polar Openings

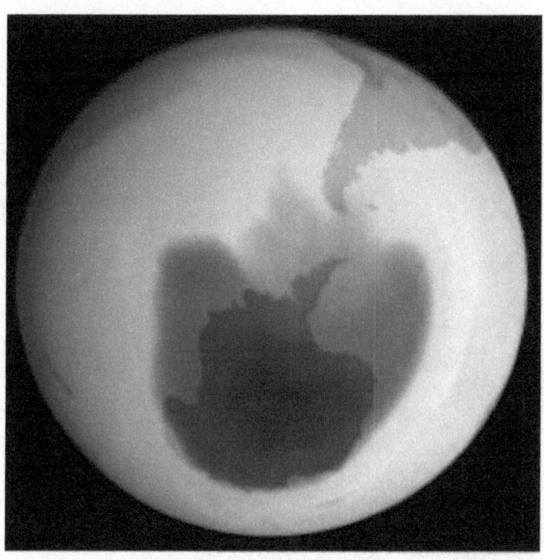

NASA ozone hole images, for example this one released on October 6, 1999, show the area of least ozone (darkest blue) in the same area the South Polar Opening is located in Antarctica.

The ozone layer that covers any sunlit area of the earth is a layer of oxygen isotope (O^3) in the upper atmosphere caused when ultraviolet light from the sun creates ozone from oxygen, and is only a few pennies thick but serves as a protection to life on earth from too much ultraviolet radiation from the sun. Scientists originally considered that the ozone holes that appear in the Arctic and Antarctic each year could be caused by an *"upwelling"* of air that pushes the ozone layer apart creating the "hole," but they had no mechanism that would cause this upwelling of air. The Hollow Earth solves this mystery with ozone free air from the earth's interior that blows out of the polar openings come spring in the Arctic, and fall in the Antarctic, pushing the ozone layer apart creating the ozone holes. The ozone holes have nothing to do with chlorofluorocarbon pollution. The scam to ban the chlorofluorocarbon *"freon"* was cooked up by the DuPont Corporation so they could maintain their monopoly on refrigerants as soon as their patent on Freon ran out.

Rodney M. Cluff

The Lost Greenland Viking Colony Migrated to Our Hollow Earth through the North Polar Opening

An article in the December 1923 issue of Popular Science Monthly magazine reported an Eskimo tradition regarding the fate of the lost Greenland Viking Colonies which were established there in about the 8th century, but disappeared sometime before Greenland was repopulated in 1721. According to the Eskimos, the Viking colonies did not die off as Europeans at that time thought, but instead lured by the more plentiful game, wildlife and driftwood towards the northern shores of Greenland and Northern Canada, upon the return of a hunting party reporting they had found a paradise in the north, all the colonists packed up and singing songs departed suddenly toward the northward across the ice and never returned.

Did this lost Viking colony go to the Hollow Earth via the North Polar Opening?

Mostly likely they did.

Lieutenant-Commander Fitzhugh Green of the US Navy in 1923 at the time this article was published in Popular Science Monthly sincerely believed they would soon find a large lost Arctic continent to which this Viking colony may have migrated, which studies of Arctic Ocean tides and currents indicated exists. The Navy even outfitted a dirigible named the ZR-1 to search out this hidden land, but unfortunately it crashed!

Nevertheless, Lt. Green wrote a novel published in 1925, the ZR WINS, in which he flies in the ZR dirigible across the Arctic from Point Barrow, Alaska, in which his character discovers a large island between Alaska and the pole where the lost Viking Colonists had migrated to. The novel is considered by some as even better than Jules Verne's, JOURNEY TO THE CENTER OF THE EARTH.

The Disney movie, The Island at the Top of the World (30th Anniversary Edition), most likely was also based on Lt. Green's theory that there was an island up near the North Pole where the lost Viking colonies migrated to. But, as revealed by Olaf Jansen, it is not an island, but the inner continent within the North Polar Opening.

You can read Lt. Green's Popular Science Monthly article, *"Will the ZR-1 Discover a Polar Paradise"* at:

http://www.ourhollowearth.com/Page%201%20-%20Popular%20Science%20Article%20-%20December%201923.htm

An Ancient Mercator Map Depicts the Lost Garden of Eden inside the Earth Reached through an Opening in the Arctic

Gerardus Mercator drew this map of Our Hollow Earth's Inner Continent using the best reports he could obtain from the explorers of his day. The map shows the Lost Garden of Eden on a high mountain in the Inner Continent reached through an opening in the Arctic. The abundant waters that flow from the Garden of Eden's artesian fountain divide into four rivers that then flow to the four cardinal points of the compass within Our Hollow Earth. The River Pison flows to our South Pole (their North Pole) towards the left. The River Hiddekel is the river that flows toward the right where it empties into the Arctic Ocean within the North Polar Opening near Northland, Russia.

This Mercator's map of the Arctic first appeared as a vignette in his 1569 world map atlas. Mercator's map inspired explorers like the Englishman Martin Frobisher to seek a northwest passage from Canada to China.

An Ancient Mercator Map Depicts the Lost Garden of Eden inside the
Earth Reached through an Opening in the Arctic

The prophet Isaiah spoke of the Garden of Eden when he
wrote that in the Last Days God, *"...shall set up an ensign for
the nations, and shall assemble the outcasts of Israel, and
gather together the dispersed of Judah from the FOUR
CORNERS of the earth."* (Isaiah 11:12) The only *"four corners"*
that the earth has is the navel of the earth in the Garden of
Eden inside Our Hollow Earth where a great natural artesian
fountain of water issues forth and then divides into four rivers
that flow towards the four cardinal points of the compass. No
such location exists on the earth's exterior, so it must be
located within Our Hollow Earth, as reported by Arctic explorer
Olaf Jansen and confirmed by Colonel Billie F. Woodard.

Charles A. Lindbergh Discovered the North Polar Opening

Another interesting story was told to me by our expedition member who met with Hank Krastman. He told me that his friend Don Tolman, a holistic healer from Utah had been told by the great-grandson of Charles A. Lindbergh that after his great-grandfather's famous first-ever solo flight across the Atlantic Ocean, that Lindbergh made several flights into the Arctic to map out possible northwest passages when he discovered the North Polar Opening.

I estimate this discovery probably happened in 1931.

When he returned to New York, he contacted the U.S. military to tell them of his discovery, but they told him to be silent. Lindbergh didn't agree with this, and so he was about to take his story to the New York Times and publish his discovery of the North Polar Opening to the world. That night his baby boy was kidnapped and he was told his baby would be killed if he didn't remain silent about his discovery of the North Polar Opening.

Interesting Hollow Earth Emails

I am including a few of the interesting emails I have received over the years. The rest you can read on my website at:

https://www.ourhollowearth.com/Emails.htm

From: Jack
Date: February 17, 2010
To: rodneycluff
https://www.bodyguards.com/
Hi,

A little over 5 Months ago gentleman contacted us and requested if we can arrange 2 security personnel for a 4 month travel contract.

The person agreed to meet with me in NYC, he told me that the security guys will travel with a group of 6 Kabbalist Rabbis and 6 Priests and he cannot give specifics of the travel but introduced me to a pilot and someone with a ship captain that both had extreme credentials, so we agreed to send 2 of our top guys with them.

We were told from the start that there will be little or no contact for 4 months as they are exploring a very spiritual thing but rest assured everyone will return safe, I am not a spiritual person at all but I respect those who are and they seemed very legitimate.

On Friday Feb 5th 2010 the 2 security guys came back to NYC and asked to meet me ASAP, what happened next is not a dream or fantasy but rather the most bizarre thing I think I will ever encounter.

They walk into our meeting location, both wearing a religious hat, and they obviously didn't shave. They sit down with me and they tell me that what they are about to share with me is not a practical joke but rather something very serious.

They tell me they converted to Judaism and they will be leaving early as the sabbat starts soon, they rather not continue working on locations but rather only do consulting work as after what they saw in the past 4 months had a serious impact on their life.

Both of these guys like myself where not spiritual people, def. not anything with Judaism.

447

I begged them to give me any info of whats up but they seriously didn't want to share much, then as they were about to leave one of them turns to me and asks "what did the rabbis tell you the sambatyon is?", I seriously never heard the word sambatyon in my life, so I told him I think they should both go see a doctor.

I drove them to some synagogue in NYC, on the way one broke down and cried, so you understand this guy is 6'5 320lbs, tough as nails, now im starting to really question is he went insane.

I demand he tell me whats up.

He brakes down even thou as the other guy says he shouldn't break his promise, he tells me that the past 4 weeks as eye opening as it was for him was literally hell, that he questions if he would have been better off not knowing.

He told me he was in a whole other world, but unlike some hole you mention, it was some travel to a river that from there had a gateway, he said it was the most perfect thing he saw, he did say the people are 10ft plus, he told me there is no thing as money there, and they follow the bible to a T. He said thou they are all jews, no Christmas there and the new year is not when we have new year.

He told me they said that we are getting close to the end of days, that there will be a final war soon and they will come to defend Israel, they explained that in the coming months we will slowly see more countries against Israel and that like when the jews crossed the red sea they had no one to cry to but to god, it will be the same thing, and god once again will help.

He said there is no force on this planet that would be able to remotely win over these people, he saw flying saucers, no gas no electric just based on some secret about levitation.

2 hours he they sat in my car and told me things that where just unreal, but obviously had a effect on them as they changed to a way I never thought these 2 guys will change to.

I don't know if these guys where hypnotized or what, but wanted to share this info.

Good luck in all your work
Jack

After receiving this email, I had one of our expedition members contact Jack and verify what he told me. Our expedition member went to New York and verified everything, and was even invited by the Rabbis to go on their next trip to Sambatyon and said they would contact him when they were ready. But the call never came. Needless to say, our expedition member was very disappointed.

From what our expedition member learned, apparently the Rabbi expedition with these security guards went to Israel, and from there they went by air and boat and then walked into a remote location guarded by an indigenous tribe who let them through where they entered a cavern with artificial lighting. They descended far into the earth to a cavern city of some lost tribes of Israel, but who, like the people that went to the hollow earth, were now very tall.

They stayed with them several months before returning to the surface. They were told that the country of Israel here on our surface world in Palestine are in grave danger of extermination, as more and more countries turn against them, and that the Lost Tribes are prepared to help them. On this trip, the lost tribe colony living in this cavern gave the Rabbis invisibility technology to help the Israelis. It is the same invisibility technology that they use to hide their cavern entrances and is described by Bill, the flying saucer captain who met Larry Foreman outside Los Angeles in 1960. What they do is spray a field of iodide crystals over an area and then display an image on it of what they want people to see. For invisibility, the Israelis could spray a field of iodide crystals in front of a tank and display on it a view from the rear of the tank, so that anybody in front of the tank would only see the view from the tank's rear.

From: Dean D
Date: Friday, November 20, 2009 2:26 PM
To: allplanets-hollow@yahoogroups.com
Subject: [allplanets-hollow] Dennis Crenshaw Interview with Hank Krastman

List members,

This was an interview by Dennis Crenshaw with Dr. Hank Krastman, done way back in 1994.

Dean

JOURNEY TO PALATKWAPI
From: THE HOLLOW EARTH INSIDER, Vol. 2, No. 5: [Exclusive Interview with Dr. Hank Krastman Ph.D] Dr. Hank Krastman

has spent the last 23 years researching and reporting on the Earth's Ley Lines, Vortexes, Men in Black, Crystals and Crystal Skulls and Hopi entrances to the Inner Earth. He has written and published many books and videos on these and other subjects, and until recently published UNEXPLAINED magazine. His continuing work will soon be found in INTERNATIONAL UFO LIBRARY magazine.

Dennis: Dr. Krastman, how did you become involved in your research, particularly in your study of the Hopi Indians and the Inner Earth?

Hank: Please call me Hank. In 1961, while attending classes at the University of Northern Arizona I met a young, blue eyes, blond headed fellow student named Karl Kopavi Waltz [of the Waltz clan of the Hopi nation]. Through conversation I learned that he was a Hopi Indian with a Dutch ancestor and was planning on becoming a teacher.

The name *"Karl Waltz"* came from his Great Great Grandfather *"Jacob Waltz"* who, was from Holland. As I was also from Holland this became our bond. I learned that Jacob Waltz was well known to western history buffs as the *"lost Dutchman,"* holder of the secret of The Lost Dutchman Mine in Arizona's Superstition Mountains. I will go into this in detail in my upcoming book *"Kopavi,"* but in 1875 at the age of 65, Jacob married a 16 year old Hopi girl who was named -- because of her light complexion and blue eyes -- MUHA, the *"Fire One."*

Kopavi and I became close friends and as time went on he told me many things that made such good sense that I wanted to know more. One of the most fascinating things he told me was the real secret of *"The Lost Dutchman Mine."*

Kopavi told me that his grandmother had told him that his Great Great Grandfather, Jacob Waltz had lived with the Pima Indians at the Superstition Mountains. The Indians grew to trust the 64 year old Dutchman and since the Pima's were not allowed to enter the sacred entrances to the underground, they asked him to take sacks of salt to the underground people. Salt is scarce in the inner world and gold is not. Jacob in turn would receive sacks of gold as his reward.

Dennis: So the *"Lost Dutchman Mine"* never existed?

Hank: That's right. It's actually an entrance to the inner lands and, as I learned later, just one of many entrances found around the world. I also was amazed to learn that Kopavi was actually from the underground city of PALATKWAPI.

One day Kopavi asked if I'd like to actually see an entrance to the underground... I was thrilled.

[Note: the Hopis, and Cherokee believe that the Pueblo Indians had in the most ancient times taken up residence within an underground 'cavern' world and after a long period of time emerged once again onto the surface. Some say that reptilian humanoids drove them to the surface, other versions state that some of the underground Hopi turned to practicing sorcery and made things bad for the rest, while other versions state that an underground *"flood"* forced them to seek refuge topside. Some versions say that they emerged through a cavern, other through a mound, and still other versions state that the portal of emergence, the sipapu or sipapuni, lies hidden under a pool of yellowish water. Most of the Hopi versions however agree with the LOCATION of the sipapu... somewhere along the Little Colorado River, a 'short' ways upstream from its confluence with the Colorado River. The sipapu is considered one of the most sacred sites of the Hopi people. - Branton]

First we went to the Grand Canyon where we hired two donkeys. Karl assured me [that] the donkeys were sure-footed, but I wasn't so sure once we started to climb, the trail became small and full of rocks and from time to time I could feel the donkeys hind legs slip on the smooth, loose slate. All I could do was hold on and try to shut out the sound of loosened rock as it clattered down the cliff side to the ground below. Finally we came to a wide spot on the trail and I was able to breathe again. We took our breather on a small plateau from which the panoramic view of the Grand Canyon spread out below displaying the rich earth colors found only in the American southwest. Karl brought out a blindfold and explained that, while he had secured permission from the Council of Nine for me to enter these sacred grounds, he had been requested to cover my eyes from here on out. The rest of the ride into the Canyon was terrifying to say the least. How much time had passed I did not know, I was too busy hanging on to the saddle of the small brave beast.

After what seemed to be an eternity, we came to a stop and Karl removed my blindfold. As my eyes adjusted to the bright sunlight again, I looked around me. We were in a large clearing flush against the canyon wall. Karl motioned for me to climb down from my now, uncomfortable perch, as he did the same. After stretching my tensed up body I turned to Karl who was standing next to the sheet cliff wall. *"Put your hand against the wall right here,"* he said pointing to the solid rock face. I did as he asked, and my hand didn't stop. It continued right through the wall. It seemed that the cliff wall was only a created illusion to keep unwanted people away. Actually it was the entrance to a large secret cave called *"PUPOVI."* As we walked inside I was amazed to find the cave was brightly lit. Before us was a sort of escalator type device which you lay down on. There was some sort of transparent bubble which covered you called a HAWIOVI. My heart was beating fast, so I stood looking at this strange sight as I caught my breath.

Now I have a confession to make. For years I have been telling people that this is as far as I went. I lied. I was told not to reveal any more than this. However, I have now been given permission to tell the rest of the story. Actually, I did proceed from this point all the way into the underground Hopi city of PALATKWAPI. I am putting the whole story in a book to be entitled *"KOPAVI,"* which I am currently working on.

Dennis: Wow! This is quite a revelation. So you took this Hawiovi to Palatkwapi?

Hank: Only part of the way. At the bottom of the escalator, which travels at the speed of light, we boarded a PATUWVOTA or flying machine which moved through subterranean tunnels on energy fields. I will go into detail about all of this in my book.

Dennis: Yes, I realize that your time is almost up and you have another engagement. But, can you answer a few more questions?

Hank: I do have to go... but yes I'll answer a few.

Dennis: Upon exiting the PATUWVOTA what did you see?

Hank: We first went into a large cave room called the PUSIVI and my friend and guide said, *"Welcome to TUWANASAVI, the center of the world."* The large room had many doors all around with strange markings on them. Karl explained that the doors were an added precaution, in case an unauthorized person penetrated the area they wouldn't know which door leads to the city. I understood without him saying anything that you didn't want to open the wrong door. Karl walked up to one of the doors, placed his hand over a symbol in the center and we entered a clean white room with a light purple glow called the POWAMU WUWUCHPI or purification room where all the negativities and bacteria were removed from our bodies, both outside and inside. Only then could we enter the city of Palatkwapi.

Dennis: There is so much I want to ask. I don't know where to start. Is Palatkwapi located inside the subterranean world of tunnels or is it located on the underside of the mantle, or the inner-world?

Hank: It is located on the surface of the inner-world.

Dennis: What color was the sky? Was it red?

Hank: No. The sky is a beautiful blue, no clouds and the sun was stationary in the sky. The air was the freshest I'd breathed.

Dennis: Can you describe what the city looks like?

Hank: Sure. The houses were all built in the Greek style with columns and red tile roofs. The rooms are all open and airy with no glass in the open windows. There are gardens with bright blooming flowers and plants everywhere. It is very quiet, other than the soft rush of wind which was constant. I learned this wind is called APONIVI and is pulled in from the North and South polar openings by machines which are powered by the inner sun, then blown back through tunnels to the surface of the earth in hundreds of places. This system was designed to compensate for the gravitational pull and centrifugal force of the earth and the heat of the sun in order to maintain the gravity and a constant temperature of 76 degrees.

-- end of email –

Expedition Member (left) visiting Hank Krastman (right) in his home in Los Angeles, 2008

One of our expedition members visited Hank Krastman in his home in Los Angeles, in 2008. He learned that while Hank was in our Hollow Earth he fell in love with a hollow earth girl, whom he said was a very beautiful woman. In fact, Hank said that the women of the hollow earth are so beautiful that any single man could not help but fall in love with them. He wanted to marry her, but decided first he wanted to go back to the surface world to get some personal belongings. So his Hopi friend took him back to the surface and told him that when he returned to the Grand Canyon, to just yell out his name and he would come get him. But when Hank returned with his belongings to the Grand Canyon and yelled out his friend's name, his friend never came. Hank was so disappointed that he didn't marry until his 70's. Even then he only married so he could take care of a homeless lady. He retired, moved to the Philippines and passed away in March 2010.

Another interesting side-line on this story is from Billie Woodard. Before I learned of this story of Hank Krastman, Billie told me that one time that he had gone back to visit his step-parent's home in Apache Junction, Arizona. He was visiting with one of the Indian Elders when the Elder invited him to go for a walk up on the Superstition Mountains. After walking past sundown on the mountain up a canyon to a certain place, Billie was so tired he was ready to drop, when the Indian Elder said, *"Wait here."* The Elder then went on ahead and soon came back and said, *"Let's go."* They then walked a short distance up this canyon and came up to the canyon wall. The Elder walked right into the canyon wall leading Billie by the hand. A holographic image of the canyon wall was hiding the cavern entrance.

It was the Lost Dutchman cavern where many years ago Jacob Waltz would enter with his donkeys loaded with salt to take to a cavern city of people living within the shell of the earth, deep down in the earth beneath the mountain, which they paid for with gold. Billie followed the Indian Elder further into the cavern and came to a cavern room filled with cubes of gold, each about 4"x4"x4" in size. It was the Lost Dutchman's stash of gold. After examining the gold, they returned to Apache Junction.

When I told Billie about Hank's story, Billie said he had never heard Hank's story before, and wanted to speak with him. So I put them in contact with each other. Billie later told me that he gave Hank a tip for locating one of these camouflaged entrances: To take a compass and if you manage to get in the proximity of the entrance, the compass needle will start to spin. Hank thanked Billie for the tip.

July 29, 2009 at 07:07:25
Name: Patrick
comments: Hello Friend of the Truth !

I read your site and its very nice and informative, most of the stuff i already known.
I looked for more infos about the shown *"german letter of u 209"* and i found some infos in an german submarine archive :

http://www.u-boote-online.de/dieboote/u0209.html

The u 209 sub is missed since 07.05.1943:
Aus unbekannter Ursache verlorengegangen
"missed in unknown reasons"

the last operation-area was
*"Operationsgebiet: Nordatlantik südlich von Island und südöstlich von Grönland
07.05.1943 Verlust des Bootes"*
north-atlantic, south of Iceland and southeast of Greenland
Position:(52°00'N-38°00'W)
the commander was Mr. Heinrich Brodda, like you said too.
There are 4 years between last contact and the letter from Mexico.
the letter is dated on 20.04 - the birthday of Hitler. Its courios and i think its real. Like the radio-call which was received by british army *"our submarines found an paradise"*
I know that the earth is hollow, i said i KNOW, not i believe.
My Grandfather war in german marine too, he is missed too, no one knows for what reason, like other dozens of german subs.
I'm 36, i never met him. But i dream very often of him, my whole life. He grows older in any dream i have, slowly but older.
He talk to me in my dreams since i was an little boy, he told me that he isn't dead, he told me that he lives inside the earth.
Believe it, or not. its real, i'm not joking or i ain't a gracy idiot.

My live i'm searching for the truth about it, i was in India an talked to an guru bevore dome years, he showed me the entry to hollow earth and he told me that there is an labyrinth in this cave, wich isn't easy to get. He told me that there are barries inside, magical barries, wich cant *"get troght"* by anyone, only the ones which true heart can get inside.

He said there was an guru that lived in the 1930, and at the age of 60 he was going inside this cave and he came out in the year 2001 - not older then before. Its an *"wonder"* or miracle in India, but i forget his name.

He comes back to leave a message for mankind, like thousand messiahs before him, like byrd and others and he get back to inner earth.

Ich will go back to India next year, i will go in this gave and i will find the way through it. Im free at heart, im mixture from christian and buddism and i understand the messages of the old ones, of the gods.

Thanks for your work, and sorry for my bad english, i never learned it in school, i ever understand it like the most languages too. don't know why, there is much stuff i don't know.

but i will find it, that's the reason why i'm here.

From: Dan
Date: December 26, 2004, 4:19 AM
Subject: Good Book

Hi Rodney.

I heard you on George Noory the other night. For some reason your manner and what you said appealed to me.

I just got your book. I thought it would be just some entertaining fiction. (I thought it would be maybe 40 pages long. I surely didn't think 555.) But I can see from looking through it just a little bit that it is really something good.

I have a Ph.D. from the University of Illinois at Chicago in continuum mechanics. I can already see that you put a lot of science into it. The combination of science with religion is very compelling.

One question that occurred to me when I heard you talk, was why the oceans wouldn't drain into the hole at the North Pole. I'm sure you have an answer in the book. (Maybe the ice goes down into the hole, effectively forming a wall. But then if global warming melts the ice, I would speculate that the ocean would flow into the hole.)

I hope that in the future you can publish your work as a regular book. I think you would sell a lot of copies. I would buy one for myself, and others to give as gifts. I think you have a great combination with the subject - along with the old stories, text, calculations, pictures, graphics and photographs.

I think you have devoted yourself to a very interesting and also hopeful topic. And I think it takes courage to propose a theory that 99.9% of people would reject without investigation as being absolutely crazy.

Best wishes, and good luck in the future.

Dan Baron.

1/17/2003

Dear Rodney,

I have just recently finished reading your book and must tell you that it is an excellent work.

You have a rare overview of the actual state of our planet and its inhabitants, and have taken the effort and the risk to communicate it to the public. Your message is vitally important to the welfare of humanity and in a sense this makes you a prophet.

What I find most unique about your book is your ability to identify Heaven as an actual physical location and not just as some ethereal realm. Now, everyone who reads your book, and hopefully others like it, will know what the bible really means when it speaks of *"heaven and earth"*.

Heaven is always an inner, central, and solar location, and there are many heavens, from the sub-atomic to the Heaven of heavens at the center of the Wheel of Creation itself. Our cosmos is in fact a hierarchy of suns or heavens, and the pattern of one is always replicated in the other.

The other thing I find unique about your book is your identification of the lost ten tribes of Israel as the leading power within the hollow earth, and the United States government as their leading nemesis. It is sad but true that most of our outer world leaders have sold out to Satan.

I am also enthralled by your description of where we go to at death. For years I have suspected the same thing but have never actually heard it voiced until I read your book.

I find your description of the center of our galaxy very intriguing, and your geographical understanding of Hell very accurate. LOCATION is so important in the clear understanding of any concept. May I add that the actual dividing line between Hell and Eden-Paradise is the central gravity sphere deep within the earth's shell, and that the interior of the earth's shell in general is popularly referred to as Middle Earth. As you know, it is here, in the first few hundred miles beneath and above our outer surface where the powers of evil have made their home, and where poor humanity is trapped. Happily, I know that these evil beings will one day soon be removed or relocated so that humanity can finally flourish to its glorious potential, unimpeded.

I am waiting anxiously for that day when I will be strolling through Eden once again with not a worry in the world.

Note: There are two important authors or books I have not seen mentioned in your book.

One is Theodore Fitch, who has written extensively on the bible and the hollow earth. His classic *"A Mansion is Built For You in Paradise"* is available at Health Research Books.

The other is a book called *"The Hollow Earth"* by Rudy Rucker, which describes an amazing journey to the south polar entrance and into the hollow earth, and gives a fascinating description of the inner sun.

If you have not read them I urge you to do so.

see you in Eden,

Nick

Date: Mon, 08 Oct 2001 11:37:59 -0600
Subject: THE HOLLOW EARTH/ADMIRAL RICHARD BYRD!
From: Jack
To: Rodney Cluff

DEAR SIR,

I'M 50 YEARS OLD AND HAVE STUDIED THE HOLLOW EARTH THEORY SINCE ABOUT 1974.

I KNOW THE BOOK VERY WELL; THE AUTHOR BEING RAYMOND BERNARD, AND THE ACTUAL EXPEDITION AND MATERIAL PUT TOGETHER BY ADMIRAL RICHARD E. BYRD. I'M ALSO MORMON WITH A BACKGROUND IN SPECIAL FORCES, SEABEES, U.S. NAVY. THE NAVY MAINTAINS THE ANTARTIC DEPT WHICH GUARDS THE OPENING TO THAT PARTICULAR

REGION. IN MY MANY YEARS IN THE NAVY, I'VE MET FOUR CAPTAINS, ONE LIEUTENANT, ONE WARRANT OFFICER AND ONE RETIRED GREEN BERET WHO HAVE ALL SEEN THE OPENING PERSONALLY. THEY HAVE ALL TOLD THEIR PARTICULAR STORY TO ME. THE REASON I AM WRITING YOU THIS E-MAIL IS TO LET YOU KNOW THAT THERE ARE MANY PEOPLE OUT THERE THAT KNOW OF THE HOLLOW EARTH AND THAT ADMIRAL BYRD IS THE TRUE EXPLORER OF BOTH OF THE OPENINGS TO THE INTERIOR OF THE EARTH. I THOUGHT IT PECULIAR COMING ACROSS YOUR INTERNET SITE AND TALKING ABOUT THE HOLLOW EARTH THAT YOU NOT ONCE MENTIONED ADMIRAL BYRD. I PERSONALLY MET HIS DAUGHTER IN LAW, THE WIFE OF THE YOUNGER ADMIRAL BYRD. I WAS ALSO GIVEN A COPY OF ONE OF THE ORIGINAL MANUSCRIPTS. IT'S GREAT THAT YOU'RE PROMOTING THE CHURCH THROUGH THE BOOK OF MORMON AND IT'S TIES WITH THE LOST 10 TRIBES. IN CLOSING, IT IS ONLY A SUGGESTION THAT THE MAN WHO ACTUALLY DISCOVERED THE HOLLOW EARTH SHOULD BE GIVEN HIS JUST DUE AND BE RECOGNIZED FOR HIS GREAT DISCOVERY!

FROM ANOTHER BELIEVER! JACK A.

Jack,

I think I do give Admiral Byrd credit in my book and on my website of his discoveries of our hollow earth and his flights beyond the poles through both north and south polar openings, which discoveries he wanted to tell the world, but which his superiors in the Navy told him not to. Why?

The reason is: The world controllers which control our outer world governments, the scientific and education establishments, and who own the news media outlets have kept the discovery that our earth is hollow as THE WORLD'S TOP SECRET because they do not want to lose their slaves on this prison planet, and do not want all of us fleeing to the terrestrial paradise that Our Hollow Earth is.

Rod

From: (Name removed by request)
Sent: Tuesday, January 1, 2019 11:45 PM
To: rodneycluff
Subject: Suzie talked about you
Attachments: Suzie's Q&A.docx

Hi, Bro. Cluff,

Allow me to give you some background before I go into the main topic.

I'm Chinese from Taiwan and was baptized into the church in Taipei in 1984. I came to the States in 1987 as a graduate student. In early 1990s I got married and raised a family in the San Francisco Bay Area. Towards the end of 2013 while visiting Utah with a former visiting teaching companion, I learned lots of things regarding the last days upon us. I was introduced to a small group of members who sometimes got together to share their insights about the Latter-Day Gadianton Robbers and taught each other preparedness. Though I enjoyed the spirit listening to them in one such meeting, I wasn't convinced the Lord's return was on the doorstep. Several personal dreams and visions were shared with me, and unspeakable evils were shown to me on this website http://itccs.org/

This marked the beginning of my awakening I suppose. After the trip I joined AVOW, an on-line LDS forum where members shared "another voice of warning." I'm not sure if you are aware of this forum or have been a member yourself. In 2014 while reading their weekly digest, I noticed an old post about a woman that claimed to be from the Hollow Earth. That's when I learned about your work. I did copy the post into a word file for further investigation. I'm attaching it here in case you've never read this before.

I discontinued my AVOW membership after a year to focus on my own studies and preparations. I vaguely remember the woman's name was Suzie, so let's just call her that. Following her recommendation, I visited your website to learn about the amazing place she said her people lived.

I've since read a lot, though not everything, from your website. Since I came from a family of aviators--father, brother, brothers-in-law were all trained pilots, I quite enjoyed Admiral

460

Byrd's story. While contemplating moving to Utah to further my preparations for the Lord's return, I read Larry's book downloaded from your website. I now live in Eagle Mountain Utah, Eagle being the symbol of freedom and liberty. The Lord very subtly feeds me pieces of information to teach me the mysteries of His kingdom, as Father has promised me in my patriarchal blessing. Bro. Cluff, I'm now a firm believer of the Hollow Earth. Thank you for your awesome work!

Suzie's Q&A.pdf from AVOW forum.

On Mar 2, 2025, at 12:47□PM, N__ R________
<_____@gmail.com> wrote:
To: rodneycluff
Subject: HE account from my neighbor

Maryanne Ryan my neighbor told me her uncles mission companions brother had a patriarchal blessing that he would teach the lost tribes. In his late 20s he was a pilot and flying by the north pole and disappeared. The last thing anyone heard hi say was "wow it's so beautiful it's so green up here!"

Rodney M. Cluff

THE ORIGIN, CAUSE
and
CONTROL of GRAVITY -- FOUND!

Gravity is an Ether Flow

"The gravity of the centre of the earth, the gravity of the global earth, the solar flood, the air force, the force emanating from the planets and stars, the sun's and moon's gravitational forces, and the gravitational force of the Universe, all together enter the layers of the earth in the proportion of 3,8,11,5,2,6,4,9 and aided by the heat and moisture therein, cause the origin of metals, of various varieties, grades and qualities." -- Vymaanika-Shaastra Aeronautics, Or the Science of Aeronautics, by Maharishi Bharadwaaja, circa 400 B.C., translated and published by G. R. Josner, Mysore, India, 1979.

"God, the holiest of all, through Jesus Christ his Son—He that ascended up on high, as also he descended below all things, in that he comprehended all things, that he might be in all and through all things, the light of truth; which light shineth. This is the LIGHT OF CHRIST. As also he is in the sun, and the light of the sun, and the power thereof by which it was made. As also he is in the moon, and is the light of the moon, and the power thereof by which it was made; As also the light of the stars, and the power thereof by which they were made; and the earth also, and the power thereof, even the earth upon which you stand. And the light which shineth, which giveth you light, is through him who enlighteneth your eyes, which is the same light that quickeneth your understandings; which light proceedeth forth from the presence of God TO FILL THE IMMENSITY OF SPACE -- The LIGHT WHICH IS IN ALL THINGS, which giveth life to all things, which is the law by which all things are governed, EVEN THE POWER OF GOD who sitteth upon his throne, who is in the bosom of eternity, who is in the midst of all things." (D&C 88:5-13)

"The earth rolls upon her wings, and the sun giveth his light by day, and the moon giveth her light by night, and the stars also give their light, as they roll upon their wings in their glory, IN THE MIDST OF THE POWER OF GOD." (D&C 88:45)

"There is no such thing as immaterial matter. All spirit is matter, but it is more fine or pure, and can only be discerned by purer eyes; We cannot see it; but when our bodies are purified we shall see that it is all matter." -- D&C 131:7, 8

Because it is claimed that the Michelson/Morley light experiment at the turn of the 19th century failed to measure the presence of the Ether "wind" blowing past the earth as the earth passed through it, scientists today say they reject the Ether theory. At that time, Nicola Tesla had just tried to build a generator that would power the world with free electricity, but was stopped because his banker, J. P. Morgan said that if he couldn't charge for the electricity produced, he was going to put an end to Nicola's invention. Part of that effort was to put into the school text books that the ether of space does not exist.

The truth is that the Michelson/Morley light experiment did not result in a null value, as is mistakenly claimed. Subsequent to the Michelson/Morley experiment, Edward Morley continued extensive interferometer measurements with Dayton Miller in which an ether wind WAS detected.

In an article at:

http://www.orgonelab.org/miller.htm

titled, Critical Review of the Shankland, et al, Analysis of Dayton Miller's Ether-Drift Experiments, by James DeMeo, Ph.D. of the Orgone Biophysical Research Lab, PO Box 1148, Ashland, Oregon 97520, Dr. DeMeo points out that Dayton Miller conducted a total of over 200,000 individual readings made in over 12,500 individual turns of the interferometer, at many different months of the year, starting in 1902 with Edward Morley at Case University in Cleveland, and ending in 1927 with his Mt. Wilson experiments. Dr. DeMeo insists that the Michelson/Morley experiment of 1887 should cease being referred to as the most definitive work on the question of the existence of the ether and ether-drift. Instead the definitive status belongs to Dayton Miller, as summarized in his paper of 1933 in Reviews of Modern Physics.

Not only did the Michelson/Morley experiment show a slight positive result (and not a "null" or "zero" result as is chronically misreported in the physics literature), but it encompassed only six hours of data-gathering from a grand total of 36 turns with the interferometer -- this was less than 1% of the work undertaken by Miller over more than two decades, which encompassed over 12,500 individual turns of the interferometer at different seasonal epochs and at different altitudes and times of day.

Dr. DeMeo points out that the Michelson/Morley experiment was conducted in a basement with thick stone and brick walls which shielded the experiment from the ether wind. Miller, however, found a greater detection of the ether wind at high altitudes using glass and canvas walls in his observational hut. Miller's measurements showed the ether to be moving along a generally southerly-to-northerly vector relative to the ecliptic. Miller concluded that the Earth was drifting at a speed of 208 km/sec. towards an apex in the Southern Celestial Hemisphere, towards Dorado, the sword-fish, right ascension 4 hrs. 54 min., declination of -70 deg. 33', in the middle of the Great Magellanic Cloud and 7 deg. from the southern pole of the ecliptic. (Miller 1933, p.234)

This is based upon the assumption that the Earth was pushing through a stationary ether in that particular direction. However, if ether is considered to have the attributes of a prime mover, which pushes the planets along on their paths, then we would logically view the detected ether-drift as an expression of a movement of the cosmic ether from Dorado generally towards the northern pole of the ecliptic carrying the Earth along with it as it moved, and this would put Miller's ether-drift measurements in reasonably good agreement with current estimates that the Earth-Sun system is moving towards the star Vega, which is generally close to the northern pole of the ecliptic, almost 180 degrees opposite of the constellation of Dorado.

Dr. DeMeo concludes that in view of the fact that the ether can be entrained by the Earth, that it moves slower at lower altitudes and can be blocked by metals and dense materials, this strongly suggests it has a substantial interaction with matter. And then if ether has mass, then momentum would be imparted from its movement to the stars and planetary bodies imbedded in it. Such a mechanism for planetary movements has a long history, but has its most clear contemporary expression in the theory of Cosmic Superimposition of Wilhelm Reich (1951). Reich argued he had discovered an ether-like energetic continuum in the atmosphere and high vacuum, which was more active at higher altitudes, interactive with matter, and specifically reflected by metals.

Not only does the Ether wind blow past the earth, as Dayton Miller's interferometer measurements show, pushing the earth along with the whole solar system like a giant wheel in vortex fashion in the direction of the star Vega, but it is the assertion of this author that the ether wind also blows INTO the earth producing the effect known as gravity.

Michelson and Miller set up their interferometer experiments with two light beams at right angles to each other so that if the Ether wind wasn't blowing toward one light beam, it might be blowing toward the other. Scientists, as definitively proven by Dayton Miller's extensive work, have been mistaken in their assertion there was no detection of the ether wind from the Michelson/Morley experiment with 36 turns of their interferometer, or by Miller's subsequent 12,500 individual turns of his interferometer.   (See also Dr. James DeMeo's book, THE DYNAMIC ETHER OF COSMIC SPACE, Correcting a Major Error in Modern Science.)

It is suggested by this author that in addition to the two horizontal light beams used in the interferometer experiments that a third vertical light beam be used to detect the ether wind flowing into the earth. Had Michelson or Miller set up a third light beam straight up, they would have discovered that it would have been slowed by the downward pressure of gravity. It is the prediction of this author that a third vertical light beam in the interferometer experiment should detect an ether wind blowing into the Earth producing the effect we know as gravity.

Such an experiment has recently been carried out. On August 14, 2009, Martin Grusenick of Germany performed an interferometer experiment using a laser light beam in a vertical position confirming my prediction of detecting the ether wind flowing into the earth. Upon going up, the interference patterns shift toward the left and when going down the interference patterns shift towards the right which indicates that gravity is an ether wind blowing into the earth in vortex fashion. See his youtube video of the experiment here: https://www.youtube.com/watch?v=7T0d7o8X2-E&feature=youtu.be

Scientists have detected black holes in space, where gravity is so powerful that even light cannot escape it. Instead of being a hole, most likely black holes are planets with such a high gravity that even light cannot escape them. These are probably the prisons of the galaxies, positioned at their centers, where devils and the sons of perdition are imprisoned for eternity for refusing to obey God's laws. These black hole planets also seem to be hollow with polar openings, because astronomers have noticed giant vortexes of auroral energy spewing out at their extremities so powerful that they extend high above and below the galactic plane.

NASA photo of galaxy Centaurus A taken by the Chandra X-Ray Observatory showing a black hole at its center spewing out brightly glowing gases above the galactic plane.

It is a well-known fact that gravity will bend a beam of light. Astronomers report that if a star is just out of sight on the far side of the sun, it can still be seen because the light beam from that star is bent around the sun by the sun's gravity. However, this is not evidence, as orthodox scientists would maintain, of Einstein's Theory of Relativity and his claim that space is curved, and that it is the mass of the sun "pulling" on the light beam. Instead, it could be the ether wind of gravity accelerating into the sun that bends the light beam.

The ether of space flowing into the sun pushing on the light beam toward the sun as the light beam passes the sun could be what gravity is -- an Ether wind flowing into the mass of the sun.

Another indication that gravity is an Ether flow is the fact that a spinning object falls faster than when not spinning. According to present physics, all objects, no matter what shape or size, should all fall at the same rate in a vacuum. The fact that a spinning object falls faster than a non-spinning object indicates that the spin helps the object pass through the Ether easier than when it is not spinning. In a Gyro Drop Experiment, (see Appendix) it was found that a spinning gyroscope fell .333% faster than when it was not spinning.

What present day scientists have misunderstood about gravity is that it isn't a pull toward earth, but a push from space that holds us to the surface of the planet. Among the great scientists of the 20th century who rejected the Ether theory was Albert Einstein. However, it has been noted that shortly before his death, he changed his mind. Why? Was it because he discovered evidence for the existence of the Ether?

The power source of gravity is this: All space is full of an etheric, or spiritual gas called the Ether. Many 19th Century scientists believed the Ether, sometimes spelled "aether," exists, including Sir Isaac Newton, the father of gravitation theory.

In the scriptures, God revealed to his prophets that the Ether that fills all space is called the *Spirit of Christ*, *The Light of Truth*, or *the Power of God*. Joseph F. Smith, in his **Doctrines of Salvation**, Vol I, p. 52, called it a *"substance that fills the immensity of space and emanates from God. It is by this power that man is able to think clearly. Without it, no vegetation would grow, the worlds would not stay in their orbits. We know it is matter, only more refined and pure."* Hebrews 11:1 says that it is a substance, *"Now faith is the substance of things hoped for, the evidence of things not seen."*

Gravity, that mysterious force which holds us to the earth's surface, is the primary evidence for the existence of the Ether. To understand gravity, one must realize that in space there can exist no such thing as an attractive force. How can a particle of matter "attract" another? Whether we are talking nuclear force, magnetic force, electrical force or the force of gravity, there is just no such thing as an attractive force. Particles of matter just do not attract each other.

Isaac Newton refused to believe the attractive theory of gravity as he pointed out in his third letter to Bentley (February 25, 1692) in which he wrote, *"It is inconceivable that inanimate, brute matter should, without the mediation of something else, which is not material, operate upon, and affect other matter without mutual contact, as it must be if gravitation, in the sense of Epicurus, be essential and inherent in it. And this is one reason why I desired you would not ascribe innate gravity to me. That gravity should be innate, inherent, and essential to matter, so that one body may act upon another at a distance through a vacuum, without the mediation of anything else, by and through which their action and force may be conveyed from one to another, is to me so great an absurdity that I believe no man who has in philosophical matters a competent faculty of thinking can ever fall into it."* (THE METAPHYSICAL FOUNDATIONS OF MODERN SCIENCE, by Edwin A. Burtt, 1954, p. 266)

Subsequent to publishing his book, THE FINAL THEORY, Mark McCutcheon proposed two experiments with his friend Roland Michel Tremblay that show that gravity is not an attractive force of matter exerted by the earth.

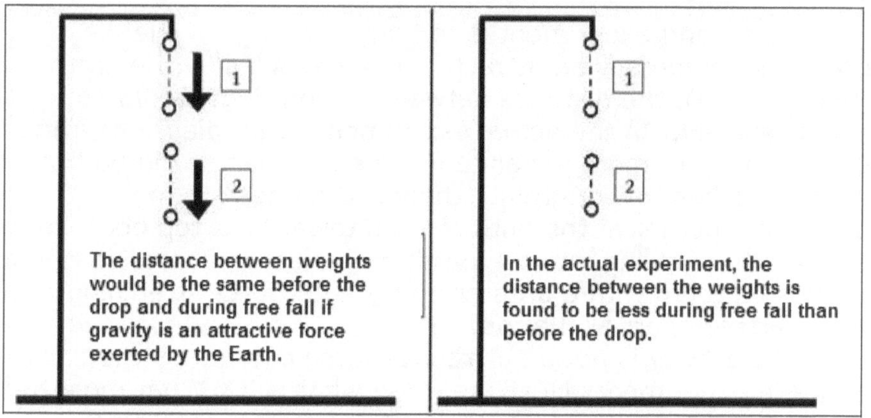

Elastic Band Separating Two Weights Drop Experiment

The first experiment is to hang two equal weights, one below the other, with an elastic band between them, and then let them fall. An attractive force of gravity from the earth (diagram on the left) would pull on each weight equally keeping their distance apart the same before the drop as while in free fall. In the actual experiment (diagram on the right), the distance between the weights is found to be less in free fall than before the drop. This can be explained by the ether flow theory of gravity because as the weights go into free fall, the resistance of the ether that the bottom weight meets as it moves through the ether causes it to slow down decreasing the distance between the weights. Additionally, the ether hitting the bottom weight shields somewhat the top weight from being hit by as much ether flow. The combined effect causes the distance between the weights to be less in free fall than before the drop.

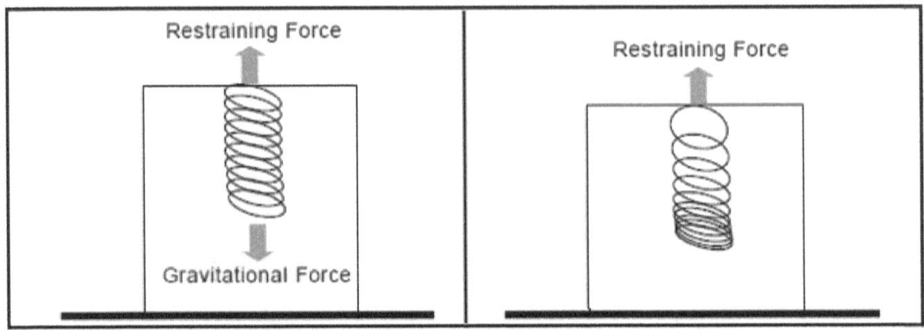

Spring Experiment

The second experiment is to hang a spring. If the gravitational force is an attractive force exerted by the earth (left diagram), the distance between spring rings would be evenly spaced. In the actual experiment (right diagram), the spring rings are more expanded at the top than at the bottom. With the ether flow of gravity theory, the rings are more bunched together at the bottom than toward the top because as the ether flowing into the earth meets each successive ring it pushes on them with a greater and greater force stretching the rings farther apart at the top.

So if gravity is not an attractive force exerted by the earth, as these experiments illustrate, then what is it? What must be realized is that the only way particles of matter can come together is if they are "pushed" together. Therefore, gravity is not a pull toward the earth, but a push from space by the Ether. The pressure of the Ether of space accelerating into the earth is what keeps our feet firmly planted on the earth's surface and is what gives weight to matter on that surface. Without the Ether of space flowing through matter and being resisted by that matter to its flow through it, all matter would be weightless everywhere. By recognizing the existence of the Ether and that its flow into and through matter and the resistance of matter to that flow is what creates the pressure force we call gravity, we can begin to understand what gravity is, where it originates, what causes it to flow into matter and how it can be controlled for the benefit of man.

The great secret to the control of gravity is that electrostatic force can be used to redirect the ether flow. This is done by displacing the positive nuclei of atoms in a non-central direction away from their orbiting electrons. In fact, electrostatic forces can be made equal to or greater than the gravitational force. Expressed mathematically, these two forces are:

$$Fg = G \frac{m \times M}{r^2} \qquad Fe = k \frac{Q_1 \times Q_2}{d^2}$$

The gravitational force Fg is measured in centimeters of acceleration, which at the earth's surface averages 980.665 cm per second squared. The pressure caused by the acceleration of the Ether into the earth is what holds us to the surface of the planet and is equal to the weight of any mass body being measured. The formula states that the gravitation force Fg is equal to the weight of one mass-body in grams times the weight of the second mass-body in grams divided by the distance between their centers in centimeters squared times the gravitational constant G.

Fe is the electrostatic force. Q_1 and Q_2 are charges of opposite sign and **d** is the distance between the surfaces of the charged bodies; **k** is a proportionality constant.

The significance of this is that an equal electrostatic force can counter the gravitation force and overcome it. To do this, it will help to understand the cause of gravity. Gravity flows into the atom and radiation flows out.

First we will consider what electromagnetic radiation is and where it originates.

The Gyroscopic Particle -- The Force of Magnetism

Scientist and inventor Joseph Newman, as he studied the original writings of James Clerk Maxwell, the father of electromagnetic theory, discovered the statement made by Maxwell that in a magnetic field there exists *"matter in motion."*

On page 125 of James Clerk Maxwell: a Biography by Ivan Tolstoy are recorded the words of Maxwell, *"The theory I propose may...be called a theory of the Electromagnetic Field because it has to do with the space in the neighborhood of the electric or magnetic bodies, and it may be called a Dynamical Theory, because it assumes that in that space there is matter in motion, by which the observed electromagnetic phenomena are produced."*

With further study Newman discovered that this electromagnetic energy consists of "matter in motion" in the magnetic field of atoms and consists of extremely small gyroscopic particles which are spinning at the speed of light and traveling at the speed of light. Expressed as a formula, it is: Energy (E) = mass (m) x speed of light squared (c^2), as developed by the famous scientist, Albert Einstein. All atoms contain a magnetic field that consists of gyroscopic particles emanating from the south pole of atoms, continuing on the outside of atoms and then returning in the north pole.

Although never seen, the effects of these gyroscopic particles indicate that they consist of spinning balls of Ether. In his book, THE ENERGY MACHINE, Joseph Newman demonstrated how a gyroscope acts the same way as these gyroscopic particles act when they produce an electric current in a copper wire when the wire is passed through the magnetic field of a horseshoe magnet.

Newman attributes the production of current when a wire is passed through the field at right angles to the flow of magnetic flux, but does not produce current if the wire is passed in parallel through the field, to the spin of these gyroscopic particles. Since the field consists of gyroscopic particles all rotating in the same direction, a current is produced in a wire when it is passed through the field lines at right angles to the lines of magnetic flux because the wire hits the gyroscopic particles as it passes through their line of orbit. When the gyroscopic particles hit electrons in the wire, the electrons are knocked down the wire in the direction of the gyroscopic particle spin thus producing an electric current. However, if the wire is passed through the magnetic field parallel to the lines of magnetic flux, the gyroscopic particles in the magnetic field lines are spinning at right angles to the wire and so cannot knock electrons down the wire, and so no current is produced.

To get current, the wire has to be passed through the magnetic field lines at right angles to the lines of magnetic flux so the spinning gyroscopic particles when hitting electrons in the wire will knock them down the wire in the direction of the gyroscopic particle spin.

The gyroscopic particle, postulated by James Clerk Maxwell and named by Joseph Newman, is the smallest particle of matter. All other particles of matter, protons, neutrons, electrons, etc., are also made of the ether of space in as yet to be defined configurations. With sufficient knowledge of the nature of the ether, and the shapes and functions of each particle of matter, it should be possible to devise a way to construct atoms with it. Instead of building a house, you could "grow" a house.

By understanding the nature of gyroscopic particles, how they originate in the nucleus of all atoms, how they produce what is known as the magnetic field of atoms, and how they can become radiant energy, we can comprehend how they are the essence of magnetic and electromagnetic energy. Indeed, the gyroscopic particle is what all electromagnetic radiation consists of, and the secret of its origin and control will lead us to the cause and control of gravity itself. To do this we will need to look into the structure of the atom and its nucleus where the cause of gravity originates.

How Atomic Nuclei Cause the Ether Flow We Call Gravity

What is the cause of the gravitational force? -- it is produced by a vacuum in the nucleus of atoms. Atoms have the same shape as the earth. They are hollow with a central nucleus radiation generator (a central sun) and polar openings in their electron shell. The radiation emitted by the nuclear radiation generator consists of spinning gyroscopic particles of ether that emit from the south polar opening of the atom and become part of the magnetic field of the atom.

Gyroscopic particles are continually being lost from the magnetic fields of atoms by the collision of free-flying electrons and other particles into the magnetic field of atoms. This causes gyroscopic particles to be knocked out of the field in the form of light and other electromagnetic radiation. It is the theory of this author that gyroscopic particles lost from the magnetic field of atoms are continually being replaced by newly created gyroscopic particles within the nucleus of atoms. If this was not so, atoms would eventually lose so many gyroscopic particles from these collisions that they would lose their magnetic field and become very cold. Our planet would become uninhabitable within a short period of time if gyroscopic particles lost from the magnetic fields of atoms in the form of heat and light were not replaced by newly created ones made from the ether of space.

Newly created gyroscopic particles are formed in the nucleus of atoms from the Ether which fills all space including the space within atoms. The effect of gravity, then, is produced when newly created gyroscopic particles are formed within the nucleus of atoms from the Ether of space. In fact, the ether of space is the prime construct of the universe. Its flow into matter creates the effect we know as gravity. It's formation into gyroscopic particles in the nucleus of atoms creates the magnetic field of the atom.

The secret to the creation of gravity is this: As the nucleus of an atom forms gyroscopic particles from the ether of space and ejects the gyroscopic particles out the south polar opening of the atom, a vacuum is created in the ether in the nucleus of the atom. The ether of space then rushes in to fill this void. As the ether rushes into the nucleus to fill the vacuum created there by the creation of gyroscopic particles from the ether, it exerts a pressure on all matter it passes through. This pressure of the ether rushing into all the atoms of the earth is what keeps our feet firmly planted on the ground. It also keeps the electrons in orbit about the nucleus of the atom, and the gyroscopic particles in the magnetic field of the atom. And when gyroscopic particles are knocked out of the atom's magnetic field by loose flying electrons and other particles, they fly through space in little bunches called photons and create light, heat and radio waves.

After studying Isaac Newton's writings regarding his thoughts on the cause of gravity, Henry C. Warren Jr., on page 9 in his Paper 1, *The Entrained Spatial Medium: Gravitation* (about 2000-2001) came to basically the same conclusion that I have as to what causes gravity, in which hewrote that gravity was the result of *"a condensation causing a flow of ether with a corresponding thinning of the ether density associated with the increased velocity of flow."* (See: https://www.olypen.com/hcwarren/)

In my theory of gravity, I have added that the condensation of the ether occurs in the nucleus of all atoms where the nucleus condenses the ether into gyroscopic particles that then become part of the magnetic field of the atom. The condensation causes a vacuum which Warren and Newton called a *"thinning of the ether density"* in the nucleus of the atom causing what Warren called an *"increased velocity of flow"* of the ether into the atom. The increased flow of the ether into the nucleus of all atoms results in a pressure force we call gravity.

In fact, the ether of space is the prime candidate for solving the illusive Unified Field Theory of physics. Maybe we could take a look at how The Unified Field Theory could be developed using our knowledge of the ether, its flow in and around particles of matter and its transformation into gyroscopic particles of matter.

The Unified Field Theory

The Ether of space, its formation into matter and its flow in and around matter provides the foundation for the development of the elusive Unified Field Theory in which all forces of nature can be linked to one underlying cause.

The premises of this theory are:

A.   Energy is matter in motion.

B.   How that matter moves determines which kind of force it exerts.

C.   The matter that is the underlying cause of all the forces of nature is the Ether of space.

D.   The Ether consists of matter in the form of a tenuous spiritual gas that fills the universe.

E.   All physical matter is made of the Ether.

F.   The Ether's passage through, around and into physical matter is what causes all the forces of nature.

The Unified Field Theory should be able to predict all the observable forces of nature in terms of the ether.

1.   **The force of Gravity** is the pressure of the ether of space flowing in to fill the vacuum in the nucleus of atoms when gyroscopic particles are created by the nucleus from the ether of space.

Atoms are hollow with a nucleus radiation generator, orbiting electrons and a magnetic field:

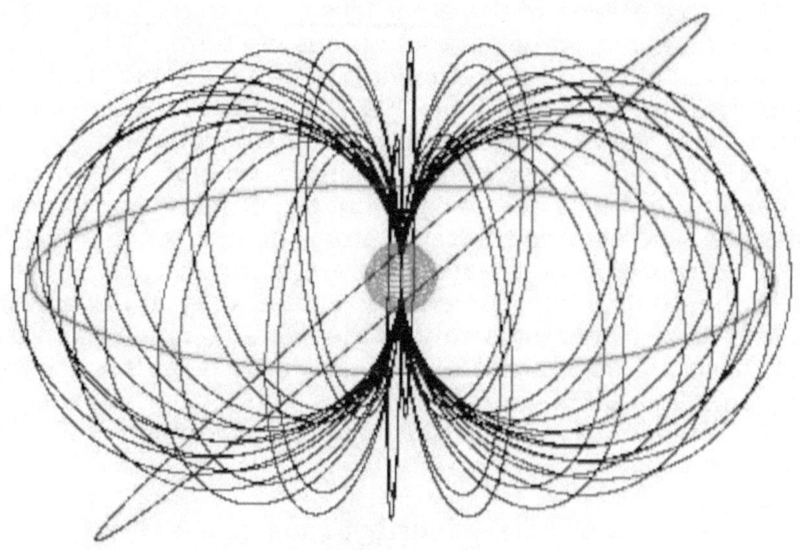

In this diagram, the yellow nucleus radiation generator emits gyroscopic particles that follow the black magnetic field lines as they emit from the south polar opening of the atom. Met by inflowing ether, they then bend around into a polar orbit traveling between the layers of electrons (blue and purple lines). As they approach the north polar opening, the gyroscopic particles are then pushed inside the atom into the nucleus where they are then again pushed out the south polar opening.

All atoms are hollow bodies consisting of a central nucleus radiation generator, electron shells with polar openings and a magnetic field. The nuclei of atoms create the atomic magnetic field by concentrating the ether of space by spinning it into gyroscopic particles which then become the magnetic field of the atom. This concentration of the ether into gyroscopic particles in the nuclei of atoms causes a vacuum in the ether because the gyroscopic particles take up less space than the ether they were made from. Since the ether is a gas, and nature abhors a vacuum, the ether gas surrounding the atom rushes in to fill the void in the nucleus caused by the concentration of the ether into gyroscopic particles. As it rushes in from space to fill the void, the ether exerts a pressure in the direction of the void in the nucleus of the atom. This is the gravitational force: a pressure caused by the acceleration of the ether of space to fill the void caused by the creation of gyroscopic particles in the nuclei of atoms.

The acceleration of the ether falls off as to the inverse square of the distance from the nucleus of atoms. The acceleration of the ether in the vicinity of the atomic nuclei is traveling at the speed of light, 299,796 km/sec. At the surface of the Earth, the average acceleration of the ether is only 980.665 cm/sec². Close-by atoms are causing an acceleration in the ether at the speed of light, but the greatest portion of the ether is accelerating towards atoms deeper in the earth. Because of the greater distance of most atoms in the earth from the surface, the ether is accelerating slower towards those farther away atoms so that the average acceleration of the ether towards all the atoms in the Earth at the surface of the planet is 980.665 cm/sec².

For a planet of radius r, the surface acceleration a of gravity is,

a = K / r²

where K = a planetary constant equal to a * r².

The force of gravity on an object on the Earth's surface is determined by the pressure of the ether accelerating into the earth on that object. The weight of an object is defined as the force of gravity on that object and is calculated as the mass of the object in question times the acceleration of gravity:

w = ma

where:

w = force of gravity on the object, expressed in Newtons

In International System of Units of force (SI), one Newton is equal to the force that would give a mass of one kilogram an acceleration of one meter per second per second, and is equivalent to 100,000 dynes.

1 Newton (N) = 1 kg of mass times the acceleration of gravity, thus 1 kilogram of mass exerts a downward force of gravity of 9.8 Newtons on the Earth's exterior surface.

a = the acceleration of gravity, expressed as meters/second²

For example, on the surface of the Earth, the acceleration of gravity is 9.8 meters/second². The acceleration of gravity on the surface of the Earth is the rate at which the Ether of space is flowing into the Earth. So the force of gravity acting on my body weight of 70 kilograms = 70 x 9.8 = 686 Newtons – the gravity force that keeps my body firmly planted on the exterior ground surface of Earth.

The force of gravity between planets is determined from Physicist Al Snyder's modified gravitation force formula using relative orbital velocities:

F = Sqrt(Sqrt(m/M))/R

More on this will be discussed later. See: A More Correct Gravitation Force Formula

2. **The Magnetic force** is the flow of gyroscopic particles (which are spinning balls of ether) in the magnetic field of the atom, which is kept in the toroidal vortex of the atom by the gravitational ether flowing towards the nuclei of the atom. The gyroscopic particles are created from the ether in the nucleus of the atom and then ejected out the south polar opening of the atom. As the gyroscopic particles emit from the south polar opening of the atom, they are met by inflowing ether which bends the trajectory of the outflowing gyroscopic particles around the atom in a path that takes them in and out of the nucleus of the atom through its polar openings. At the north polar opening of the atom, the gyroscopic particles are pushed back in the atom by the inflowing ether. Depending on their angle of emission from the south polar opening of the atom, the gyroscopic particles flow between layers of orbiting electrons or outside the electron shell.

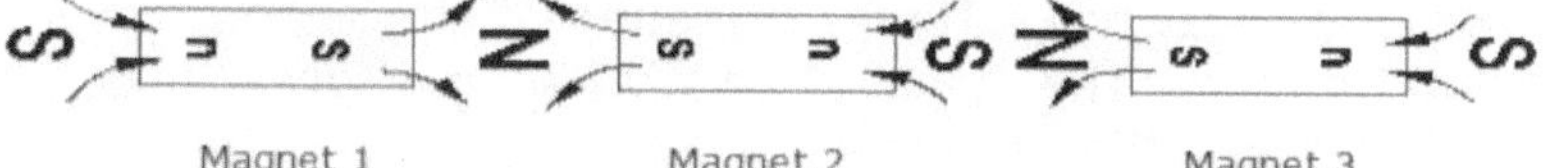

A material that has been magnetized has some or all of its atoms aligned in the same direction. In a magnetic bar, for example, the end commonly designated as the North Pole is actually the end that the south poles of atoms in the bar are pointing towards. Gyroscopic particles emerge from the south polar openings of atoms and are what causes a magnet to repel another magnet when placed North Pole to North Pole. In magnets, the placement of North Pole to North Pole repels because gyroscopic particles emitting from the south poles of the atoms in the magnet push atoms away in the opposing magnet. The gyroscopic particles after emerging from the south polar opening of the atom (North Pole of the magnet) continue to travel around the atom from south to north and enter the north polar opening of the atoms (South Pole of the magnet).

The gyroscopic particles entering the north polar opening of atoms is what causes magnets to attract each other when placed North Pole to South Pole (see diagram). As mentioned before, there exists no such thing as an attractive force. The apparent attraction of magnets is actually a push toward the magnet by gyroscopic particles returning to the nuclei of atoms as they continue their path in and out of atoms. After entering the north polar opening of atoms, the particles then enter the nuclei again where they are again propelled out the south polar opening in the same path as before. This is how the magnetic field of an atom is created.

3. **The Nuclear force** is actually the same as gravity. It is the pressure of the ether flowing into the nucleus of atoms. As the ether flow accelerates towards the vacuum created in the nucleus of all atoms, the ether approaches the speed of light. At the surface of the earth, the acceleration of the ether into the earth is only 980.665 cm/sec^2 and exerts a pressure equal to the weight of mass. But as it approaches the nucleus of atoms, the ether is traveling at the speed of light, 299,796 km/sec. This great speed causes the great pressures found in the so-called Nuclear force. This great velocity of the ether is then redirected by the nucleus into the spin of the gyroscopic particles when the nucleus transforms the ether into gyroscopic particles. Once created from the incoming ether by the nucleus into gyroscopic particles, the gyroscopic particles continue to spin at the velocity of light. The velocity of the ether in the vicinity of the nucleus is also imparted to the velocity with which gyroscopic particles travel in the magnetic field of the atom. The gyroscopic particles continue to travel at the velocity of light, whether in the magnetic field of the atom or whether they are knocked out of the magnetic field of the atom into space in the form of heat, light or radio waves. This velocity of the gyroscopic spin also contributes to the speed of

light with which electricity flows when the spinning gyroscopic particles knock electrons down a conductor.

The Nuclear force is what holds the atom together. It is none other than the gravitational force of the atom as the ether of space accelerates towards the vacuum created in the nuclei of atoms by the creation by the nuclei of gyroscopic particles from the ether of space. As the ether rushes towards the nuclei, it exerts a pressure on orbiting electrons keeping them in orbit. The centrifugal force of the orbiting electrons exactly counterbalances the inward gravitational flow of the ether into the nucleus of the atom. The inflow of the ether towards the nuclei of atoms also keeps the gyroscopic particles in the toroidal vortex in the magnetic field of atoms. Without the inflowing ether, all atoms would disintegrate and disappear. Thus the Nuclear force keeps the atom together.

In Einstein's formula for energy, $E = mc^2$, **m** = the mass of atom, and **c** is the speed of light squared, and refers to the gyroscopic particles in the magnetic field of the atom or after they are knocked out of the magnetic field in the form of light, heat or radio waves. The speed of light is squared because, as Joseph Newman explains in his book, THE ENERGY MACHINE, the gyroscopic particles are traveling at the speed of light and at the same time are spinning at the speed of light.

4. **The Electrical force** is the flow of electrons between outer electron shells of atoms. These outer electrons are knocked out of their orbit about the nucleus by gyroscopic particles in the magnetic field of a neighboring atom which has invaded the space of the other atom. The spin on the gyroscopic particles in the magnetic field is what knocks the electrons out of orbit. The electrons are then propelled in the direction of the gyroscopic particle spin. When the gyroscopic particles in a magnetic field hit electrons in a wire, for example, the electrons are knocked down the wire in the direction of the gyroscopic particle spin thus producing an electric current.

Superconductivity is a method of reducing the size of the magnetic field of atoms to facilitate the flow of electrons in a conductor. Atomic magnetic fields are reduced either by cooling or locking their orientation in a ceramic so that electrons can pass more easily between atoms without passing through orbiting gyroscopic particles in the magnetic fields of neighboring atoms.

This breakthrough of recent years in improving superconductivity was in consultation with Joseph Newman. His theory of gyroscopic particles explains how superconductivity works. In a conductor, electrons pass from atom to atom when a voltage is placed on a circuit. Heat is produced in the conductor because as electrons jump from atom to atom they have to pass first through their own atom's electromagnetic field and then through the field of the atom they are jumping to. As these electrons pass through the fields lines, they impact with gyroscopic particles in the magnetic fields and knock gyroscopic particles out of the field in the form of heat, light or radio waves.

Superconductivity comes into play when the magnetic fields of atoms are reduced in size by the extraction of gyroscopic particles in the cooling process. If the atoms are cooled down enough to eliminate several layers of field lines, then electrons passing from atom to atom in a current don't have to pass through or hit gyroscopic particles in the magnetic field of neighboring atoms. Electricity is thus allowed to transmit without generating heat which makes those circuits super conductive -- the advantage being that lower voltages and smaller wires are needed to move more current, and in appliances like computers no fans are needed to keep the heat from burning the wires and components in a circuit. An advantage is also produced when superconductivity is reached because without each atom's magnetic field interfering with the orientation of surrounding atoms, all atoms align as in a magnet and their combined fields produce a much greater overall magnetic field of the object the atoms are in. It is also by reaching the super conductive threshold that gravity control is achieved.

5. **Electromagnetic radiation** -- radio, heat and light radiation consist of groups of gyroscopic particles that have been knocked out of the magnetic fields of atoms by loose flying electrons, neutrons or protons that hit the atom.

In the magnetic field of atoms, the gyroscopic particles are orbiting in and out of the nucleus of atoms in streams of single entities as opposed to group configurations as they are found in protons and other particles. Heat, light and radio waves also are composed of groups of gyroscopic particles that are more loosely grouped than as found in protons, neutrons, etc. Heat, light and radio waves are produced when atoms are struck by free flying electrons, protons, or other particles. As a free flying particle enters the magnetic field of an atom, it passes through the field lines at more or less right angles. As the invading electron or particle passes through the magnetic field lines, gyroscopic particles in each field line layer are knocked out of the atom by the invading particle. These gyroscopic particles are then knocked out of the magnetic field of the atom in small, loosely configured bunches, called photons. One photon is emitted from each field line passed through by the invading particle.

Therefore, electromagnetic waves consist of bunches of gyroscopic particles which have been knocked out of the magnetic fields of atoms by invading particles.

All the characteristics of heat, light and radio waves can thus be explained by the spin of the gyroscopic particles traveling in bunches through space. For instance, the particle and wave characteristics of light which have eluded explanation by scientists for so long can now be explained by this gyroscopic particle theory. The frequency of light waves is determined by the distance one bunch of gyroscopic particles is from the next bunch. The intensity of the wave is determined by how many gyroscopic particles are in each bunch. Each bunch of spinning gyroscopic particles in a light beam is a photon. The wavelength of light is the distance one photon is from the next one. The frequency is the quantity of photons passing a certain point per unit of time.

Other characteristics of heat, light and radio waves can also be explained by the spinning nature of the gyroscopic particles. The diffraction of light is caused by the spinning nature of the gyroscopic particles which when they enter a transparent or translucent material are deflected when the impact of the particles throws them away from their direction of spin. Polarization of light occurs when only those gyroscopic particles that are spinning parallel to a slit pass through. Those spinning at right angles and other angles to the slit are deflected away from the opening because of their spin orientation. Interference patterns are produced by particles being deflected at different angles because of the different angles of spin when the particles contact an opening.

Heat is electromagnetic radiation of a wavelength between radio waves and light, called infrared. Heat is produced in the same manner as is light -- electrons passing more or less at right angles through the magnetic field of atoms knock gyroscopic particles out of the field with the frequency of infra-red light, which is heat. When heat is generated by rubbing, electrons in the outer shell of neighboring atoms are pushed through the magnetic field of neighboring atoms causing gyroscopic particles in the magnetic fields of both atoms to be knocked out of their magnetic fields with the frequency of heat. The lower frequency waves like heat waves result from knocking gyroscopic particles out of the outermost magnetic field lines of the atoms because the field lines are farther apart on the outer reaches of the atom. Higher frequency radiation is generated by knocking gyroscopic particles out of the inner field lines of atoms that are closer together. Heat produced by a fire results from electrons lost in the chemical reaction of the fire impacting other atoms causing light and heat to be emitted with varying frequencies all the way from radio frequencies to heat and light.

Different frequencies are emitted because as the invading particles pass down through the layers of the atom's magnetic field, it first passes through the outer layers which are spaced farther apart than layers closer to the nucleus. So radio waves with large wavelengths would be generated first, then heat waves with shorter wavelengths and as particles reach deeper layers, light with even shorter wavelengths and higher frequencies are emitted -- the different frequencies and wavelengths caused by the distance the layers of magnetic field lines in the atoms being invaded are from each other.

6. **The Electrostatic force** is produced by the redirection of the Ether in the proximity of electrons and protons by virtue of their spin. Joseph Newman illustrates in his book that particles spinning in the same direction repel each other as they come in contact with each other's periphery because balls spinning in the same direction are actually moving in opposite directions at their point of contact thus knocking each other away from each other. Their spin also causes the Ether to flow in the direction of their spin in the proximity of their surfaces. Thus same spinning particles repel each other because their spin sets up a high pressure zone in the ether between them which pushes them apart.

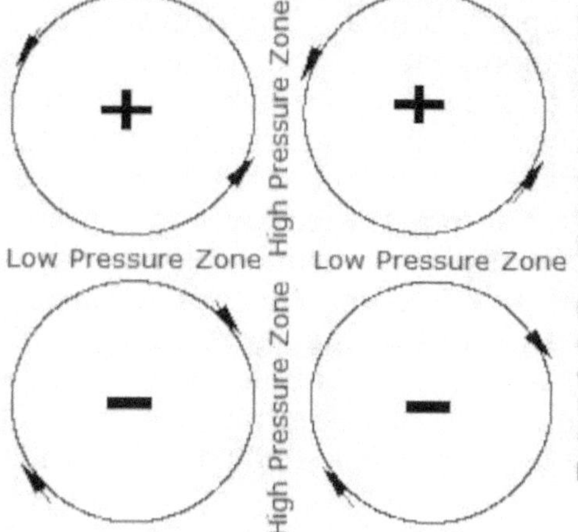

Particles spinning in opposite directions are actually moving in the same direction at their point of contact. Similar to a venturi, the more rapidly moving ether flowing between oppositely spinning particles creates a low pressure zone between them causing ether on either side of the particles to push the particles together towards the low pressure zone between them. Therefore, opposite electric charges do not "attract" each other, they are actually pushed together by the Ether as it pushes on the electrons as the Ether flows towards the low pressure zone between the oppositely spinning electrons.

The formula for electrostatic force is,

$$Fe = k\,(\,q1 * q2\,) / d^2$$

Fe is the electrostatic force. q1 and q2 are charges of opposite sign and **d** is the distance between the surfaces of the charged bodies; **k** is a proportionality constant.

7. **The force of Inertia** is the resistance of mass to movement and its tendency to continue to move once put in motion. This is a direct result of the gravitational directional flow of the ether of space into the nuclei of all atoms. Since the ether is flowing into the atom from all directions, any movement of the atom in any direction will be resisted by the inflowing ether, but once the atom is put into motion or any matter made of atoms is placed into motion, that matter has a tendency to continue to move forward because the matter is

dragging ether along with it as it moves. This becomes more obvious when we put on the brakes as we are driving our cars. When the brakes are applied, the ether that we have been dragging along with us as we move still continues to flow through us as we put on the brakes. The pressure of the ether that has been moving along with us then pushes us forward as we put on the brakes of the vehicle we are in. When going around a curve, or accelerating very fast, the resistance to the turn or to the acceleration is due to the resistance of the mass attempting to move through the ether. A high pressure zone in the ether is set up in front of the moving vehicle that resists its forward movement. If we could get the ether we are moving through to move along with us at the same velocity, such as when we are falling in an elevator, then movement would be facilitated instead of opposed by the ether. By controlling the directional flow of the ether of space, which is accomplished with electrostatic forces, the limitless energy of the ether can be at our command.

8. **Centrifugal force** is caused when an object rotates. On the inside of a rotating cylinder, it can create the same effect as gravity. Centrifugal force is directed away from the axis of rotation. By rotating an object, ether is thrown away from the axis of rotation causing the Centrifugal force. It is a substantial force that can easily be felt by taking a rope, attaching a bottle of water and swinging it around your head. The closer you have it to yourself, the less the force. The farther it swings from the axis of rotation the slower it swings, which is similar to the planets orbiting the sun. At 4 times the distance its orbital velocity is cut in half and its orbital period is twice as long as the radius units. A formula to describe this scenario is:

$$R = V*T$$
Where:
R = the radius units
V = the velocity
T = the orbital period

Variations of this formula are:
$$V = 1/Sqrt(R)$$
$$T = Sqrt(R)*R$$
For example, 4 = ½ * 8. Here, if the orbit radius is extended to 4 times the distance, its orbital velocity decreases

to one-half and its orbital period is doubled. If we should extend the bottle of water 9 times the distance, its orbital velocity would be 1/3 and its orbital period 27.

If, however, the rope is traded for a merry-go-round, where the bottle of water is not allowed to slow down the farther it is placed from the center of rotation, then Centrifugal force increases on the bottle of water effectively making it weigh more, because the ether is pressing the bottle of water against the outer edge of the merry-go-round.

This Centrifugal force is expressed with the formula:

$F_c = mv^2/r$,

Where:

F_c = the Centrifugal force

m = the mass of the bottle of water

v = the speed of rotation

r = the radius distance the bottle of water is from the axis of rotation.

With the merry-go-round scenario, the object being the bottle of water, the more massive the object, the greater the force; the faster the rotation, the greater the force; and the greater the distance from the center, the greater will be the force.

9. **The Tidal force** is caused by a low gravitational pressure zone in the ether of space flowing into a planet when a nearby planetary body passes by. For example, in the Earth-Moon system, the ether flowing into the Earth decreases when the Moon passes into the path of the ether as it flows into the Earth. The tide stands higher under the Moon because there is less gravitational pressure of the ether pressing onto the surface of the ocean in that area. The mass of the Moon is blocking some of the inflowing ether in the area of the high tide by shielding the Earth from some of the gravitational pressure of the ether flowing into the Earth.

The tidal force is determined from Physicist Al Snyder's modified gravitation force formula,

$F = Sqrt(v*V))/R$

Where,

v = the relative orbital velocity of the secondary body

V = the relative orbital velocity of the primary body

R = the orbital radius of the secondary body about the primary body

In fact, all forces, whether they be gravitational, magnetic, nuclear, electrical, electromagnetic, electrostatic, inertial, centrifugal, mechanical, or tidal, all have their origin in the ether of space. The Ether of space is the underlying foundational premise of the Unified Field Theory. It is the prime construct of the universe. Everything is made of it. It's ever constant flow into the earth holds us to the planet. It governs the motions of both solar systems and the electrons orbiting the nucleus of atoms, their magnetic fields, electrostatic and centrifugal forces. It is the force of inertia. The ether is a tenuous etheric substance or spiritual gas that fills the universe. It gives life to all things. It is the medium that the communication of prayer uses to transmit messages. It's resistance to the passage of light is what gives light a speed limit. It is the "Light of Christ" that fills every soul to give us life and understanding. It is the Power of God which he uses to govern the universe. It is so all encompassing that almost no one is aware of its existence.

The Formulas of Gravity

When Galileo performed his experiments on gravity over 300 years ago, he found that the acceleration of gravity above the surface of the earth varies as to the earth's radius distance squared. Expressed in a formula this is:

$K = a * r^2$ **Acceleration of Gravity formula**

Where:

K = a planetary constant of gravity

r = the planetary radius

a = the surface acceleration of gravity

With this formula, the acceleration of gravity can be determined at any distance from a planet's surface.

The gravitation formula was developed from the Acceleration of Gravity formula. Solving for **a**:

$a = K / r^2$

This was substituted into the momentum force formula,

$F = M * a$ **Momentum Force Formula**

making it:

$F = M * K / r^2$

Since K is a planetary constant, a universal gravitation constant was necessary to calculate the gravitation force of any body in space. K was then replaced in the formula by G, the universal gravitation constant to make it:

$F = G M / r^2$

For any planet of radius r, F = a, the surface acceleration of gravity, therefore,

$a = G M / r^2$ **Surface Acceleration of Gravity Formula**

Since GM has replaced K in the Acceleration of Gravity formula, then:

$K = GM$ **Planetary Mass Formula**

The gravitation constant is the gravity force exerted by a unit amount of matter.

The gravity force exerted by a whole planet would be:

$F = GM$

Since the acceleration of gravity varies as to the distance squared that bodies are from each other, then it is assumed that the gravitation force between two bodies in space varies as to their inverse distance squared.

Let's return to the Momentum formula. For a small body falling towards the surface of a planet the momentum force is,

$F = m * a$

If we substitute **a** for the Acceleration of Gravity formula we get,

$F = m * K / r^2$

Since we have determined that the gravity force for a planet, $K = GM$, we can substitute GM for K to arrive at the gravity force acting between the small mass and the gravity of the whole planet,

$F = m * GM / r^2$

Thus, the final gravitation force formula is:

$F = GmM / r^2$ **The Gravitation Force Formula**

How does the gravity force relate to orbital velocity? To understand this, we must analyze how the acceleration of gravity relates to the momentum of a mass in orbit. You can learn how orbital velocity relates to orbital radius by observing a bucket whirled about yourself on the end of a rope. It is found that as the radius distance increases, the velocity of the bucket decreases.

In fact, it has been found that squaring the orbital velocity of a satellite in orbit about a planet divided by the orbital radius equals the acceleration of gravity towards that planet at the orbital radius distance. This keeps the satellite in orbit. Like the bucket, if the orbital radius is increased, the orbital velocity of the satellite decreases. The first satellites put up into orbit orbited the earth in about 90 minutes. Later telecommunication satellites were placed further out, at the distance called the geosynchronous orbit at 22,243.528 miles. At that orbital distance from the earth, the satellite stays directly over one spot on the ground below giving it an orbital period of about 24 hours.

So the relationship of the acceleration of gravity to orbital velocity is,

$a = v^2 / R$

When we substitute this into the momentum formula,

$F = m * a$

we get the orbital centrifugal force formula:

$F = m * v^2 / R$ **The Orbital Centrifugal Force Formula**

When we substitute $a = v^2 / R$ into the Acceleration of Gravity formula, $K = a * r^2$, for a theoretical surface orbiting satellite (ignoring air friction) where the planetary radius $r = R$, the orbital radius, we get:

$K = R^2 * v^2 / R$, which reduces to:

$K = v^2 * R$ **The Orbital Velocity Formula**

By substituting the Planetary Mass Formula for **K** into the Orbital Velocity Formula, we can calculate the mass of any planet with an orbiting satellite body:

$GM = v^2 * R$

solving for M:

$M = v^2 * R / G$

Or solving for v,

$v = Sqrt(GM/R)$

To summarize, in our study of gravity, we have to work with, three planetary constant formulas:

$K = a * r^2$ The Acceleration of Gravity Formula
$K = GM$ The Planetary Mass Formula
$K = v^2 * R$ The Orbital Velocity Formula

and three force formulas:

$F = m * a$ The Momentum Force Formula
$F = GmM / r^2$ The Gravitation Force Formula
$F = m v^2 / R$ The Orbital Centrifugal Force Formula

From the geosynchronous formula, $R^3 = GMT^2 / 4\ pi^2$,

$G = 4\ pi^2 R^3 / T^2 M$

And since

$G = K / M$,

Then also,

$K = 4\ pi^2 R^3 / T^2$

In order to calculate the mass of the planets, it was necessary to obtain a unit value for gravity. How much gravitational force does one gram of matter exert? This was answered when the gravitation constant was directly measured in the Cavendish experiment.

Calculating the Gravitation Constant

The purpose of the gravitation constant is to calculate the amount of gravitation force a given amount of matter exerts. If the only cause of gravity is the vacuum in the nucleus of atoms, then any given amount of matter should exert the same gravitation force anywhere in the universe. Spectroscopic analysis of light emitted by all bodies of matter in space indicates that matter is the same everywhere as we find it here on earth.

In recent decades, Scientists have performed many experiments relative to measuring the gravitation constant first devised by Lord Cavendish. They have noticed variations in the surface acceleration of gravity when measured on high poles and in deep wells. However, they have not taken into account the existence of the ether and its flow which would be different both on a high pole and deep in a well than on the earth's surface. On a high pole, the ether that is entering the earth hasn't assumed a fully vertical orientation as it enters the earth in the form of a vortex from space. That vortex flow is directed towards a planet at a great distance, but as it nears the earth, the direction of the flow begins to move laterally with the earth's rotation then gradually assuming a vertical direction as the surface is approached. Then after entering the surface of the planet, the ether spreads out to form a hollow sphere at the central sphere of gravity located at a depth of 700 miles from the outer surface of the planet, according the Inner Guide in ETIDORPHA.

In a deep well, ether is flowing in all directions towards matter that is all around the experiment. The proportionality of the ether flow pressure would be dependent upon the amount of matter in any direction from the experiment. In the shell of a hollow planet, that flow would be greater towards the sides than either above or below one's position in the shell because at any location within the shell there would be more mass to the sides than above or below. Ideally, the gravitation constant experiment should be conducted as far from any matter as possible with only two mass bodies, and the measuring equipment built into the masses.

The original Cavendish experiment was conducted on the earth's surface with a bar and two sets of masses. In that experiment, two "moons" of equal mass are attached to the ends of a bar. In the exact center of the bar, equidistant from the ends, the bar is suspended in still air with a fiber with no twist in it. Two "earths" of equal mass are then placed perpendicular to the rod on the counterclockwise side of the "moons" at equal distances from the "moon" masses. The downward pressure of gravity on the two moons would be balanced to zero by the balancing fiber in the center of the rod, leaving the gravity force free to act between each "earth" and it's "moon." The bar is allowed to rotate with its "moons" in the direction of their respective "earths." The distance that the "moons" accelerate toward their "earths" is measured as well as the angle and oscillation period of the bar with the "moons." Since the bar is connected and suspended by the fiber, as the gravitational interaction of the "moons" with the "earths" causes the "moons" to accelerate towards the "earths," a torsion force is exerted by the fiber to keep the "moons" from touching the "earths." This causes the bar with the "moons" to oscillate back and forth towards and then away from the "earths." This oscillation is compared to the oscillation of a pendulum and the torsion force exerted by the fiber is assumed to be equal to the gravitation force exerted by the "earth-moon" pairs. The gravitation constant is then computed from Newton's gravitation formula.

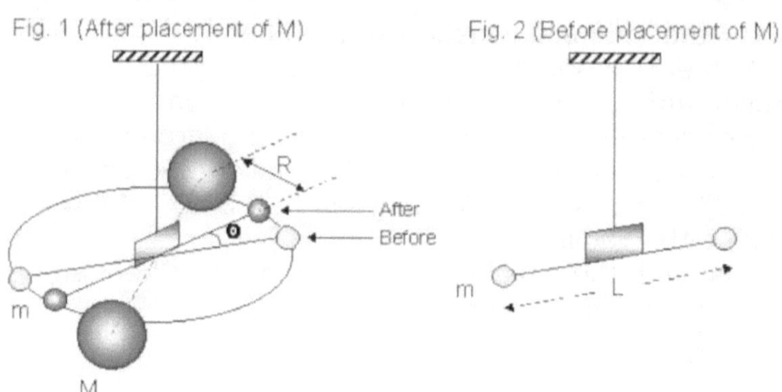

We have a textbook example of a Cavendish experiment in FUNDAMENTALS OF PHYSICS, by David Halliday and Robert Resnick. Values given for the experiment are given as,

M = 12,700 gm   Mass of large spheres

m = 9.85 gm   Mass of small spheres

L = 52.4 cm   Length of the rod from the center of one small sphere to the center of the other small sphere

T = 769 sec   Oscillation period of the small spheres

R = 10.37281 cm   Distance between center of the large and small spheres

2Ø = 0.516 degrees   The angle between the two equilibrium positions of the small spheres before and after placement of the large spheres

In this experiment, the Gravitational torque of the small spheres acting through the bar towards the large spheres equals the Torsional torque of the fiber resisting the gravitational interaction between the large and small spheres. Therefore,

Torsional Torque = Gravitational Torque

The Torsional Torque is expressed as,

t = k Ø

Where

k = the torsional constant of the fiber

Ø (theta) = one-half the angle between the two equilibrium positions of the small spheres on the bar as they oscillate back and forth towards the large spheres.

Theta is converted to radians. 1 radian = 360/2pi, or 57.29577951, therefore, .258 degrees / 57.29577951 = 4.502949×10^{-3} radian

The bar with the small spheres twisting the fiber can be compared to a mass on a spring as,

$$T = 2 * \pi * \sqrt{\frac{I}{k}}$$

where I is the moment of inertia, which accounts for the mass of the small spheres rotating about the central axis of the fiber suspension point on the bar, and replaces m in the original spring formula. T is the period of the pendulum.

The moment of inertia of the small spheres on the rod is calculated as,

$$I = 2 \, m \, d^2$$

Where,

2 represents the two small spheres
m is the mass of a small sphere
d = one-half L, the distance from one small sphere to the fiber suspension point on the bar
Solving for the moment of inertia,
$I = (2*0.00985kg*(0.524m/2)^2) = 1.352287 \times 10\text{-}3 \ kg \ m^2$
Solving for the k torsional constant of the suspending fiber,

$$k = \frac{4\pi^2 I}{T^2}$$

$= (4 \ pi^2)(1.352287 \times 10^{-3} \ kg \ m^2) / (769 \ s)^2$
$= 9.02767 \times 10^{-8} \ N \ m$
N = Newtons. 1 Newton is the force necessary to give an acceleration of one meter per second per second to one kilogram of mass.

The magnitude of the torsional torque exerted by the fiber therefore is,
$t = k \ \varnothing$
$= (9.02767 \times 10^{-8} \ N \ m)(4.502949 \times 10^{-3} \ radians)$
$= 4.06512 \times 10^{-10} \ N \ m$

This torsional torque is balanced by the gravitational torque that the large spheres exert on the small spheres. The gravitational force is from the Newtonian gravitation formula,
$F = GmM / R^2$

The gravitational torque is the gravitational force acting through the bar through the distance of 1/2 L, half the length of the rod. This is given by,

$$\frac{GMm}{R^2} * \frac{L}{2} + \frac{GMm}{R^2} * \frac{L}{2}$$

which reduces to,

$$\frac{GmML}{R^2}$$

Therefore,

$$t = \frac{GmML}{R^2}$$

Solving for G yields,

$$G = \frac{tR^2}{mML}$$

$=$

$(4.06512 \times 10^{-10} \ N \ m)(0.1037281 \ m)^2/(12.7 \ kg)(9.85 \times 10^{-3})(0.524 \ m)$

$= 6.67259 \times 10^{-11} \ N \ m^2 / kg^2,$

The value for G used with centimeters-grams is,

6.67259 x 10⁻⁸ Dynes   **The Gravitation Constant**

What this Gravitation Constant means, is that for every gram of matter in the universe, an acceleration in the ether of space is generated to the value of 6.67259×10^{-8} centimeters per second squared towards the center of that gram of mass.

Calculating the Mass of the Earth

In determining the mass of the earth, it is assumed that the momentum of a small mass accelerating towards the earth near its surface is equal to the earth's gravitational force acting on that small mass:

F = m * a   The Momentum Force (Newton's Second Law)

F = GmM/r^2 The Gravitation Force

m * a = GmM/r^2

Solving for **a**, the small mass **m** cancels out leaving:

a = GM/r^2

We can now solve for M, the mass of earth, using the gravitation constant:

M = a * r^2 /G

Where:

a = 980.665 cm/sec^2, average acceleration of gravity at the earth's surface

r = 6.378 x 10^8 cm, equatorial radius of the earth

G = 6.67259 x 10^{-8}, the Gravitation Constant

(980.665 cm/sec^2)(6.378 x 10^8)^2 / (6.67259 x 10^{-8})

 = **5.97854 x 10^{27} gm, Mass of the Earth**

From the Density formula,

D = M/V

we obtain the density of the earth.

From the volume of a sphere formula,

V = $PiD^3/6$

= the volume of the earth is 1.082 x 10^{27} cc.

The density of the earth then is:

5.97854 x 10^{27} gm / 1.082 x 10^{27} cc

= **5.525 gm/cc**

Since on average, earth surface rocks are 2.7 gm/cc (water is 1 gm/cc), then if the earth's overall average density is 5.5, then the interior of the earth would have to be at least as dense as steel (8 gm/cc). For example, (8.3 + 2.7) / 2 = 5.5. If the earth were full of water (no land), then the earth would have a density of 1. If it had the density of surface rocks, it would be 2.7. Therefore, if the earth has a density of 5.5, then the interior has to contain more dense material than surface rocks.

For a hollow earth, if the shell is assumed to be 10% of the earth's diameter, 800 miles thick, this would give a shell volume of about 50% of the Earth's volume which is 5.41 x

10^{26} cc, and since most of the earth's mass would be in its shell, the density of the earth's shell would be **11.05 gm/cc**, which is close to the density of lead (11.3). So 0 + 11.05 gm/cc divided by 2 = an average density of 5.525 gm/cc for the whole earth.

We do not know how many layers the shell of the earth consists of, but there are certainly more layers than two. At a minimum there should be three layers, one for the outer surface layer, another for the inner surface layer, and then a layer between them.

Surface rocks have an average density of 2.7 gm/cc. If we divide the 50% of the volume of the earth that the shell comprises into three layers, each layer would be 16.67% of the volume of the shell. The outer surface layer would have a density of 2.7 gm/cc. If the inner surface layer also has a density of 2.7 gm/cc, then the middle layer would have a density of **5.65 gm/cc**: 2.7 + 2.7 + **5.65** = 11.05 gm/cc. If the outer and inner layers are thinner in volume than 16.67%, then the middle layer would have a density greater than 5.65 grams per cubic centimeter.

This assumes that most of the earth's mass is located in its shell. As you can see, Newtonian physics requires an average shell density almost as dense as lead (11.3). And since surface rocks average 2.7 gm/cc in density, then the interior of the shell would have to be greater than the average density.

So we can say that a shell density of 11.05 gm/cc could be in the realm of possibility. After all, the earth DOES ring like a bell after a rather large earthquake. A bell is hollow and is made of metal, just as a hollow earth must be.

You may say that the interior sun could contain some of the mass of the earth which would lower the density of the shell. But an interior sun of an estimated diameter of 600 miles would contain very little of the mass of the earth.

Assuming the interior sun has a density of glass, which evidence indicates all stars are actually hollow crystal balls instead of burning gas, its mass would be only .01% of the mass of the Newtonian mass of the earth.

$$V = \frac{\pi D^3}{6}$$

    pi * (600 mi * 1.60934722 km * 100,000 cm)^3 / 6
    = 4.714130881 x 10^{23} cc   Volume of the Inner Sun

Let's assume that the Inner Sun is also hollow and has a shell 10% of its diameter, or 60 miles. This would give the Inner Sun's hollow a volume of 2.413635011 x 10^{23} cc. So the volume of its shell would be 2.30049587 x 10^{23} cc multiplied by 2.6, the density of glass gives,

    Mass = Volume * Density

    = 5.981289262 x 10^{23} gm, Mass of inner sun

    divided by mass of earth of 5.95877 x 10^{27} gm

    = .0001003779 * 100 = .01%

By far, most of a hollow earth's mass would be located in its shell.

Another possibility, you may say, is that the earth's shell is thicker giving a lower average density. This could be a possibility. Some method of determining the shell's thickness needs to be devised. This could easily be determined by entering the hollow of the earth through a polar opening and bouncing radar waves off the opposite side of the hollow interior.

In all, I see nothing in the Newtonian mass of the earth that would completely exclude the earth from being hollow. Earthquake waves have been noticed to bend as they descend into the earth causing them to curve back up to the surface before hitting the discontinuity inside the earth scientists claim is the outer core. This indicates the earth does increase in density with depth which is consistent with a hollow earth with a shell using the Newtonian mass of the earth. In fact, if the earth is hollow and the Newtonian mass of the earth requiring an increased density with depth is correct, then that in itself would exclude the claim to a molten interior. That discontinuity inside the earth could be the inner surface.

If we apply hollow planet theory to the masses and densities of the planets and sun and assume that most of their mass is located in a shell with a hollow interior containing an interior sun, this would allow them all to have solid surfaces. This is in contrast to present scientific belief that the sun is gaseous as well as the big outer planets. Depending on the thickness of their shells, all the planets, moon's and even the sun could easily have solid surfaces at least as dense as earth's (2.7). The scriptures indicate that the sun is a crystal. Glass has a density of 2.6. The scientific work of Michael Mozina from observations of the Sun provided by solar observing satellites has provided indication that the Sun has a solid surface as well as the fact that it has a magnetic field which could not be generated by swirling gases.

With the gravitation constant **G**, we can calculate the masses and densities of the planets, moons and sun of our solar system if we can arrive at a value for the surface acceleration of gravity for each body. We can do this by using the geosynchronous orbit formula using data from orbiting satellites.

The Geosynchronous Orbit Formula

The geosynchronous formula was developed from Johannes Kepler's Third Law of Planetary Motion. In his own words Kepler wrote, "The ratio which exists between the periodic times of any two planets is precisely the ratio of the 3/2 power of the mean distances." (The Attractive Universe, E. G. Vales, pg. 42)

Expressed as a formula, this is:

$R^{3/2} = T$

The accepted geosynchronous formula is based on Newton's squaring of Johannes Kepler's third law of planetary motion. Newton squared this formula giving:

$R^3 = T^2$

Newton's squared version of Kepler's formula can easily be seen in the accepted geosynchronous formula:

(1) $R^3 = GMT^2 / 4 \text{ pi}^2$

Since GM = K, any of the K constant formulas for a planet can be substituted for GM in this formula. So other variations of the geosynchronous formula could be:

(2) $R^3 = r^2 \, a \, T^2 / 4 \text{ pi}^2$

or

(3) $R^3 = v^2 \, r \, T^2 / 4 \text{ pi}^2$

However, since Newton squared Kepler's original formula, the correct geosynchronous formula is the square root of the accepted formula. Even though either formula works, reducing the equation to its lowest form is more correct mathematically. For example, taking geosync formula (2), and reducing it to lowest terms we get:

$R^{3/2} = r * \text{Sqrt}(a) * T / 2 \text{ pi}$

The other variations of the formula are:

$R^{3/2} = \text{Sqrt}(GM) * T / 2 \text{ pi}$

$R^{3/2} = v * \text{Sqrt}(r) * T / 2 \text{ pi}$

Kepler's original Third Law formula $R^{3/2} = T$, can be easily seen in the reduced geosynchronous formula.

The geosynchronous formula can be developed in the following manner:

A geosynchronous orbit has very little to do with the mass of an orbiting satellite as discovered by Galileo when he demonstrated that all non-spinning objects, no matter their mass, all fall at essentially the same rate. This is because the mass of falling objects are so small that they exert a negligible acceleration as compared to the mass of the earth.

The acceleration of gravity formula is:

$a = K/r^2$

Where:

$a = 980.665$ cm/sec^2, average acceleration of gravity at the earth's surface

$r = 6.378 \times 10^8$ cm, average radius of the earth

$K = a * r^2$

the K planetary constant for earth is $3.989235778 \times 10^{20}$ gm cm

The formula for an orbiting satellite is:

$v = $ Sqrt(K) / Sqrt(R)   **Orbital Velocity Formula**

Where:

v = the orbital velocity

R = the orbital radius from the planet's center

K = the planetary constant

Another way of arriving at the orbital velocity would be from the circle circumference formula divided by the orbital time:

$v = $ 2PiR / T   **Circumference Velocity Formula**

Therefore,

Circumference Velocity Formula = Orbital Velocity Formula:

2PiR / T = Sqrt(K) / Sqrt(R)

Solving for **R**:

$R^{3/2} = $ Sqrt(K) * T / 2 Pi

Substituting **K** from the Acceleration of Gravity Formula:

$K = r^2 a$

we arrive at the corrected Geosynchronous Formula:

$R^{3/2} = $ r Sqrt(a) T / 2 Pi

This is the same formula as the corrected orthodox formula (2) mentioned above:

$R^{3/2} = $ r Sqrt(a) T / 2 Pi

Solving this formula for Earth we get:

r = 6,378 km (radius of the earth)

T = 86,164.09 sec (one sidereal revolution of the Earth is 23 hrs 56 min)

a = .00980665 km/sec (acceleration of gravity at the surface)

Pi = 3.141592654

Then

6,378 * 86,164.09 * .099028531 / (2 * 3.141592654) =

54,421,581.38 / 6.283185307 = 8,661,463.687$^{3/2}$ =

42,175.56 km - r (radius of Earth) =

35,797.56 km to the geosynchronous orbit from the surface / 1.60934722 mi =

22,243.528 miles, the standard known geosynchronous orbit of Earth.

Surface Gravities of the Planets

The surface gravities of the planets can be calculated two ways, with the same results.

One, with the Geosynchronous Orbit Formula,

$R^{3/2}$ = Sqrt(a) * r * T / 2pi

where:

R is the mean orbital radius on an orbiting satellite

r is the radius of the planet you are seeking the surface acceleration for

T is the orbital period of an orbiting satellite in seconds

Solving for a, the surface acceleration of gravity of the primary body:

a = $(R^{3/2}$ * 2pi / (r * T)$)^2$

and Two, with the Orbital Velocity Formula for an orbiting satellite

v = Sqrt(K/R)

To get the Gravity Constant **K** of the planet:

K = R * v^2

then using this constant to arrive at the surface acceleration of gravity using the Acceleration of Gravity Formula:

a = K/r^2

For example, solving for **a** for the Sun using the acceleration of gravity variation of the Geosynchronous Formula:

a = $(R^{3/2}$ * 2pi / (r * T)$)^2$

R = 148,736,000 km, orbital radius of the Earth, a satellite of the Sun

r = 696,041.28 km, radius of the Sun

T = 31,484,358 sec, Earth's orbital time in seconds

a = .270488217 km/sec, is the surface acceleration of gravity for the Sun, or **27,048.8217** cm/sec^2

Now, using the Orbital Velocity formula in conjunction with the Acceleration of Gravity formula:

v = Sqrt(K/R)

or

K = R * v^2

First we calculate the orbital velocity of the Earth using the Circumference Velocity formula:

v = 2piR/T

2 * 3.141592654*148,736,000 km/31,484,358 sec =

v = 29.68254426 km/sec
Solving for **K** of the Sun from the Orbital Velocity formula we get:
$K = Rv^2$
148,736,000 km * 29.68254426^2 =
$K = 1.310443635 \times 10^{11}$
then using the Acceleration of Gravity formula:
$a = K/r^2$
$a = 1.310443635 \times 10^{11} / (696,041.28 \text{ km})^2$
$a = .270488217$ km/sec^2, or **27,048.8217 cm/sec^2**, the surface acceleration of gravity for the Sun
The surface acceleration of gravity for the Earth can be obtained using the Geosynchronous Orbit formula with the Moon, satellite of Earth. Solving for the Earth's average surface acceleration of gravity,
$a = (R^{3/2}*2pi/r*T)^2$
Where,
r = the average radius of earth, 6.36745×10^8 cm
R = the average orbital radius of the Moon, 3.821672×10^{10} cm
T = the Moon's orbital period, 2.354149431×10^6 sec
Solving,
$((3.821672 \times 10^{10}$ cm$)^{3/2} * 2 * 3.141592654/6.36745 \times 10^8$ cm $* 2.354149431 \times 10^6$ sec$)^2 =$ **980.665 cm/sec^2**, the average outer surface acceleration of gravity of the Earth.
For a table of Surface Gravities of the Planets, Moon and Sun see The Solar System table in the CONTENTS.

Centrifugal Force Equals the Gravitation Force Acting on an Orbiting Body

For any orbiting satellite, the centrifugal force exactly balances the gravitation force acting on that satellite to keep it in orbit.   The Gravitation formula with its Gravitation Constant should be consistent with and equal to the Centrifugal force formula.

Consider a simulated scenario where my body weight is put into a theoretical surface orbit ignoring any air friction.

MYWEIGHT = 160 lbs * 453.592 gm = 72,574.72 gm

Calculating the surface orbital velocity:

$K = v^2 * R$

$v = Sqrt(K/R)$

$= 790,865.4$ cm/sec

Calculating the centrifugal force on me at surface orbital velocity:

$F = v^2 * MYWEIGHT / R$

$790,865.4^2 * 72,574.72 / 6.378 * 10^8$ cm

= **7.117149 x 10^7 gm** centrifugal force

Now calculating the Gravitation force on MYWEIGHT at surface orbit:

Mass of Earth:

$M = K / G$

$= 5.97854173 \times 10^{27}$ gm

$F = G * MYWEIGHT * M / R^2$

= **7.117149 * 10^7 gm** gravity force, which is the same as the centrifugal force necessary to keep the satellite in orbit.

Using the centrifugal force we can also calculate the mass of the earth with this scenario. Since the centrifugal force acting on my body weight at surface orbital velocity (ignoring air friction) equals the gravity force keeping me in orbit, the centrifugal force = the gravity force:

$F = v^2 * m / R$ Centrifugal force formula

$F = G * m * M / R^2$ Gravitation Force Formula

Then,

$v^2 * m / R = G * m * M / R^2$

Solving for **M**, mass of the Earth,

$M = R * v^2 / G$

$= 5.97854173 \times 10^{27}$ gm Mass of the Earth

So the value of the Gravitation Constant **G** with the Newtonian Gravitation formula gives a mass for the earth

Centrifugal Force Equals the Gravitation Force Acting on an Orbiting
Body

consistent with a hollow earth, and provides gravity forces that
equal the centrifugal force on an orbiting satellite.

Squaring Kepler's Third Law Orbital Velocities Gives the Relative Masses of the Planets

What Johannes Kepler discovered was that if a planet is 4 times the distance from the Sun as the Earth, then it's orbital velocity is cut in half and it's orbital period is twice as long as the orbital radius in Astronomical Units.

The earth's average distance to the sun is one Astronomical Unit (AU). Expressed in a simplified formula this is:

R = V * T   **The RIVET Formula**

Earth 1 AU = 1 * 1
Planet 2 AU 4 = 1/2 * 8
Planet 3 AU 9 = 1/3 * 27
Planet 4 AU 16 = 1/4 * 64
Planet 5 AU 25 = 1/5 * 125
Planet 6 AU 36 = 1/6 * 216
Planet 7 AU 100 = 1/10 * 1,000

From this table, a pattern can be seen in which V = 1/Sqrt(R) Therefore,

R = 1/Sqrt(R) * T

or

Sqrt(R) * R = T

or

$R^{3/2}$ = T, which is **Kepler's Third Law formula**.

The relative orbital velocities of secondaries (moon's) orbiting primaries (planets) can be obtained with the following Keplerian formula:

$R^{3/2}$ / T = Relative Orbital Velocity

Where:

R = the relative orbital radius, or R_1 of one orbiting body / R_2 of a second orbiting body

T = the relative orbital period, or T_1 of the same orbiting body / T_2 of the same second orbiting body

For example, the ratio of the orbital radius of Io, a moon of Jupiter to the orbital radius of Earth's moon is:

261,942 mi / 238,857 mi = 1.096647785

The ratio of the orbital times of the two moons are:

152,854 sec / 2,360,580 sec = .06475273

Plugging these values into Kepler's formula, we get:

$1.096647785^{3/2}$ / .06475273 = 17.73546851 times faster that Io orbits Jupiter than our Moon orbits the earth.

Since the gravitation force is equal to the centrifugal force in an orbiting satellite, at the orbital distance, the gravitational force or mass of the planet is proportional to the velocity squared of any of its orbiting satellites.

Since the gravitational force is directly proportional to the planet's mass, squaring the relative orbital velocities of a planet's moon gives the relative mass of that planet, relative to earth being 1.

For example, since Io, a moon of Jupiter orbits Jupiter 17.73546851 times faster than our Moon orbits Earth, squaring that relative orbital velocity gives the relative mass of Jupiter,

$17.73546851^2 = 314.5$ times more massive than earth.

$F = v^2 * m / R$ Centrifugal Force

$F = GmM/R^2$ Gravitation Force

Centrifugal Force = Gravitation Force

$v^2 m / R = GmM/R^2$

The mass of the satellite cancels out showing that the acceleration of gravity acting on an orbiting satellite is equal to its velocity squared divided by its orbital radius:

$a = v^2 / R$

$v^2 / R = GM/R^2$

Solving for the mass of the planet,

$M = v^2 * R / G$

Isaac Newton obtained the relative masses of the planets by squaring Kepler's Third Law, because the relative masses of the planets are directly proportional to the orbital velocities squared of their orbiting satellites.

$R^{3/2} / T$ = Relative Orbital Velocity - Kepler

$R^3/T^2 =$ Relative planetary mass - Newton

The original Keplerian ratio gives the relative orbital velocities of the planets' moons. Newton's squaring of the Keplerian ratios gives the relative mass of the planets because the centrifugal force acting on an orbiting body is defined as the orbital velocity squared.

Because centrifugal force acting on an orbiting satellite equals the gravity force acting upon it, the relative mass of the planet is equal to the relative orbital velocity squared of its orbiting satellites as defined in the Centrifugal Force Formula.

The formulas we are dealing with here are:

$F = v^2 * m / R$ Centrifugal Force Formula

$F = GmM / R^2$ Gravity Force Formula at the orbital radius

$F = m * a$ Momentum Force Formula

$R^{3/2} = T$ Kepler's Third Law Formula

Let's take for example, an orbiting body of the planet Uranus to illustrate.

From the surface acceleration of gravity variation of the geosynchronous formula,

$a = (R^{3/2} * 2pi / (r * T))^2$

we can obtain the surface acceleration of gravity of Uranus using Titania, a moon of Uranus

R of Titania = 4.377424438E+10 cm

T of Titania = 752,160 sec

r of Uranus = 2.54E+9 cm

$(4.377424438E+10^{3/2} * 2pi / (2.54E+9 * 752,160))^2$

= 907.25 cm/sec^2 surface acceleration of gravity of Uranus.

For a small object accelerating towards a planet near the surface,

$m * a = GmM/r^2$

$a = GM/r^2$

Solving for **M** of Uranus:

$M = a * r^2 / G$

907.25 * 2.54E+9^2 / 6.67259 x 10^{-8}

= **8.772027 E+28 gm** mass of Uranus

Now let's compare our moon orbiting earth to our moon theoretically orbiting Uranus at the Moon's same orbital radius distance to Earth.

From the **K** Constant Formula,

$K = a * R^2$

K of Uranus is 5.8532141E+21

For Earth K = 3.976049315E+20

From the Orbital Velocity Formula

$K = v^2 * R$

our Moon's orbital velocity is,

$v = Sqrt(K/R)$

Sqrt(3.976049315E+20 / 3.821672E+10)

= 101,999.766 cm/sec   orbital velocity of our Moon.

Calculating the moon's orbital velocity about Uranus at the same orbital radius,

Sqrt(5.8532141E+21 / 3.821672E+10)

= 391,354.651 / 101,999.766 = **3.8368**, the Keplerian ratio of Uranus to Earth. So hereby is proof that the Keplerian ratio is a relative orbital velocity ratio and NOT a relative mass ratio as Al Snyder supposed in his book, NEWTON'S LAWS ARE FULL OF FLAWS.

We previously determined that the planetary gravity force is equal to K, the planetary gravity constant, when we determined that K=GM.

The Keplerian ratio is therefore,
Sqrt(K/K) = Relative Orbital velocity
And,
K/K = Relative mass
For Uranus and Earth this is,
Sqrt(5.8532141E+21/3.976049315E+20) = **3.8368**, the Keplerian ratio or the relative orbital velocity of a moon of Uranus compared to the orbital velocity of our Moon about Earth.
And,
5.8532141E+21/3.976049315E+20 = **14.72**, the relative mass of Uranus to Earth, Earth = 1.
To prove this we can compare the mass of the two planets. The relative mass of Uranus to Earth is:
Earth: $a * r^2$ / G
980.665 * $6.36745E+8^2$ / 6.67259 x 10^{-8}
= 5.958779597 E+27 gm
Uranus:
907.25 * $2.54 E+9^2$ / 6.67259 E-8
= 8.772027 E+28 gm / 5.958779597 E+27 = **14.72**, the same as the Newtonian relative mass for Uranus.
Now let's calculate the centrifugal and gravity forces on a satellite in Uranus orbit at the Moon's average orbital distance.
Let's take a satellite of my body weight,
MYWEIGHT = 160 lbs * 453.592 gm = 72,574.72 gm
Calculating the orbital velocity:
$K = v^2 * R$
v = Sqrt(K/R)
Sqrt(5.8532141E+21 / 3.821672E+10)
= 391,354.65 cm/sec. orbital velocity at moon radius from Uranus
Calculating the centrifugal force on me at this orbital velocity:
$F = v^2 *$ MYWEIGHT / R
$391,354.65^2$ * 72,574.72 / 3.821672E+10
= **290,852.6 gm** centrifugal force
Now calculating the Gravitation force on MYWEIGHT at this orbital distance:
Mass of Uranus:
M = K / G
5.8532141 E+21 / 6.67259 E-8
= 8.772027204 E+28 gm mass of Uranus
$F = G *$ MYWEIGHT $* M / R^2$

6.67259 E-8 * 72,574.72 * 8.772027 E+28 /
3.821672E+10^2

= **290,852.6** gm gravity force, which is the same as the centrifugal force necessary to keep the satellite of MYWEIGHT in orbit about Uranus.

Now to show that squaring the Keplerian ratio or orbital velocities gives the relative mass of the planet.

We know that the centrifugal force acting on an orbiting satellite equals the gravity force acting on it at the orbital distance,

Fc = Fg

Fc = v^2 * m / R   Centrifugal Force

Fg = GmM / R^2 Gravitation Force at the orbital distance

v^2 m / R = GmM / R^2

The mass of the satellite cancels out showing that the gravitational force on an orbiting satellite is equal to its orbital velocity squared divided by its orbital radius:

v^2 / R = GM/R^2

Solving for mass of the planet,

M = v^2 * R / G

Our moon orbiting Uranus would have a velocity of 391,354.65 cm/sec. Plugging this into the above orbital mass formula,

391,354.65^2 * 3.821672E+10 / 6.67259 E-8

= **8.772027 E+28 gm**, mass of Uranus.

Therefore, Newton squared the Keplerian ratio to arrive at the relative mass of the planets because in the orbital mass formula the orbital velocity is squared. This is proof that the original Keplerian ratio squared is equal to relative masses of the planets.

Equal Gravisphere Distances Are Calculated from Relative Orbital Velocities

Physicist Al Snyder, in his book, <u>NEWTON'S LAWS ARE FULL OF FLAWS</u>, determined that relative orbital velocities are a more correct measure of the gravity force between planets or particles of matter than the acceleration of gravity that Newtonians use. After extensive study of Snyder's claims, I can see no argument that would nullify Snyder's claim that relative orbital velocities are a better measure of the force of gravity between bodies in space. My study of centrifugal force clearly demonstrated that the relative masses of the planets are the square of their relative orbital velocities.

Relative orbital velocities as a more correct measure of the gravity force between planets are used to determine the equal gravisphere distances between bodies because no body in space is stationary. The ether flow of gravity into planets enters in the form of a vortex. Relative Orbital velocities are better in determining the equal gravisphere distance because they take into account both the horizontal flow of gravity in the vortex and the vertical flow into both bodies, which gives a measure of the relative gravity force between planets. The acceleration of gravity is a measure of the vertical flow of gravity. The orbital velocity is a measure of both the horizontal flow and the vertical flow. That is why velocity is squared in the orbital velocity formula.

The formula for calculating the equal gravisphere distance from the secondary, as given by Al Snyder, is:

$Eg = R * Sqrt(v_2) / Sqrt(v_2) + Sqrt(v_1)$,

Where

v_2 is Kepler's relative orbital velocity of the secondary

v_1 is Kepler's relative orbital velocity of the primary (Earth = 1)

R is the orbital radius of the secondary

See the Equal Gravisphere Distances table in Contents for the equal gravisphere distances of all the planets.

An example calculation is the equal gravisphere distance between the sun and the earth. Since we have previously determined that the relative orbital velocity is equal to the square root of the relative mass, for the sun this is $v_1 = Sqrt(329,831.7975) = 574.3098445$. Therefore,

v_2 (earth) = 1
v_1 (sun) = 574.3098445
R = 92,960,000 mi, or 1.496049176E+13 cm (earth's average orbital radius)
92,960,000 * (Sqrt(1) / Sqrt(1) + Sqrt(574.3098441))
= 3,723,648.475 miles is the equal gravisphere distance between the earth and the sun, the distance from earth.

A More Correct Gravitation Force Formula

Several inconsistencies between calculations with the Classical Gravitation formula and observed phenomenon indicate the need for a more correct gravitation formula.

Of these inconsistencies we will first discuss the inconsistency in Classical Gravitation regarding the behavior of a pendulum during a solar eclipse.

Inconsistency Regarding the Behavior of a Pendulum during a Solar Eclipse

It has been noted that as a total solar eclipse begins, a pendulum will rotate slightly and then start swinging faster as if gravity flowing into the earth has been increased by the Moon eclipsing the Sun. The slight rotation of the pendulum seems to indicate that gravity flows into the earth in vortex fashion. After the eclipse, the pendulum returns to its normal swing.

This observation of the pendulum swinging faster during an eclipse is called the Allais Effect. It was named for Maurice Allais, a French economist who was also an amateur astronomer who believed in the existence of the Ether of space. In 1959 Allais wrote an article in *Aero/Space Engineering* titled, "*Should the Laws of Gravitation be Reconsidered?*" In 1954, and again in 1959, Allais performed an experiment with a Paraconical pendulum to see if the ether flowing into the Earth would be affected when the Moon covers the Sun's disk during a solar eclipse.

What Allais found was that during a solar eclipse, the acceleration of gravity increases into the surface of the Earth beneath the eclipse. This increase in gravity causes the pendulum to swing faster during the eclipse. Let's examine the behavior of a pendulum when gravity increases.

The formula of the Period of a Pendulum is:

$$T = 2\pi \sqrt{\frac{L}{a}}$$

Where:
T is the time of the period, the period being the point where the pendulum starts its swing until the pendulum returns to that point
L is the length of the pendulum, and
a is the acceleration of gravity

T is inversely proportional to the square root of ***a***, therefore as ***a*** increases, such as when the pendulum enters a stronger gravitational field, the period ***T*** decreases. Therefore, the pendulum swings faster in greater gravity.

What Allais' experiment showed was that gravity flowing into the Earth INCREASED during a solar eclipse. Here is a graph showing the increase in the velocity of the swings of the pendulum. During the eclipse, the pendulum swung faster because the acceleration of gravity into the Earth beneath the eclipse increased:

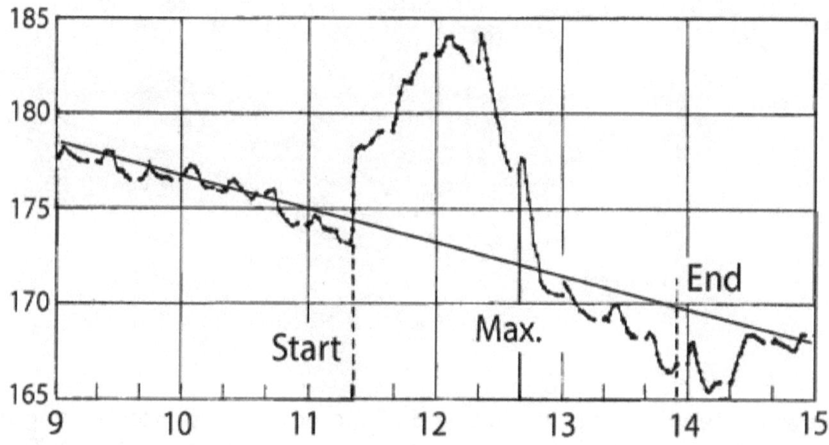

Graph showing the precession rate shift during an eclipse.   ALLAIS, 1959

So why does the acceleration of gravity into the Earth increase beneath the Moon during a solar eclipse? This can be understood by understanding that gravity is a pressure force caused by an ether wind accelerating into the surface of the Earth. This acceleration of gravity flowing into the Earth is partially blocked by the mass of the Moon causing the Earth's tides to rise.

Every time the Moon comes up, Earth's ocean tides rise. The Moon comes up on average 50 minutes later each day, and so the ocean tides rise 50 minutes later each day. We have known for centuries that the Moon, and the Sun to a lesser extent, causes the tides to rise. The rise in ocean tides is caused by the mass of the Moon blocking some of the gravity flowing into the Earth beneath the Moon which causes a partial vacuum in the Ether of space between the Earth and the Moon which causes less pressure of the ether pressing onto the surface of the ocean resulting in the ocean surface rising.

In his book, THE HIGH BRIDGE INCIDENT, THE STORY BEHIND THE STORY, contactee Howard Menger, explained that the mass of the Moon blocks the ether of space flowing into the Earth as gravity and causes the tides to rise beneath the Moon. Menger used this knowledge of the behavior of the ether of space in his efforts to control gravity in the flying saucers he built.

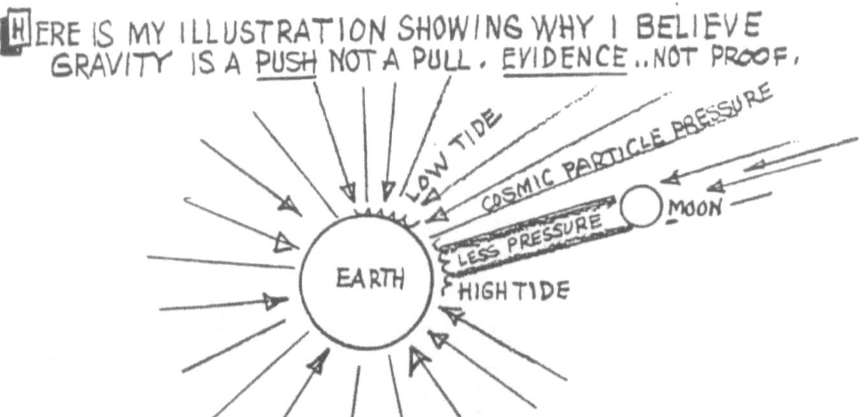

Howard Menger explained in this drawing how gravity accelerating into the Earth is blocked by the mass of the Moon to create a High Tide beneath the Moon because there is less gravity beneath the Moon pressing down on the ocean surface.

There exists a partial vacuum in the Ether of space between the Earth and the Moon that causes less gravitational pressure to be exerted upon the surface of the ocean beneath the Moon. Less gravitational pressure on the ocean surface beneath the Moon allows normal pressure on the ocean surface elsewhere not immediately beneath the Moon to push up on the ocean surface beneath the Moon towards that partial vacuum in the ether between the Earth and the Moon -- which causes the tide to go up beneath the Moon.

So what happens during a solar eclipse? In a solar eclipse, gravity flowing into the Moon towards the Sun from the Earth side of the Moon added to the gravity flowing into the Sun offsets the gravity flowing toward the earth on the far side of the Moon which causes that partial vacuum between the Earth and the Moon during high tide to disappear, which then allows normal gravity pressure to flow into the Earth beneath the

Moon during a solar eclipse. This could be tested by measuring the tide height beneath the Moon during a solar eclipse. The tide height beneath the Moon should go down somewhat because of more gravity flowing into the ocean surface during the eclipse.

An experiment similar to that performed by Maurice Allais was carried out by Erwin J. Saxl and Mildred Allen with a Torsion Pendulum during the 1970 Solar Eclipse in Boston, Massachusetts (bottom line in Figure 1).

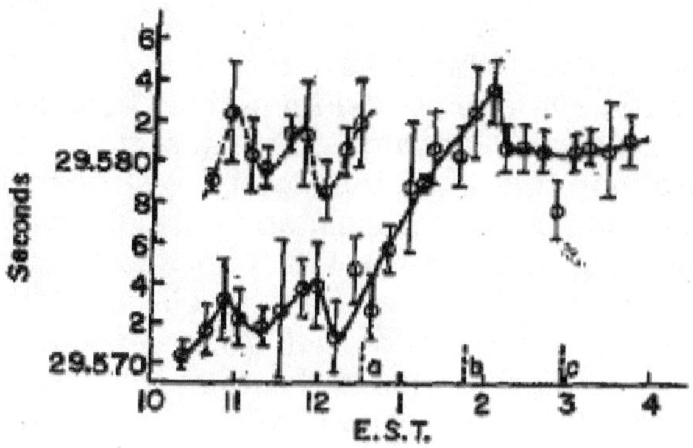

Figure 1.
> Times required to traverse the fixed part of the path of oscillation (ordinates) vs. the hour at which the observations were made, from about 10 a.m. until nearly 4 p.m. (abscissas). The full line shows the observations made on 7 March 1970, the day of the total eclipse. The short vertical dashed lines, *a*, *b*, and *c*, show the times of onset, midpoint and endpoint of the eclipse. The curved dashed line shows the data taken two weeks later, 21 of March, when the sun and moon were on opposite sides of the earth.

This graph shows that as the eclipse began at 12:31 pm Eastern Standard Time in Boston, the pendulum started swinging faster, which from our previous analysis of the behavior of a pendulum means that a greater gravitational acceleration into the Earth begins to occur during the solar eclipse. The report by Saxl and Allen stated that, *"This agrees qualitatively with the work of Allais with a paraconical pendulum, where the change of azimuth increased substantially in the first half of the eclipse of 30 June 1954. Both these*

519

effects would seem to have a gravitational basis which cannot be explained by accepted classical (gravitation) theory."

Because Classical Gravitation theory rejects the existence of the ether of space, this behavior of the pendulum during a total solar eclipse is inexplicable, but which can be explained when acknowledging the existence of the ether and that the acceleration of the ether into the Earth is what we call gravity.

The more I study gravity, the more I am finding that I agree with Isaac Newton, the father of gravity theory. Newton believed the ether of space exists.

Note that in the graph of Saxl and Allen, the bottom line representing the average pendulum periods before the eclipse started shows slower pendulum speeds which would be the case during a high tide when the Moon is overhead. The Moon is overhead during a solar eclipse, and normally would be causing a high tide with less gravitational acceleration into the Earth, but since we are now entering a solar eclipse, the eclipse somehow negates the high tide with an increased acceleration of gravity into the Earth. This is shown in the graph. When the line enters the eclipse, the pendulum speeds increase indicating an increase in the gravitational acceleration into the Earth beneath the Moon and the Sun. An increased acceleration of gravity into the ocean surface normally would cause a low tide because ether accelerating into the ocean surface pushes the ocean surface down with an increased acceleration of gravity. What is happening during the eclipse is a disappearance of the partial vacuum in the ether between the Earth and the Moon which normally causes a high tide.

Notice the dashed lines above the solid line in the graph of of Saxl and Allen which represents the normal gravitational acceleration into the Earth two weeks later when the Moon was on the opposite side of the Earth during a low tide in Boston harbor. The graph shows that during low tide, the gravitational pressure on the ocean surface is approximately the same as during the solar eclipse. Thus during low tide and an eclipse, the gravitation pressure on the ocean surface is approximately the same and is greater than during high tide.

Inconsistency with the Density of the Moon

In addition to the inconsistency in Classical Gravitation regarding the behavior of a pendulum during a solar eclipse there are several other inconsistencies between calculations with the Classical Gravitation formula and observed phenomena that indicate the need for a more correct gravitation formula.

Orbital periods and radii of Moon orbiting satellites as reported by NASA result in a surface gravity of 162 cm/sec^2 for the Moon.

For example, NASA reported the Lunar Prospector spacecraft of April 23, 1998 in orbit 1225 to have a Periselene Altitude (closest distance) above the moon's surface of 81.8 km and an Aposelene (farthest distance above the moon's surface) altitude of 118 km for an average orbital altitude of 99.9 km. Adding the moon's radius, r, of 1,738 km, or 173,800,000 cm, gives an orbital radius, R, of the spacecraft of 1,837.9 km, or 183,790,000 cm. The orbital period given was 118 minutes, or 7,080 sec. Using the geosynchronous orbit formula and solving for the Moon's surface acceleration of gravity we get,

$a = (R^{3/2} * 2pi / (r * T))^2$

$(1.8379E{+}8^{3/2} * 2\ pi / (1.738E{+}8 * 7,080))^2$

= 162 cm/sec^2

A drop experiment of objects dropped on the Moon by Apollo 15 Astronaut David Scott in 1971 as seen on NASA's video of that occasion also gives the surface acceleration rate on the Moon.

In the drop experiment on the Moon, a feather and a hammer were dropped simultaneously from a height of about 1.6 m (160 cm) from the Moon's surface. See: The Apollo 15 Hammer-Feather Drop - NASA Science

So how long would it take on Earth for an object to fall 160 centimeters? The formula is,

$d = at^2 / 2$

Where:

d = distance

t = seconds traveled

a = average Earth surface acceleration of gravity of 980.665 cm/sec^2

Solving for **t**,

t = Sqrt(2d / a)

Sqrt(2 * 160 cm / 980.665 cm/sec^2)
t = .57 seconds
On Earth it would take .57 seconds for an object to fall 160 centimeters.

On the Moon, how long did it take for the Astronauts' objects to fall 160 cm? ChatGPT reports that the frame-speed of the television camera used for the Apollo 15 hammer-feather drop experiment was 29.97 frames per second (standard NTSC broadcast equivalent). On the video, the drop started at time-frame 59 and ended on the ground at 100 for a frame span of 41 frames. So 29.97x = 41 * 1, Solving for x = 41 * 1 / 29.97, x = 1.368 seconds.

Taking our distance formula, and solving for **a**,
a = 2d / t^2
2 * 160 / 1.368^2
= 170.99 cm/sec^2

On the Moon, the acceleration of gravity at the location where Apollo 15 astronauts conducted the drop experiment was 170.99 cm/sec^2, which was higher than the average Moon surface acceleration of gravity of 162 cm/sec^2 because the location of the drop experiment was on the edge of the Imbrium mascon, an area of higher rock density. The Apollo 15 landing location was 26.132° N, 3.633° E in the Hadley-Apennine area between the Imbrium mascon and the Apennine highlands. Mascons on the Moon are locations with higher density materials that have been extruded from the interior to the surface. Because of the higher density of the rock in the mascons, there exists a higher acceleration of gravity in those areas. The higher density of the mascon areas cause orbiting satellites to dip and accelerate over them.

An interesting discovery by Apollo astronauts that confimed a higher density for the Moon's interior were the results of seismic tests conducted on the Moon. Apollo 12 astronauts set up very sensitive seismometers in November, 1969, on the Moon's Sea of Storms and after returning from the lunar surface to the orbiting command module, sent the lunar module ascent stage crashing back onto the Moon creating a moon quake. The vibrations set up by the impact, as recorded by the seismometers, were very similar to the vibrations of a bell when struck. A bell is made out of metal and is hollow. This indicates that the interior of the Moon must contain a substantial amount of metal and must be hollow. At first the vibrations were small, but gained in size to a peak, then tapered off and finally died off after 55 minutes, indicating a high metallic content in the Moon's shell.

The Saturn rocket booster of Apollo 13 was sent to the Moon where it hit causing the Moon to vibrate more than 3 hours and 20 minutes. Apollo 14's Saturn booster rocket was also crashed into the lunar surface causing the Moon to ring like a gong causing the Moon to vibrate for more than three hours.

Also, with the seismometers set up at different locations on the Moon's surface, by the Apollo 14 and 15 missions, they were able to triangulate the vibrations of subsequent bombardments of the Moon's surface traveling down 15 miles into the crust whereupon they picked up speed and traveled at the speed they would travel through metal another 45 miles down at which point they bounced back to the surface, indicating that they had reached the inside surface of a 60-mile thick shell. (OUR SPACESHIP MOON, pp. 99-103, 114)

Classical Gravitation gives a density of the Moon of 3.349687 gm/cc. So what would the density of the Moon's 60-mile thick shell be? It would be more dense than Earth's shell. The Apollo missions to the Moon proved that the Moon's shell contains a much greater percentage of dense and strong metals than does the Earth. If the Moon is hollow as is the Earth, and the metallic content of its surface rocks are denser than Earth's surface rocks, the specific gravity of the Moon's shell should be greater than Earth's shell.

Classical Gravitation gives the Earth an overall density of 5.5 and the Moon 3.3.   And yet the high density of the moon rocks brought back from the Moon and the results from the seismic tests showing the Moon to "ring" like a bell with a period of over three hours when hit by a large meteor, whereas the Earth "rings" with a period of 54 minutes when a large earthquake strikes, indicates that the Moon's shell is much denser and thinner than the Earth's shell.

If we consider that our hollow earth has a shell 10% of the Earth's 8,000 mile diameter for a shell thickness of 800 miles and a shell volume of 50% of Earth's total volume, the density of the Earth's shell would be 11.05 gm/cc. Since NASA discovered that the Moon has a shell thickness of only 60 miles, which is only 2.78% of the Moon's diameter, this gives the Moon's shell a density of 21.256 gm/cc, which is much denser and thinner than Earth's shell. This is consistent with the seismic tests carried out by NASA astronauts on the Moon and the dense moon rocks brought back to Earth. The average density of the Moon therefore is 0 gm/cc (for the hollow) + 21.256 gm/cc (for the shell) divided by 2 = **10.6 gm/cc** whereas the Earth's average density is 0 gm/cc (for the hollow) + 11.05 gm/cc (for the shell) divided by 2 = **5.525 gm/cc**. Isaac Newton in his *Principia Book III*, also determined the Moon to have a greater density than the Earth. My estimate of the Moon to Earth relative densities is thus 10.6 to 5.5, as compared to that given by Newton of 9 to 5 which he based on his tidal studies.

It would appear that the gravity force is not only related to mass, but also to density. In determining the density of the Moon, scientists must take into account that the Moon is hollow with a shell that is relatively thinner and denser than earth's shell. Apparently, Classical Gravitation is useful in determining the mass of a planet, but cannot calculate the correct gravity force acting between planets, nor how gravity works within a planet.

For hollow planets, it would appear that a more correct gravitation formula would relate the force of gravity and the surface acceleration of gravity to the volume and density of the shell of a planet rather than insisting that it varies only as to the inverse distance squared to the center.

Inconsistency with the Equal Gravisphere Distances

Another inconsistency with Classical Gravitation is with the Equal Gravisphere Distances. The Equal Gravisphere Distance between a primary body and its orbiting secondary is the location where the gravity force between them is equal.

For example, if we calculate the Equal Gravisphere Distance between the Earth and the Sun, we find that the Classical Gravitation formula gives an Equal Gravisphere Distance of about 160,000 miles from Earth. This is not possible because the Moon's average orbital radius is over 250,000 miles from Earth. The Equal Gravisphere Distance between the Earth and the Sun would need to be much greater than the Moon's orbital radius about the Earth, otherwise the Sun's gravitation force would draw the Moon out of its orbit about the Earth. If the Moon were in actuality outside the Equal Gravisphere Distance between the Earth and the Sun, the Moon could not orbit the Earth. Since the Moon obviously does orbit the Earth, we must conclude that the Classical Gravitation formula does not correctly calculate the Equal Gravisphere Distance.

Using Physicist Al Snyder's Equal Gravisphere Distance formula, (which is just a simple percentage calculation formula)
$$Eq = R * Sqrt(m_2) / (Sqrt(m_2) + Sqrt(m_1))$$
to calculate the Equal Gravisphere Distance using the relative masses of the Earth and the Sun (Earth = 1, Sun = 329,831.7975), we get:
m_2 (Earth) = 1
m_1 (Sun) = 329,831.7975 (times greater than Earth in mass)
R = 92,960,000 mi, or $1.496049176 \times 10^{13}$ cm (Earth's average orbital radius)
92,960,000 * (1 / 1 + 574.31)
= 161,582.45 miles, the Classical Gravitation Equal Gravisphere Distance from Earth, 161,582.45 miles x 1.609344 km x 100,000 cm = $2.600423391 \times 10^{10}$ cm

When we plug these values into the Classical Gravitation formula, we find that at this distance from earth, the acceleration of gravity is .589 cm/sec^2 flowing into both the Sun and the Earth.

For the Sun:
$F = GM / R^2$
Where:
6.67259 x 10^{-8} Gravitation Constant
1.9698095 x 10^{33} Mass of the Sun
1.496049176 x 10^{13} Average orbital radius of the Earth
Solving,
6.67259 x 10^{-8} * 1.9698095 x 10^{33} / (1.493448752609 x 10^{13})^2
= .589 cm/sec^2

For the Earth:
Where:
5.9785417 x 10^{27} Average radius of the Earth
2.600423391 x 10^{10} Classical Gravitation Equal Gravisphere Distance from Earth
6.67259 x 10^{-8} * 5.9785417 x 10^{27} / (2.600423391 x 10^{10} cm)^2
= .589 cm/sec^2

We see that at 161,582.45 miles from Earth, the acceleration of gravity into the Earth and the Sun are equal. Thus we see that the Classical Gravitation formula is actually just an acceleration of gravity formula, not a gravity FORCE formula.

We can clearly see that the Classical Gravitation formula does not provide an Equal Gravisphere Distance compatible with the facts. This is because the Equal Gravisphere Distance between the Earth and the Sun CANNOT be located between the Earth and the Moon. If it were, the Sun's gravity would cause the Moon to leave its orbit about the Earth.

If we take the Acceleration of Gravity formula,
$K = a \times r^2$
Where,
a = 980.665 cm/sec^2 Average surface acceleration of gravity for the Earth
r = 6.371 x 10^8 Average radius of the Earth
Solving,
$a \times r^2$
= K, 3.980484029 x 10^{20}
and solving for **a**:

$a = K / R^2$

and using the Classical Gravitation derived Equal Gravisphere Distance of 161,582.495 miles from Earth (2.600423391 x 10^{10} cm), we find that at this distance the acceleration of gravity into the Earth is:

3.980484029 x 10^{20} / (2.600423391 x 10^{10} cm)^2

= .589 cm/sec^2

For the Sun:

$K = a \times r^2$

Where:

a = 27,048.8217 cm/sec^2 Surface Acceleration of Gravity of the Sun

r = 6.957 x 10^{10} cm   Average Radius of Sun

Solving for K of the Sun:

27,048.8217 x (6.957 x 10^{10})^2

= 1.3091588859 x 10^{26}

And using the Classical Gravitation derived Equal Gravisphere Distance from the Sun of 92,960,000 miles AU – 161,582.45 miles = 92,798,417.55 miles x 1.609344 km x 100,000 cm = 1.493 x 10^{13} cm Equal Gravisphere Distance from the Sun and solving for a:

$a = K / R^2$

Solving for the Sun

1.3091588859 x 10^{26}/ (1.493 x 10^{13})^2

= .587 cm/sec^2

We can therefore conclude that Newton's gravitation formula is nothing more than a variation of the acceleration of gravity formula. It is not a gravity force formula that will calculate the force of gravity between planets. It only calculates the acceleration of gravity into each body.

The Equal Gravisphere Distance between the Earth and the Moon was confirmed by the Apollo moon shots. Prior to the Apollo missions, scientists believed that the Equal Gravisphere Distance between the Earth and the Moon was 1/81th of the distance from the Moon to the Earth, or about 3,000 miles from the Moon. That is why the first probes sent to the Moon missed it altogether or crashed into the Moon -- mission control was aiming at passing the Equal Gravisphere Distance between the

Moon and the Earth at a supposed distance of only 3,000 miles from the Moon.

In Physicist Al Snyder's book, <u>NEWTON'S LAWS ARE FULL OF FLAWS</u>, is a copy of NASA's radar readings for the Apollo 16 moon shot which discovered by radar the distance of 54,828.7 nautical miles from the Moon that the rocket stopped decelerating away from the Earth and started to accelerate towards the Moon. This is 63,095.7 statute miles from the Moon that the Apollo 16 moon shot discovered the Equal Gravisphere Distance. At that time, the distance from the Earth to the Moon was 252,477.1 statute miles.

Al Snyder discovered that it is better to use relative orbital velocities of each secondary body orbiting about its primary body to calculate the gravity force between bodies at the Equal Gravisphere Distance.

We can calculate the Equal Gravisphere Distance between Earth and the Moon with relative orbital velocities, as Snyder described in his book,

$Eg = R * Sqrt(v_2) / Sqrt(v_2) + Sqrt(v_1)$
Where:
v_1 = relative orbital velocity of the Earth about the Moon (1)
v_2 = relative orbital velocity of the Moon about the Earth (.111)
R = average orbital radius of the Moon about the Earth, but which in this case of the Apollo 16 moon shot = 252,477.1 miles
For the Moon, the Equal Gravisphere Distance is,

252,477.1 miles * Sqrt(.111) / Sqrt(.111) + Sqrt(1)

= 63,103.48 miles from the Moon, or 1.0155×10^{10} cm, the Equal Gravisphere Distance, which is pretty close to the actual radar reading given above of the Equal Gravisphere Distance

The Equal Gravisphere Distance from Earth then is 252,477.1 miles - 63,103.48 = 189,373.62 miles or 3.04767×10^{10} cm.

With this data we can use the reciprocal of Snyder's equal gravisphere formula to arrive at the relative mass of the Moon, the mass of the Moon and the surface acceleration of gravity of the Moon.

Snyder's equal gravisphere equation is:
$Eq = R * Sqrt(v_2)/Sqrt(v_2) + Sqrt(v_1)$

The reciprocal of this (in order to solve for the relative orbital velocity of the Moon) is:
$Er = R * Sqrt(v_1) / Sqrt(v_2) + Sqrt(v_1)$

We can then solve for v_2, the relative orbital velocity of the Moon (which Snyder claimed, incorrectly, to be the relative mass of the Moon). The relative mass of the Moon then is the square of the relative orbital velocity of the Moon.

The Relative Orbital Velocity of the Moon^2 = relative mass of the Moon:
$.111^2 = .012321$

Solving for v_2:
$Er * (Sqrt(v_2) + Sqrt(v_1)) = R * Sqrt(v_1)$
Moving the Er to the other side of the equation:
$Sqrt(v_2) + Sqrt(v_1) = R * Sqrt(v_1) / Er$
Moving the $Sqrt(v_1)$ to the other side of the equation:
$Sqrt(v_2) = (R * Sqrt(v_1) / Er) - Sqrt(v_1)$

The reciprocal of the Equal Gravisphere Distance is:
252,477.1 miles - 63,103.48 = 189,373.62 miles = Er

Solving the equation:
$(252,477.1 * 1 / 189,373.62) - 1$
$= .333 = Sqrt(v_2)$
$v_2 = .111$ = the relative orbital velocity of the Moon about the Earth

And since the relative mass is equal to the relative orbital velocity squared, then
$v_2^2 = .012321$ = the relative mass of the Moon to Earth. Multiply this by the mass of the Earth to get the mass of the Moon,
$.012321 * 5.958779597E+27 =$ **7.3418E+25** gm for mass of the Moon,

then plugging that into the surface acceleration of gravity mass formula,

$a = GM / r^2$

and solving for a:

6.67259E-8 * 7.3418E+25 / 1.738E+8^2 (Moon's radius squared)

= **162 cm/sec^2**, the average surface acceleration of gravity of the moon.

Snyder's work showed that relative orbital velocities give a more accurate measure of the gravity force between bodies in space than the acceleration of gravity that Classical Gravitation uses. He showed that at the Equal Gravisphere Distance, the gravity force using relative orbital velocities is equal for both the primary and the secondary.

At the Equal Gravisphere Distance,

for the Moon,
$Fm = Sqrt(v_2)/R$

for the Earth,
$Fe = Sqrt(v_1)/R$

$Fm = Fe$

for the Moon:
$Sqrt(.111)/ 1.0155 \times 10^{10}$ cm
= **3.28 x 10^{-11}**

for the Earth:
$Sqrt(1)/ 3.04767 \times 10^{10}$ cm.
= **3.28 x 10^{-11}**

We can also arrive at the gravity force between planets using orbital velocities calculated from the centrifugal force formula and the gravity force formula.

Since the centrifugal force of an orbiting body is equal to the acceleration of gravity force at the orbital distance,

$Fc = v^2m/R$ Centrifugal Force
$Fg = GmM/R^2$ Gravity Force
$Fc = Fg$
Then,

$v^2m/R = GmM/R^2$
Solving for **v**,
$v = Sqrt(GM/R)$

Solving for the orbital velocity of the Moon about the Earth, where the mass of the Earth is used:

$G = 6.67259 \times 10^{-8}$ The Gravitation Constant
$M = 5.9785417 \times 10^{27}$ gm   The Mass of earth
$R = 3.844 \times 10^{10}$ cm   Moon's average orbital radius
$Sqrt(6.67259 \times 10^{-8} * 5.9785417 \times 10^{27} / 3.844 \times 10^{10})$
$= 101,871.61$ cm/sec   the orbital velocity of the Moon about the Earth

Now solving for the orbital velocity of the Earth about the Moon, where the mass of the Moon is used (7.3418×10^{25} gms):
$V = Sqrt(Gm/R)$
$Sqrt(6.67259 \times 10^{-8} * 7.3418 \times 10^{25} / 3.844 \times 10^{10})$
$= 11,289.03$ cm/sec   orbital velocity of the Earth about the Moon

Now solving for the relative orbital velocity of the Moon to the Earth,

V_{earth} / V_{moon}
$11,289.03 / 101,871.61$
$= 0.111$ the relative orbital velocity of the Moon about the Earth, Earth = 1
So therefore,
$v_2 = Sqrt(Gm/R) / Sqrt(GM/R)$
$v_2 = Sqrt(m/M)$
Solving,
$Sqrt(7.3418 \times 10^{25} / 5.9785417 \times 10^{27}$ gm$)$
$= 0.111$ the relative orbital velocity of the Moon about the Earth

Now using Snyder's relative orbital gravity force formula, we can show that at the Equal Gravisphere Distance, the gravity force of the earth and the moon are equal.

Solving for Snyder's gravity force at the Equal Gravisphere Distance from Earth:
$v_1 = 1$, relative orbital velocity of Earth about the Moon

Ee = 3.04767 x 10^{10} cm, Equal Gravisphere Distance from Earth
F = Sqrt(v_1)/Ee
Sqrt(1)/ 3.04767 x 10^{10} cm
= 3.28 x 10^{-11}

Now solving for Snyder's gravity force at the Equal Gravisphere Distance from the Moon,
v_2 = .111, Relative Orbital Velocity of Moon about the Earth
Em = 1.01543 x 10^{10} cm, Equal Gravisphere Distance from Moon
F = Sqrt(v_2)/Em
Sqrt(.111)/ 1.0155 x 10^{10} cm
= 3.28 x 10^{-11}

So therefore,
Sqrt(v_1)/Ee = Sqrt(v_2)/Em
And substituting v_2 = Sqrt(m/M):

F = Sqrt(Sqrt(m/M))/R   **Corrected Gravitation Formula**
Where,
m = the mass of the secondary, and
M = the mass of the primary, and
R = the orbital radius

Thus we see that by using relative orbital velocities or the Corrected Gravitation Formula using masses, at the Equal Gravisphere Distance between the Earth and the Moon, as confirmed by radar to exist at about 63,000 miles from the Moon, during the Apollo 16 Moon shot, the gravity force of the Earth and the Moon are equal.

Using masses in the Moon / Earth system:
m = 7.3418 x 10^{25} Mass of the Moon
M = 5.9785417 x 10^{27} gm   The Mass of earth
R = 1.0155 x 10^{10} cm   Equal Gravisphere Distance from Moon

Solving for the Moon,
F = Sqrt(Sqrt(m/M))/R   Corrected Gravitation Formula
Sqrt(Sqrt(7.3418 x 10^{25} / 5.9785417 x 10^{27} gm))/ 1.0155 x 10^{10} cm
= 3.28 x 10^{-11}

Solving for the Earth,
Where:
R = Equal Gravisphere Distance from the Earth
Sqrt(Sqrt(1))/ 3.04767 x 10^{10} cm
= **3.28 x 10^{-11}**

Inconsistency of the Tides

The fourth inconsistency with Classical Gravitation we will discuss is the inconsistency of the tides. When we use the Classical Gravitation formula to calculate the gravity force of the Moon and the Sun on Earth's tides, we find again, that the Classical Gravitation force on Earth's tides is not consistent with observable facts.

It has been known for centuries that the Moon exerts the greater force of gravity on Earth's tides than does the Sun. Yet, even with our highest estimate of the Moon's mass, the Classical Gravitation formula tells us that the Sun exerts 99.44% of the gravity force on the tides and the Moon only 0.56%. This is impossible. We KNOW the moon exerts the greater force because every time the Moon comes up, the tides go up even when the Sun is down. The tides are higher when both the Sun and the Moon are on the same side of the Earth, but the tide still goes up on the side of the Earth that the Moon is located when the Sun and the Moon are on opposite sides of the Earth. We also know that the tides are principally caused by the Moon, because the Moon comes up on average 50 minutes later every day and the tides also rise 50 minutes later every day.

So to calculate the Classical Gravitation force for the tides:

From the geosynchronous formula we arrive at the surface acceleration of gravity of the Sun:
$R^{3/2} = Sqrt(a) * r * T / 2pi$
Solving for a:
$a = (R^{3/2} * 2pi / (r * T))^2$
$= 27,048.8217$ cm/sec^2 , surface acceleration of gravity of the Sun (see Surface Gravities of the Planets in Contents)

The mass formula is:
$M = a * r^2 / G$
$G = 6.67259 \times 10^{-8}$
$r = 6.957 \times 10^{10}$ cm   Average radius of the Sun

Solving for the mass of the Sun,
$27,048.8217 * (6.957 \times 10^{10}$ cm$)^2 / 6.67259 \times 10^{-8}$
$= 1.961995 \times 10^{33}$ gm   Mass of the Sun

From the geosynchronous formula we arrive at the surface acceleration of gravity of the Moon:
$$a = (R^{3/2} * 2pi / (r * T))^2$$
$$(1.8379E+8^{3/2} * 2\ pi / (1.738E+8 * 7{,}080))^2$$
= 162 cm/sec²

Assuming an average surface acceleration of gravity for the Moon of 162 cm/sec², we arrive at a mass of the Moon of,
$$a = GM/r^2$$
Solving for M, mass of the Moon:
$$M = a * r^2 / G$$
Where:
$$G = 6.67259 \times 10^{-8}$$
$$r = 1.7374 \times 10^8\ cm \qquad \text{Average radius of the Moon}$$
Solving for M, mass of the Moon
$$162\ cm * (1.7374 \times 10^8)^2 / 6.67259 \times 10^{-8}$$
= 7.328586 x 10²⁵ gm

For the Sun:
$$F = GM / R^2$$
$$R = 1.496 \times 10^{13}\ cm \qquad \text{Earth's average orbital radius}$$
$$6.67259 \times 10^{-8} * 1.961995 \times 10^{33}\ gm / (1.496 \times 10^{13}\ cm)^2$$
$$= 0.58496\ cm/sec^2$$

For the Moon:
$$R = 3.84399 \times 10^{10}\ cm \qquad \text{Moon's average orbital radius}$$
$$6.67259 \times 10^{-8} * 7.328586 \times 10^{25} / (3.84399 \times 10^{10}\ cm)^2$$
$$= .003309\ cm/sec^2$$

This gives a total for the two forces of 0.588269 cm/sec². The Sun's percent of the force is **99.44%** and the Moon **0.56%**.

We can therefore conclude that the Classical Gravitation formula does not give the correct gravitational force values acting on the Earth's tides by the Sun and the Moon, because we KNOW that the Moon exerts the greater force on Earth's tides because when the Moon comes up, the tides rise.

Now using relative orbital velocities to calculate the gravity force of the Moon and the Sun on Earth's tides:

In physicist Al Snyder's observation of the tides in Los Angeles harbor, he arrived at the relative force of the Sun and the Moon on earth's tides at quarter moon position. His formula for the tidal force is:

$F = Sqrt(v)/R$
Where:
R is the average orbital radius, and
v is the relative orbital velocity

Compared to Earth, the Moon's mass is .012321 of Earth's mass. The relative orbital velocity of the Moon is the square root of the relative Moon mass as compared to Earth:

$Sqrt(.012321) = .111$ the relative orbital velocity of the Moon

Using Snyder's tidal force formula,

For the Moon,
$F = Sqrt(v) / R$
Where:
$v = .111$ The relative orbital velocity of the Moon
$R = 3.844 \times 10^{10}$ cm Moon's average orbital radius
$= Sqrt(.111) / 3.844 \times 10^{10}$ cm
$= 8.667 \times 10^{-12}$ Tidal force of the Moon

For the Sun,
$F = Sqrt(V) / R$
Where:
$V = 574.31$ The relative orbital velocity of the Sun
$R = 1.496 \times 10^{13}$ cm   Earth's average orbital radius
$Sqrt(574.31) / 1.496 \times 10^{13}$ cm
$= 1.6019 \times 10^{-12}$ Tidal force of Sun

The total force is 1.02689×10^{-11}. The Moon's percent of the total is **84.41%** and the Sun is **15.59%**.

This is very close to the values arrived at by Al Snyder in which he observed that the average highest high tide in Los Angeles harbor to be 6.4 ft at New Moon position. At Quarter Moon position, the high-high tide is **4.4** ft.

At New Moon position, the Moon Force + Sun Force = 6.4 ft, actual observed tide height.

84.41% + 15.59% = 100%

100% * 6.4 ft = 6.4 ft calculated tide height

Now at Quarter Moon position, the Moon Force - Sun Force = 4.4 ft actual observed tide height.

84.41% - 15.59% = 68.82%

68.82% * 6.4 ft = **4.4 ft** calculated tide height.

Using relative orbital velocities to determine the gravity force, the calculated tide height at Quarter Moon position is the same as the observed tide height. Snyder's tidal force formula calculates the actual tide height whereas the Classical Gravitation formula does not.

It would appear that the Snyder tidal force formula using relative orbital velocity values could be considered a more correct gravitation formula.

Inconsistency in the Location of the Center of Gravity

On August 3, 1988, an article appeared in The Charlotte Observer newspaper titled, *Deep-In-The Earth Experiments Question Newton Gravity Theory.*

In this article, it was reported that physicists have assumed every push or pull in the universe, from the flex of a muscle to an atomic explosion, can be explained as the work of four fundamental forces. They are *gravity*, *electromagnetism* (which manifests itself as heat, light, radio waves and other things), the "*strong force*" that holds atomic nuclei together and the "*weak force*" involved in radioactive decay.

Experiments a mile below the surface of the Greenland ice sheet have yielded evidence that Isaac Newton's 301-year-old theory may not fully explain gravity. Mark Ander, a Los Alamos physicist said,

"We're saying we appear to have the cleanest evidence to date of something that cannot be explained by Newtonian gravity."

The experiment involved lowering a sensitive gravity meter into a hole bored more than a mile into the ice and seeing how the strength of gravity changed as the meter moved closer to the center of the earth. Newtonian gravity theory says it should change in a certain way because, as the meter goes down, there is less matter below it to exert a pull. The Charlotte Observer of Wednesday, August 3, 1988 reported, *"What the researchers found was that THE PULL DIMINISHED FASTER THAN EXPECTED"* as the meter descended down the hole.

Although physicists postulated a need for a more complex form of gravity and others contended that there must be an unknown *"fifth"* fundamental force in the universe to account for this anomaly, the results of this experiment are exactly what would be expected if the Earth was not full of matter throughout, but hollow within -- a Hollow Earth.

If the earth is full of matter throughout as present day scientists claim, and it were somehow possible to drill a hole through the center of the earth and a gravimeter were then lowered to the earth's center of gravity at the earth's center, it would be observed that at first the acceleration of gravity would increase a certain distance inside the earth and then there would be a very gradual decrease in resultant gravitational acceleration as it was lowered the thousands of miles to the earth's center where the resultant gravitational acceleration would be zero.

However, the Greenland experiment measured and thereby proved that earth's gravitational acceleration actually diminishes faster towards the center of the earth than if the earth were full of matter. The physicists, therefore, think they are seeing a flaw in Newton's famous 300+-year-old gravitation formula.

Classical Gravitation assumes that the earth contains matter throughout because it calculates an earth density of 5.5 times an equal weight of water. Surface rocks being on the average 2.7 times as dense as water, Classical Gravitation's interior earth would have to have at least the density of steel, 8.3, to give an overall earth density of 5.5 (8.3 + 2.7 / 2 = 5.5).

However, the Classical Gravitation formula does not necessarily require that the earth to be solid throughout. If the earth is actually hollow with perhaps a shell thickness of 10% of the earth's diameter, say 800 miles, applying the Classical Gravitation mass of the earth to that shell would give the shell a density of 11.02, which is close to the density of lead (11.3). This is not beyond the realm of possibility.

The Greenland Ice Hole experiment indicates that the center of gravity is not in the center of the earth, but very much closer to the earth's surface as it would be if the earth is hollow. If the center of gravity is located closer to the earth's surface, as the Greenland Ice Hole experiment discovered, then this means that the earth is a hollow sphere instead of containing matter all the way through the earth.

As the gravimeter was lowered down the shaft, a decrease in gravitational acceleration was actually observed to occur more rapidly than if the earth consists of matter all the way through to the center of the earth. This indicates that the earth's center of gravity is not in the center of the earth as formerly supposed, but is closer to the surface of the earth.

If the center of gravity is closer to the earth's surface, then the earth must be hollow. This is because the center of gravity in any body of matter is located at the center of the mass of that body. If the earth's center of mass is closer to the earth's surface as this experiment indicates, then there must be a considerable volume within the earth that contains little or NO matter, a HOLLOW sphere within the earth.

From the February 27, 1989 Physical Review Letters journal, p. 986, we can review the empirical data provided by the Ice Hole Experiment in the TABLE 1:

Table 1.

z	$\wedge z$	$g^{\wedge}z$	g_{ice}	g_r	g_m	g_{obs}	$g_{obs} - g_m$	δ_q
213.00	0.00	0.00	0.00	0.00	0.00	0.00	0.00	0.00
396.12	182.90	56.37	-14.06	-0.25	42.06	42.39	0.33	0.25
579.00	365.78	112.74	-28.13	-0.52	84.06	84.72	0.63	0.23
761.83	547.61	169.09	-42.21	-0.82	126.06	127.13	1.07	0.20
944.63	731.41	225.45	-56.31	-1.14	168.00	169.48	1.48	0.19
1,309.40	1,096.18	337.91	-84.44	-1.92	251.55	254.10	2.55	0.15
1,491.18	1,277.96	393.97	-98.47	-2.40	293.10	296.32	3.22	0.13
1,673.23	1,4670.01	450.11	-112.52	-2.95	334.64	338.51	3.87	0.25

TABLE 1 Variables:
Where:

z	the absolute observation depths from the surface of the Greenland Ice sheet
$\wedge z$	the depths relative to the shallowest observation point
$g^{\wedge}z$	the theoretical free-air term (Newtonian calculated change in acceleration of gravity)
g_{ice}	gravitational effect due to the ice (approximately the second term in Equation (1))
g_r	the gravitational attraction of the sub-ice terrain
g_m	$= g^{\wedge}z + g_{ice} + g_r$, the theoretical gravity differences (based on the assumption the center of gravity is located in the center of the earth)
g_{obs}	the observed gravity differences
$g_{obs} - g_m$	the anomalies
δ_q	the uncertainties

The modeled and observed values are offset to make them both zero at $z = 213$ m, which is permitted since all the gravity observations are relative. All distances are in meters and all gravity values are in mGal (1 mGal = 10^{-3} cm/s^2).

The scientists reported,
"Some unified field theories raise the possibility that forces exist in nature with ranges on the order of 10^2 - 10^5 m and coupling strengths close to that of gravity. If they exist, these new forces would be apparent violations of Newton's inverse-square law. Recent geophysical measurements in a mine and

on a tall television antenna have reported small deviations from the classical law."

"The Newtonian prediction of the gravity profile in the borehole (2033 m deep located at Dye 3 Greenland), based on a density model of the ice and the topographic relief of the bedrock developed from geophysical measurements, was compared with measured values. Differences in gravity g were measured at several depths z and modeled by":

Equation (1) $g_m(z) = g(z) - 4Gp_i(z) + g_r(z)$

where:

g   = the theoretical free-air gravity gradient

G   = the Newtonian gravitational constant

p_i = the ice density

g_r = a correction to the gravity differences based on the attraction of the sub-ice terrain

Conclusion of the experimental scientists:

"After all these conventional adjustments are applied, there remains an unexplained gravity difference of 3.87 +/- 0.36 mGal between the gravity value at a depth of 213 m and the one at a depth of 1673 m."

"We have found an anomalous gravity gradient that could be taken as evidence for non-Newtonian gravity."

Whereas these scientists feel they have perhaps discovered an indication of an error in Newton's inverse square law of gravitation, this anomaly is better explained by a hollow earth. In a hollow earth, the mass is mostly located in the planet's shell and a central sphere of gravity would be located somewhere between the outer and inner surfaces -- 700 miles down from the outer surface in an 800 mile thick shell, according to the Inner Earth Guide in ETIDORPHA.

In the following simplified Table 2 of the Greenland Ice Hole Experiment, it can be more clearly seen that the Observed acceleration of gravity is decreasing more rapidly than the Calculated Newtonian acceleration as the gravimeter is lowered down the ice hole.

Table 2.

LEVEL km	Earth Radius km	mGal Calculated a	Calculated Change in a	Observed Change in a	mGal Observed a
Surface	6371.00	981601.00			981601.00
0.213	6370.79	981666.64	0.00	0.00	
0.39612	6370.60	981723.07	56.37	42.39	981723.07
0.579	6370.42	981779.44	112.72	84.72	981765.46
0.76183	6370.24	981835.80	169.07	127.13	981807.79
0.94463	6370.06	981892.15	281.54	169.48	981850.20
1.3094	6369.69	982004.61	337.59	254.10	981892.55
1.49118	6369.51	982060.66	393.73	296.32	981977.17
1.67323	6369.33	982116.80	505.92	338.51	982019.39
2.037	6368.96	982229.00			982061.58

$K = r^2 * a$
Where:
  r = 6371 km
  a = 981601 mGal   (1 mGal = 10^{-3} cm/sec^2)
  K = 3.98428322 E+13

A MUCH GREATER decrease in the gravitational acceleration was OBSERVED in the Greenland Ice Hole gravity experiment as the gravimeter was lowered down the ice hole (compare the mGal Calculated to the mGal Observed columns) than the Newtonian CALCULATED acceleration (based on the assumption that the center of gravity is located in the center of the earth 4000 miles away) -- which indicates that the earth's center of gravity is closer to the surface than if it were located in the center of the earth.

The difference between the theoretical gravitational gradient and the observed acceleration of gravity in this Greenland Ice hole experiment is illustrated in this chart:

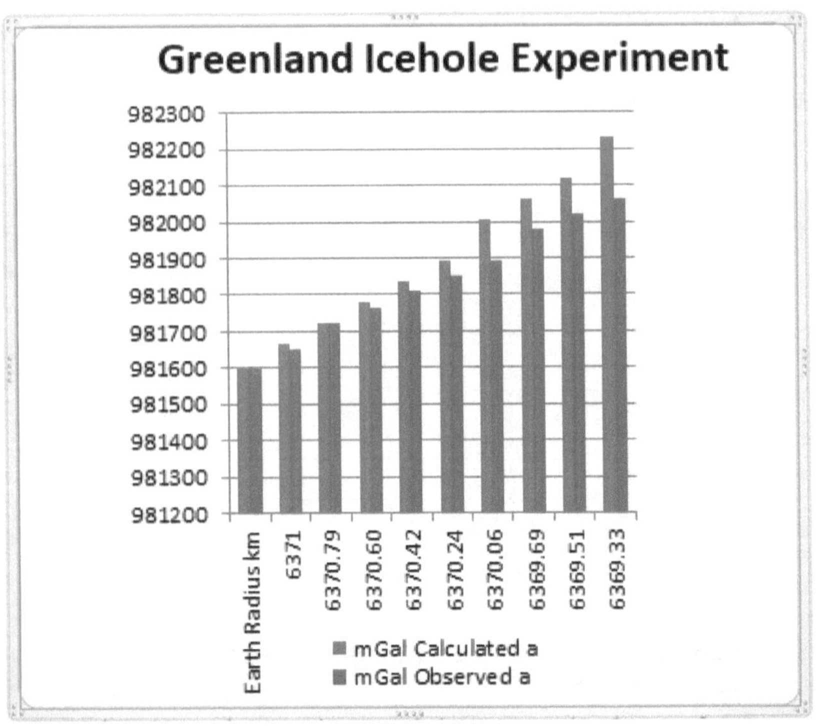

A gravity meter lowered toward that central sphere of gravity located 700 miles down in the shell of the hollow earth would require a much more rapid decrease (red in chart – mGal Observed) in the acceleration of gravity in order to reach zero acceleration at the center of gravity in only a few hundred miles down. Whereas, a much more gradual decrease in acceleration (blue in the chart – mGal Calculated) would be required to reach zero acceleration in the far more distant 4,000 miles to the center of the earth having matter throughout.

At a depth of 2.037 km from the surface of the ice, the acceleration of gravity was calculated at 982.229 cm/sec^2, compared to the observed acceleration of 982.06158 cm/sec^2 -- a difference of 167.42 LESS mGals for the Observed acceleration than for that Calculated by the inverse square law when assuming the center of gravity is located in the center of the earth (6,371 km from the surface of the Greenland ice sheet).

If sufficient data could be obtained on the acceleration of gravity from space to the center of gravity, a graph of this data would take the shape of a bell curve, where acceleration starts

544

at zero (or close to zero) many miles in space and ascends to the highest acceleration somewhere between the surface of the earth and the center of gravity, and then descends back to zero at the center of gravity. The Greenland Ice Hole Experiment shows that the apex of that graph would most likely be reached sooner within a few hundred miles depth such as would be expected in a hollow earth, instead of the thousands of miles to the center of the earth as required by a solid earth.

Thus the gravimeter observations of the Greenland Ice Hole experiment discovered that the center of gravity in the Earth is closer to the surface such as would be the case if the Earth is hollow than the calculations of the Classical Gravitation formula which assumes the center of gravity is in the center of the Earth.

Inconsistency of the Gravitation Formula

As previously explained, Al Snyder pointed out another inconsistency in the Classical Gravitation formula. It's the scenario of two sets of magnets, one set 10 times more power than the first. Using the Classical Gravitation formula, he showed that for the first set of magnets of power 1, separated by a distance of 1 unit, the force is 1, but for the second set of magnets 10 times more powerful than the first, look at the force:

$$100 = 10 * 10 / 1^2$$

Classical Gravitation would maintain that the second set of magnets is 100 times more powerful than the first set, instead of the actual 10 times more powerful that we KNOW they are. Therefore, Snyder concluded that in the Classical Gravitation formula, F is actually squared,

$$F^2 = G \, m * M / R^2$$

Snyder maintained that the planets are hollow and thus less massive. In fact, his calculations gave the Earth about 1/4 the mass that Classical Gravitation gives. I have been unable to verify his claims to a less massive Earth. But Snyder's work does show that relative orbital velocities give a more accurate measure of the gravity force than the acceleration of gravity that Classical Gravitation uses.

In his book, NEWTON'S LAWS ARE FULL OF FLAWS, Al Snyder determined that to be consistent with his magnet scenario described above, the corrected gravitation formula should be,

$$F^2 = G * m * M \ / R^2$$

and solving for F,

$$F = Sqrt \, (G * m * M \,) / R$$

but since relative orbital velocities result in gravity forces more consistent with Equal Gravisphere Distances and the tides, Snyder claimed that relative orbital velocities of the secondary and primary should be used in the corrected gravitation formula.

But since the relative masses = relative orbital velocities squared, and the square root of the relative mass = the relative orbital velocity, then relative masses can also be used, so that:

$V^2 = M$, so that
$V = Sqrt(M)$

And,
$v^2 = m$, so that
$v = Sqrt(m)$

Where:
$F = Sqrt(v/V)/R = Sqrt(Sqrt(m/M))/R$

So then,
Corrected Gravitation Force Formula:
$F = Sqrt(Sqrt(m/M))/R$
Where,
m = the mass of the secondary
M = the mass of the primary
R = the distance between their centers

For example, the gravity force between the Earth and the Sun at the Equal Gravisphere Distance between the Earth and the Sun can be determined with the Equal Gravisphere Distance formula using relative masses, which are the square root of the relative orbital velocities:

From the geosynchronous formula we arrive at the surface acceleration of gravity of the Sun:
$R^{3/2} = Sqrt(a) * r * T / 2pi$
Solving for a:
$a = (R^{3/2} * 2pi / (r * T))^2$
= 27,048.8217 cm/sec^2 , surface acceleration of gravity of the Sun
(see Surface Gravities of the Planets in Contents)

The mass formula is:
$M = a * r^2 / G$
$r = 6.957 \times 10^{10}$ cm   Average Radius of Sun

Solving for the mass of the Sun,
27,048.8217 * (6.957 x 10^{10} cm)^2 / 6.67259 x 10^{-8}
= 1.961995 x 10^{33} gm   Mass of the Sun

Solving for the mass of earth:
M = a * r^2 /G
Where:
a = 980.665 cm/sec^2, average acceleration of gravity at the earth's surface
R = 6.371 x 10^8 cm   Average radius of the earth
G = 6.67259 x 10^{-8} Gravitation Constant
(980.665 cm/sec^2)(6.371 x 10^8)^2 / (6.67259 x 10^{-8})
  = 5.96542576 x 10^{27} gm   Mass of the earth

Calculating the relative mass of the Sun,
M/m   The Relative Mass of the Sun, Earth = 1
1.961995 x 10^{33} gm / 5.96542576 x 10^{27} gm
= 328,894   Relative Mass of the Sun, Earth = 1

The relative orbital velocity is the square root of the relative mass:
Sqrt(328,894) = 573, Relative Orbital Velocity of Sun, Earth = 1
92,960,000 miles   1 AU, average orbital radius of the earth

Calculating the Equal Gravisphere Distance between the Earth and the Sun,
Eq = R * Sqrt(v2) / (Sqrt(v2) + Sqrt(v1))
Where,
v1 = Relative Orbital Velocity of the Sun = 573
v2 = Earth = 1

Solving,
92,960,000 miles * Sqrt(1) / (Sqrt(1) + Sqrt(573))
= 3,727,794 miles from Earth, the Equal Gravisphere Distance = 5.9993 x 10^{11} cm
Now to show that at the Equal Gravisphere Distance the gravity force of the Earth and the Sun are equal using the Corrected Gravitation Formula:
F = Sqrt(Sqrt(m/M))/R   Corrected Gravitation Formula

92,960,000 miles - 3,727,794 miles = 89,232,206 miles distance to Equal Gravisphere from Sun * 160,934.4 cm per mile = 1.436 x 10^{13} cm
Solving for the Earth:
F = Sqrt(Sqrt(m/M)) / R   for the Earth
Where:
5.96542576 x 10^{27} gm   Mass of the Earth
1.961995 x 10^{33} gm   Mass of the Sun
5.9993 x 10^{11} cm   Equal Gravisphere Distance from the Earth

Solving for the Earth:
Sqrt(Sqrt(5.96542576 x 10^{27} gm / 1.961995 x 10^{33} gm))/ 5.9993 x 10^{11} cm
= **6.96 x 10^{-14}**, Gravity Force of the Earth at the Equal Gravisphere Distance from the Earth

Solving for the Sun:
1.436 x 10^{13} cm   Equal Gravisphere distance from Sun
Sqrt(Sqrt(1)) / 1.436 x 10^{13} cm
= **6.96 x 10^{-14}**, Gravity Force of the Sun at the Equal Gravisphere Distance from the Sun

At the Equal Gravisphere Distance from both the primary mass body and the secondary mass body, the gravity force is equal, as we have just determined between the Earth and the Sun.

The Equal Gravisphere Distance between the Earth and the Moon was confirmed by the Apollo moon shots as previously mentioned.

So let's take the Equal Gravisphere Distance between the Moon and Earth in the Apollo 16 radar reading, and use Snyder's Corrected Gravitation Formula to calculate to see if at that distance the gravity force is equal.

We can calculate the Equal Gravisphere Distance between Earth and the Moon with relative orbital velocities, as Snyder described in his book,
Eg = R * Sqrt(v_2) / Sqrt(v_2) + Sqrt(v_1)
Where:

v_1 = relative orbital velocity of the Earth about the Moon (1)
v_2 = relative orbital velocity of the Moon about the Earth (.111)
R = average orbital radius of the Moon about the Earth
For the Moon, the Equal Gravisphere Distance is,
252,477.1 miles * Sqrt(.111) / Sqrt(.111) + Sqrt(1)
= 63,103.48 miles from the Moon, the Equal Gravisphere Distance, pretty to close the actual radar reading given above of the Equal Gravisphere distance, or 1.01555×10^{10} cm

The reciprocal Equal Gravisphere Distance from Earth then is 252,477.1 mi - 63,103.48 mi = 189,373.62 miles or 3.04767×10^{10} cm.

Using the Corrected Gravitation Force Formula,
F = Sqrt(Sqrt(m/M))/R   Corrected Gravitation Formula,
we calculate the gravitation force at the Equal Gravisphere Distance for the Moon and the Earth:

Using masses in the Moon / Earth system:
m = 7.3418×10^{25} Mass of the Moon
M = 5.9785417×10^{27} gm   The Mass of earth
R = 1.0155×10^{10} cm   Equal Gravisphere Distance from Moon

Solving for the Moon,
F = Sqrt(Sqrt(m/M))/R   Corrected Gravitation Formula
Sqrt(Sqrt(7.3418×10^{25} / 5.9785417×10^{27} gm))/ 1.0155×10^{10} cm
= 3.28×10^{-11}

Solving for the Earth,
Where:
R = Equal Gravisphere Distance from the Earth
Sqrt(Sqrt(1))/ 3.04767×10^{10} cm
= 3.28×10^{-11}

Besides obtaining the relative orbital velocity by square rooting the relative mass of each body, we could also use the orbital velocities of each body about the other to show that the gravity force is equal at the Equal Gravisphere Distance.

Since the centrifugal force of an orbiting body is equal to the acceleration of gravity force at the orbital distance,

$F = v^2m/R$ Centrifugal Force
$F = GmM/R^2$ Gravity Force
Then,
    $v^2m/R = GmM/R^2$
Solving for **v**,
$v = Sqrt(GM/R)$

Now solving for the orbital velocity of the Moon about the Earth,
Where:
$G = 6.67259 \times 10^{-8}$ The Gravitation Constant
$M = 6.044031 \times 10^{27}$ gm The Mass of earth
$R = 3.843990 \times 10^{10}$ cm   Moon's average orbital radius
$Sqrt(6.67259 \times 10^{-8} * 5.9785417 \times 10^{27} / 3.843990 \times 10^{10})$
$= 101,871.74$ cm/sec   orbital velocity of the Moon about the Earth

Now solving for the orbital velocity of the Earth about the Moon, where the mass of the Moon is used (7.320058×10^{25} gms):
$Sqrt(6.67259 \times 10^{-8} * 7.320058 \times 10^{25} / 3.843990 \times 10^{10})$
$= 11,272.32$ cm/sec   orbital velocity of the Earth about the Moon

Solving for the relative orbital velocity of the Moon to the Earth,
v_{earth} / v_{moon}
$11,272.32 / 101,871.74$
$= 0.111$
Using Snyder's orbital gravity force formula, we can show that at the Equal Gravisphere Distance, the gravity force of the Earth and the Moon are equal.

Solving for Snyder's gravity force for the moon,
$v_2 = .111$, Relative Orbital Velocity of Moon about the Earth
$Em = 1.012576 \times 10^{10}$ cm   Equal Gravisphere Distance from Moon
$F = Sqrt(v_2)/Em$
$Sqrt(.111)/ 1.012576 \times 10^{10}$ cm
$= \mathbf{3.28 \times 10^{-11}}$

Solving for Snyder's gravity force for the Earth,
$Ee = 3.048995 \times 10^{10}$ cm, Equal Gravisphere Distance from Earth

$v_1 = 1$, relative orbital velocity of Earth about the Moon
$F = \mathrm{Sqrt}(v_1)/Ee$
$\mathrm{Sqrt}(1)/\ 3.048995 \times 10^{10}$
$= 3.28 \times 10^{-11}$

Thus we see that by using relative orbital velocities or the Corrected Gravitation Formula, at the Equal Gravisphere Distance, the gravity force of the Earth and the Moon are equal as confirmed by the radar readings of the Equal Gravisphere location between the Earth and the Moon in the Apollo 16 moon shot, as also the tidal forces of the Sun and the Moon on Earth's tides are correctly calculated as confirmed by the actual observation of the tide heights.

And when gravimeter readings are taken in a hole in the Earth, such as was done in the Greenland Ice Hole experiment, it was found that the Earth's center of gravity is closer to the surface such as would be found in a hollow earth rather than in the center of the earth as has been assumed by Classical Gravitation.

Additionally, seismograph readings on the Moon confirmed that the Moon's shell of 60 miles thick is relatively thinner at 2.78% of the diameter of the Moon giving a density of 21.256 gm/cc, that is greater than Earth's 800-mile thick shell, when assuming that the Earth is hollow having a shell that is 10% of the diameter of Earth with a density of 11.02 gm/cc which results in a greater average density for the Moon of 10.6 gm/cc than the Earth's average density of 5.5 gm/cc.

Shell Densities of the Planets and their Moons

When all bodies in our solar system are assumed to be hollow with shells having a small percentage of the planetary diameter, we discover that all of the planets, including the Sun can have solid surfaces. And since nearly all the planets have been found to have magnetic fields and auroras indicating the existence of interior suns within their hollow interiors, we conclude that most probably all the planets, moons and even the Sun are inhabitable within and may even now contain perfect climates giving life to plants, animals and even human life within their interiors.

The Solar System

$F = G \, m * M / r^2$ Gravitation Formula
$F = \mathrm{Sqrt}(\mathrm{Sqrt}(m/M))/R$ Corrected Gravitation Formula
$G = \mathbf{6.67259 \times 10^{-8}}$ cm/sec^2
1 km = 100,000 cm
pi = 3.14159265
1 mi = 1.609344 km
Density = Mass/Volume
Surface acceleration of gravity: $a = ((R^{3/2} * 2pi / (r * T))^2$
Planetary Mass: $M = a * r^2 / G$
Volume of sphere = $PiD^3/6$

	Shell Density gm/cc	Surface Acceleration of Gravity cm/sec²	Planet Densities gm/cc	Planet's Diameter km	Surface Gravity as multiple of Earth's	Relative Planetary Mass Earth = 1	Average Orbital Radius mi
Sun	2.85766	27,129.93	1.34945	1,400,000	27.66	329,831.797	
Mercury	11.085	368.76678	5.4095	4,878	.376	.05499	36,000,000
Venus	10.698	882.80	5.2206	12,100	.9	.81	67,230,000
Moon	21.256	162.18	3.349687	3,476	.4516	.01232	238,855
Earth	11.05	980.665	5.525	12,756	1.0	1	92,960,000
Mars	7.795869	361	.175275	6,790	.368	.02733	141,700,000
Jupiter	2.4887	2,422	1.214502886	142,700	2.4698	309.08	483,700,000
Saturn	1.26	1,033	.0615979629	120,000	1.05	93.22	885,200,000
Uranus	2.577596	893	1.261966	50,800	.910	14.442	1,781,000,000
Neptune	3.44855	1,143	5.286969	48,600	1.17	16.91878	2,788,000,000
Pluto	25.04	512.395	12.2217	3,000	.5224	.02886	3,660,000,000

Planetary shell is assumed to be 10% of the planet's diameter except the moon which NASA determined was 60 miles thick (using seismometers placed on the moon and crashing an Apollo lunar module into the moon to make it ring like a bell).

Most of the planets except the Sun have a surface gravity very close to that of earth's surface gravity, the average of all being .92178 that of the Earth. Additionally, the sun, planets and moons if considered hollow with a shell thickness 10% of the planet's diameter would allow them all to have solids shells (not gaseous), except Saturn (density of 1.26 gm/cc) which has a shell density closer to water (assuming a shell thickness of 10% of the planet's diameter). However, if we assume a shell thickness for Saturn of 5% of its planetary diameter, its shell density would be 2.27 gm/cc, which would be solid.

The Density of Earth's Shell

Earth's average shell thickness: 800 mi, 1,287.4752 km
Earth's shell x 2 = 2,574.9504 km
6,356.752 km   Earth's polar radius
6,378.137 km   Earth's equatorial radius
6367.445 km   Earth's average radius
6367.445 x 2 = 12,734.889 km   Diameter of Earth =
1.2734889×10^9 cm

Volume of Earth $PiD^3/6$
$(3.14159265 \times (1.2734889 \times 10^9)^3)/6 = 1.081394 \times 10^{27}$ cc

Diameter of Earth's hollow =
Diameter of Earth – Diameter of Earth's shell x 2
12,734.889 km – 2,574.9504 km = 10,159.9386 km * 100,000
cm = 1.01599386×10^9 cm = Diameter of Earth's hollow

Volume of Hollow: $PiD^3/6$
$(3.14159265 \times (1.01599386 \times 10^9)^3)/6 = 5.491258296 \times 10^{26}$ cc

Volume of Earth - Volume of Hollow = Volume of Shell
$1.082 \times 10^{27} - 5.491258296 \times 10^{26} = 5.32268 \times 10^{26}$

So on average, half the volume of the Earth consists of the volume of the shell and half the volume of the Earth consists of the volume of the hollow.

Given that the volume of the earth is 1.082×10^{27} cc, the density of the earth is:

Density = Mass / Volume
Earth Density = Earth Mass / Earth Volume

5.97854×10^{27} gm / 1.082×10^{27} cc

= **5.525 gm/cc**

(Hollow Density + Shell Density)/2 = Earth Density
(0 + 11.05)/2 = 5.525 gm/cc,

Density of Shell = 11.05 gm/cc

The Densities of the Planets

$D = M/V$

$M = a * r^2 / G$

$a = ((R^{3/2} * 2pi / (r * T))^2$

$G = \mathbf{6.67259 \times 10^{-8}}$ cm/sec^2

D = Density of planet in grams/cubic centimeter

M = Mass of planet in grams

V = Volume of planet in cubic centimeters

Volume of sphere = Pi $* D^3 / 6$

Shell thickness assumed to be 10% of planet's diameter (except the Moon shell thickness = 60 miles)

The planets are hollow worlds containing interior suns that emit solar winds through polar openings that light up their auroras, having shell densities with solid surfaces and most probably containing interior climates, surface gravities and environments ideal for plant, animal and human life.

Even the sun has a shell density (2.85766 gm/cc) that would make it a solid. The scriptures indicate the sun is a giant crystal. The only planet that shows a shell density less than solid is Saturn (1.26 gm/cc). If its shell were thinner than 10% of the planet's diameter, it may also have a solid shell. None of the planets, not even the sun are entirely gaseous.

	Density gm/cc	Shell Density gm/cc	Surface gravity cm/sec^2	Diameter cm	Volume cc	Mass gm
Sun	1.3945	2.8576	27,129.93	1.392082E+11	1.412516E+33	1.969E+33
Mercury	5.4095	11.085	368.766	4.878E+8	6.077486E+25	3.2876E+26
Venus	5.2206	10.698	882.80	1.21E+9	9.275871E+26	4.8426E+27
Moon	3.3496	21.256	162	3.476E+8	2.199064E+25	7.3661E+25
Earth	5.525	11.05	980.665	1.2756E+9	1.082E+27	5.97854E+27
Mars	3.8043	7.7958	361	6.79E+8	1.639109E+26	6.2358E+26
Jupiter	1.2144	2.4887	2,422	1.427E+10	1.521495E+30	1.8478E+30
Saturn	.06159	1.26	1,033	1.2E+10	9.047786E+29	5.5732E+29
Uranus	1.2578	2.5775	893	5.08E+9	6.864197E+28	8.6342E+28
Neptune	1.6828	3.4485	1,143	4.86E+9	6.010456E+28	1.0114E+29
Pluto	12.221	25.044	512.395	3.0E+8	1.413716E+25	1.7277E+26

Equal Gravisphere Distances

$Eg = R * Sqrt(v_2) / Sqrt(v_2) + Sqrt(v_1)$
Eg = Equal gravisphere (point of equal gravity between primary & secondary -- miles from secondary planet)
R = Average orbital radius of secondary
v_1 = Relative orbital velocity of primary
v_2 = Relative orbital velocity of secondary

	Relative Orbital Velocity	Equal Gravisphere Distance Miles from Secondary	Average Orbital Radius (Miles)
Sun	574.31		
Mercury	.2345	713,074	36,000,000
Venus	.9	2,560,190	67,230,000
Moon	.111	63,103	238,855
Earth	1.0	3,723,648	92,960,000
Mars	.3257	3,296,147	141,700,000
Jupiter	17.58	72,185,919	483,700,000
Saturn	9.66	101,721,820	885,200,000
Uranus	3.8	133,980,056	1,781,000,000
Neptune	4.113	218,124,457	2,788,000,000
Pluto	.16988	47,669,110	3,660,000,000

Relative orbital velocities are probably better to use to determine the equal gravisphere distances between bodies because no body in space is stationary. The ether flow of gravity enters the planets in the form of a vortex, as viewed looking down on the planets above their poles. Thus, relative orbital velocities are better in determining the equal gravisphere distance because they take into account both the horizontal flow of gravity in the vortex and the vertical flow into both bodies, as well as the gravity force between them. The acceleration of gravity is a measure of the vertical flow of gravity. The orbital velocity is a measure of both the horizontal flow and the vertical flow.

The Gravity of Inner Earth

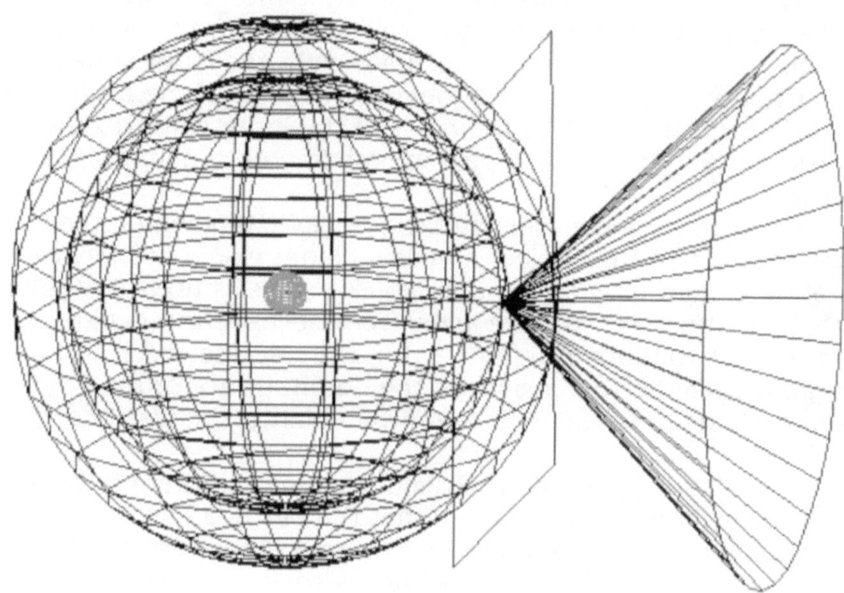

Orthodox Science claims that the center of gravity of the earth is ONLY in the center of the earth. However, they ignore that the earth is hollow. If gravity accelerates towards where the mass of the earth is located, in a hollow planet, there are TWO masses, 1) the inner sun and 2), the shell of the planet. So gravity should be accelerating both towards the inner sun, and also towards the shell, both towards the outer surface and towards the inner surface. This is confirmed by those that have been to Our Hollow Earth and have returned to report that their feet are as firmly planted on the inner surface as our feet are on the outer surface of the planet.

In a hollow earth, the greatest portion of the mass of the earth is located in the shell. The inner sun would have very little mass in comparison to the mass of the shell. In Chapter Two of this book, I calculated that the mass of a 600 mile diameter inner sun would have a mass of $5.981289262 \times 10^{23}$ gm, which is only .01% of the mass of the earth assuming the inner sun is a hollow crystal having approximately the density of glass of 2.6 gm/cc and a shell thickness of 10% of its diameter.

$5.981289262 \times 10^{23}$ gm Mass of inner sun

divided by mass of earth of 5.978541732 x 10^{27} gm
= .000100046 * 100 = .01%

Therefore, gravity will be accelerating into the inner sun, but an even greater acceleration will be into the shell of the Earth, both into the outer surface AND into the inner surface.

Gravity accelerating into the inner surface of the shell WILL be affected somewhat by the mass of the inner sun AND the mass of the earth above a person standing on the inner surface.

Orthodox Science claims that a person standing on the inner surface of the shell of a hollow planet would actually be floating in zero gravity. In their college physics texts it is called the Shell Theorem. But we know from hollow earth explorers that the inhabitants of Inner Earth have a surface acceleration of gravity similar to our own on the outer surface of the planet. Gravity keeps their feet firmly planted on the inner surface.

So obviously, Orthodox Science is wrong. Where have they gone wrong? First, they ignore that the earth is hollow and contains a central sphere of gravity somewhere between the inner and outer surfaces of the Earth's shell. Second, they claim the Earth has only one center of gravity and that all gravity accelerates towards the center of the earth. Third, they claim that gravity is an attraction of matter and deny the existence of the ether of space.

The Greenland Ice Hole experiment proved that the center of gravity in Earth's shell is not in the center of the earth. Because the gravimeter measured a greater decrease in the acceleration of gravity as it was lowered down the ice hole than if the center of gravity were located in the center of the Earth, this showed that the center of gravity is closer to the surface, which would be the case if the Earth is hollow. Indeed, the Greenland Ice Hole experimental results proved that the center of gravity is located in the Earth's shell, somewhere between the inner and outer surfaces.

The 25 years of interferometer readings performed by Dayton Miller and Edward Morley between 1902 and 1927 definitively detected the existence of the Ether, and was unfairly brushed off by subsequent scientists as described by Dr. James DeMeo, PhD in his book, THE DYNAMIC ETHER OF COSMIC SPACE.

I have provided quotes from the father of gravity theory, Isaac Newton, who believed the ether of space exists and that he did not believe that gravity was a force of attraction, but rather is a flow of the ether caused by a condensation of the ether in the nucleus of all atoms. This condensation of the ether in the nucleus of all atoms creates a vacuum in the ether in the nucleus of the atom causing surrounding ether to accelerate into the nucleus of the atom creating the pressure force we call gravity.

The ether condenses into gyroscopic particles in the nucleus of the atom that then become part of the magnetic field of the atom and are continually being knocked out of the magnetic field of the atom by particles colliding into the atom. Therefore, there is a continual flow of the ether into the atom and a continual emission of radiation flowing out of the atom. The same happens for planets. There is a continuous flow of the ether into the earth creating the pressure force we call gravity, and scientists have discovered that all the planets emit more radiation than they receive from the Sun.

This correction in gravity theory allows for an inner surface gravity. As the ether of space flows into the inner surface, the pressure of the flow is directed towards the inner surface as the ether flows towards the central sphere of gravity in the shell of the planet.

Realistically, the vector forces of mass gravity acceleration acting on an inner earth inhabitant would fall off towards his sides, so that the actual mass causing an acceleration of gravity on his person would be acting on him in the shape of a cone, rather than a plane. Mass towards his sides would not be causing any downward acceleration, but would increase as the angle increases towards a position directly beneath his feet. Mass directly beneath his feet would cause the greatest component of the acceleration of gravity acting on him, and would decrease to zero for mass off to the sides as the angle directly beneath his feet reduces to horizontal. At a certain angle off to the sides, gravity acceleration acting upon him would cease. Therefore the gravity acceleration acting upon him to hold his feet to the inner surface would thus describe a cone with his feet at the apex of the cone, as depicted in the diagram. The mass in that cone down to the central sphere of gravity would be the only mass causing a gravitational acceleration on the inner earth person towards the inner surface beneath his feet.

From the acceleration of gravity formula,

$$a = GM / r^2$$

we can calculate the mass in the earth that produces the outer surface acceleration of gravity, by solving for M,

$M = r^2a/G$

Assuming the earth's shell to be of uniform density to simplify our calculations, the center of gravity in a 798 mile thick shell at the equator where the acceleration of gravity is 978 cm/sec^2 would be 399 miles (6.42128×10^7 cm) down from either the outer or inner surfaces.

Solving for M, we get: $(6.42128 \times 10^7$ cm$)^2$ * 978 cm/sec^2 / $6.67259 \times 10^{-8} = 6.04349 \times 10^{25}$ gm

My DesignCAD drawing gives me a cone beneath the feet of our Hollow Earth person standing on the inner surface at the equator with a depth of 798 miles and a volume of 936,494,497.53 cubic miles, or 3.90348×10^{24} cm^3, at an average density of 11.05 gm/cc, gives a mass of 4.31334×10^{25} gm.

My intuition tells me that we should double the size of the cone beneath our inner person that would include all the possible mass that would affect our inner surface person pressing him towards the central sphere of gravity in the shell of the planet. Doubling the size of our cone would give us a mass of 8.63×10^{25} gm.

Calculating for the inner surface acceleration of gravity, using the mass in this double sized cone, minus the mass in the shell that gives the outer surface acceleration of 978 cm/sec^2 at the equator, $8.63 \times 10^{25} - 6.04349 \times 10^{25}$ gm we get, 2.583197×10^{25} gm for the mass in the shell that produces the inner surface acceleration of gravity,

$a = GM / r^2$

6.67259×10^{-8} * 2.583197×10^{25} gm / $(6.42128 \times 10^7$ cm$)^2$

= 418.03 cm/sec^2

Granted, that the mass of the Earth above the head of an inner earth inhabitant will counteract the acceleration of gravity into the Earth's shell beneath his feet. How much could that be?

Since only the mass on the inside of the central sphere of gravity on the opposite side of the earth will be exerting an acceleration on our inner surface person from above his head ignoring the acceleration due to the mass of the inner sun and the acceleration on our inner earth person due to centrifugal force, we can then calculate the acceleration due to the mass above his head as, (r = distance from inner surface to center of M above his head)

6.67259×10^{-8} * 2. 583197×10^{25} gm / $(1.05 \times 10^9$ cm$)^2$

= 1.56 cm/sec^2

When we subtract this from the acceleration of gravity caused by the mass to the central sphere of gravity in the cone beneath our inner person's feet of 418.03 cm/sec^2, it results in an inner surface acceleration of 416.47 cm/sec^2.

To this we would need to add the centrifugal force exerted by the rotating earth, which is calculated from the Centrifugal force formula,

$F = m\,v^2 / r$

For one gram of matter,

$a = v^2 / r$

r being the distance from the inner surface to the center of the earth, and

v being the velocity of the earth on the inner surface at the equator. Earth's shell at the equator in my model is 798 miles thick.

From the Circumference Velocity Formula,

$v = 2PiR / T$

$R = 5.093843488 \times 10^8$ cm, which is 3165.167601 miles from the inner surface at the equator to center of the Earth

$T = 86{,}164.09$ sec, sidereal time of one Earth day

$= 37144.89688$ cm/sec

plugging this into:

$a = v^2 / r$

$= 2.7086489$ cm/sec^2 acceleration, centrifugal force

Adding the 2.7086489 cm/sec^2 acceleration of centrifugal force, we get an inner surface acceleration of gravity at the equator of 419.18 cm/sec^2.

How much acceleration would our inner sun's gravity exert on our inner earth person? It is negligible, only 0.154 cm/sec^2. The calculations are as follows:

$a = GM / r^2$

$6.67259 \times 10^{-8} * 5.981289262 \times 10^{23}$ gm mass of inner sun $/ (5.093843488 \times 10^8$ cm$)^2$ (distance from center of earth to inner surface at the equator)

$= .154$ cm/sec^2, acceleration of gravity of inner sun exerted on our inner surface person. Subtracting this amount from the previously calculated inner surface acceleration of gravity gives an inner surface acceleration of gravity at the equator of **419.03 cm/sec^2**.

If you weigh 100 lbs. at the equator on our outer surface, on the inner surface with this scenario you would weigh 42.85 lbs., somewhat more than 2/5[ths] of your weight on the outer surface of the earth, or 42.85% of your weight on the outer surface.

Can Gravity Alone Keep the Inner Sun Positioned in the Center of the Earth?

Granted, that gravity alone cannot account for the stable suspension of the inner sun within the hollow of our earth. The reason for this is because with the interaction of any outer earth bodies such as the moon, sun or planets on that inner sun, may displace it somewhat towards one side or other of the inner surface. With any slight displacement of the inner sun toward one side or other of the inner surface, the inner sun would be caught in a greater acceleration of gravity flowing towards the inner surface it was closest to, while at the same time the acceleration of gravity on the opposite side of the inner sun flowing towards that side of the inner surface would decrease. This would cause the inner sun to start to fall towards the inner surface it was closest to, and it would crash into that inner surface -- if gravity was the only acting force here.

In order for the inner sun to maintain a centrally located position in the hollow of any planet, it must do so by some force other than just gravity.

Perhaps the force that could keep inner suns centrally located in hollow planets is the electrostatic force and ion emission of the inner suns.

In this respect, the inner sun is very similar to Prof. Robert R. Searl's flying saucer invention he calls an Inverse-G craft. The engine he invented gives off a very strong ion emission. This ion emission he directs to the periphery of his craft where he redirects it off the top and bottom hulls of his craft. By directing the ion emission at differing strengths on different parts of the hull of his craft, he is able to control its flight speed and direction. The ion emission also causes his craft's hull to be charged electrostatically, which in turn charges anything that comes in its proximity to the same charge. Like charges repel and so the craft travels through the air or water in a vacuum at incredible speeds.

Our inner sun emits a tremendous ion emission, strong enough to light up the auroras above the polar openings by the impact of these highly energetic particles impacting on air particles in the atmosphere. The inner sun's ion emission is emitted in all directions, which helps to keep it stationarily positioned in the hollow. The ion emission also impacts upon the inner atmosphere and sets up an electrostatic charge that repels the inner sun towards its centrally located position in the hollow.

Additionally, the interaction of the magnetic fields of the Inner Sun and the Shell of the Earth could keep the Inner Sun centrally positioned in the center of the hollow of our Earth.

The North Magnetic Pole of the Earth's shell is located at our South Magnetic Pole. The Inner Sun's South Magnetic Pole would be located towards the Earth's shell's North Magnetic Pole. The South Magnetic Pole of the Earth's shell is located at our North Magnetic Pole. The Inner Sun's North Magnetic Pole would be located towards the Earth's shell's South Magnetic Pole and thus the apparent attractions of magnetic poles would also keep the Inner Sun centrally positioned in the center of the Earth.

The Discovery of Flying Saucer Technology and Gravity Control

Flying Saucer technology was discovered in the 1950's by several independent inventors.

Thomas Townsend Brown (see https://www.thomastownsendbrown.com/) documented his discoveries in several United States patents (Nos. 2,949,550, 3,017,394, 3,022,430, and 3,187,206). He actually built working models.

Another inventor, Horace C. Dudley, filed a patent in 1960 (No. 3,095,167) for the enhancement of vehicular flight by the counteraction of gravity with very high electrostatic charges on the hull of the craft.

UFO contactee, Howard Menger describes in his book, THE HIGH BRIDGE INCIDENT: THE STORY BEHIND THE STORY, how he was befriended by extraterrestrials from Venus at a very young age as he was playing with his brother on his father's farm. From his life-long association with these Venusians and flights on their craft, he came to the conclusion that gravity is a push from space and can be countered by electro-static forces. Using this knowledge he constructed a three-foot-in-diameter model flying saucer in 1951. He flew it by remote radio control from the ground. However, it flew out of range and he lost it.

Later FBI agents visited him with parts of his crashed flying saucer and expressed interest in the propulsion system. From this contact, in 1961 the Pentagon set up a high tech laboratory installation near Colorado Springs where Howard Menger helped the government and participating big industry build a full size flying saucer craft which Menger test flew successfully. For this the government promised him a tax-free $1,500 check each month for the rest of his life. However, after one year the checks quit coming. He tried to check back with the project, but they had gone underground into the deep black projects.

These pioneers together with more recent scientists have contributed to our knowledge of gravity that will help us achieve electrogravitic propulsion. It is just a matter of putting the details together for a flying saucer to be actually constructed. This craft will be a silent electrogravitic propelled saucer shaped craft capable of underwater, air and space travel.

The advantages of this saucer craft will be its unlimited range, record breaking speed, the elimination of costly depletable fuels, and its ability to fly underwater and into space. This would make it more difficult for enemy missiles to hinder a mission. With this craft, missions could easily be undertaken to the moon and other planets. Perhaps the surfaces of the other planets are uninhabitable on their surfaces, however, with all probability, their interiors are gardens of Eden.

Saucer technology is based on the discoveries of several scientists, Townsend Brown, Horace C. Dudley, Joseph Newman, inventor John R. R. Searl of England and Howard Menger of Vero Beach, Florida, in combination with additional knowledge concerning the Ether as taught by 19th Century scientists.

Control of the Gravitational Ether Flow

Flying Saucer technology, then, is based on the control of the gravitational force, which is the directional flow of the Ether, by electrical forces. The secret of Saucer technology is the ability to direct the flow of the Ether in any direction by electro-static forces. This conclusion is supported by patents obtained by Townsend Brown on August 16, 1960 (#2,949,550) and by Horace C. Dudley on June 25, 1963 (#3,095,167).

Horace C. Dudley's patent #3,095,167 for an *"Apparatus for the Promotion and Control of Vehicular Flight,"* states that he discovered that by placing a positive static charge on the hull of a rocket, the charge counteracts gravitational force on the rocket.

Dudley wrote,

"It has been demonstrated that a missile, such as a ball of conductive material, can be projected upwardly against the action of gravity by placing a relatively high charge of the order of 400,000 to 500,000 volts on the ball by means of a suitable electrostatic generator. With this procedure, altitudes of as much as 10 centimeters were attained without the use of ANY propelling charge and PROVES THAT GRAVITY CAN NOT ONLY BE COUNTERBALANCED BY ELECTROSTATIC CHARGES, BUT ALSO THAT SUCH CHARGES CAN ACTUALLY PROPEL A MISSILE FROM THE EARTH'S SURFACE."

In Dudley's experiments, he took a model rocket made of high dielectric plastic material weighing 53 grams and when fired, traversed an unstable, erratic trajectory of relatively low altitude under 100 feet. When the identical rocket was coated on the inside with a conductive varnish, stable flight was attained to an altitude of approximately 300 feet. Upon coating of both surfaces, inner and outer, with a conductive varnish, stable flight was obtained to about 600 feet. Conductive surfaces increased flight stability and height 600%. If the rocket is given an appropriate electrostatic charge prior to firing, further increased altitudes will be attained. It was found that if the charges are significant, the effect of gravity can be completely counterbalanced.

The electrostatic force field also creates a vacuum around the vehicle enhancing its mobility.

Again Dudley wrote,

"Still another effect occurs which results in the reduction of friction between the rocket and the atmosphere...In positively charging the outer shell of a rocket or other aerial vehicle, it follows that the adjacent molecules of gas will become positively ionized and take on substantially the same charge as that of the vehicle. Inasmuch as molecules of gas adjacent to the surface of the vehicle assume the same charge as the vehicle, they will be repelled and the vehicle will therefore move in what may be termed a 'self-generated vacuum' induced by the charge on the vehicle itself."

Dudley describes how the outer surface of the vehicle should be devoid of all sharp edges, points, fins, and trailing wires that would result in the loss of the electrostatic charge on the surface of the vehicle. Surfaces should be smooth and have the greatest possible radius of curvature and the surface should be large relative to the mass. The conductive surface should then be coated with a dielectric material to aid in the retention of the charge. During flight, appropriate means must be carried for the maintenance of the electrostatic charges on the surface of the missile. Vehicular control can be accomplished by controlling the electrostatic charge.

A similar discovery was made by Townsend Brown.

From the smallest atom to the largest galaxy, the Universe operates on three basic forces -- electricity, magnetism, and gravitation. Taken separately, electricity and magnetism are not of much practical use. However, when coupled to work in combination with each other, almost endless technical applications arise. To date, our total electrical development has been based on coupling electricity with magnetism -- electromagnetism.

The coupling of electricity and gravitation is a whole new field of scientific endeavor with presently unimaginable practical applications yet future to be discovered. This coupling of electricity with gravitation is called electrogravitation.

Townsend Brown was one of the few known experimental scientists involved in discovering the uses of electrogravitation.

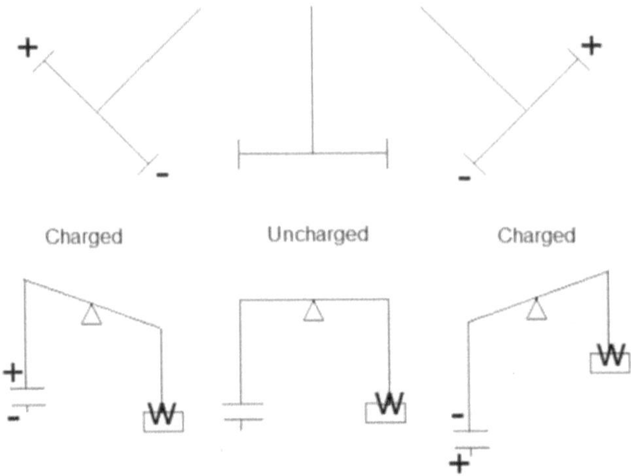

In 1923, Professor Biefeld of Denison University suggested to his protegé, Townsend Brown, certain experiments which led to the discovery of the Biefeld-Brown Effect, and ultimately, to the discovery of electrogravitational energy.

The first empirical experiments by Townsend Brown concerned a charged condenser which when hung in free suspension with horizontal poles exhibited a forward thrust toward the positive pole! A reversal of polarity caused a reversal of the direction of thrust. Further development of this phenomenon illustrated an "antigravity" effect. When balanced on a beam balance, and then charged, the condenser moves. If the positive pole is up, the condenser moves UP, that is, becomes "lighter"; if the positive pole is pointed down, it moves DOWN (becomes "heavier").

These two simple experiments demonstrate what is now known as the Biefeld-Brown effect. The intensity of the effect is determined by five factors.

1. The separation of the plates of the condenser -- the closer the plates, the greater the effect.

2. The higher the "K" factor, the greater the effect. ("K" is a measure of the ability of a material to store electric energy in the form of elastic stress.)

3. The greater the area of the condenser plates, the greater the effect.

4. The greater the voltage (potential) difference between the plates, the greater the effect.

5. The greater the mass of the material between the plates (dielectric), the greater the effect.

$$G = 6.67259E^{-8}\left(\frac{\frac{(km)AE}{d}}{m}\right)N - 1$$

Where:

G = the acceleration of the ether flow (Earth = 1)
m = the mass of the dielectric
k = the dielectric constant of the dielectric (air = 1)
A = the area of one plate
d = the distance between plates
E = the voltage across the plates
N = the number of plates

Formula from THE ANTI-GRAVITY HANDBOOK, compiled by D. Hatcher Childress, p. 48.

On the basis of further experimental work, in 1926 Townsend Brown described what he called a "space car." It had no moving parts; its motion controlled by simply varying the orientation of the positive charge on its periphery. A scale model flying around a stationary pole tethered to its power supply seemed to have no limit to the speed possible, and even when run in a vacuum, the machine flew so fast it had to be shut off before it developed enough inertia to fly apart.

After working with the problem of horizontal thrust, Brown discovered that a saucer shape for the craft was the most efficient shape to assist the gravitational field for maximum lift. The craft consisted of two charged plates, the upper charged positive and lower negative, imparting lift to the craft, and directional control was achieved by the use of a segmented structure on the periphery which would be switched in the direction of desired flight. The two condenser plates were separated by dielectric material.

As is evident in Townsend Brown's patents, the shape and position of the capacitor plates is of utmost importance in obtaining the maximum ether thrust. In general, one of the plates must be large in comparison to the other. For example, there is no resultant gravitational force on an atom or a planet, because the ether is flowing from all directions towards the positively charged nucleus at the center. So positioning the plates of a capacitor is all important. To get a resultant motive force by the ether flow on an atom, the nucleus would have to be displaced away from the center. The reason resultant ether flow has generally not been discovered in capacitors is because both plates are always the same size as each other, and usually both are rolled up in roll. It was only by making one big and the other small and repositioning the plates away from each other that Brown discovered the thrust of the ether flow. However, even Brown never suspected the thrust was caused by the ether, but assumed the thrust was caused by an ion flow. Even if Brown had suspected the thrust was an ether flow, he would not have been able to obtain a patent on his discoveries since modern science disclaims the existence of the ether.

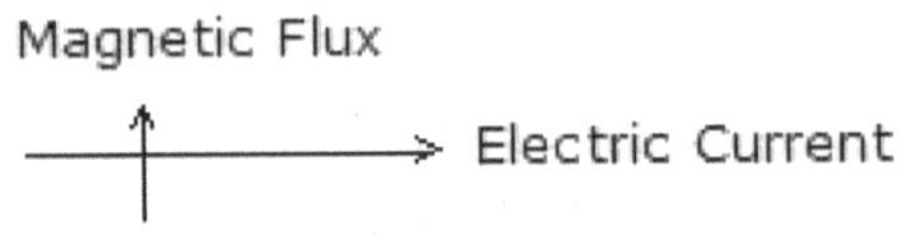

The Ether flow would be INTO the page.

If we take the example of the atom, the gravitational ether appears to flow at right angles to the lines of force of the magnetic field and the electrical current, and towards the positive electrode. In the above diagram, assume that the positive electrode is located beneath the page. The magnetic flux is northward; the electric current to the right. The ether flow would then be into the page toward the positive electrode.

With these discoveries by Townsend Brown and Horace Dudley the characteristics of flying saucers that have baffled present-day scientists as inexplicable with orthodox science can now be explained:

1. Method of propulsion. We now know that flying saucers consist of very highly charged condensers which produce such a strong electrostatic field that the gravitational flow of the Ether is redirected through the craft in the direction of the positive charge on the hull of the craft. As the Ether flow passes through the craft, it pushes the craft in the direction of the positive charge on the hull of the craft.

2. Tremendous accelerations and changes of direction. According to orthodox science, both machine and occupants would endure unbearable stresses given the observed tremendous accelerations and changes of direction flying saucers are seen to perform. "Not so," says Brown. No stresses would be felt, since craft, occupants, and load respond equally to distortion of the local gravitational field as a unit. The nearest analogy would be like going DOWN in an elevator. When the elevator starts down, both the elevator and its occupants flow with the gravitational ether flow without any shoving or stresses. The electrostatic field produced by the extremely high charges on the condenser plates of the flying

saucer redirect the gravitational ether flow and the saucer FALLS in the direction of the positive charge.

3. Flying saucers are surrounded by a glowing corona. The high-charged capacitors ionize the surrounding air letting off a bluish-violet glow.

4. Independence of aerodynamic effects. The ionized air surrounding the highly charged capacitor disks of the flying saucer create a vacuum "buffer zone" about the craft permitting the craft to travel through the air or under water without any resistance. The air or fluid in the proximity of the craft's highly charged capacitor disks become charged with the same charge as the surface of the disks. Since like charges repel, a vacuum is created between the surface of the flying saucer and the surrounding medium whether it be air or water allowing the craft to travel equally well through the atmosphere or under water. The lack of resistance to the movement of the craft allows the redirected gravitational ether flow to push the craft with greater acceleration and speed. Since the resistance of inertia is the resistance to the movement of matter through the ether, a flying saucer experiences no such resistance because the ether flows with the Saucer. This explains why Saucers can accelerate at unheard of velocities and change directions without being affected by G-forces on the craft or the occupants.

5. The problem of the generation of the high electrostatic charge for the saucer capacitor disks. In order to redirect the ether flow, capacitor disks need to be charged to extremely high voltages. Townsend Brown charged his disks up to 300 KV DC to achieve the electrogravitic propulsion he obtained and observed that the thrust appeared to be approximately linear with the voltage. A full scale saucer craft would need voltages in the millions in order to redirect the Ether flow with enough power to propel it.

A Powerful Electrogravitic Energy Source:
The Searl Effect Generator

The problem Townsend Brown encountered with his capacitor disks was he had no onboard high voltage source capable of producing the required high electrostatic charges necessary to redirect the ether flow to propel his craft. The solution to the onboard high electrostatic charge generator was solved by an inventor in England named John R. R. Searl.

The following is a commentary on the work of John Roy Robert Searl from John A. Thomas, Jr.'s book, ANTIGRAVITY: THE DREAM MADE REALITY, THE STORY OF JOHN R. R. SEARL.

ANTIGRAVITY: THE DREAM MADE REALITY, THE STORY OF JOHN R. R. SEARL, by John A. Thomas, Jr. is a primer on Prof. Searl's technology and inventions. Some of Prof. Searl's technology has been documented in 10 books on the Law of the Squares also published by John A. Thomas, Jr. Bradley Lockerman produced an excellent DVD documentary on the extraordinary life of John Roy Robert Searl. Professor John R. R. Searl has given his technology to Fernando Morris in San Diego, California to bring his inventions and technology to market. Their website is: https://segmagnetics.com/

John R. R. Searl (May 2, 1932 to December 4, 2018) was born May 2, 1932 at the Downs, Newbury Road, Wantage, England. At 4 years of age, John's father abandoned his mother and children and from her inability to support the family, the government placed the children in foster homes where John was often mistreated. From age 4.5 to 10, John had 2 dreams twice a year from which he claims he developed his Law of the Squares -- the basis of the technology that he developed producing free-energy electric generators and flying saucer-type craft. From his revolutionary work on free-energy technology and devices, John R. R. Searl was awarded an honorary degree as Professor of Mathematical Structures of Creation and Energy, he was listed in the International Registry of Who's Who in the World and he had many years of study at schools and universities coupled with the practical experience in the design and construction of electric motors, generators, medicine, navigation, electronics and computers.

John's first Searl Effect Generator was assembled when he was 14. The most notable effect produced by the generator was that when its circuit was overloaded, it would suddenly lose its gravity and become airborne. From what John learned over the years from his SEG, he and his friends built about 40 flying saucers operated by radio remote control. John also powered his home for 30 years with one of his SEG generators. The first SEG generators John built would become airborne and would shoot off into space and were lost. After discovering that his SEG generators can become airborne, John and his friends gave it a body and it became a flying saucer craft.

In 1968, they were within three months of having a completed manned craft in the space race to the moon, when the electric utility company had John thrown in jail and his work destroyed for supposedly *"stealing"* electricity for his home. At that time, a person could not produce his own electricity. His home was connected to the grid, but he wasn't using the power company's electricity. He had been powering his home with his SEG generator. The police came and hacked through his home until they found his SEG generator and took it. They also burned his papers and sold his tools. The flying saucer he was building with his friends in a field was dismantled and sold for scrap. His wife also disappeared.

John R. R. Searl made a great portion of his technological knowledge available to the world in a series of books, THE LAW OF THE SQUARES. In his 80's, with his health failing, John was working as a technical advisor to SEG Magnetics in Spring Valley, California where he died of complications of pneumonia on December 4, 2018. At SEG Magnetics they are attempting to recreate his inventions hoping to finally get this wonderful free-energy technology out to the people. The SEG generators they hope to produce will produce all the free electricity a homeowner can use. Besides generators, and flying saucer-type space craft, there are endless applications for this 21st Century technology.

The following is a description of the Searl Effect Generator and how it operates:

1. It is a magnetic motor, electric generator and gravity controller. It is silent, makes no noise and has no touching moving parts. As such it will never wear out. Its energy source is the ether of space interacting with the magnetic field of the magnetic parts and the electrostatic field generated by the moving parts. The Searl Effect is the prime mover. The SEARL EFFECT is the EFFECT of four forces acting in two sets of two forces, acting at right angles to each other.

2. It can consist of any number of a special design of magnetic rings, with a minimum of three -- one inside the other and all on the same plane. Each ring (and the rollers than orbit between the rings) consists of four layers that are hydraulically pressed together in an Argon atmosphere (to prevent combustion). The inner layer consists of a rare earth such as Neodymium which acts as a reservoir of electrons. The second layer out consists of nylon which restricts the flow of electrons between the inner layer and the outer layers so that there is an even flow of electrons within the ring. Without the nylon layer, the flow of electrons would be pulsated. The third layer consists of magnetizable metal such as nickel or iron. This is covered by a thin layer of metal that acts as an electrical conductor and passes the electrons from the rings to the rollers and on to other rings.

3. The magnetic rings are stationary.

4. Between each magnetic ring are magnetic rollers that rotate on their own axis as they rotate around the rings suspended in the space between the rings by their interacting

magnetic fields. The magnetic fields of the inner rings reinforce the field in the outer rings to make the rollers in the outer rings move faster. The speed of the rollers is 2.5 times the speed of the next inner ring of rollers. The rollers maintain themselves at right angles to the rings. The magnetic rollers are 1/4 inch shorter at the ends than the rings to keep the rollers from jumping out of the space between the rings. Each magnetic roller consists of 8 segments that are stacked on top of each other and are held together only by their magnetic attraction.

5. The rings are magnetically imprinted on their outer side with a Sine wave with the south pole on the top and north pole on the bottom which gives an AC type field around the rings. The magnetic imprint placed on the rollers is at right angles to the imprint on the rings. The roller magnetic imprint is circular, such as the magnetic field that exists around a wire that is carrying electric DC current. The magnetic imprint on the rings interacts with the magnetic field of the rollers and produces a magnetic wave in which the rollers move. The rollers move faster when the imprint on the rings is at a greater angle than 90 degrees. A greater angle of the imprint is needed for flight rather than for home power generation.

6. The rollers on the outer ring may pass through coils and induce a current in the coils.

7. A circuit is connected with the positive lead to the inner ring and the negative leads to the outer coils. The current induced in the coils is alternating current.

8. A capacitor-type electrostatic field is produced as electrons are extracted from the inner ring leaving it positively charged and the outer coils negatively charged.

9. When a load is placed in the circuit, the SEG's temperature lowers.

10. As greater loads are placed in the circuit, the SEG's rollers will jump to higher speeds incrementally.

11. As greater loads are placed in the circuit, the SEG's electrostatic field expands continuously outward.

12. As the electrostatic field expands continuously outward, gyroscopic particles in the field are lost as they get farther and farther away from the rings and when they hit atmospheric atoms or other particles. At very high electrostatic charges, the gyroscopic particles in the field may even produce light on the top south polar area of the SEG's rings where gyroscopic particles are emitting from the SEG hitting air particles above the SEG.

13. As the electrostatic field expands continuously, air, water and other approaching objects become electrostatically charged with the same charge and are repelled away. The SEG eventually is enveloped in a vacuum.

14. As gyroscopic particles in the continuously expanding field are lost, the SEG gets colder and colder. As atomic magnetic field layers are reduced by the loss of gyroscopic particles, at 4 degrees Kelvin, the atoms in the SEG suddenly all align and become superconductive. The overall magnetic field of the SEG then suddenly becomes super powerful because of the aligning of all the atoms. The magnified magnetic field combined with the electrostatic field redirects the ether flow and the SEG levitates.

15. As gyroscopic particles are lost in the continuously expanding electrostatic field, more particles are created in the nucleus of the SEG's atoms from the ether of space.

16. As gyroscopic particles are created in the nucleus of the SEG's atoms, a vacuum is produced in the nucleus of the SEG's atoms because the spinning gyroscopic particles take up less space than the ether they are made of.

17. The ether of space rushes in to fill the vacuum in the nucleus of the SEG's atoms exerting a gravitational pressure in the direction of their greatest flow.

18. Gyroscopic particles emitting from the south polar top side of the rings cause a lower ether pressure in the inflowing ether than at the north polar bottom of the SEG.

19. Gyroscopic particles entering the atoms of the SEG rings at the north polar bottom causes a greater ether flow/gravitational pressure. The greater ether pressure on the bottom of the SEG and less on the top causes it to levitate.

20. A shell is built around the SEG with a cabin in the center ring to give it a vehicle capability.

21. The placement of plungers in flight control cells on the periphery of the SEG craft gives directional control to the ether flow and the craft. When a plunger is activated, it momentarily touches the rotating outer rollers and the craft immediately tips down on that side similar to how a gyroscope tips when its rotating periphery is touched.

22. Increasing the load on the circuit of the SEG in the craft gives control to the proportionality of the ether flow to provide craft speed control. Electricity in an SEG used for electric power is tapped with coils spaced around the outer ring and its rollers. The rollers passing through the U shaped coils induce an alternating electric current in the coils. For an SEG used for

flight, instead of coils, the outer perimeter consists of wire brushes that slightly touch the orbiting rollers to extract electrostatic electricity which is then directed to the top and bottom surfaces of the craft.   Directional flight control can also be obtained by redirecting the emission of electrostatic ions to different sections of the craft hull.

23. A comfortable one-half G is constant in the center ring/cabin area because the ether entering the SEG from above is 1/2 that entering from below.

24. The craft is inertia-free since the craft carries its own ether along with it and controls it.

25. The SEG has an energizing effect on life forms, wounds heal faster, and a greater resistance to disease is produced by coming within its proximity because of the abundantly emitted free electrons discharged by the generator into the surrounding medium.

26. The SEG's south polar top emitting gyroscopic particles energizes material and will put out fires because more energy is pumped into the burning mass than leaves it.

27. Electrons emitting from the SEG places a charge on pollution particles in the air and the particles drop to the ground to clean the air. The electrons placed in the air by the SEG also energize life forms improving health and energy.

28. Innumerable applications of this technology could be developed including: flying cars by putting SEG's in the wheels, floating beds, cordless appliances, home power generation, space craft and space cities, cities under the ocean, Antarctica, the Sahara desert, levitating cranes, a pollution-free, free-energy planet. Its energy could be used to heal disease, and is especially effective in healing burn victims. It can be used to clean the air of pollution. It can power water purifiers. It could be used to extinguish fires. Homes designed with SEG technology would be resistant to fire, earthquakes, tornados, hurricanes, and burglars. SEG craft could be used to protect its inhabitants from solar flares, repel asteroids, disperse hurricanes and tornadoes and control the weather.

The Law of the Squares

The Searl Effect is what drives the Searl Effect Generator or Inverse-G disk and was developed from the Law of the Squares as discovered and developed by Prof. John R. R. Searl. A square is a series of sequentially unique numbers forming a square and is balanced when the horizontal, vertical and two diagonal line values all equal. Only whole numbers are used. Each number in the square represents the measure in grams of the compounds which are used to make the SEG. Starting with the highest number, this specifies the grams of a rare earth -- an element which has a greater number of electrons than other powders used. This element goes in the inner ring. Layers going out from the inner ring use atoms with less and less electrons. Elements are selected from the Periodic Table starting at the bottom of the Table with the inner ring and continuing up the table as layers in the rings are added from the center. Each ring consists of several layers. Layers are separated by a nylon layer to slow down the electron flow in the SEG. The Square of the SEG is selected by calculating the volume of the rings necessary to give room to the required number of rollers and then selecting a GROUP TWO (even numbered square) that will satisfy this volume.

The LAW OF NATURE states that no two bodies or particles can share the same SPACE FRAME within the same TIME FRAME, but that two bodies or particles CAN share the same space frame at different time frames. These two states must exist at all times; they cannot exist independent of each other. Within each square are TWO states, a Space Frame and a Time Frame. Each square = a space frame; each level = a time frame. The level number of a square begins with the lowest number in the square minus 1. The SPACE FRAME (square value, or line value) on a series of different squares may be the same, but the TIME FRAME (level value) for each will be different. Conversely, the TIME FRAME (level value) on a series of different squares may be the same, but the SPACE FRAMES (square value, or line value) will be different -- this is the LAW OF NATURE and cannot be broken. If a Time Frame is set, the space gets smaller as you move down the squares.

The number of a square = the number of digits on any one side of a square. There are three groups of squares. A GROUP ONE square (odd numbered squares) has a center point which is the line value divided by the number of the square. Group One numbers ROTATE. GROUP TWO and THREE squares (even numbered squares) have a center block of four numbers at the center of the square. Because group one squares do not have a center block, a group one square cannot be used in making a Searl Effect Generator. <u>Only GROUP TWO squares can be used in making an SEG</u>. Its center block is shared by the total square. Group Two squares OSCILLATE. Group Two squares are half of all even numbered squares -- those that can be divided by 4. A GROUP THREE cannot be used in making a SEG because its center block of numbers acts as a point and is not shared by the whole square. Although the group three center appears like a group two center, it only belongs to the rotating cross of the center of the square. But there is a way through this problem by applying it to flight. Group Three squares are half of all even numbered squares -- those that cannot be divided by 4. A Group Three square OSCILLATES, except its center cross ROTATES.

When a value can be used in more than just one square, it gives much more choice in the design structure than just one square can offer. The SEG needs 8 squares to be in the offer, one of which must be the value of the square which you intend to use. The SEARL EFFECT is the EFFECT of four forces acting in two sets of two forces, acting at right angles to each other. The Searl Effect is a true prime mover and through its action can produce electricity or motive power.

In addition, the following about nature must be understood: There are TWO PRIME STATES in nature, ENERGY and MATTER and can be converted from one into the other. Nature is BINARY in its actions, Low/High, Yes/No, 0/1, Go/No Go, many of which are the mirror image of the other, such as, Left/Right, Day/Night, Male/Female, North/South, In/Out, Latitude/Longitude. Sometimes Nature makes a triangle agreement in its actions or functions, such as, One Input/Two Outputs, Two Inputs/One Output. There are THREE PRIME STATES in Nature: Gas, Liquid, and Solid. In SOLIDS we have Conductors, Semi-Conductors, Non-Conductors. The ELEMENTS are the building blocks of Nature and consist of Non-Metals, Light Metals and Heavy Metals. Materials are Brittle, Ductile and Elastic. Conditions to consider are Pressure, Temperature, and Vacuum. Forces are Electricity, Magnetism, and Gravity. Chemical states are Acid, Alkaline or Neutral. Temperature points are Melting Point, Boiling Point, and Superconductive Point. Other important items of nature are Density, Amphoteric, Crystaline Structure, Electron Configuration, Atomic Weight, Atomic Volume, Atomic Radii, and Covalent Radius.

Important questions to ask are 1) What material you are going to use? 2) How much material are you going to use? 3) What AC are you going to energize the material with? 4) How are you going to apply the DC?

When you make a circle engine, you must have no less than three rings and two states must be involved, 1) stationary -- plates, and 2) motion -- the runners. The lowest value that can form a square of uniform nature is 12, and square 3 is the smallest square possible. So we need no less than 12 runners upon the first plate for smooth operation. Square 3, level 1 begins with 0 with a line value of 12. A square 4 is the smallest square that can be used to build an SEG. The size of the SEG is determined by the size of the Level used (how big of numbers in the square).

In the layout of numbers of a square, you can either have UNIFORM numbering in which the line values do not add up to the diagonals, or you can have RANDOM numbering in which the line values DO line up to the diagonals. The SEG uses random numbered squares in which the lines values, column values and diagonals all equal. Since random can switch to uniform and uniform to random, this is taken to mean that mass can switch to energy and energy can switch to mass. The center square for both random and uniform states remains the same. For group two, the two diagonals remain the same in both random and uniform states and form the line value. For FREE MOTION, there must be four forces acting, in which the horizontal force must always have the difference of 2 units and this is what groups 2 and 3 produce. In the center squares of group 2, the difference between the top and bottom will always be 2. One-half the difference between the left and right vertical of the center square is also the number of the square.

When designing an SEG, the volume of the rings for the size of machine wanted must be calculated. Prof. Searl then has a level chart that will give the number of the square needed for that volume. The frequency for the squares at this level is then determined by the number of magnetically imprinted poles placed on the rings. A heavier rare-earth element from the lower part of the periodic table is used in the inner layer of each ring with lighter elements from higher in the periodic table in the outer layers of each ring. This makes the more abundant electrons in the outer orbits of bigger atoms available for the circuit. The powders of the elements are assembled according to the numbers in each line of the square in grams for each layer with the heaviest (highest number) for the element in the inner layer of a ring and lighter (lower numbers) for element layers towards the outside layer. For example, for a square of 4, top row, the highest number would go to measure the grams of the inner layer. The next lower number is assigned to the next layer, the next lower number to the next layer, etc. The exception to this order is that the nylon layer takes the smallest number. The next row would be assigned to the next outer ring, etc. The rollers would be done similarly with the total mass of each segment of the roller being the line value with each layer of the roller taking the same assignment of values as a ring.

SQUARE 4 (Smallest square that can be used to build an SEG)

Level 2 (Level 1 is 0):

							The metal layers are assembled from high to low from the inside out with Nylon carrying the smallest number. If the materials we are using are neodymium, nylon, nickel and copper, then the grams for Ring 1 would be:
					34		
16	9	5	4		34	Ring 1	Neodymium 16 grams Inner Layer 1
3	6	10	15		34	Ring 2	Nylon 4 grams Layer 2
2	7	11	14		34	Ring 3	Nickel 9 grams Layer 3
13	12	8	1		34	Ring 4	Copper 5 grams Outer Layer 4
34	34	34	34		34		

Square 4, Level 202:

				834
216	**209**	**205**	**204**	834
203	**206**	**210**	**215**	834
202	**207**	**211**	**214**	834
213	**212**	**208**	**201**	834
834	834	834	834	834

You can download an Excel spreadsheet to calculate any level of Square 4 here:

https://www.ourhollowearth.com/square4.zip

Constructing a Flying Saucer Space Craft

Space craft could be developed utilizing the limitless power of the Ether to power the electrogravitic field of the craft using the Searl generator design to produce the required high electrostatic force field and the segmented periphery design of Townsend Brown disks to allow directional flight of the craft.

An embodiment of this flying saucer technology perhaps could assume the following construction. The space craft would be saucer shaped, which Townsend Brown discovered was the optimum shape for maximum upward thrust. In the center it would have a cylindrical cabin made of dielectric material. A metal coating on the outside of the cabin would protect the passengers from the high electro-magnetic fields. Surrounding the cabin a Searl Effect Generator could be attached. The rotating rollers between several concentric rings in the SEG would reinforce each subsequently outer ring's magnetic field and the rollers speed. An electrostatic charge can be collected from the rollers in the periphery ring with wire brushes which can then be discharged to the segmented hull. The top and bottom hulls of the craft could be segmented into several segments that could be charged separately. They would be insulated from each other, and would function as electrical condenser plates. Opposite charges could be placed on opposite segments of the craft's hull so the craft would fly in the direction of the positive charge. The hull plates would be connected with dielectric connectors to spokes which would radiate from the axis from above and below the cabin and connect at the periphery to a dielectric rim so the top and bottom hull plates and each hull segment would be insulated from each other in order to hold their charge.

The directional thrust would be obtained by having cables feed the electricity extracted from the Searl Effect Generator wire brushes through a joy stick in the cabin which would allow switching charges between each segment on the top and bottom hulls. Switching would be a rheostat type so clean-cut switching would not occur to prevent sparking. A vertical position of the joy-stick would allow all top segments to be positive and all bottom segments negative. Any other direction would charge the top plates in the direction the joystick was pointed positive and in the opposite direction negative. A separate lever-type rheostat would increase or decrease the amount of electricity allowed to discharge to the plates allowing the field to be increased or decreased which would increase or decrease the velocity of the craft.

Craft Layout

The diagram below depicts a version of this craft with the following layout: It consists of a central cabin based on the triangle construction of the house I built in Laveen, Arizona that used 5-sided star plates at each corner. This would be the living room. Above the living room would be the control room with windows on all sides allowing a semi-spherical view. Surrounding the ring of bedrooms, kitchen and bath, will be a hall that will go completely around the craft. The entrance to the craft is through a drop down portion of the hall. Air conditioning ducts are installed over the hall ring. Over the bedrooms, kitchen and bath ring will be water tanks. A black water tank could be installed beneath the floor. Outside the hall ring will be another ring of storage spaces and rooms for washer, dryer, air-conditioning equipment, water pump, water heater, water purification, and a Searl SEG electric generator to power the living area. Outside the storage ring will be the Searl craft engine which consists of three rings with orbiting rollers between each ring and brushes on the periphery to gather the electrostatic electricity to power the craft.

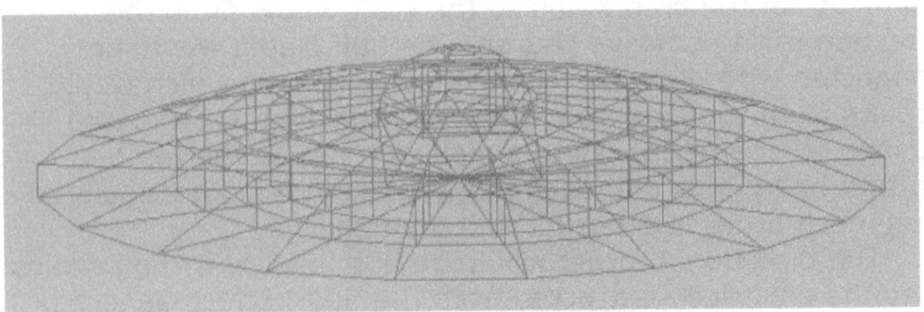

Water Purification System

The craft water purification system I propose could be based on John Nellis' Living Water Machine (see http://www.johnellis.com/), modified to use Browns gas.

John Nellis' Living Water Machine distills and purifies the water from germs with heat, then gives the purified water an electric charge that energizes the body. He has been drinking his purified water for 30 years, is actually 70 years old, but looks only 40 and he never gets sick. His doctor reports that John has the strongest heart he has ever tested, and claims that John will live forever if he continues using his living water in his diet.

The purification process of the Living Water Machine removes all inorganic minerals from the water which are not good for the human body (we need organic minerals from plant source, not inorganic minerals), kills any viruses and bacteria in the water with high heat, then gives the water an electric charge.

In the craft, recycling water will be necessary. So the water purification system must remove all materials from the water to make it absolutely pure. The electrical charge then placed on the water helps to energize the bodies of the craft occupants and also helps keep the piping and water tanks clean of debris and germs.

A modification of the Living Water purifier using Browns Gas would make the water absolutely pure. It takes 4 kwh of electricity to make 1,000 liters of Browns Gas. One liter of water makes 1,860 liters of Browns Gas. Therefore, it would take 7.44 kwh of electricity to make 1 liter of purified water. The water is turned into Browns Gas by electrolysis. Direct Current electricity supplied to two steel electrodes causes the water to separate into Hydrogen and Oxygen, which are then stored together in the exact proportion that the gases came out of the water during electrolysis. This is called Browns Gas. When this gas is then recombined, it turns back into water. This is done when a spark is introduced into the gas. The gas implodes rather than explodes and reduces immediately from 1,860 liters of gas back into 1 liter of water. The resulting vacuum can be used to power the purifier and to pump the water.

After the water is converted to gas, the gas can be passed through fine air filters just to make sure no germs or impurities are transported into the resulting purified water. The air filters and turning the water into Browns Gas should eliminate any possible contamination of the resulting water with minerals or germs. This would replace the heating element in the Living Water Machine which is used in that machine to kill the germs.

The Browns Gas generated can be changed back into water at the same time it runs a motor that pumps the water so the craft will have pressurized water. The motor could be a turbine engine where the Browns Gas is introduced and ignited with an electrical spark. The resulting vacuum of the implosion of the

Browns Gas would cause water to flow through the turbine because of differential pressure caused by the creation of the vacuum in one end of the turbine. Only purified water would be allowed to flow in the turbine cycle. The pressurized water flow would be connected to the crafts water storage tanks and piping to keep them pressurized. The Browns Gas turbine could be attached by a common shaft to a gas pump that would pressurize the Browns Gas as it comes out of the electrolysis chamber and so can be injected into the turbine. The turbine shaft could also power an electric generator that would place the electric charge on the purified water as well as supply DC current to the electrolysis chamber.

So the craft's water purification system would be both a water recycler, and a water purifier. Water should be able to recycle indefinitely with no loss if excess water in the air is extracted and placed back into the cycle. The recycler would need to have any wastes generated on the craft be liquefied before being put back into the cycle. The recycler would also need a method of removing solid waste after the water is removed. The solid waste could then be used for fertilizer in a craft's on-board food garden. This craft water purification system would thus allow for extended space flight times and for self-contained self-sufficiency.

The Virtual City of Light

I propose that a city be built using Searl flying saucer technology. The City would be circular in shape with a public sector in the center surrounded by the home sector surrounded by the business/agricultural/industrial sector. All homes would be flying saucer type construction using Searl SEG generators and craft engines to make them solar flare proof, fire proof, hurricane/tornado proof, earthquake proof and burglar proof. As such they would be mobile homes allowing the citizens to come and go at will by flight.

The monetary system of our City of Light would be my Asset Based Receipt Monetary System that provides a stable monetary system with no inflation or deflation. The benefits of all money created will go to the citizen tax payers that will cost them nothing more than their participation and yet more than double their income in every ten year period. See my Asset Based Receipt Monetary System which I have proposed for the United States of America here:
https://ourhollowearth.com/vcl/VirtualCityofLight/other.html

I have proposed a layout and city charter for The City of Light. At first, our city will be a virtual city on the Internet. When enough funds have come in, we will select some land and build a City of Light on the ground. One of our first projects as the Virtual City of Light is to sponsor an expedition to Our Hollow Earth. If successful, perhaps our first city built on the ground could be in Our Hollow Earth. You are invited to join our Virtual City of Light.

Gravity Bibliography

Burtt, Edwin A. THE METAPHYSICAL FOUNDATIONS OF MODERN SCIENCE, 1954, Garden City, New York, Doubleday.

Childress, David Hatcher. THE ANTIGRAVITY BOOK, 1993, published by Adventures Unlimited Press, 303 Main Street, PO Box 74, Kempton, Illinois 60946-0074, (815) 253-6390, Fax (815) 253-6300.

Cluff, Rodney M. WORLD TOP SECRET: OUR EARTH IS HOLLOW! https://www.ourhollowearth.com/

DeMeo, Dr. James. THE DYNAMIC ETHER OF COSMIC SPACE, 2019, Natural Energy Works, PO Box 1148, Ashland, Oregon 97520, United States of America, http://www.naturalenergyworks.net

DOCTRINE AND COVENANTS, published by the Church of Jesus Christ of Latter-Day Saints, Deseret Book Co., Salt Lake City, Utah.

Halliday, David and Robert Resnick. FUNDAMENTALS OF PHYSICS.

Hamilton, William F. CENTER OF THE VORTEX, 1979, published by Nexus & Nexus News, printed by Wilcopy, Los Angeles, California.

Menger, Howard & Connie. THE HIGH BRIDGE INCIDENT: The Story Behind the Story...Released After 35 Years of Silence, 1991, published by Howard Menger, PO Box 1405, Vero Beach, Florida 32961 (407) 562-1153.

Newman, Joseph. THE ENERGY MACHINE, 1984, published by Joseph Westley Newman, Route 1, Box 52, Lucedale, Mississippi 39452 (601) 947-7147.

Newton, Isaac, letters quoted in detail in THE METAPHYSICAL FOUNDATIONS OF MODERN PHYSICAL SCIENCE by Edwin Arthur Burtt, Double Day Anchor Books.

Palmer, Ray. FLYING SAUCERS MAGAZINE, Palmer Publications, Inc., Amherst, Wisconsin 54406.

Rensberger, Boyce, of the Washington Post. "Deep-In-The Earth Experiments Question Newton Gravity Theory," THE CHARLOTTE OBSERVER, newspaper, August 3, 1988.

Sigma, Rho. ETHER-TECHNOLOGY: A Rational Approach to Gravity Control, 1977, CSA Printing & Bindery, Lakemont, Georgia 30552.

Snyder, Al. NEWTON'S LAWS ARE FULL OF FLAWS, 1973, Snyder Institute of Research, 508 N. Pacific Coast Hwy., Redondo Beach, California 90277.

Snyder, Al. SATAN'S SAUNA AND THE DEVIL'S TRIANGLE, 1975, Snyder Institute of Research, 508 N. Pacific Coast Hwy., Redondo Beach, California 90277.

Smith, Joseph Fielding. DOCTRINES OF SALVATION, 3 Vols., 1954, Deseret Book Co., 40 E. South Temple, Salt Lake City, Utah 84104.

Thomas, John A. Jr. ANTIGRAVITY: THE DREAM MADE REALITY, THE STORY OF JOHN R. R. SEARL, 373 Rock Beach Road, Rochester, New York.

McCutcheon, Mark. THE FINAL THEORY, Rethinking Our Scientific Legacy, 2004. Universal Publishers, Boca Raton, CA.

Tolstoy, Ivan. JAMES CLERK MAXWELL: A BIOGRAPHY.

Warren, Henry C. *The Entrained Spatial Medium: Gravitation*, Paper 1, 2001, https://www.olypen.com/hcwarren/.

Wilson, Don. OUR MYSTERIOUS SPACESHIP MOON, 1975, Dell Publishing Co., Inc., 1 Dag Hammarskjold Plaza, New York, NY 10017, http://www.scribd.com/doc/40405065/Secrets-of-Our-Spacecship-Moon.

Wilson, Don. SECRETS OF OUR SPACESHIP MOON, 1979, Dell Publishing Co., Inc., 1 Dag Hammarskjold Plaza, New York, NY 10017, http://www.scribd.com/doc/40405065/Secrets-of-Our-Spacecship-Moon.

ADDENDUM

Websites on gravity and ether detection:

https://www.electrogravityphysics.com/

http://www.orgonelab.org/cart/naturalenergy.htm

Rodney M. Cluff

Gyro Drop Experiment

Performed by Kenneth Gerber, M.D. and Richard F. Merritt
Analysis by Edward Delvers

In this experiment, a fully enclosed, electrically driven gyroscope is released to fall freely under the influence of gravity. The elapsed time taken to fall a measured distance of 10.617 feet was measured, with the rotor stopped and also with the rotor spinning at approximately 15,000 RPM.

Data was gathered on a Chronometrics Digital Elapsed Time Clock measuring 1/10,000 second, actuated by two photo transistor sensors placed in the paths of two light beams which were consecutively interrupted by the edge of the casing of the falling gyroscope.

The gyroscope, of total weight 7.23 lbs (rotor weight 4.75 lbs, case weight 2.48 lbs) was released to fall along its axis. Electrical leads supplying power to the 4 1/4" diameter rotor were disconnected just prior to release.

Experimental Set-up

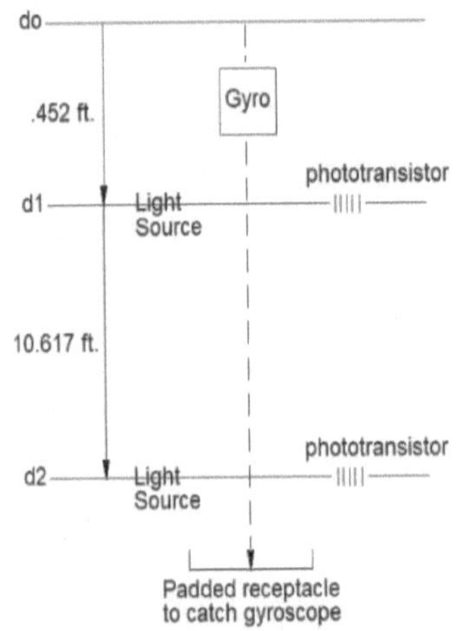

release, t0 = 0 seconds

Fully enclosed, electrically driven gyroscope released to fall along its axis. Rotor speed approximately 15,000 rpm, rotor weight 4.75 lbs, case weight 2.48 lbs, rotor diameter 4 1/4" (total weight 7.23 lbs)

elapsed time measurement begins, t = t1

Chronometric Digital Elasped Time Clock measuring 1/10,000 second, actuated by signals from phototransistors

elapsed time measurement ends, t = t2

Summary of Experimental Results

	Non-Rotating (NR)	Rotating (R)
t_0	0.0 sec	0.0 sec
t_1	.1677 sec	.1674 sec
t_2	.82973 sec	.82837 sec
Measured elapsed time, with +/- standard deviation	(.66203 +/- .000996 sec)	(.66097 +/- .000824 sec)
Number of runs	13	7
Acceleration	32.1549* ft/sec²	32.2619 ft/sec²

Change in acceleration: $\Delta a = (a_R - a_{NR}) = .1070$ feet/second²

*NOTE: Value for gravitational acceleration at sea level, 39° Latitude (Washington, D.C.) based on the formula of the U.S. Coast and Geodetic Survey. The data for the non-rotating gyroscope is normalized to this value, and the data for the rotating gyroscope is compared to it.

Fictitious Force Increment

A hypothetical, fictitious force increment which would have to be applied to the non-rotating gyroscope to impart the increased acceleration noticed in its rotating mode, was calculated for comparison purposes.

Force increment: $\Delta F = (F_R - F_{NR}) = .024$ lbs = .38 oz

Data

	Non-Rotating	Run Number	Rotating
Run Number	Time		Time
1	.6604 seconds	1	.6617 seconds
2	.6603 seconds	2	.6616 seconds
3	.6614 seconds	3	.6605 seconds
4	.6630 seconds	4	.6618 seconds
5	.6623 seconds	5	.6613 seconds
6	.6623 seconds	6	.6601 seconds
7	.6622 seconds	7	.6598 seconds
8	.6618 seconds		
9	.6627 seconds		
10	.6615 seconds		
11	.6639 seconds		
12	.6627 seconds		
13	.6619 seconds		
Mean +/- Standard Deviation	= .66203 +/- .000996 sec		= .66097 +/- .000824 sec

Statistical Analysis

Value for Student's "t" Test: t = 2.3980, p = .0275355685 (18 degrees of freedom)

On the basis of the Standard Deviations of the data from this experiment, one can say with a 97% level of confidence that a fully encased, spinning gyroscope drops faster than the identical gyroscope non-spinning, when released to fall along its axis.

Appendix

The following are calculations performed on the measured data to arrive at the values given in the Summary of Experimental Results (above).

1. Calculation to find velocity v_1 at the beginning of elapsed time measurement for the Non-Rotating gyroscope, using the equation,

$$d = v_i t + 1/2\ at^2$$

where:

$d = (d_2 - d_1) = 10.617$ feet (measured)
$t = (t_2 - t_1) = .66203$ seconds (measured)
$a = 32.1549$ ft/sec^2 (normalized value)
$v_i =$ unknown, velocity v_1 at time t_1
    Substituting values: $v_i = 5.393$ feet/sec

2. Calculation to find the distance between release position and beginning of elapsed time measurement segment for the Non-Rotating gyroscope.

$$v_f^2 = v_i^2 + 2ad$$

where:

$v_f = 5.393$ ft/sec (from 1. above)
$v_i = 0$ ft/sec (initial velocity)
$a = 32.1549$ ft/sec^2 (normalized value)
$d = (d_1 - d_0) =$ unknown.
    Solving the equation: $d = (d_1 - d_0) = .4522$ feet

3. Calculation to find the time already spent falling when the elapsed time measurement begins for the Non-Rotating condition of the gyroscope.

$$v_f = v_i + at$$

where:

$v_f = v_1$ at t1 $= 5.393$ ft/sec (from 1. above)
$v_i = 0$ ft/sec
$a = 32.1549$ ft/sec^2 (normalized value)
$t = (t_1 - t_0) =$ unknown
    Solving the equation: $t = (t_1 - t_0) = .1677$ seconds

4. Calculation to find the total time taken to fall total distance for the Non-Rotating condition of the gyroscope.
$t_{totalNR} = (t_2 - t_1)_{NR} + (t_1 - t_0)_{NR} = .66203 + .1677 = .82973$ seconds
$d_{totalNR} = (d_2 - d_1)_{NR} + (d_1 - d_0)_{NR} = 10.617 + .4522 = 11.0692$ feet

5. Calculation to find time already spent falling by the Rotating gyroscope when elapsed time measurement begins. This assumes the acceleration of the Rotating gyroscope is constant. It is found by comparing the ratio or the initial time interval to measured elapsed time interval for the Non-Rotating gyroscope, to that of the Rotating gyroscope.

$$(t_1 - t_0)_{NR} / (t_2 - t_1)_{NR} = (t_1 - t_0)_R / (t_2 - t_1)_R$$

where:

$(t_1 - t_0)_{NR}$ = .1677 sec. (calculated)
$(t_2 - t_1)_{NR}$ = .66203 sec. (measured)
$(t_1 - t_0)_R$ = unknown
$(t_2 - t_1)_R$ = .1674 seconds

6. Calculation to find acceleration (a_R) of the Rotating gyroscope using total time and total distance values, using the equation:

$$d = v_i t + 1/2 \, at^2$$

where:

d = 11.069 ft (from 4. above)
v_i = 0 ft/sec
$a = a_R$ = unknown
$t = t_{totalR} = (t_2 - t_1)_R + (t_1 - t_0)_R$ = .66097 + .1674 = .62837 seconds

Solving the equation: $a = a_R$ = 32.2619 feet/second^2

7. Change in acceleration:

$\Delta a = a_R - a_{NR}$ = 32.2619 ft/sec^2 - 32.1549 ft/sec^2 = .1070 ft/sec^2
Percent change in acceleration: $\Delta a / a_{NR}$ = .00333 = .333%

8. Fictitious Force Increment: Calculation to find a hypothetical, fictitious force increment which would have to be applied to the Non-Rotating gyroscope to cause the increased acceleration observed for the Rotating gyroscope. The mass (m) of the gyroscope is assumed not to have changed for the purposes of this calculation.

Using the equation: F = ma
a ratio is set up:

$$F_{NR} / F_R = m_{NR} a_{NR} / m_R a_R$$

where:

F_{NR} = 7.23 lbs (measured gyro and case weight)
F_R = unknown

$a_{NR} = 32.1549$ ft/sec^2 (normalized value)
$a_R = 32.2619$ ft/sec^2 (from 6. above)
 Solving the equation: F = 7.254 lbs.

The fictitious force increment is: $\Delta F = F_R - F_{NR} = 7.254 -$ 7.23 = .024 lbs, or converted to ounces: .024 lbs x 16 oz/lb = .38 ounces

References

1. "The Effect of Gravity on Rotating Objects," Edward C. Delvers and Bruce E. DePalma, March 18, 1974, Simularity Institute reprint.

2. "Is God Supernatural," Robert L. Dione, Bantam Books, NY, 1976, 553-02723-150.

3. "Gyro Drop Experiment," by Kenneth Gerber, M.D., U.S. Department of Health, Education and Welfare, Public Health Services, National Institutes of Health, National Heart, Lung and Blood Institute, Bethesda, MD 20014, Richard F. Merritt, and Edward Delvers, 1977.

4. "The Cause of Gravitation," A. Bernard Rendle, Modal Research, 51 Dorking Rd, Gt. Bookham, Surrey, England, 1971.

5. Unpublished "Elastic Collision Experiment Data," Simularity Institute report.